Summary of Single-Sample Confidence Interval Procedures

Case	Problem Type	Point Estimate	Type of Confidence Interval	$100(1-\alpha)\%$ Confidence Interval
1.	Mean μ, variance σ^2 known	$\bar{x}$	Two-sided	$\bar{x} - z_{\alpha/2}\sigma/\sqrt{n} \leq \mu \leq \bar{x} + z_{\alpha/2}\sigma/\sqrt{n}$
			One-sided lower	$\bar{x} - z_{\alpha}\sigma/\sqrt{n} \leq \mu$
			One-sided upper	$\mu \leq \bar{x} + z_{\alpha}\sigma/\sqrt{n}$
2.	Mean μ of a normal distribution, variance σ^2 unknown	$\bar{x}$	Two-sided	$\bar{x} - t_{\alpha/2,n-1}s/\sqrt{n} \leq \mu \leq \bar{x} + t_{\alpha/2,n-1}s/\sqrt{n}$
			One-sided lower	$\bar{x} - t_{\alpha,n-1}s/\sqrt{n} \leq \mu$
			One-sided upper	$\mu \leq \bar{x} + t_{\alpha,n-1}s/\sqrt{n}$
3.	Variance σ^2 of a normal distribution	s^2	Two-sided	$\dfrac{(n-1)s^2}{\chi^2_{\alpha/2,n-1}} \leq \sigma^2 \leq \dfrac{(n-1)s^2}{\chi^2_{1-\alpha/2,n-1}}$
			One-sided lower	$\dfrac{(n-1)s^2}{\chi^2_{\alpha,n-1}} \leq \sigma^2$
			One-sided upper	$\sigma^2 \leq \dfrac{(n-1)s^2}{\chi^2_{1-\alpha,n-1}}$
4.	Proportion or parameter of a binomial distribution p	$\hat{p}$	Two-sided	$\hat{p} - z_{\alpha/2}\sqrt{\dfrac{\hat{p}(1-\hat{p})}{n}} \leq p \leq \hat{p} + z_{\alpha/2}\sqrt{\dfrac{\hat{p}(1-\hat{p})}{n}}$
			One-sided lower	$\hat{p} - z_{\alpha}\sqrt{\dfrac{\hat{p}(1-\hat{p})}{n}} \leq p$
			One-sided upper	$p \leq \hat{p} + z_{\alpha}\sqrt{\dfrac{\hat{p}(1-\hat{p})}{n}}$

Engineering

Statistics

Engineering
Statistics

Second Edition

Douglas C. Montgomery
George C. Runger
Norma Faris Hubele
Arizona State University

John Wiley & Sons, Inc.

New York   Chichester   Weinheim
Brisbane   Singapore   Toronto

EDITOR Wayne Anderson
ASSISTANT EDITOR Penny Perrotto
MARKETING MANAGER Katherine Hepburn
ASSOCIATE PRODUCTION DIRECTOR Lucille Buonocore
SENIOR PRODUCTION EDITOR Monique Calello
SENIOR DESIGNER Kevin Murphy
ILLUSTRATION COORDINATOR Gene Aiello
ILLUSTRATION STUDIO Wellington Studios

This book was set in 10/12 Times Roman by Monotype Composition Company and printed and bound by R.R. Donnelley & Sons Company, Crawfordsville. The cover was printed by Phoenix Color Corporation.

This book is printed on acid-free paper. ∞

The paper in this book was manufactured by a mill whose forest management programs include sustained yield harvesting of its timberlands. Sustained yield harvesting principles ensure that the numbers of trees cut each year does not exceed the amount of new growth.

To order books or for customer service call
1-800-CALL-WILEY (225-5945).

Library of Congress Cataloging-in-Publication Data
Montgomery, Douglas C.
    Engineering statistics / Douglas C. Montgomery, George C. Runger, Norma Faris Hubele.—2nd ed.
      p.   cm.
    Includes index.
    ISBN 0-471-38879-3 (cloth: alk. paper)
    1. Statistics. 2. Probabilities. I. Runger, George C. II. Hubele, Norma Faris, III. Title.

QA276.12.M6453 2000
519.5′024′62—dc21   00-039241

Printed in the United States of America

10 9 8 7 6 5 4 3 2 1

To
Meredith, Neil, Colin, and Cheryl
Rebecca, Elisa, George, and Taylor
Norman and Michelle

Preface

Engineers play a significant role in the modern world. They are responsible for the design and development of most of the products that our society uses, as well as the manufacturing processes that make these products. Engineers are also involved in many aspects of the management of both industrial enterprises and business or service organizations. Fundamental training in engineering develops skills in problem formulation, analysis, and solution that are valuable in a wide range of settings.

Solving many types of engineering problems requires an appreciation of variability and some understanding of how to use both descriptive and analytical tools in dealing with variability. Statistics is the branch of applied mathematics that is concerned with variability and its impact on decision making. This is an introductory textbook for a first course in engineering statistics. Although many of the topics we present are fundamental to the use of statistics in other disciplines, we have elected to focus on meeting the needs of engineering students by allowing them to concentrate on the applications of statistics to their disciplines. Consequently, our examples and exercises are engineering based, and in almost all cases, we have used a real problem setting or the data either from a published source or from our own consulting experience.

Engineers in all disciplines should take at least one course in statistics. Indeed, the Accreditation Board on Engineering and Technology is requiring that engineers learn about statistics and how to use statistical methodology effectively as part of their formal undergraduate training. Because of other program requirements, most engineering students will take only one statistics course. This book has been designed to serve as a text for the one-term statistics course for all engineering students.

ORGANIZATION OF THE BOOK

The book is based on a more comprehensive text (Montgomery, D. C. and Runger, G. C., *Applied Statistics and Probability for Engineers,* Second Edition, John Wiley & Sons, New York, 1999) that has been used by instructors in a one- or two-semester course. We have taken the key topics for a one-semester course from that book as the basis of this text. As a result of this condensation and revision, this book has a modest mathematical level. Engineering students who have completed one semester of calculus should have no difficulty reading nearly all of the text. Our intent is to give the student an understanding of statistical methodology and how it may be applied in the solution of engineering problems, rather than the mathematical theory of statistics.

Chapter 1 introduces the role of statistics and probability in engineering problem solving. Statistical thinking and the associated methods are illustrated and contrasted with other engineering modeling approaches within the context of the engineering problem-solving method. Highlights of the value of statistical methodologies are discussed using simple examples. Simple summary statistics are introduced.

Chapter 2 illustrates the useful information provided by simple summary and graphical displays. Computer procedures for analyzing large data sets are given. Data analysis methods such as histograms, stem-and-leaf plots, and frequency distributions are illustrated. Using these displays to obtain insight into the behavior of the data or underlying system is emphasized.

Chapter 3 introduces the concepts of a random variable and the probability distribution that describes the behavior of that random variable. We concentrate on the normal distribution, because of its fundamental role in the statistical tools that are frequently applied in engineering. We have tried to avoid using sophisticated mathematics and the event–sample space orientation traditionally used to present this material to engineering students. An in-depth understanding of probability is not necessary to understand how to use statistics for effective engineering problem solving. Other topics in this chapter include expected values, variances, probability plotting, and the central limit theorem.

Chapters 4 and 5 present the basic tools of statistical inference: point estimation, confidence intervals, and hypothesis testing. Techniques for a single sample are in Chapter 4, and two-sample inference techniques are in Chapter 5. Our presentation is distinctly applications oriented and stresses the simple comparative-experiment nature of these procedures. We want engineering students to become interested in how these methods can be used to solve real-world problems and to learn some aspects of the concepts behind them so that they can see how to apply them in other settings. We give a logical, heuristic development of the techniques, rather than a mathematically rigorous one.

Empirical model building is introduced in Chapter 6. Both simple and multiple linear regression models are presented, and the use of these models as approximations to mechanistic models is discussed. We show the student how to find the least squares estimates of the regression coefficients, perform the standard statistical tests and confidence intervals, and use the model residuals for assessing model adequacy. Although this chapter makes some modest use of matrix algebra, we emphasize the use of the computer for regression model fitting and analysis.

Chapter 7 formally introduces the design of engineering experiments, although much of Chapters 4 and 5 was the foundation for this topic. We emphasize the factorial design and, in particular, the case in which all the experimental factors are at two levels. Our practical experience indicates that if engineers know how to set up a factorial experiment with all factors at two levels, conduct the experiment properly, and correctly analyze the resulting data, they can successfully attack a large majority of the engineering experiments that they will encounter in the real world. Consequently, we have written this chapter to accomplish these objectives. We also introduce fractional factorial designs and response surface methods.

Statistical quality control is introduced in Chapter 8. The important topic of Shewhart control charts is emphasized. The $\overline{X}$ and R charts are presented, along with some simple control charting techniques for individuals and attribute data. We also discuss some aspects of estimating the capability of a process.

The students should be encouraged to work problems to master the subject matter. The book contains an ample number of problems of different levels of difficulty. The end-of-section exercises are intended to reinforce the concepts and techniques introduced in that section. These exercises are more structured than the end-of-chapter supplemental exercises, which generally require more formulation or conceptual thinking. We use the supplemental exercises as integrating problems to reinforce mastery of concepts, as opposed to analytical technique. The Team Exercises challenge the student to apply chapter methods and concepts to problems requiring data collection. As noted below, the use of statistics software in problem solution should be an integral part of the course.

USING THE BOOK

We strongly believe that an introductory course in statistics for undergraduate engineering students should be, first and foremost, an *applied course.* The primary emphasis should be on data description, inference (confidence intervals and tests), model building, designing engineering experiments, and statistical quality control, *because these are the techniques that they will need to know how to use as practicing engineers.* There is a tendency in teaching these courses to spend a great deal of time on probability and random variables (and, indeed, some engineers, such as industrial and electrical engineers, do need to know more about these subjects than students in other disciplines) and to emphasize the mathematically oriented aspects of the subject. This can turn an engineering statistics course into a "baby math-stat" course. This type of course can be fun to teach and much easier on the instructor, because it is almost always easier to teach theory than application, but it does not prepare the student for professional practice.

In our course taught at Arizona State University, students meet twice weekly, once in a large classroom and once in a small computer laboratory. Students are responsible for reading assignments, individual homework problems, and team projects. In-class team activities include designing experiments, generating data, and performing analyses. The supplemental problems and team exercises in this text are a good source for these activities. The intent is to provide an active learning environment with challenging problems that foster the development of skills for analysis and synthesis.

USING THE COMPUTER

In practice, engineers use computers to apply statistical methods in solving problems. Therefore, we strongly recommend that the computer be integrated into the course. Throughout the book, we have presented output from Minitab as typical examples of what can be done with modern computer software. In teaching, we have used Statgraphics, Minitab, Excel, and several other statistics packages or spreadsheets. We did not clutter the book with examples from many different packages, because *how* the instructor integrates the software into the class is ultimately more important than *which* package is used. All text data and the instructor manual are available in electronic form.

In our large-class meeting times, we have access to computer software. We show the student how the technique is implemented in the software as soon as it is discussed in class. We recommend this as a teaching format. Low-cost student versions of many popular software packages are available, and many institutions have statistics software available on a local area network, so access for the students is typically not a problem.

Computer software can be used to do many exercises in this text. Some exercises, however, have small computer icons in the margin. We highly recommend using software in these instances.

The second icon is meant to represent a computer disk. This icon marks problems for which summary statistics are given, but the complete sample of data is available in the Instructors Resources CD-ROM. Some instructors may wish to have the students use the data rather than the summary statistics for problem solutions.

WEB SITE

Current supporting material for instructors and students is available at the Web site www.wiley.com/college/montgomery. We will use this site to communicate our latest information about innovations and recommendations for effectively using this text and we hope to elicit your feedback. In-depth case studies that illustrate an integration of several analysis methods will be posted as they are developed. Electronic versions of select data from the text are posted there for your convenience.

ACKNOWLEDGMENTS

We would like to express our appreciation for support in developing some of this text material from the Course and Curriculum Development Program of the Undergraduate Education Division of the National Science Foundation. We also thank Teri Reed Rhoads, Lora Zimmer, and Sharon Lewis for their work as graduate assistants in the development of the course based on this text. We are very thankful to Dr. Connie Borror for her thorough work on the instructor's volume. We appreciate the staff support and resources provided by the Department of Industrial Engineering at Arizona State University, and our department chair, Dr. Gary Hogg.

Several reviewers provided many helpful suggestions, including Dr. Hongshik Ahn, SUNY, Stony Brook; Dr. James Simpson, Florida State/FAMU; Dr. John D. O'Neil, California Polytechnic University, Pomona; Dr. Charles Donaghey, University of Houston; Professor Gus Greivel, Colorado School of Mines; Professor Arthur M. Sterling, LSU; and Professor David Powers, Clarkson University. We are also indebted to Dr. Smiley Cheng for permission to adapt many of the statistical tables from his excellent book (with Dr. James Fu), *Statistical Tables for Classroom and Exam Room*. John Wiley & Sons, Prentice-Hall, the Biometrika Trustees, The American Statistical Association, the Institute of Mathematical Statistics, and the editors of *Biometrics* allowed us to use copyrighted material, for which we are grateful.

This project was supported, in part, by the
National Science Foundation
Opinions expressed are those of the authors and not necessarily those of the Foundation

Douglas C. Montgomery
George C. Runger
Norma Faris Hubele

Contents

1

The Role of Statistics in Engineering

CHAPTER OUTLINE

1-1 THE ENGINEERING METHOD AND STATISTICAL THINKING

Engineers solve problems of interest to society by the efficient application of scientific principles. The **engineering** or **scientific method** is the approach to formulating and solving these problems. The steps in the engineering method are as follows:

1. Develop a clear and concise description of the problem.

2. Identify, at least tentatively, the important factors that affect this problem or that may play a role in its solution.

3. Propose a model for the problem, using scientific or engineering knowledge of the phenomenon being studied. State any limitations or assumptions of the model.

4. Conduct appropriate experiments and collect data to test or validate the tentative model or conclusions made in steps 2 and 3.

5. Refine the model on the basis of the observed data.

1

6. Manipulate the model to assist in developing a solution to the problem.

7. Conduct an appropriate experiment to confirm that the proposed solution to the problem is both effective and efficient.

8. Draw conclusions or make recommendations based on the problem solution.

The steps in the engineering method are shown in Fig. 1-1. Note that the engineering method features a strong interplay between the problem, the factors that may influence its solution, a model of the phenomenon, and experimentation to verify the adequacy of the model and the proposed solution to the problem. Steps 2–4 in Fig. 1-1 are enclosed in a box, indicating that several cycles or iterations of these steps may be required to obtain the final solution. Consequently, engineers must know how to efficiently plan experiments, collect data, analyze and interpret the data, and understand how the observed data are related to the model they have proposed for the problem under study.

The field of **statistics** deals with the collection, presentation, analysis, and use of data to make decisions and solve problems. Because many aspects of engineering practice involve working with data, obviously some knowledge of statistics is important to any engineer. Specifically, statistical techniques can be a powerful aid in designing new products and systems, improving existing designs, and designing, developing, and improving production processes.

Statistical methods are used to help us describe and understand **variability.** By variability, we mean that successive observations of a system or phenomenon do not produce exactly the same result. We all encounter variability in our everyday lives, and **statistical thinking** can give us a useful way to incorporate this variability into our decision-making processes. For example, consider the gasoline mileage performance of

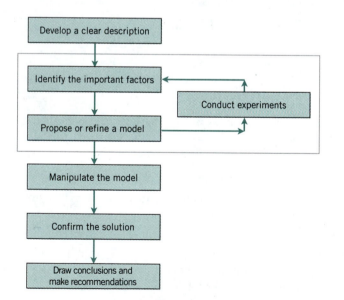

Figure 1-1   The problem-solving method.

your car. Do you always get exactly the same mileage performance on every tank of fuel? Of course not—in fact, sometimes the mileage performance varies considerably. This observed variability in gasoline mileage depends on many factors, such as the type of driving that has occurred most recently (city versus highway), the changes in condition of the vehicle over time (which could include factors such as tire inflation, engine compression or valve wear), the brand and/or octane number of the gasoline used, or possibly even the weather conditions that have been experienced recently. These factors represent potential **sources of variability** in the system. Statistics gives us a framework for describing this variability and for learning about which potential sources of variability are the most important or which have the greatest impact on the gasoline mileage performance.

We also encounter variability in dealing with engineering problems. For example, suppose that an engineer is designing a nylon connector to be used in an automotive engine application. The engineer is considering establishing the design specification on wall thickness at 3/32 inch but is somewhat uncertain about the effect of this decision on the connector pull-off force. If the pull-off force is too low, the connector may fail when it is installed in an engine. Eight prototype units are produced and their pull-off forces measured, resulting in the following data (in pounds): 12.6, 12.9, 13.4, 12.3, 13.6, 13.5, 12.6, 13.1. As we anticipated, not all of the prototypes have the same pull-off force. We say that there is variability in the pull-off force measurements. Because the pull-off force measurements exhibit variability, we consider the pull-off force to be a **random variable.** A convenient way to think of a random variable, say X, that represents a measurement, is by using the model

$$X = \mu + \epsilon$$

where μ is a constant and ϵ is a random disturbance. The constant remains the same with every measurement, but small changes in the environment, test equipment, differences in the individual parts themselves, and so forth change the value of ϵ. If there were no disturbances, then ϵ would always equal zero and X would always be equal to the constant μ. However, this never happens in the real world, so the actual measurements X exhibit variability. We often need to describe, quantify and ultimately reduce variability.

Figure 1-2 presents a **dot diagram** of these data. The dot diagram is a very useful plot for displaying a small body of data—say, up to about 20 observations. This plot allows us to see easily two features of the data; the **location,** or the middle, and the **scatter** or **variability.** When the number of observations is small, it is usually difficult to identify any specific pattern in the variability, although the dot diagram is a convenient way to see any unusual data features.

The need for statistical thinking arises often in the solution of engineering problems. Consider the engineer designing the connector. From testing the prototypes, he knows that the average pull-off force is 13.0 pounds. However, he thinks that this may be too

Figure 1-2 Dot diagram of the pull-off force data when wall thickness is 3/32 inch.

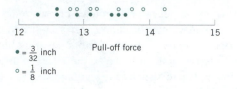

Figure 1-3   Dot diagram of pull-off force for two wall thicknesses.

low for the intended application, so he decides to consider an alternative design with a greater wall thickness, 1/8 inch. Eight prototypes of this design are built, and the observed pull-off force measurements are 12.9, 13.7, 12.8, 13.9, 14.2, 13.2, 13.5, and 13.1. The average is 13.4. Results for both samples are plotted as dot diagrams in Fig. 1-3. This display gives the impression that increasing the wall thickness has led to an increase in pull-off force. However, there are some obvious questions to ask. For instance, how do we know that another sample of prototypes will not give different results? Is a sample of eight prototypes adequate to give reliable results? If we use the test results obtained so far to conclude that increasing the wall thickness increases the strength, what risks are associated with this decision? For example, is it possible that the apparent increase in pull-off force observed in the thicker prototypes is only due to the inherent variability in the system and that increasing the thickness of the part (and its cost) really has no effect on the pull-off force?

Often, physical laws (such as Ohm's law and the ideal gas law) are applied to help design products and processes. We are familiar with this reasoning from general laws to specific cases. However, it is also important to reason from a specific set of measurements to more general cases to answer the previous questions. This reasoning is from a sample (such as the eight connectors) to a population (such as the connectors that will be sold to customers), and is referred to as **statistical inference.** See Fig. 1-4. Clearly, reasoning based on measurements from some objects to measurements on all objects can result in errors (called sampling errors). However, if the sample is selected properly, these risks can be quantified and an appropriate sample size can be determined.

Engineers and scientists are also often interested in comparing two different conditions to determine whether either condition produces a significant effect on the response that is observed. These conditions are sometimes called "treatments." The connector pull-off

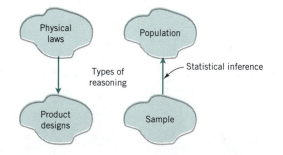

Figure 1-4   Statistical inference is one type of reasoning.

force problem illustrates such a situation; the two different treatments are the two wall thicknesses, and the response is the pull-off force. The purpose of the study is to determine whether the thicker wall results in a significant effect—increased pull-off force. We can think of each example of 8 prototype parts as a random and representative sample of all parts that will ultimately be manufactured. The order in which each part was tested was also randomly determined. This is an example of a **completely randomized experiment.**

When statistical significance is observed in a randomized experiment, the experimenter can be confident in the conclusion that it was the difference in treatments that resulted in the difference in response. That is, we can be confident that a cause-and-effect relationship has been found.

Sometimes the objects to be used in the comparison are not assigned at random to the treatments. For example, the September 1992 issue of *Circulation* (a medical journal published by the American Heart Association) reports a study linking high iron levels in the body with increased risk of heart attack. The study, done in Finland, tracked 1931 men for five years and showed a statistically significant effect of increasing iron levels on the incidence of heart attacks. In this study, the comparison was not performed by randomly selecting a sample of men and then assigning some to a "low iron level" treatment and the others to a "high iron level" treatment. The researchers just tracked the subjects over time. This type of study is called an **observational study.** Designed experiments and observational studies are discussed in more detail in the next section.

It is difficult to identify causality in observational studies, because the observed statistically significant difference in response between the two groups may be due to some other underlying factor (or group of factors) that was not equalized by randomization and not due to the treatments. For example, the difference in heart attack risk could be attributable to the difference in iron levels, or to other underlying factors that form a reasonable explanation for the observed results—such as cholesterol levels or hypertension.

The difficulty of establishing causality from observational studies is also seen in the smoking and health controversy. Numerous studies show that the incidence of lung cancer and other respiratory disorders is higher among smokers than nonsmokers. However, establishing cause and effect here has proven enormously difficult. Many individuals had decided to smoke long before the start of the research studies, and many factors other than smoking could have a role in contracting lung cancer.

1-2 COLLECTING ENGINEERING DATA

In the previous section, we illustrated some simple methods for summarizing data. In the engineering environment, the data are almost always a sample that has been selected from some population. Generally, these data are collected in one of two ways.

The first way that engineers often collect data is from an **observational study.** In this situation, the process or system that is being studied can only be observed by the engineer, and the data are obtained as they become available. For example, suppose that an engineer is evaluating the performance of a manufacturing process producing plastic components by injection molding. He or she can observe the process, select components as they are manufactured, and measure important characteristics of interest, such as the wall thickness, the shrinkage, or the strength of the part. The engineer can also measure and record potentially important process variables, such as the mold temperature, the moisture content of the raw material, and the cycle time. Often, in an observational study the engineer is interested in using the data to construct a model of the system or process. These models are often called empirical models, and they are introduced and illustrated in more detail in the next section. Another way that observational data are obtained is through the analysis of historical data on the system or process. For example, in semiconductor manufacturing it is fairly common to keep extensive records on each batch or lot of wafers that have been produced. These records would include test data on physical and electrical characteristics of the wafers as well as the processing conditions under which each batch of wafers was produced. If questions arise concerning a change in an important electrical characteristic, the process history may be studied in an effort to determine the point in time where the change occurred and to gain some insight regarding which process variables might have been responsible for the change. Often these studies involve very large data sets and require a firm grasp of statistical principles if the engineer is to be successful.

The second way that engineering data are obtained is through a **designed experiment.** In a designed experiment the engineer makes deliberate or purposeful changes in the controllable variables of some system or process, observes the resulting system output data, and then makes an inference or decision about which variables are responsible for the observed changes in output performance. The nylon connector example in the previous section illustrates a designed experiment; that is, a deliberate change was made in the wall thickness of the connector with the objective of discovering whether or not a greater pull-off force could be obtained. Designed experiments play a very important role in engineering design and development, and in the improvement of manufacturing processes. Generally, when products and processes are designed and developed with designed experiments, they enjoy better performance, higher reliability, and lower overall costs. Designed experiments also play a crucial role in reducing the lead time for engineering design and development activities. In Section 1-4 we will illustrate several types of designed experiments for the connector example.

The ability to think and analyze sample data statistically will enable us to answer questions about the system or process under study. For example, consider the problem involving the choice of wall thickness for the nylon connector. An approach that could be used in solving this problem is to compare the mean pull-off force for the 3/32-inch design, to the mean pull-off force for the 1/8-inch design, using the technique of statistical **hypothesis testing.** Chapters 4 and 5 discuss hypothesis testing and other related techniques. Generally, a hypothesis is a statement about some aspect of the system we are interested in. For example, the engineer might be interested in knowing whether the mean pull-off force of a 3/32-inch design exceeds the typical maximum load expected to be encountered in this application—say, 12.75 foot-pounds. Thus we would be interested in testing the hypothesis that the mean strength of the 3/32-inch design exceeds 12.75. This is called a single-sample hypothesis testing problem. Chapter 4 presents techniques for this type of problem. Alternatively, the engineer might be interested in testing the hypothesis that increasing the wall thickness from 3/32 to 1/8 inch results in an increase in mean pull-off force. This is an example of a two-sample hypothesis testing problem. Two-sample hypothesis testing problems are discussed in Chapter 5.

1-3 MECHANISTIC AND EMPIRICAL MODELS

Models play an important role in the analysis of nearly all engineering problems. Much of the formal education of engineers involves learning about the models relevant to specific fields and the techniques for applying these models in problem formulation and solution. As a simple example, suppose we are measuring the current flow in a thin copper wire. Our model for this phenomenon might be Ohm's law:

$$\text{Current} = \text{voltage/resistance}$$

or

$$I = E/R \tag{1-1}$$

We call this type of model a **mechanistic model,** because it is built from our underlying knowledge of the basic physical mechanism that relates these variables. However, if we performed this measurement process more than once, perhaps at different times or even on different days, the observed current could differ slightly because of small changes or variations in factors that are not completely controlled, such as changes in ambient temperature, fluctuations in performance of the gauge, small impurities present at different locations in the wire, and drifts in the voltage source. Consequently, a more realistic model of the observed current might be

$$I = E/R + \epsilon \tag{1-2}$$

where ϵ is a term added to the model to account for the fact that the observed values of current flow do not perfectly conform to the mechanistic model. We can think of ϵ as a term that includes the effects of all of the unmodeled sources of variability that affect this system.

Sometimes engineers work with problems for which there is no simple or well-understood mechanistic model that explains the phenomenon. For instance, suppose we are interested in the number average molecular weight (M_n) of a polymer. Now we know that M_n is related to the viscosity of the material (V), and that it also depends on the amount of catalyst (C) and the temperature (T) in the polymerization reactor when the material is manufactured. The relationship between M_n and these variables is

$$M_n = f(V, C, T) \tag{1-3}$$

say, where the *form* of the function f is unknown. Perhaps a working model could be developed from a first-order Taylor series expansion, which would produce a model of the form

$$M_n = \beta_0 + \beta_1 V + \beta_2 C + \beta_3 T \tag{1-4}$$

where the β's are unknown parameters. As in Ohm's law, this model will not exactly describe the phenomenon, so we should account for the other sources of variability that may affect the molecular weight by adding another term to the model, thus

$$M_n = \beta_0 + \beta_1 V + \beta_2 C + \beta_3 T + \epsilon \qquad (1\text{-}5)$$

is the model that we will use to relate molecular weight to the other three variables. This type of model is called an **empirical model;** that is, it uses our engineering and scientific knowledge of the phenomenon, but it is not directly developed from our theoretical or first-principles understanding of the underlying mechanism.

To illustrate these ideas with a specific example, consider the data in Table 1-1. This table contains data on three variables that were collected in a semiconductor manufacturing plant. In this plant, the finished semiconductor is wire bonded to a frame. The variables reported are pull strength (a measure of the amount of force required to break the bond), the wire length, and the height of the die. We would like to find a model relating pull strength to wire length and die height. Unfortunately, there is no physical mechanism that we can easily apply here, so it doesn't seem likely that a mechanistic modeling approach

Table 1-1   Wire Bond Pull Strength Data

Observation Number	Pull Strength y	Wire Length x_1	Die Height x_2
1	9.95	2	50
2	24.45	8	110
3	31.75	11	120
4	35.00	10	550
5	25.02	8	295
6	16.86	4	200
7	14.38	2	375
8	9.60	2	52
9	24.35	9	100
10	27.50	8	300
11	17.08	4	412
12	37.00	11	400
13	41.95	12	500
14	11.66	2	360
15	21.65	4	205
16	17.89	4	400
17	69.00	20	600
18	10.30	1	585
19	34.93	10	540
20	46.59	15	250
21	44.88	15	290
22	54.12	16	510
23	56.63	17	590
24	22.13	6	100
25	21.15	5	400

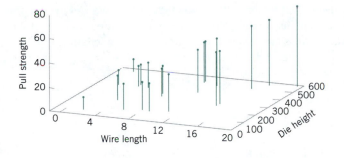

Figure 1-5 Three-dimensional plot of the wire and pull strength data.

will be successful. Note that this is an example of an observational study (refer to Section 1-2).

Figure 1-5 presents a three-dimensional plot of all 25 observations on pull strength, wire length, and die height. From examination of this plot, we see that pull strength increases as both wire length and die height increase. Furthermore, it seems reasonable to think that a model such as

$$\text{Pull strength} = \beta_0 + \beta_1(\text{wire length}) + \beta_2(\text{die height}) + \epsilon$$

would be appropriate as an empirical model for this relationship. In general, this type of empirical model is called a **regression model.** In Chapter 6 we show how to build these models and test their adequacy as approximating functions. Chapter 6 presents a method for estimating the parameters in regression models, called the method of least squares, that traces its origins to work by Karl Gauss. Essentially, this method chooses the parameters in the empirical model (the β's) to minimize the sum of the squared distances between each data point and the plane represented by the model equation. It turns out that applying this technique to the data in Table 1-1 results in

$$\widehat{\text{Pull strength}} = 2.26 + 2.74(\text{wire length}) + 0.0125(\text{die height}) \qquad (1\text{-}6)$$

where the "hat" or circumflex over pull strength indicates that this is an estimated quantity.

Figure 1-6 is a plot of the predicted values of pull strength versus wire length and die height obtained from equation 1-11. Note that the predicted values lie on a plane

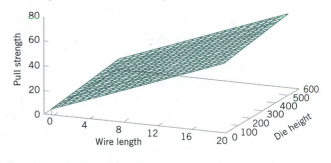

Figure 1-6 Plot of estimated values of pull strength from the empirical model in equation (1-6).

above the wire length–die height space. From the plot of the data in Fig. 1-5, this model does not appear unreasonable. The empirical model in equation 1-6 could be used to predict values of pull strength for various combinations of wire length and die height that are of interest. Essentially, the empirical model can be used by an engineer in exactly the same way that a mechanistic model can be used.

1-4   DESIGNING EXPERIMENTAL INVESTIGATIONS

Much of what we know in the engineering and physical-chemical sciences is developed through testing or experimentation. Often, engineers work in problem areas in which no scientific or engineering theory is directly or completely applicable, so experimentation and analysis of the resulting data constitute the only way that the problem can be solved. Even when there is a good underlying scientific theory that we may rely on to explain the phenomena of interest, it is almost always necessary to conduct tests or experiments to confirm that the theory is indeed operative in the situation or environment in which it is being applied. Statistical thinking and statistical methods play an important role in planning, conducting, and analyzing the data from engineering experiments.

Section 1-1 contained a brief example involving an engineer who was investigating the impact of increasing the wall thickness of a connector on the pull-off force. Recall that the engineer built eight prototypes of each design (3/32 and 1/8 inch), tested each unit and computed the sample average and sample standard deviation of pull-off force for each design. We noted that statistical hypothesis testing was a possible framework within which to investigate whether increasing the wall thickness in the design would lead to higher levels of mean pull-off force. This is an illustration of using statistical thinking to assist in analyzing data from a simple comparative experiment.

Statistical thinking can also be applied to more complex experimental problems. To illustrate, reconsider the connector wall thickness problem. Suppose that when the connector is assembled in the application, it is first immersed in an adhesive, and then the assembly is cured by applying heat over some time period. The pull-off force is measured on the final assembly. The engineer suspects that in addition to wall thickness, the cure time and cure temperature could have some effect on the performance of the connector. Therefore, it is necessary to design an experiment that will allow us to investigate the effect of all three factors on pull-off force.

When several factors are potentially important, the best strategy of experimentation is to design some type of **factorial experiment.** A factorial experiment is an experiment in which factors are varied together. To illustrate, suppose that in the connector experiment, the cure times of interest are 1 and 24 hours, and the temperature levels are 70°F and 100°F. Now since all three factors have two levels, a factorial experiment would consist of the eight test combinations shown at the corners of the cube in Fig. 1-7. Two trials, or **replicates,** would be performed at each corner, resulting in a 16-run factorial experiment. The observed values of pull-off force are shown in parentheses at the cube corners in Fig. 1-7. Note that this experiment uses eight 3/32-inch prototypes and eight 1/8-inch prototypes, the same number used in the simple comparative study in Section 1-1, but

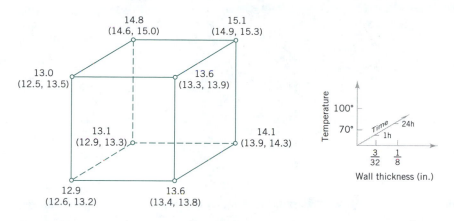

Figure 1-7 The factorial experiment for the connector wall thickness problem.

we are now investigating *three* factors. Generally, factorial experiments are the most efficient way to study the joint effects of several factors.

Some very interesting tentative conclusions can be drawn from this experiment. First, compare the average pull-off force of the eight 3/32-inch prototypes with the average pull-off force of the eight 1/8-inch prototypes (these are the averages of the eight runs on the left face and right face of the cube in Fig. 1-7, respectively), or $14.1 - 13.45 = 0.65$. Thus, increasing the wall thickness from 3/32 to 1/8 inch increases the average pull-off force by 0.65 pounds. Next, to measure the effect of increasing the cure time, compare the average of the eight runs in the back face of the cube (where time = 24 h) with the average of the eight runs in the front face (where time = 1 h), or $14.275 - 13.275 = 1$. The effect of increasing the cure time from 1 to 24 hours is to increase the average pull-off force by 1 pound; that is, cure time apparently has an effect that is larger than the effect of increasing the wall thickness. The cure temperature effect can be evaluated by comparing the average of the eight runs in the top of the cube (where temperature = 100°F) with the average of the eight runs in the bottom (where temperature = 70°F), or $14.125 - 13.425 = 0.7$. The effect of increasing the cure temperature is to increase the average pull-off force by 0.7 pounds. Thus, if the engineer's objective is to design a connector with high pull-off force, there are apparently several alternatives, such as increasing the wall thickness and using the "standard" curing conditions of 1 hour and 70°F or using the original 3/32-inch wall thickness but specifying a longer cure time and higher cure temperature.

There is an interesting relationship between cure time and cure temperature that can be seen by examination of the graph in Fig. 1-8. This graph was constructed by calculating the average pull-off force at the four different combinations of time and temperature, plotting these averages versus time, and then connecting the points representing the two temperature levels with straight lines. The slope of each of these straight lines represents the effect of cure time on pull-off force. Note that the slopes of these two lines do not appear to be the same, indicating that the cure time effect is *different* at the two values of cure temperature. This is an example of an **interaction** between two factors. The

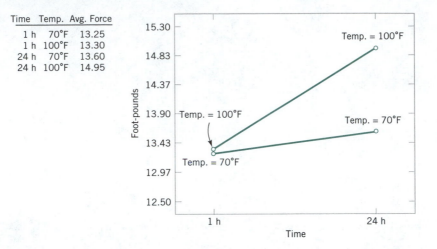

Time	Temp.	Avg. Force
1 h	70°F	13.25
1 h	100°F	13.30
24 h	70°F	13.60
24 h	100°F	14.95

Figure 1-8   The two-factor interaction between cure time and cure temperature.

interpretation of this interaction is very straightforward; if the standard cure time (1 h) is used, cure temperature has little effect, but if the longer cure time (24 h) is used, increasing the cure temperature has a large effect on average pull-off force. Interactions occur often in physical and chemical systems, and factorial experiments are the only way to investigate their effects. In fact, if interactions are present and the factorial experimental strategy is not used, incorrect or misleading results may be obtained.

We can easily extend the factorial strategy to more factors. Suppose that the engineer wants to consider a fourth factor, type of adhesive. There are two types: the standard adhesive and a new competitor. Figure 1-9 illustrates how all four factors, wall thickness, cure time, cure temperature, and type of adhesive, could be investigated in a factorial design. Since all four factors are still at two levels, the experimental design can still be represented geometrically as a cube (actually, it's a *hypercube*). Note that as in any factorial design, all possible combinations of the four factors are tested. The experiment requires 16 trials.

Generally, if there are k factors and they each have two levels, a factorial experimental design will require 2^k runs. For example, with $k = 4$, the 2^4 design in Fig. 1-9 requires 16 tests. Clearly, as the number of factors increases, the number of trials required in a

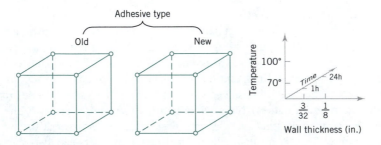

Figure 1-9   A four-factorial experiment for the connector wall thickness problem.

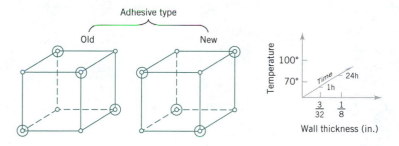

Figure 1-10 A fractional factorial experiment for the connector wall thickness problem.

factorial experiment increases rapidly; for instance, eight factors each at two levels would require 256 trials. This amount of testing quickly becomes unfeasible from the viewpoint of time and other resources. Fortunately, when there are four to five or more factors, it is usually unnecessary to test all possible combinations of factor levels. A **fractional factorial experiment** is a variation of the basic factorial arrangement in which only a subset of the factor combinations are actually tested. Figure 1-10 shows a fractional factorial experimental design for the four-factor version of the connector experiment. The circled test combinations in this figure are the only test combinations that need to be run. This experimental design requires only 8 runs instead of the original 16; consequently it would be called a **one-half fraction.** This is an excellent experimental design in which to study all four factors. It will provide good information about the individual effects of the four factors and some information about how these factors interact.

Factorial and fractional factorial experiments are used extensively by engineers and scientists in industrial research and development, where new technology, products, and processes are designed and developed and where existing products and processes are improved. Since so much engineering work involves testing and experimentation, it is essential that all engineers understand the basic principles of planning efficient and effective experiments. Chapter 7 focuses on these principles, concentrating on the factorial and fractional factorials that we have introduced here.

1-5   OBSERVING PROCESSES OVER TIME

Whenever data are collected over time it is important to plot the data over time. Phenomena that might affect the system or process often become more visible in a time-oriented plot and the concept of stability can be better judged.

Figure 1-11 is a dot diagram of concentration readings taken periodically from a chemical process. The large variation displayed on the dot diagram indicates a possible problem, but the chart does not help explain the reason for the variation. Because the data are collected over time, they are plotted over time in Fig. 1-12. A shift in the process mean level is visible in the plot and an estimate of the time of the shift can be obtained.

The quality guru W. Edwards Deming stressed that it is important to understand the nature of variation over time. He conducted an experiment in which he attempted to drop

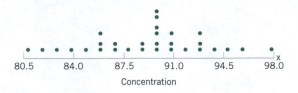

Figure 1-11   A dot diagram illustrates variation but does not identify the problem.

marbles as close as possible to a target on a table. He used a funnel mounted on a ring stand and the marbles were dropped into the funnel. See Fig. 1-13. The funnel was aligned as closely as possible with the center of the target. Deming then used two different strategies to operate the process. (1) He never moved the funnel. He simply dropped one marble after another and recorded the distance from the target. (2) He dropped the first marble and recorded its location relative to the target. He then moved the funnel an equal and opposite distance in an attempt to compensate for the error. He continued to make this type of adjustment after each marble was dropped.

After both strategies were completed Deming noticed that the variability in the distance from the target for strategy 2 was approximately two times larger than for strategy 1. The adjustments to the funnel increased the deviations from the target. The explanation is that the error (the deviation of the marble's position from the target) for one marble provides no information about the error that will occur for the next marble. Consequently, adjustments to the funnel do not decrease future errors. Instead, they tend to move the funnel farther from the target.

This interesting experiment points out that adjustments to a process based on random disturbances can actually *increase* the variation of the process. This is referred to as **overcontrol** or **tampering.** Adjustments should be applied only to compensate for a nonrandom shift in the process—then they can help. A computer simulation can be used

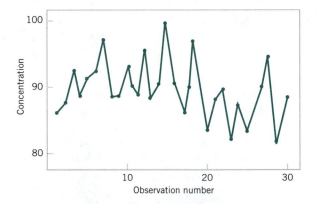

Figure 1-12   A time plot of concentration provides more information than the histogram.

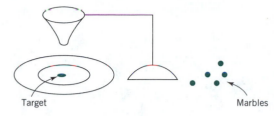

Target   Marbles

Figure 1-13   Deming's funnel experiment.

to demonstrate the lessons of the funnel experiment. Figure 1-14(a) displays a time plot of 50 measurements (denoted as y) from a process in which only random disturbances are present. The target value for the process is 10 units. Figure 1-14(b) displays the same data after adjustments are applied to the process mean in an attempt to produce data closer to the target. Each adjustment is equal and opposite to the deviation of the previous measurement from target. For example, when the measurement is 11 (one unit above target), the mean is reduced by one unit before the next measurement is generated. The overcontrol has increased the deviations from target.

Figure 1-15(a) displays the same data as in Fig. 1-14(a) except that the measurements after observation number 25 are increased by two units to simulate the effect of a shift in the mean of the process. When there is a true shift in the mean of a process, an adjustment can be useful. Figure 1-15(b) displays the data obtained when an adjustment (decrease two units) is applied to the mean after the shift is detected (at observation number 28). Note that this adjustment decreases the deviations from target.

The question of when to apply adjustments (and by what amounts) begins with an understanding of the types of variation that affect a process. A **control chart** is an invaluable way to examine the variability in time-oriented data. Figure 1-16 presents a control chart for the chemical process concentration data from Fig. 1-12. The *center line* on the control chart is just the average of the concentration measurements for the first 20 samples (91.5 g/l) when the process is stable. The *upper control limit* and the *lower*

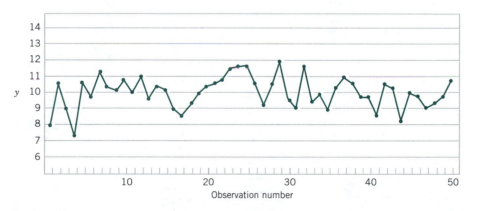

Figure 1-14(a)   Process data with only random disturbances.

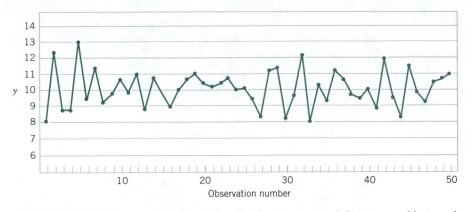

Figure 1-14(b) Adjustments applied to random disturbances overcontrol the process and increase the deviations from target.

control limit are a pair of statistically derived limits that reflect the inherent or natural variability in the process. These limits are located at appropriate values above and below the center line. If the process is operating as it should, without any external sources of variability present in the system, the concentration measurements should fluctuate randomly around the center line and almost all of them should fall between the control limits.

In the control chart of Fig. 1-16, the visual frame of reference provided by the center line and the control limits indicates that some upset or disturbance has impacted the process around sample 20, because all of the following observations are below the center line and two of them actually fall below the lower control limit. This is a very strong signal that corrective action is required in this process. If we can find and eliminate the underlying cause of this upset, we can improve process performance considerably.

Control charts are critical to engineering analyses for the following reason. In some cases, the data in our sample is actually selected from the population of interest. The

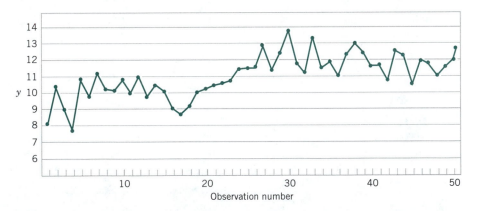

Figure 1-15(a) Process mean shifts (upward by two units) following observation number 25.

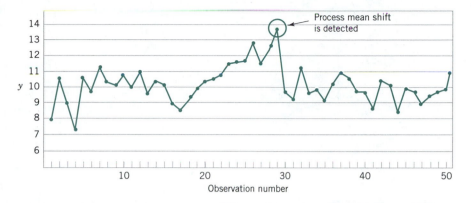

Figure 1-15(b) Process mean shift is detected at observation number 28 and the following measurements are decreased by two units.

sample is a subset of the population. For example, a sample of three wafers might be selected from a production lot of wafers in semiconductor manufacturing. Based on data in the sample, we want to conclude something about the lot. For example, the average of the resistivity measurements in the sample is not expected to exactly equal the average of the resistivity measurements in the lot. However, if the sample average is high we might be concerned that the lot average is too high.

In many other cases, we use the current data to make conclusions about the future performance of a process. For example, not only are we interested in the concentration measurements in Fig. 1-12. We also want to make conclusions about the concentration of future production that will be sold to customers. This population of future production does not yet exist. Clearly, this analysis requires some notion of stability as an additional assumption. For example, it might be assumed that the current sources of variability in production are the same as those for future production. A control chart is the fundamental tool to evaluate the stability of a process.

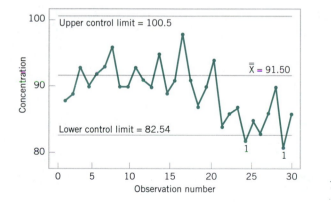

Figure 1-16 A control chart for the chemical process concentration data.

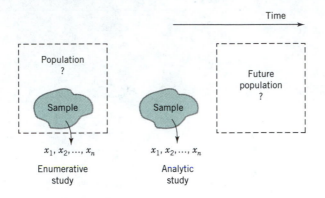

Figure 1-17 Enumerative versus analytic study.

The wafers-from-lots example is called an **enumerative** study. A sample is used to make an inference to the population from which the sample is selected. The concentration example is called an **analytic** study. A sample of data is used to make an inference to a future population. The statistical analyses are usually the same in both cases, but an analytic study should begin with a control chart to evaluate stability. Figure 1-17 provides an illustration.

Control charts are a very important application of statistics for monitoring, controlling, and improving a process. The branch of statistics that makes use of control charts is called **statistical process control** or **SPC.** We will discuss SPC and control charts in Chapter 8.

Data Summary
and Presentation

2-1   DATA SUMMARY AND DISPLAY

Well-constituted data summaries and displays are essential to good statistical thinking because they can focus the engineer on important features of the data or provide insight about the type of model that should be used in problem solution.

The computer has become an important tool in the presentation and analysis of data. Although many statistical techniques require only a hand-held calculator, much time and effort may be required by this approach, and a computer will perform the tasks much more efficiently.

Most statistical analysis is done using a prewritten library of statistical programs. The user enters the data and then selects the types of analysis and output displays that are of interest. Statistical software packages are available for both mainframe machines and personal computers. Among the most popular and widely used packages are SAS (Statistical Analysis System) for both servers and personal computers (PCs) and Statgraphics and Minitab for the PC. We will present some examples of output from several statistics

19

packages throughout the book. We will not discuss the hands-on use of the packages for entering and editing data or using commands. You will find these packages, or similar ones, available at your institution, along with local expertise in their use.

We can describe data features numerically. For example, we can characterize the location or central tendency in the data by the ordinary arithmetic average or mean. Because we almost always think of our data as a sample, we will refer to the arithmetic mean as the **sample mean.**

Definition

If the n observations in a sample are denoted by $x_1, x_2, \ldots, x_n$, then the **sample mean** is

$$\bar{x} = \frac{x_1 + x_2 + \cdots + x_n}{n}$$

$$= \frac{\displaystyle\sum_{i=1}^{n} x_i}{n} \qquad (2\text{-}1)$$

EXAMPLE 2-1

Consider the prototype connection described in Chapter 1. The data are shown in Fig. 2-1. The sample mean pull-off force for the eight observations on pull-off force is

$$\bar{x} = \frac{x_1 + x_2 + \cdots x_n}{n} = \frac{\displaystyle\sum_{i=1}^{8} x_i}{8} = \frac{12.6 + 12.9 + \cdots + 13.1}{8}$$

$$= \frac{104}{8} = 13.0$$

A physical interpretation of the sample mean as a measure of location is shown in Fig. 2-2, which is a dot diagram of the pull-off force data. Note that the sample mean $\bar{x} = 13.0$ can be thought of as a "balance point." That is, if each observation represents 1 pound of mass placed at the point on the x-axis, then a fulcrum located at $\bar{x}$ would exactly balance this system of weights.

The sample mean is the average value of all the observations in the data set. Usually, these data are a **sample** of observations that have been selected from some larger **population** of observations. Here the population might consist of all the connectors that will be

Figure 2-1 Dot diagram of the pull-off force data when wall thickness is 3/32 inch.

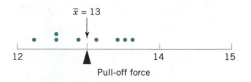

Figure 2-2 The sample mean as a balance point for a system of weights.

sold to customers. Sometimes there is an actual physical population, such as a lot of silicon wafers produced in a semiconductor factory. We could also think of calculating the average value of all the observations in a population. This average is called the **population mean,** and it is denoted by the Greek letter μ (mu).

When there are a finite number of observations (say, N) in the population, then the population mean is

$$\mu = \frac{\sum\limits_{i=1}^{N} x_i}{N} \qquad (2\text{-}2)$$

The sample mean, $\bar{x}$, is a reasonable estimate of the population mean, μ. Therefore, the engineer designing the connector using a 3/32-inch wall thickness would conclude, on the basis of the data, that an estimate of the mean pull-off force is 13.0 pounds.

Although the sample mean is useful, it does not convey all of the information about a sample of data. The variability or scatter in the data may be described by the **sample variance** or the **sample standard deviation.**

Definition

If the n observations in a sample are denoted by $x_1, x_2, \ldots, x_n$, then the **sample variance** is

$$s^2 = \frac{\sum\limits_{i=1}^{n} (x_i - \bar{x})^2}{n-1} \qquad (2\text{-}3)$$

The **sample standard deviation,** s, is the positive square root of the sample variance.

The units of measurements for the sample variance are the square of the original units of the variable. Thus, if x is measured in pounds, the units for the sample variance are (pounds)^2. The standard deviation has the desirable property of measuring variability in the original units of the variable of interest, x.

How Does the Sample Variance Measure Variability?

To see how the sample variance measures dispersion or variability, refer to Fig. 2-3, which shows the deviations $x_i - \bar{x}$ for the connector pull-off force data. The greater the amount of variability in the pull-off force data, the larger in absolute magnitude some of the deviations $x_i - \bar{x}$ will be. Since the deviations $x_i - \bar{x}$ always sum to zero, we must use a measure of variability that changes the negative deviations to nonnegative quantities. Squaring the deviations is the approach used in the sample variance. Consequently, if s^2 is small, then there is relatively little variability in the data, but if s^2 is large, the variability is relatively large.

EXAMPLE 2-2

Table 2-1 displays the quantities needed for calculating the sample variance and sample standard deviation for the pull-off force data. These data are plotted in Fig. 2-3. The numerator of s^2 is

$$\sum_{i=1}^{8} (x_i - \bar{x})^2 = 1.60$$

so the sample variance is

$$s^2 = \frac{1.60}{8-1} = \frac{1.60}{7} = 0.2286 \text{ (pounds)}^2$$

and the sample standard deviation is

$$s = \sqrt{0.2286} = 0.48 \text{ pounds}$$

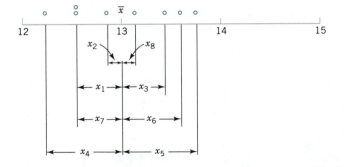

Figure 2-3   How the sample variance measures variability through the deviations $x_i - \bar{x}$.

Table 2-1   Calculation of Terms for the Sample Variance
and Sample Standard Deviation

i	x_i	$x_i - \bar{x}$	$(x_i - \bar{x})^2$
1	12.6	-0.4	0.16
2	12.9	-0.1	0.01
3	13.4	0.4	0.16
4	12.3	-0.7	0.49
5	13.6	0.6	0.36
6	13.5	0.5	0.25
7	12.6	-0.4	0.16
8	13.1	0.1	0.01
	104.0	0.0	1.60

The computation of s^2 requires calculation of $\bar{x}$, n subtractions, and n squaring and adding operations. If the original observations or the deviations $x_i - \bar{x}$ are not integers, the deviations $x_i - \bar{x}$ may be tedious to work with, and several decimals may have to be carried to ensure numerical accuracy. It can be shown that a more efficient computational formula for the sample variance is

$$s^2 = \frac{\sum_{i=1}^{n} (x_i - \bar{x})^2}{n - 1}$$

$$= \frac{\sum_{i=1}^{n} (x_i^2 + \bar{x}^2 - 2\bar{x}x_i)}{n - 1}$$

$$= \frac{\sum_{i=1}^{n} x_i^2 + n\bar{x}^2 - 2\bar{x} \sum_{i=1}^{n} x_i}{n - 1}$$

and since $\bar{x} = (1/n) \sum_{i=1}^{n} x_i$, this last equation reduces to

$$s^2 = \frac{\sum_{i=1}^{n} x_i^2 - \frac{\left(\sum_{i=1}^{n} x_i\right)^2}{n}}{n - 1} \tag{2-4}$$

Note that equation 2-4 requires squaring each individual x_i, then squaring the sum of the x_i, subtracting $(\Sigma x_i)^2/n$ from Σx_i^2, and finally dividing by $n - 1$. Sometimes this is called the shortcut method for calculating s^2 (or s).

EXAMPLE 2-3

We will calculate the sample variance and standard deviation using the shortcut method, equation 2-4. The formula gives

$$s^2 = \frac{\sum\limits_{i=1}^{n} x_i^2 - \frac{\left(\sum\limits_{i=1}^{n} x_i\right)^2}{n}}{n-1} = \frac{1353.6 - \frac{(104)^2}{8}}{7} = \frac{1.60}{7} = 0.2286 \text{ (pounds)}^2$$

and

$$s = \sqrt{0.2286} = 0.48 \text{ pounds}$$

These results agree exactly with those obtained previously.

Analogous to the sample variance s^2, there is a measure of variability in the population called the **population variance.** We will use the Greek letter σ^2 (sigma squared) to denote the population variance. The positive square root of σ^2, or σ, will denote the **population standard deviation.** When the population is finite and consists of N values, we may define the population variance as

$$\sigma^2 = \frac{\sum\limits_{i=1}^{N} (x_i - \mu)^2}{N} \tag{2-5}$$

A more general definition of the variance σ^2 will be given later. We observed previously that the sample mean could be used as an estimate of the population mean. Similarly, the sample variance is an estimate of the population variance.

Note that the divisor for the sample variance is the sample size minus one $(n - 1)$, whereas for the population variance it is the population size N. If we knew the true value of the population mean μ, then we could find the *sample* variance as the average squared deviation of the sample observations about μ. In practice, the value of μ is almost never known, and so the sum of the squared deviations about the sample average $\bar{x}$ must be used instead. However, the observations x_i tend to be closer to their average, $\bar{x}$, than to the population mean, μ. Therefore, to compensate for this we use $n - 1$ as the divisor rather than n. If we used n as the divisor in the sample variance, we would obtain a measure of variability that is, on the average, consistently smaller than the true population variance σ^2.

Another way to think about this is to consider the sample variance s^2 as being based on $n - 1$ **degrees of freedom.** The term *degrees of freedom* results from the fact that the n deviations $x_1 - \bar{x}, x_2 - \bar{x}, \ldots, x_n - \bar{x}$ always sum to zero, and so specifying the values of any $n - 1$ of these quantities automatically determines the remaining one. This was illustrated in Table 1-1. Thus, only $n - 1$ of the n deviations, $x_i - \bar{x}$, are freely determined.

EXERCISES FOR SECTION 2-1

2-1. Eight measurements were made on the inside diameter (in mm) of forged piston rings used in an automobile engine. The coded data are 1, 3, 15, 0, 5, 2, 5, and 4. Calculate the sample average and sample standard deviation, construct a dot diagram, and comment on the data.

2-2. In *Applied Life Data Analysis* (Wiley, 1982), Wayne Nelson presents the breakdown time of an insulating fluid between electrodes at 34 kV. The times, in minutes, are as follows: 0.19, 0.78, 0.96, 1.31, 2.78, 3.16, 4.15, 4.67, 4.85, 6.50, 7.35, 8.01, 8.27, 12.06, 31.75, 32.52, 33.91, 36.71, and 72.89. Calculate the sample average and sample standard deviation.

2-3. Seven oxide thickness measurements of wafers are studied to assess quality in a semiconductor manufacturing process. The data (in angstroms) are: 1264, 1280, 1301, 1300, 1292, 1307, and 1275. Calculate the sample average and sample standard deviation. Construct a dot diagram of the data.

2-4. An article in the *Journal of Structural Engineering* (Vol. 115, 1989) describes an experiment to test the yield strength of circular tubes with caps welded to the ends. The first yields (in kN) are 96, 96,

102, 102, 102, 104, 104, 108, 126, 126, 128, 128, 140, 156, 160, 160, 164, and 170. Calculate the sample average and sample standard deviation. Construct a dot diagram of the data.

2-5. An article in *Human Factors* (June 1989) presented data on visual accommodation (a function of eye movement) when recognizing a speckle pattern on a high-resolution CRT screen. The data are as follows: 36.45, 67.90, 38.77, 42.18, 26.72, 50.77, 39.30, and 49.71. Calculate the sample average and sample standard deviation. Construct a dot diagram of the data.

2-6. The following data are direct solar intensity measurements (watts/m^2) on different days at a location in southern Spain: 562, 869, 708, 775, 775, 704, 809, 856, 655, 806, 878, 909, 918, 558, 768, 870, 918, 940, 946, 661, 820, 898, 935, 952, 957, 693, 835, 905, 939, 955, 960, 498, 653, 730, and 753. Calculate the sample average and sample standard deviation.

2-7. For each of Exercises 2-1 through 2-6, discuss whether the data result from an observational study or a designed experiment.

2-2   STEM-AND-LEAF DIAGRAM

The dot diagram is a useful data display for small samples, up to (say) about 20 observations. However, when the number of observations is moderately large, other graphical displays may be more useful.

For example, consider the data in Table 2-2. These data are the compressive strengths in pounds per square inch (psi) of 80 specimens of a new aluminum-lithium alloy undergoing evaluation as a possible material for aircraft structural elements. The data were recorded in the order of testing, and in this format they do not convey much information about compressive strengths. Questions such as "What percent of the specimens fail below 120 psi?" are not easy to answer. Because there are many observations, constructing a dot diagram of these data would be relatively inefficient; more effective displays are available for large data sets.

Table 2-2   Compressive Strength of 80 Aluminum-Lithium Alloy Specimens

105	221	183	186	121	181	180	143
97	154	153	174	120	168	167	141
245	228	174	199	181	158	176	110
163	131	154	115	160	208	158	133
207	180	190	193	194	133	156	123
134	178	76	167	184	135	229	146
218	157	101	171	165	172	158	169
199	151	142	163	145	171	148	158
160	175	149	87	160	237	150	135
196	201	200	176	150	170	118	149

A **stem-and-leaf diagram** is a good way to obtain an informative visual display of a data set $x_1, x_2, \ldots, x_n$, where each number x_i consists of at least two digits. To construct a stem-and-leaf diagram, we divide each number x_i into two parts: a **stem,** consisting of one or more of the leading digits, and a **leaf,** consisting of the remaining digits. To illustrate, if the data consist of percent defective information between 0 and 100 on lots of semiconductor wafers, then we can divide the value 76 into the stem 7 and the leaf 6. In general, we should choose relatively few stems in comparison with the number of observations. It is usually best to choose between 5 and 20 items. Once a set of stems has been chosen, they are listed along the left-hand margin of the diagram. Besides each stem all leaves corresponding to the observed data values are listed in the order in which they are encountered in the data set.

EXAMPLE 2-4

To illustrate the construction of a stem-and-leaf diagram, consider the alloy compressive strength data in Table 2-2. We will select as stem values the numbers 7, 8, 9, . . . , 24. The resulting stem-and-leaf diagram is presented in Fig. 2-4. The last column in the diagram is a frequency count of the number of leaves associated with each stem. Inspection of this display immediately reveals that most of the compressive strengths lie between 110 and 200 psi and that a central value is somewhere between 150 and 160 psi. Furthermore, the strengths are distributed approximately symmetrically about the central value. The stem-and-leaf diagram enables us to determine quickly some important features of the data that were not immediately obvious in the original display in the table.

In some data sets, it may be desirable to provide more classes or stems. One way to do this is to modify the original stems as follows: Divide the stem 5 (say) into two new stems, 5L and 5U. The stem 5L has leaves 0, 1, 2, 3, and 4, and stem 5U has leaves 5, 6, 7, 8, and 9. This will double the number of original stems. We could increase the number of original stems by four by defining five new stems: 5z with leaves 0 and 1, 5t (for twos and three) with leaves 2 and 3, 5f (for fours and fives) with leaves 4 and 5, 5s (for six and seven) with leaves 6 and 7, and 5e with leaves 8 and 9.

Stem	Leaf	Frequency
7	6	1
8	7	1
9	7	1
10	5 1	2
11	5 8 0	3
12	1 0 3	3
13	4 1 3 5 3 5	6
14	2 9 5 8 3 1 6 9	8
15	4 7 1 3 4 0 8 8 6 8 0 8	12
16	3 0 7 3 0 5 0 8 7 9	10
17	8 5 4 4 1 6 2 1 0 6	10
18	0 3 6 1 4 1 0	7
19	9 6 0 9 3 4	6
20	7 1 0 8	4
21	8	1
22	1 8 9	3
23	7	1
24	5	1

Figure 2-4 Stem-and-leaf diagram for the compressive strength data in Table 2-2.

EXAMPLE 2-5

Figure 2-5 illustrates the stem-and-leaf diagram for 25 observations on batch yields from a chemical process. In Fig. 2-5*a* we used 6, 7, 8, and 9 as the stems. This results in too few stems, and the stem-and-leaf diagram does not provide much information about the data. In Fig. 2-5*b* we divided each stem into two parts, resulting in a display that more adequately displays the data. Figure 2-5*c* illustrates a stem-and-leaf display with each stem divided into five parts. There are too many stems in this plot, resulting in a display that does not tell us much about the shape of the data.

Figure 2-6 shows a stem-and-leaf display of the compressive strength data in Table 2-2 produced by Minitab. The software uses the same stems as in Fig. 2-4. Note also that the computer orders the leaves from smallest to largest on each stem. This form of the plot is usually called an **ordered stem-and-leaf display.** This ordering is not usually done when the plot is constructed manually because it can be time consuming. The computer adds a column to the left of the stems that provides a count of the observations at and above each stem in the upper half of the display and a count of the observations at and below each stem in the lower half of the display. At the middle stem of 16, the column indicates the number of observations at this stem.

The ordered stem-and-leaf display makes it relatively easy to find data features such as percentiles, quartiles, and the median. The **median** is a measure of central tendency that divides the data into two equal parts, half below the median and half above. If the number of observations is even, the median is halfway between the two central values.

(a)	
Stem	Leaf
6	1 3 4 5 5 6
7	0 1 1 3 5 7 8 8 9
8	1 3 4 4 7 8 8
9	2 3 5

(b)	
Stem	Leaf
6L	1 3 4
6U	5 5 6
7L	0 1 1 3
7U	5 7 8 8 9
8L	1 3 4 4
8U	7 8 8
9L	2 3
9U	5

(c)	
Stem	Leaf
6z	1
6t	3
6f	4 5 5
6s	6
6e	
7z	0 1 1
7t	3
7f	5
7s	7
7e	8 8 9
8z	1
8t	3
8f	4 4
8s	7
8e	8 8
9z	
9t	2 3
9f	5
9s	
9e	

Figure 2-5   Stem-and-leaf displays for Example 2-5.

From Fig. 2-6 we find the 40th and 41st values of strength as 160 and 163, so the median is $(160 + 163)/2 = 161.5$. If the number of observations is odd, the median is the central value. The **range** is a measure of variability that can be easily computed from the ordered stem-and-leaf display. It is the maximum minus the minimum measurement. From Fig. 2-6 the range is $245 - 76 = 169$.

We can also divide data into more than two parts. When an ordered set of data is divided into four equal parts, the division points are called **quartiles.** The *first* or *lower quartile, q_1,* is a value that has approximately one-fourth (25%) of the observations below it and approximately 75% of the observations above. The *second quartile, q_2,* has approximately one-half (50%) of the observations below its value. The second quartile is exactly equal to the median. The *third* or *upper quartile, q_3,* has approximately three-fourths (75%) of the observations below its value. As in the case of the median, the quartiles may not be unique. The compressive strength data in Fig. 2-6 contains $n = 80$ observations. Minitab software calculates the first and third quartiles as the $(n + 1)/4$ and $3(n + 1)/4$ ordered observations and interpolates as needed. For example, $(80 + 1)/4 = 20.25$ and $3(80 + 1)/4 = 60.75$. Therefore, Minitab interpolates between the 20th and 21st ordered observation to obtain $q_1 = 143.50$ and between the 60th and

Character Stem-and-Leaf Display

Stem-and-leaf of Strength   N = 80
Leaf Unit = 1.0

```
   1     7    6
   2     8    7
   3     9    7
   5    10    1 5
   8    11    0 5 8
  11    12    0 1 3
  17    13    1 3 3 4 5 5
  25    14    1 2 3 5 6 8 9 9
  37    15    0 0 1 3 4 4 6 7 8 8 8 8
 (10)   16    0 0 0 3 3 5 7 7 8 9
  33    17    0 1 1 2 4 4 5 6 6 8
  23    18    0 0 1 1 3 4 6
  16    19    0 3 4 6 9 9
  10    20    0 1 7 8
   6    21    8
   5    22    1 8 9
   2    23    7
   1    24    5
```

Figure 2-6 A stem-and-leaf diagram from Minitab.

61st observation to obtain $q_3 = 181.00$. The **interquartile range** is the difference between the upper and lower quartiles, and it is sometimes used as a measure of variability. In general, the $100k$th **percentile** is a data value such that approximately $100k\%$ of the observations are at or below this value and approximately $100(1 - k)\%$ of them are above it.

Many statistics software packages provide data summaries that include these quantities. The output obtained for the compressive strength data in Table 2-2 from Minitab is shown in Table 2-3. Note that the results for the median and the quartiles agree with those given previously. The SE mean will be discussed in a later chapter.

Table 2-3 Summary Statistics for the Compressive Strength Data from Minitab

Variable	N	Mean	Median	StDev	SE Mean
	80	162.66	161.50	33.77	3.78
	Min	Max	Q1	Q3	
	76.00	245.00	143.50	181.00	

EXERCISES FOR SECTION 2-2

2-8. An article in *Technometrics* (Vol. 19, 1977, p. 425) presents the following data on the motor fuel octane ratings of several blends of gasoline:

88.5	87.7	83.4	86.7	87.5
94.7	91.1	91.0	94.2	87.8
84.3	86.7	88.2	90.8	88.3
90.1	93.4	88.5	90.1	89.2
89.0	96.1	93.3	91.8	92.3
89.8	89.6	87.4	88.4	88.9
91.6	90.4	91.1	92.6	89.8
90.3	91.6	90.5	93.7	92.7
90.0	90.7	100.3	96.5	93.3
91.5	88.6	87.6	84.3	86.7
89.9	88.3	92.7	93.2	91.0
98.8	94.2	87.9	88.6	90.9
88.3	85.3	93.0	88.7	89.9
90.4	90.1	94.4	92.7	91.8
91.2	89.3	90.4	89.3	89.7
90.6	91.1	91.2	91.0	92.2
92.2	92.2			

Construct a stem-and-leaf display for these data.

2-9. The following data are the numbers of cycles to failure of aluminum test coupons subjected to repeated alternating stress at 21,000 psi, 18 cycles per second:

1115	1567	1223	1782	1055
1310	1883	375	1522	1764
1540	1203	2265	1792	1330
1502	1270	1910	1000	1608
1258	1015	1018	1820	1535
1315	845	1452	1940	1781
1085	1674	1890	1120	1750
798	1016	2100	910	1501
1020	1102	1594	1730	1238
865	1605	2023	1102	990
2130	706	1315	1578	1468
1421	2215	1269	758	1512
1109	785	1260	1416	1750
1481	885	1888	1560	1642

(a) Construct a stem-and-leaf display for these data.

(b) Does it appear likely that a coupon will "survive" beyond 2000 cycles? Justify your answer.

2-10. The percentage of cotton in material used to manufacture men's shirts is given here. Construct a stem-and-leaf display for the data.

34.2	33.6	33.8	34.7
33.1	34.7	34.2	33.6
34.5	35.0	33.4	32.5
35.6	35.4	34.7	34.1
36.3	36.2	34.6	35.1
35.1	36.8	35.2	36.8
34.7	35.1	35.0	37.9
33.6	35.3	34.9	36.4
37.8	32.6	35.8	34.6
36.6	33.1	37.6	33.6
35.4	34.6	37.3	34.1
34.6	35.9	34.6	34.7
33.8	34.7	35.5	35.7
37.1	33.6	32.8	36.8
34.0	32.9	32.1	34.3
34.1	33.5	34.5	32.7

2-11. The data that follow represent the yield on 90 consecutive batches of ceramic substrate to which a metal coating has been applied by a vapor-deposition process. Construct a stem-and-leaf display for these data.

94.1	87.3	94.1	92.4	84.6	85.4
93.2	84.1	92.1	90.6	83.6	86.6
90.6	90.1	96.4	89.1	85.4	91.7
91.4	95.2	88.2	88.8	89.7	87.5
88.2	86.1	86.4	86.4	87.6	84.2
86.1	94.3	85.0	85.1	85.1	85.1
95.1	93.2	84.9	84.0	89.6	90.5
90.0	86.7	78.3	93.7	90.0	95.6
92.4	83.0	89.6	87.7	90.1	88.3
87.3	95.3	90.3	90.6	94.3	84.1
86.6	94.1	93.1	89.4	97.3	83.7
91.2	97.8	94.6	88.6	96.8	82.9
86.1	93.1	96.3	84.1	94.4	87.3
90.4	86.4	94.7	82.6	96.1	86.4
89.1	87.6	91.1	83.1	98.0	84.5

2-12. Find the median and the quartiles for the motor fuel octane data in Exercise 2-8.

2-13. Find the median and the quartiles for the failure data in Exercise 2-9.

2-14. Find the median, mode, and sample average for the data in Exercise 2-10. Explain how these three measures of location describe different features in the data.

2-15. Find the median and the quartiles for the yield data in Exercise 2-11.

2-3 FREQUENCY DISTRIBUTION AND HISTOGRAM

A **frequency distribution** is a more compact summary of data than a stem-and-leaf diagram. To construct a frequency distribution, we must divide the range of the data into intervals, which are usually called **class intervals, cells,** or **bins.** If possible, the bins should be of equal width in order to enhance the visual information in the frequency distribution. Some judgment must be used in selecting the number of bins so that a reasonable display can be developed. The number of bins depends on the number of observations and the amount of scatter or dispersion in the data. A frequency distribution that uses either too few or too many bins will not be informative. We usually find that between 5 and 20 bins is satisfactory in most cases and that the number of bins should increase with n. Choosing the number of bins approximately equal to the square root of the number of observations often works well in practice.

A frequency distribution for the comprehensive strength data in Table 2-2 is shown in Table 2-4. Since the data set contains 80 observations and since $\sqrt{80} \approx 9$, we suspect that about 8 to 9 bins will provide a satisfactory frequency distribution. The largest and smallest data values are 245 and 76, respectively, so the bins must cover a range of at least $245 - 76 = 169$ units on the psi scale. If we want the lower limit for the first bin to begin slightly below the smallest data value and the upper limit for the last bin to be slightly above the largest data value, then we might start the frequency distribution at 70 and end it at 250. This is an interval or range of 180 psi units. Nine bins, each of width

Table 2-4 Frequency Distribution for the Compressive Strength Data in Table 2-2

Class Interval (psi)	Tally	Frequency	Relative Frequency	Cumulative Relative Frequency
$70 \le x < 90$	||	2	0.0250 2/80	0.0250
$90 \le x < 110$	|||	3	0.0375 3/80	0.0625
$110 \le x < 130$	⊮ |	6	0.0750	0.1375
$130 \le x < 150$	⊮ ⊮ ||||	14	0.1750	0.3125
$150 \le x < 170$	⊮ ⊮ ⊮ ⊮ ||	22	0.2750	0.5875
$170 \le x < 190$	⊮ ⊮ ⊮ ||	17	0.2125	0.8000
$190 \le x < 210$	⊮ ⊮	10	0.1250	0.9250
$210 \le x < 230$	||||	4	0.0500	0.9750
$230 \le x < 250$	||	2	0.0250	1.0000

20 psi, give a reasonable frequency distribution, so the frequency distribution in the table is based on 9 bins.

The fourth column of the table contains a **relative frequency distribution.** The relative frequencies are found by dividing the observed frequency in each bin by the total number of observations. The last column of Table 2-4 expresses the relative frequencies on a cumulative basis. Frequency distributions are often easier to interpret than tables of data. For example, from Table 2-4 it is very easy to see that most of the specimens have compressive strengths between 130 and 190 psi and that 97.5% of the specimens fail below 230 psi.

It is also helpful to present the frequency distribution in graphical form, as shown in Fig. 2-7. Such a display is called a **histogram.** To draw a histogram, use the horizontal axis to represent the measurement scale and draw the boundaries of the bins. The vertical axis represents the frequency (or relative frequency) scale. If the class intervals are of equal width, then the heights of the rectangles drawn on the histograms are proportional to the frequencies. If the class intervals are of unequal width, then it is customary to draw rectangles whose areas are proportional to the frequencies. However, histograms are easier to interpret when the class intervals are of equal width. The histogram, like the stem-and-leaf plot, provides a visual impression of the shape of the distribution of the measurements, as well as information about the scatter or dispersion of the data. Note the symmetric, bell-shaped distribution of the strength measurements in Fig. 2-7.

In passing from either the original data or stem-and-leaf diagram to a frequency distribution or histogram, we have lost some information because we no longer have the individual observations. However, this information loss is often small compared with the conciseness and ease of interpretation gained in using the frequency distribution and histogram.

Figure 2-8 shows a histogram of the compressive strength data from Minitab. The "default" settings were used in this histogram, leading to 17 bins. We have noted that histograms may be relatively sensitive to the number of bins and their width. For small data sets, histograms may change dramatically in appearance if the number and/or width

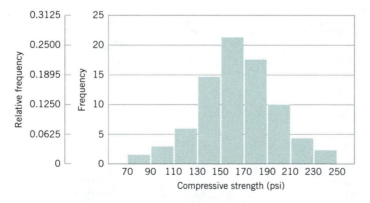

Figure 2-7   Histogram of compressive strength for 80 aluminum-lithium alloy specimens.

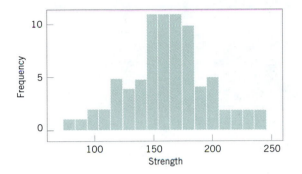

Figure 2-8   A histogram of the compressive strength data from
Minitab with 17 bins.

of the bins changes. Histograms are more stable for larger data sets, preferably of size
75 to 100 or more. Figure 2-9 shows the Minitab histogram for the compressive strength
data with 9 bins. This is similar to the original histogram shown in Fig. 2-7. Since the
number of observations is moderately large ($n = 80$), the choice of the number of bins
is not especially important, and both Figs. 2-8 and 2-9 convey similar information.

Figure 2-10 shows a variation of the histogram available in Minitab, the **cumulative
frequency plot.** In this plot, the height of each bar is the total number of observations
that are less than or equal to the upper limit of the bin. Cumulative distributions are also
useful in data interpretation; for example, we can read directly from Fig. 2-10 that there
are approximately 70 observations less than or equal to 200 psi.

Frequency distributions and histograms can also be used with qualitative or categorical
data. In some applications there will be a natural ordering of the categories (such as
freshman, sophomore, junior, and senior), whereas in others the order of the categories

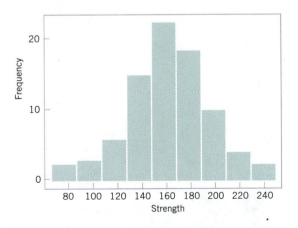

Figure 2-9   A histogram of the compressive strength data from
Minitab with 9 bins.

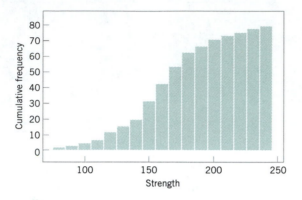

Figure 2-10 A cumulative distribution plot of the compressive strength data from Minitab.

will be arbitrary (such as male and female). When using categorical data, the bars should be drawn to have equal width.

EXAMPLE 2-6

Figure 2-11 presents a histogram showing the production of transport aircraft by the Boeing Company in 1985. Note that the 737 was the most popular model, followed by the 757, 747, 767, and 707.

In this section we have concentrated on descriptive methods for the situation in which each observation in a data set is a single number or belongs to one category. In many cases, we work with data in which each observation consists of several measurements. For example, in a gasoline mileage study, each observation might consist of a measurement

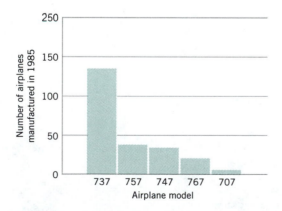

Figure 2-11 Airplane production in 1985. (*Source:* Boeing Company.)

of miles per gallon, the size of the engine in the vehicle, engine horsepower, vehicle weight, and vehicle length. This is an example of **multivariate data.** In later chapters, we will discuss analyzing this type of data.

EXERCISES FOR SECTION 2-3

2-16. Construct a frequency distribution and histogram for the motor fuel octane data from Exercise 2-8. Use 8 bins.

2-17. Construct a frequency distribution and histogram using the failure data from Exercise 2-9.

2-18. Construct a frequency distribution and histogram for the cotton content data in Exercise 2-10.

2-19. Construct a frequency distribution and histogram for the yield data in Exercise 2-11.

2-20. Construct a frequency distribution and histogram with 16 bins for the motor fuel octane data in Exercise 2-8. Compare its shape with that of the histogram with 8 bins from Exercise 2-16. Do both histograms display similar information?

2-21. **The Pareto Chart.** An important variation of a histogram is the Pareto chart. This chart is widely used in quality improvement efforts, where the data usually represent different types of defects, failure modes, or other categories. The categories are ordered so that the category with the largest frequency is on the left, followed by the category with the second largest frequency, and so forth. These charts are named after the Italian economist V. Pareto, and they usually exhibit "Pareto's law"; that is, most of the defects can be accounted for by only a few categories. Suppose that the following information on structural defects in automobile doors is obtained: dents, 4; pits, 4; parts assembled out of sequence, 6; parts undertrimmed, 21; missing holes/slots, 8; parts not lubricated, 5; parts out of contour, 30; and parts not deburred, 3. Construct and interpret a Pareto chart.

2-4 BOX PLOT

The stem-and-leaf display and the histogram provide general visual impressions about a data set, whereas numerical quantities such as $\bar{x}$ or s provide information about only one feature of the data. The **box plot** is a graphical display that simultaneously describes several important features of a data set, such as center, spread, departure from symmetry, and identification of observations that lie unusually far from the bulk of the data. (These observations are called "outliers.")

A box plot displays the three quartiles, on a rectangular box, aligned either horizontally or vertically. The box encloses the interquartile range with the left (or lower) edge at the first quartile, q_1, and the right (or upper) edge at the third quartile, q_3. A line is drawn through the box at the second quartile (which is the 50th percentile or the median), $q_2 = \tilde{x}$. A line, or **whisker,** extends from each end of the box. The lower whisker is a line from the first quartile to the smallest data point within 1.5 interquartile ranges from the first quartile. The upper whisker is a line from the third quartile to the largest data point within 1.5 interquartile ranges from the third quartile. Data farther from the box than the whiskers are plotted as individual points. A point beyond a whisker, but less than 3 interquartile ranges from the box edge, is called an **outlier.** A point more than 3 interquartile ranges from a box edge is called an **extreme outlier.** See Fig. 2-12. Occasion-

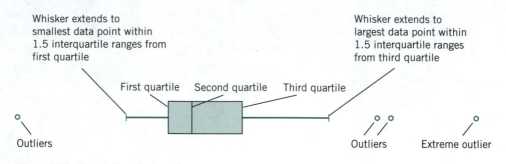

Figure 2-12 Description of a box plot.

ally, different symbols, such as open and filled circles, are used to identify the two types of outliers. Sometimes box plots are called *box-and-whisker plots*.

Figure 2-13 presents the box plot from Minitab for the alloy compressive strength data shown in Table 2-2. This box plot indicates that the distribution of compressive strengths is fairly symmetric around the central value, because the left and right whiskers and the lengths of the left and right boxes around the median are about the same. There are also two outliers on either end of the data.

Box plots are very useful in graphical comparisons among data sets, because they have high visual impact and are easy to understand. For example, Fig. 2-14 shows the comparative box plots for a manufacturing quality index on semiconductor devices at three manufacturing plants. Inspection of this display reveals that there is too much variability at plant 2 and that plants 2 and 3 need to raise their quality index performance.

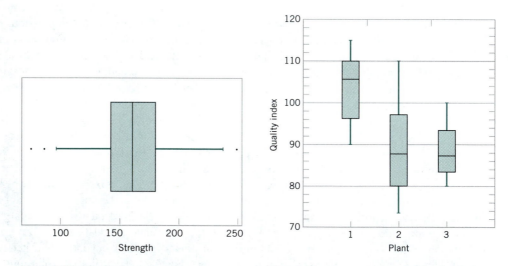

Figure 2-13 Box plot for compressive strength data in Table 2-2.

Figure 2-14 Comparative box plots of a quality index at three plants.

EXERCISES FOR SECTION 2-4

2-22. The following data are the joint temperatures of the O-rings (degrees F) for each test firing or actual launch of the space shuttle rocket motor (from *Presidential Commission on the Space Shuttle Challenger Accident,* Vol. 1, pp. 129–131): 84, 49, 61, 40, 83, 67, 45, 66, 70, 69, 80, 58, 68, 60, 67, 72, 73, 70, 57, 63, 70, 78, 52, 67, 53, 67, 75, 61, 70, 81, 76, 79, 75, 76, 58, 31.
  (a) Compute the sample mean and sample standard deviation.
  (b) Find the upper and lower quartiles of temperature.
  (c) Find the median.
  (d) Set aside the smallest observation (31°F) and recompute the quantities in parts (a), (b), and (c). Comment on your findings. How "different" are the other temperatures from this smallest value?
  (e) Construct a box plot of the data and comment on the possible presence of outliers.

2-23. An article in the *Transactions of the Institution of Chemical Engineers* (Vol. 34, 1956, pp. 280–293) reported data from an experiment investigating the effect of several process variables on the vapor phase oxidation of naphthalene. A sample of the percentage mole conversion of naphthalene to maleic anhydride follows: 4.2, 4.7, 4.7, 5.0, 3.8, 3.6, 3.0, 5.1, 3.1, 3.8, 4.8, 4.0, 5.2, 4.3, 2.8, 2.0, 2.8, 3.3, 4.8, 5.0.
  (a) Calculate the sample mean.
  (b) Calculate the sample variance and sample standard deviation.
  (c) Construct a box plot of the data.

2-24. The "cold start ignition time" of an automobile engine is being investigated by a gasoline manufacturer. The following times (in seconds) were obtained for a test vehicle: 1.75, 1.92, 2.62, 2.35, 3.09, 3.15, 2.53, 1.91.
  (a) Calculate the sample mean and sample standard deviation.

  (b) Construct a box plot of the data.

2-25. The nine measurements that follow are furnace temperatures recorded on successive batches in a semiconductor manufacturing process (units are °F): 953, 950, 948, 955, 951, 949, 957, 954, 955.
  (a) Calculate the sample mean, variance, and standard deviation.
  (b) Find the median. How much could the largest temperature measurement increase without changing the median value?
  (c) Construct a box plot of the data.

2-26. An article in the *Journal of Aircraft* (1988) describes the computation of drag coefficients for the NASA 0012 airfoil. Different computational algorithms were used at $M_\alpha = 0.7$ with the following results (drag coefficients are in units of drag counts; that is, 1 count is equivalent to a drag coefficient of 0.0001): 79, 100, 74, 83, 81, 85, 82, 80, and 84.
  (a) Calculate the sample mean, variance, and standard deviation.
  (b) Find the upper and lower quartiles of the drag coefficients.
  (c) Construct a box plot of the data.
  (d) Set aside the largest observation (100) and redo parts (a), (b), and (c). Comment on your findings.

2-27. The following data are the temperatures of effluent at discharge from a sewage treatment facility on consecutive days:

43	47	51	48	52	50	46	49
45	52	46	51	44	49	46	51
49	45	44	50	48	50	49	50

  (a) Calculate the sample mean and median.
  (b) Calculate the sample variance and sample standard deviation.
  (c) Construct a box plot of the data and comment on the information in this display.
  (d) Find the 5th and 95th percentiles of temperature.

2-5 TIME SEQUENCE PLOTS

The graphical displays that we have considered thus far such as histograms, stem-and-leaf plots, and box plots are very useful visual methods for showing the variability in data. However, we noted in Section 1-5 that time is an important factor that contributes to variability in data, and those graphical methods do not take this into account. A **time series** or **time sequence** is a data set in which the observations are recorded in the order in which they occur. A **time series plot** is a graph in which the vertical axis denotes the observed value of the variable (say, x) and the horizontal axis denotes the time (which could be minutes, days, years, etc.). When measurements are plotted as a time series, we often see trends, cycles, or other broad features of the data that we could not see otherwise.

For example, consider Fig. 2-15a, which presents a time series plot of the annual sales of a company for the last 10 years. The general impression from this display is that sales show an upward **trend.** There is some variability about this trend, with some years' sales increasing over those of the last year and some years' sales decreasing. Figure 2-15b shows the last 3 years of sales reported by quarter. This plot clearly shows that

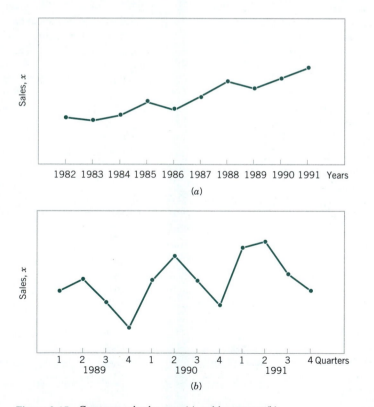

Figure 2-15 Company sales by year (a) and by quarter (b).

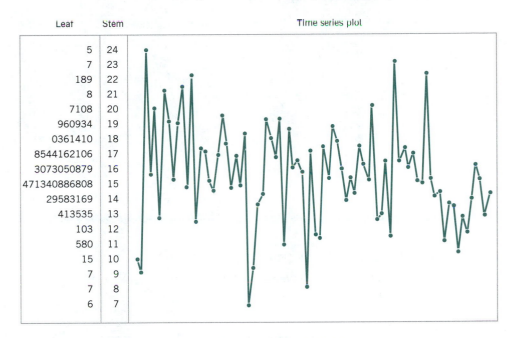

Leaf	Stem	Time series plot
5	24	
7	23	
189	22	
8	21	
7108	20	
960934	19	
0361410	18	
8544162106	17	
3073050879	16	
471340886808	15	
29583169	14	
413535	13	
103	12	
580	11	
15	10	
7	9	
7	8	
6	7	

Figure 2-16 A digidot plot of the compressive strength data in Table 2-2.

the annual sales in this business exhibit a **cyclic** variability by quarter, with the first- and second-quarter sales generally greater than sales during the third and fourth quarters.

Sometimes it can be very helpful to combine a time series plot with some of the other graphical displays that we have considered previously. J. Stuart Hunter (*The American Statistician,* Vol. 42, 1988, p. 54) has suggested combining the stem-and-leaf plot with a time series plot to form a **digidot plot.**

Figure 2-16 shows a digidot plot for the observations on compressive strength from Table 2-2, assuming that these observations are recorded in the order in which they occurred. This plot effectively displays the overall variability in the compressive strength data and simultaneously shows the variability in these measurements over time. The general impression is that compressive strength varies around the mean value of 162.67, and there is no strong obvious pattern in this variability over time.

The digidot plot in Fig. 2-17 tells a different story. This plot summarizes 30 observations on concentration of the output product from a chemical process, where the observations are recorded at 1-hour time intervals. This plot indicates that during the first 20 hours of operation this process produced concentrations generally above 85 g/l, but that following sample 20, something may have occurred in the process that resulted in lower concentrations. If this variability in output product concentration can be reduced, then operation of this process can be improved. The control chart, which is a special kind of time series plot, of these data was shown in Fig. 1-16 of Chapter 1.

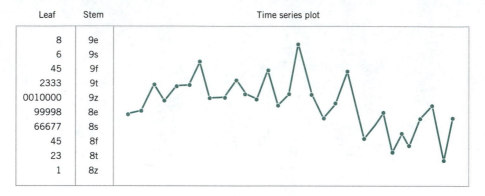

Leaf	Stem	Time series plot
8	9e	
6	9s	
45	9f	
2333	9t	
0010000	9z	
99998	8e	
66677	8s	
45	8f	
23	8t	
1	8z	

Figure 2-17   A digidot plot of chemical process concentration readings, observed hourly.

EXERCISES FOR SECTION 2-5

2-28. The College of Engineering and Applied Science at Arizona State University had a VAX computer system. Response times for 20 consecutive jobs done on the system were recorded and are shown next in order. (Read down, then left to right.)

5.3	6.2	8.5	12.4
5.0	5.9	4.7	3.9
9.5	7.2	11.2	8.1
10.1	10.0	7.3	9.2
5.8	12.2	6.4	10.5

Construct and interpret a time series plot of these data.

2-29. The following data are the viscosity measurements for a chemical product observed hourly. (Read down, then left to right.)

47.9	48.8	48.6	43.2	43.0
47.9	48.1	48.0	43.0	42.8
48.6	48.3	47.9	43.5	43.1
48.0	47.2	48.3	43.1	43.2
48.4	48.9	48.5	43.0	43.6
48.1	48.6	48.1	42.9	43.2
48.0	48.0	48.0	43.6	43.5
48.6	47.5	48.3	43.3	43.0

(a) Construct and interpret either a digi-

dot plot or a separate stem-and-leaf and time series plot of these data.

(b) Specifications on product viscosity are at 48 ± 2. What conclusions can you make about process performance?

2-30. The pull-off force for a connector is measured in a laboratory test. Data for 40 test specimens are shown next. (Read down the entire column, then left to right.)

241	220	249	209
258	194	251	212
237	245	238	185
210	209	210	187
194	201	198	218
225	195	199	190
248	255	183	175
203	245	213	178
195	235	236	175
249	220	245	190

(a) Construct a time series plot of the data.

(b) Construct and interpret either a digidot plot or a stem-and-leaf and time series plot of the data.

2-31. In their book *Time Series Analysis, Forecasting, and Control* (Holden-Day, 1976), G. E. P. Box and G. M. Jenkins present chemical process concentration readings made every 2 hours. Some of these data are shown next. (Read down, then left to right.)

17.0	16.7	17.1	17.5	17.6
16.6	17.4	17.4	18.1	17.5
16.3	17.2	17.4	17.5	16.5
16.1	17.4	17.5	17.4	17.8
17.1	17.4	17.4	17.4	17.3
16.9	17.0	17.6	17.1	17.3
16.8	17.3	17.4	17.6	17.1
17.4	17.2	17.3	17.7	17.4
17.1	17.4	17.0	17.4	16.9
17.0	16.8	17.8	17.8	17.3

Construct and interpret either a digidot plot or a stem-and-leaf and time series plot of these data.

2-32. The annual Wolfer sunspot numbers from 1770 to 1869 are shown in Table 2-5. (For an interesting analysis and interpretation of these numbers, see the book by Box and Jenkins referenced in Exercise 2-31. The analysis requires some advanced knowledge of statistics and statistical model building.)

(a) Construct a time series plot of these data.

(b) Construct and interpret either a digidot plot or a stem-and-leaf and time series plot of these data.

2-33. In their book *Forecasting and Time Series Analysis,* 2nd edition (McGraw-Hill, 1990), D. C. Montgomery, L. A. Johnson, and J. S. Gardiner analyze the data in Table 2-6, which are the monthly total passenger airline miles flown in the United Kingdom, 1964–1970 (in millions of miles).

Table 2-5   Annual Sunspot Numbers

1770	101	1795	21	1820	16	1845	40
1771	82	1796	16	1821	7	1846	62
1772	66	1797	6	1822	4	1847	98
1773	35	1798	4	1823	2	1848	124
1774	31	1799	7	1824	8	1849	96
1775	7	1800	14	1825	17	1850	66
1776	20	1801	34	1826	36	1851	64
1777	92	1802	45	1827	50	1852	54
1778	154	1803	43	1828	62	1853	39
1779	125	1804	48	1829	67	1854	21
1780	85	1805	42	1830	71	1855	7
1781	68	1806	28	1831	48	1856	4
1782	38	1807	10	1832	28	1857	23
1783	23	1808	8	1833	8	1858	55
1784	10	1809	2	1834	13	1859	94
1785	24	1810	0	1835	57	1860	96
1786	83	1811	1	1836	122	1861	77
1787	132	1812	5	1837	138	1862	59
1788	131	1813	12	1838	103	1863	44
1789	118	1814	14	1839	86	1864	47
1790	90	1815	35	1840	63	1865	30
1791	67	1816	46	1841	37	1866	16
1792	60	1817	41	1842	24	1867	7
1793	47	1818	30	1843	11	1868	37
1794	41	1819	24	1844	15	1869	74

Table 2-6   United Kingdom Airline Miles Flown

	1964	1965	1966	1967	1968	1969	1970
Jan.	7.269	8.350	8.186	8.334	8.639	9.491	10.840
Feb.	6.775	7.829	7.444	7.899	8.772	8.919	10.436
Mar.	7.819	8.829	8.484	9.994	10.894	11.607	13.589
Apr.	8.371	9.948	9.864	10.078	10.455	8.852	13.402
May	9.069	10.638	10.252	10.801	11.179	12.537	13.103
June	10.248	11.253	12.282	12.953	10.588	14.759	14.933
July	11.030	11.424	11.637	12.222	10.794	13.667	14.147
Aug.	10.882	11.391	11.577	12.246	12.770	13.731	14.057
Sept.	10.333	10.665	12.417	13.281	13.812	15.110	16.234
Oct.	9.109	9.396	9.637	10.366	10.857	12.185	12.389
Nov.	7.685	7.775	8.094	8.730	9.290	10.645	11.594
Dec.	7.682	7.933	9.280	9.614	10.925	12.161	12.772

(a) Draw a time series plot of the data and comment on any features of the data that are apparent.

(b) Construct and interpret either a digidot plot or a stem-and-leaf and time series plot of these data.

SUPPLEMENTAL EXERCISES

2-34. The pH of a solution is measured eight times by one operator using the same instrument. She obtains the following data: 7.15, 7.20, 7.18, 7.19, 7.21, 7.20, 7.16, and 7.18.

(a) Calculate the sample mean. Suppose that the desirable value for this solution was specified to be 7.20. Do you think that the sample mean value computed here is close enough to the target value to accept the solution as conforming to target? Explain your reasoning.

(b) Calculate the sample variance and sample standard deviation. What do you think are the major sources of variability in this experiment? Why

is it desirable to have a small variance of these measurements?

2-35. A sample of six resistors yielded the following resistances (ohms): $x_1 = 45$, $x_2 = 38$, $x_3 = 47$, $x_4 = 41$, $x_5 = 35$, and $x_6 = 43$.

(a) Compute the sample variance and sample standard deviation using the method in equation 2-4.

(b) Compute the sample variance and sample standard deviation using the definition in equation 2-3. Explain why the results from both equations are the same.

(c) Subtract 35 from each of the original resistance measurements and compute s^2 and s. Compare your results with those obtained in parts (a) and (b) and explain your findings.

(d) If the resistances were 450, 380, 470, 410, 350, and 430 ohms, could you use the results of previous parts of this problem to find s^2 and s? Explain how you would proceed.

2-36. The percentage mole conversion of naphthalene to maleic anhydride from Exercise 2-23 follows: 4.2, 4.7, 4.7, 5.0, 3.8, 3.6, 3.0, 5.1, 3.1, 3.8, 4.8, 4.0, 5.2, 4.3, 2.8, 2.0, 2.8, 3.3, 4.8, 5.0.

(a) Calculate the sample range, variance and standard deviation.

(b) Calculate the sample range, variance, and standard deviation again, but first subtract 1.0 from each observation. Compare your results with those obtained in part (a). Is there anything "special" about the constant 1.0, or would any other arbitrarily chosen value have produced the same results?

2-37. Suppose that we have a sample x_1, $x_2, \ldots, x_n$ and we have calculated $\bar{x}_n$ and s_n^2 for the sample. Now an $(n + 1)$st observation becomes available. Let $\bar{x}_{n+1}$ and s_{n+1}^2 be the sample mean and sample variance for the sample using all $n + 1$ observations.

(a) Show how $\bar{x}_{n+1}$ can be computed using $\bar{x}_n$ and x_{n+1}.

(b) Show that $ns_{n+1}^2 = (n - 1)s_n^2 +$
$$\frac{n}{n+1}(x_{n+1} - \bar{x}_n)^2.$$

(c) Use the results of parts (a) and (b) to calculate the new sample average and standard deviation for the data of Exercise 2-35, when the new observation is $x_7 = 46$.

2-38. **The Trimmed Mean.** Suppose that the data are arranged in increasing order, $T\%$ of the observations are removed from each end, and the sample mean of the remaining numbers is calculated. The resulting quantity is called a *trimmed mean*. The trimmed mean generally lies between the sample mean $\bar{x}$ and the sample median $\tilde{x}$. Why?

(a) Calculate the 10% trimmed mean for the yield data in Exercise 2-11.

(b) Calculate the 20% trimmed mean for the yield data in Exercise 2-11 and compare it with the quantity found in part (a).

(c) Compare the values calculated in parts (a) and (b) with the sample mean and median for the data in Exercise 2-11. Is there much difference in these quantities? Why?

(d) Suppose that the sample size n is such that the quantity $nT/100$ is not an integer. Develop a procedure for obtaining a trimmed mean in this case.

2-39. Consider the two samples shown here:

Sample 1: 10, 9, 8, 7, 8, 6, 10, 6
Sample 2: 10, 6, 10, 6, 8, 10, 8, 6

(a) Calculate the sample range for both samples. Would you conclude that both samples exhibit the same variability? Explain.

(b) Calculate the sample standard deviations for both samples. Do these quantities indicate that both samples have the same variability? Explain.

(c) Write a short statement contrasting the sample range versus the sample standard deviation as a measure of variability.

2-40. An article in *Quality Engineering* (Vol. 4, 1992, pp. 487–495) presents viscosity data from a batch chemical process. A sample of these data is presented next. (Read down the entire column, then left to right.)

13.3	14.9	15.8	16.0
14.5	13.7	13.7	14.9
15.3	15.2	15.1	13.6
15.3	14.5	13.4	15.3
14.3	15.3	14.1	14.3
14.8	15.6	14.8	15.6
15.2	15.8	14.3	16.1
14.5	13.3	14.3	13.9
14.6	14.1	16.4	15.2
14.1	15.4	16.9	14.4
14.3	15.2	14.2	14.0
16.1	15.2	16.9	14.4
13.1	15.9	14.9	13.7
15.5	16.5	15.2	13.8
12.6	14.8	14.4	15.6
14.6	15.1	15.2	14.5
14.3	17.0	14.6	12.8
15.4	14.9	16.4	16.1
15.2	14.8	14.2	16.6
16.8	14.0	15.7	15.6

(a) Draw a time series plot of all the data and comment on any features of the data that are revealed by this plot.

(b) Consider the notion that the first 40 observations were generated from a specific process, whereas the last 40 observations were generated from a different process. Does the plot indicate that the two processes generate similar results?

(c) Compute the sample mean and sample variance of the first 40 observations, then compute these values for the second 40 observations. Do these quantities indicate that both processes yield the same mean level? the same variability? Explain.

2-41. A manufacturer of coil springs is interested in implementing a quality control system to monitor his production process. As part of this quality system, it decided to record the number of nonconforming coil springs in each production batch of size 50. During production, 40 batches of data were collected and are reported here. (Read down the entire column, then left to right.)

9	12	6	9	7	14	12	4	6	7
8	5	9	7	8	11	3	6	7	7
11	4	4	8	7	5	6	4	5	8
19	19	18	12	11	17	15	17	13	13

(a) Construct a stem-and-leaf plot of the data.

(b) Find the sample average and sample standard deviation.

(c) Construct a time series plot of the data. Is there evidence that there was an increase or decrease in the average number of nonconforming springs made during the 40 days? Explain.

2-42. A communication channel is being monitored by recording the number of errors in a string of 1000 bits. Data for 20 of these strings are given here. (Read the data left to right, then down.)

3	1	0	1	3	2	4	1	3	1
1	1	2	3	3	2	0	2	0	1

(a) Construct a stem-and-leaf plot of the data.

(b) Find the sample average and sample standard deviation.

(c) Construct a time series plot of the data. Is there evidence that there was an increase or decrease in the number of errors in a string? Explain.

Team Exercises

2-43. As an engineering student you have frequently encountered data (for example, in engineering or science laboratory courses). Choose one of these data sets or another data set of interest to you. Describe the data with appropriate numerical and graphical tools.

2-44. Select a data set that is time-ordered. Describe the data with appropriate numerical and graphical tools. Discuss potential sources of variation in the data.

2-45. Consider the data on weekly waste (percent) for five suppliers of the Levi-Strauss clothing plant in Albuquerque and reported at the Web site http://lib.stat.cmu.edu/DASL/Stories/wasterunup.html. Generate box plots for the five suppliers.

2-46. Thirty-one consecutive daily carbon monoxide measurements were taken by an oil refinery northeast of San Francisco and reported at the Web site http://lib.stat.cmu.edu/DASL/Datafiles/Refinery.html. Draw a time series plot of the data and comment on features of the data that are revealed by this plot.

2-47. Consider the famous data set listing the time between eruptions of the geyser "Old Faithful" found at http://lib.stat.cmu.edu/DASL/Datafiles/differencetestdat.html from the article by A. Azzalini and A. W. Bowman, "A Look at Some Data on the Old Faithful Geyser," *Applied Statistics,* 1990, pp. 57–365.

(a) Construct a time series plot of all the data.

(b) Split the data into two data sets of 100 observations each. Create two separate stem-and-leaf plots of the subsets. Is there reason to believe that the two subsets are different?

3 Random Variables and Probability Distributions

CHAPTER OUTLINE

Earlier in this book, numerical and graphical summaries were used to summarize data. A summary is often needed to transform the data to useful information. Furthermore, conclusions about the process that generated the data are often important; that is, we might want to draw some conclusions about the long-term performance of a process based on only a relatively small sample of data. Because only a sample of data is used, there is some uncertainty in our conclusions. However, the amount of uncertainty can be quantified and sample sizes can be selected or modified to achieve a tolerable level of uncertainty if a probability model is specified for the data. The objective of this chapter is to describe these models and present some important examples.

3-1 INTRODUCTION

The measurement of current in a thin copper wire is an example of an **experiment.** However, the results might differ slightly in day-to-day replicates of the measurement because of small variations in variables that are not controlled in our experiment—changes in ambient temperatures, slight variations in gauge and small impurities in the chemical composition of the wire if different locations are selected, current source drifts, and so forth. Consequently, this experiment (as well as many we conduct) can be considered to have a **random** component. In some cases, the random variations that we experience are small enough, relative to our experimental goals, that they can be ignored. However, the variation is almost always present and its magnitude can be large enough that the important conclusions from the experiment are not obvious. In these cases, the methods presented in this book for modeling and analyzing experimental results are quite valuable.

An experiment that can result in different outcomes, even though it is repeated in the same manner every time, is called a **random experiment.** We might select one part from a day's production and very accurately measure a dimensional length. Although we hope that the manufacturing operation produces identical parts consistently, in practice there are often small variations in the actual measured lengths due to many causes—vibrations, temperature fluctuations, operator differences, calibrations of equipment and gauges, cutting tool wear, bearing wear, and raw material changes. Even the measurement procedure can produce variations in the final results.

No matter how carefully our experiment is designed and conducted, variations often occur. Our goal is to understand, quantify, and model the type of variations that we often encounter. When we incorporate the variation into our thinking and analyses, we can make informed judgments from our results that are not invalidated by the variation.

Models and analyses that include variation are not different from models used in other areas of engineering and science. Figure 3-1 displays the relationship between the model and the physical system it represents. A mathematical model (or abstraction) of a physical system need not be a perfect abstraction. For example, Newton's laws are not perfect descriptions of our physical universe. Still, these are useful models that can be studied and analyzed to quantify approximately the performance of a wide range of engineered products. Given a mathematical abstraction that is validated with measurements

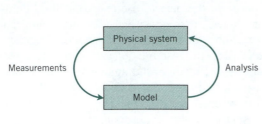

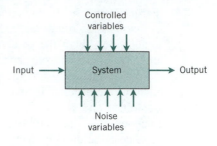

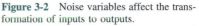

Figure 3-1 Continuous iteration between model and physical system.

Figure 3-2 Noise variables affect the transformation of inputs to outputs.

from our system, we can use the model to understand, describe, and quantify important aspects of the physical system and predict the response of the system to inputs.

Throughout this text, we discuss models that allow for variations in the outputs of a system, even though the variables that we control are not purposely changed during our study. Figure 3-2 graphically displays a model that incorporates uncontrollable variables (noise) that combine with the controllable variables to produce the output of our system. Because of the noise, the same settings for the controllable variables do not result in identical outputs every time the system is measured.

For the example of measuring current in a copper wire, our model for the system might simply be Ohm's law,

$$current = voltage/resistance$$

As described previously, variations in current measurements are expected. Ohm's law might be a suitable approximation. However, if the variations are large relative to the intended use of the device under study, we might need to extend our model to include the variation. It is often difficult to speculate on the magnitude of the variations without empirical measurements. With sufficient measurements, however, we can approximate the magnitude of the variation and consider its effect on the performance of other devices, such as amplifiers, in the circuit. We are therefore sanctioning the model in Fig. 3-2 as a more useful description of the current measurement. Consequently, the techniques presented in this text for the analysis of models including variation are often useful. (Also, see Fig. 3-3.)

As another example, in the design of a communication system, such as a computer network or a voice communication network, the information capacity available to service individuals using the network is an important design consideration. For voice communication, sufficient external lines need to be purchased from the phone company to meet the requirements of a business. Assuming each line can carry only a single conversation, how many lines should be purchased? If too few lines are purchased, calls can be delayed or lost. The purchase of too many lines increases costs. Increasingly, design and product development are required to meet customer requirements *at a competitive cost.*

In the design of the voice communication system, a model is needed for the number

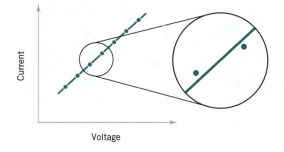

Figure 3-3 A closer examination of the system identifies deviations from the model.

of calls and the duration of calls. Even knowing that, on average, calls occur every five minutes and that they last five minutes is not sufficient. If calls arrived precisely at five-minute intervals and lasted for precisely five minutes, then one phone line would be sufficient. However, the slightest variation in call number or duration would result in some calls being blocked by others. See Fig. 3-4. A system designed without considering variation will be woefully inadequate for practical use. Our model for the number and duration of calls needs to include variation as an integral component. An analysis of models including variation is important for the design of the phone system.

3-2 RANDOM VARIABLES

In an experiment, a measurement is usually denoted by a variable such as X. In a random experiment, a variable whose measured value can change (from one replicate of the experiment to another) is referred to as a **random variable.** For example, X might denote the current measurement in the copper wire experiment. A random variable is conceptually no different from any other variable in an experiment. We use the term "random" to indicate that noise disturbances can change its measured value. An uppercase letter is used to denote a random variable.

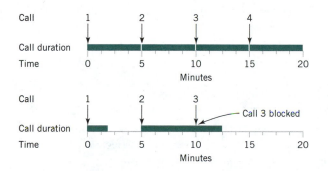

Figure 3-4 Variation causes disruptions in the system.

> **Definition**
>
> A **random variable** is a numerical variable whose measured value can change from one replicate of the experiment to another.

After the experiment is conducted, the measured value of the random variable is denoted by a lowercase letter such as $x = 70$ milliamperes. We often summarize a random experiment by the measured value of a random variable.

This model can be linked to data as follows. The data are the measured values of a random variable obtained from replicates of a random experiment. For example, the first replicate might result in a current measurement of $x_1 = 70.1$, the next day $x_2 = 71.2$, the third day $x_3 = 71.1$, and so forth. These data can then be summarized by the descriptive methods discussed in Chapter 2.

Often, the measurement of interest—current in a copper wire experiment, length of a machined part—is assumed to be a real number. Then arbitrary precision in the measurement is possible. Of course, in practice, we might round off to the nearest tenth or hundredth of a unit. The random variable that represents this measurement is said to be a **continuous** random variable.

In other experiments, we might record a count such as the number of transmitted bits that are received in error. Then the measurement is limited to integers. Or we might record that a proportion such as 0.0042 of the 10,000 transmitted bits were received in error. Then the measurement is fractional, but it is still limited to discrete points on the real line. Whenever the measurement is limited to discrete points on the real line, the random variable is said to be a **discrete** random variable.

In some cases, the random variable X is actually discrete but, because the range of possible values is so large, it might be more convenient to analyze X as a continuous random variable. For example, suppose that current measurements are read from a digital instrument that displays the current to the nearest one hundredth of a milliampere. Because the possible measurements are limited, the random variable is discrete. However, it might be a more convenient, simple approximation to assume that the current measurements are values of a continuous random variable.

> Examples of continuous random variables:
>   electrical current, length, pressure, temperature, time, voltage, weight
>
> Examples of discrete random variables:
>   number of scratches on a surface, proportion of defective parts among 1000 tested, number of transmitted bits received in error

EXERCISES FOR SECTION 3-2

Decide whether a discrete or continuous random variable is the best model for each of the following variables.

3-1. The life of a biomedical device after implant in a patient.

3-2. The number of times a transistor in a computer memory changes state in one operation.

3-3. The strength of a concrete specimen.

3-4. The number of luxury options selected by an automobile buyer.

3-5. The number of users on a computer network at a specific time of the day.

3-6. The weight of an injection-molded plastic part.

3-7. The number of molecules in a sample of gas.

3-3 PROBABILITY

Probability is used to quantify the likelihood, or chance, that a measurement falls within some set of values. Usually, a random variable is used to denote the measurement. "The chance that X, the length of a manufactured part, is between 10.8 and 11.2 millimeters is 25%" is a statement that quantifies our feeling about the possibility of part lengths. Probability statements describe the likelihood that particular values occur. The likelihood is quantified by assigning a number from the interval [0, 1] to the set of values (or a percentage from 0 to 100%). Higher numbers indicate that the set of values is more likely.

The probability of a result can be interpreted as our subjective probability, or **degree of belief,** that the result will occur. Different individuals will no doubt assign different probabilities to the same result. Another interpretation of probability can be based on repeated replicates of the random experiment. The probability of a result is interpreted as the proportion of times the result will occur in repeated replicates of the random experiment. For example, if we assign probability 0.25 to the result that a part length is between 10.8 and 11.2 millimeters, then we might interpret this assignment as follows. If we repeatedly manufacture parts (replicate the random experiment an infinite number of times), then 25% of them will have lengths in this interval. This example provides a **relative frequency** interpretation of probability. The proportion, or relative frequency, of repeated replicates that fall in the interval will be 0.25. Note that this interpretation uses a long-run proportion, the proportion from an infinite number of replicates. With a small number of replicates, the proportion of lengths that actually fall in the interval might differ from 0.25.

To continue, if every manufactured part length will fall in the interval, then the relative frequency, and therefore the probability, of the interval is one. If no manufactured part length will fall in the interval, then the relative frequency, and therefore the probability, of the interval is zero. Because probabilities are restricted to the interval [0, 1], they can be interpreted as relative frequencies.

A probability is usually expressed in terms of a random variable. For the part length example, X denotes the part length and the probability statement can be written in either of the following forms

$$P(X \in [10.8, 11.2]) = 0.25 \quad \text{or} \quad P(10.8 \leq X \leq 11.2) = 0.25$$

Both equations state that the probability that the random variable X assumes a value in [10.8, 11.2] is 0.25.

Probabilities for a random variable are usually determined from a model that describes the random experiment. Several models will be considered in the following sections. However, several general probability properties are stated here that can be understood from the relative frequency interpretation of probability.

The following terms are used. Given a set E, the complement of E is the set of elements that are not in E. The **complement** is denoted as E'. The set of real numbers is denoted as R. The sets $E_1, E_2, \ldots, E_k$ are **mutually exclusive** if the intersection of any pair is empty. That is, each element is in one and only one of the sets $E_1, E_2, \ldots, E_k$.

Probability Properties

1. $P(X \in R) = 1$, where R is the set of real numbers.
2. $0 \leq P(X \in E) \leq 1$ for any set E. $\hspace{2cm}$ (3-1)
3. If $E_1, E_2, \ldots, E_k$ are mutually exclusive, then
$$P(X \in E_1 \cup E_2 \cup \cdots \cup E_k) = P(X \in E_1) + \cdots + P(X \in E_k).$$

Property 1 can be used to show that the maximum value for a probability is one. Property 2 implies that a probability can't be negative. Property 3 states that the proportion of measurements that fall in $E_1 \cup E_2 \cup \cdots \cup E_k$ is the sum of the proportions that fall in E_1 and $E_2, \ldots,$ and E_k, whenever the sets are mutually exclusive. For example,

$$P(X \leq 10) = P(X \leq 0) + P(0 < X \leq 5) + P(5 < X \leq 10)$$

Property 3 is also used to relate the probability of a set E and its compliment E'. Because E and E' are mutually exclusive and $E \cup E' = R$, $1 = P(X \in R) = P(X \in E \cup E') = P(X \in E) + P(X \in E')$. Consequently,

$$P(X \in E') = 1 - P(X \in E)$$

For example, $P(X \leq 2) = 1 - P(X > 2)$. In general, for any fixed number x,

$$P(X \leq x) = 1 - P(X > x)$$

Let $\varnothing$ denote the null set. Because the complement of R is $\varnothing$, $P(X \in \varnothing) = 0$.

Assume that the following probabilities apply to the random variable X that denotes the life in hours of standard fluorescent tubes: $P(X \leq 5000) = 0.1$, $P(5000 < X \leq 6000) = 0.3$, $P(X > 8000) = 0.4$. The following results can be determined from the probability properties. It may be helpful to graphically display the different sets.

The probability that the life is less than or equal to 6000 hours is

$$P(X \le 6000) = P(X \le 5000) + P(5000 < X \le 6000) = 0.1 + 0.3$$
$$= 0.4$$

from Property 3. The probability that the life exceeds 6000 hours is

$$P(X > 6000) = 1 - P(X \le 6000) = 1 - 0.4 = 0.6$$

The probability that the life is greater than 6000 and less than or equal to 8000 hours is determined from the fact that the sum of the probabilities for this interval and the other three intervals must equal 1. That is, the union of other three intervals are the complement of the set $\{x \mid 6000 < x \le 8000\}$. Therefore

$$P(6000 < X \le 8000) = 1 - (0.1 + 0.3 + 0.4) = 0.2$$

The probability that the life is less than or equal to 5500 hours cannot be determined exactly. The best we can state is that

$$P(X \le 5500) \le P(X \le 6000) = 0.4 \text{ and}$$
$$0.1 = P(X \le 5000) \le P(X \le 5500)$$

If it were also known that $P(5500 < X \le 6000) = 0.15$ then we could state that

$$P(X \le 5500) = P(X \le 5000) + P(5000 < X \le 6000) - P(5500 < X \le 6000)$$
$$= 0.1 + 0.3 - 0.15 = 0.25$$

Events

A measured value is not always obtained from an experiment. Sometimes, the result is only classified (into one of several possible categories). For example, the current measurement might only be recorded as *low, medium,* or *high*; a manufactured electronic component might be classified only as defective or not; and a bit transmitted through a digital communication channel is received either in error or not. The possible categories are usually referred to as **events.** More generally, an event can also refer to a union of categories. The concept of probability can be applied to these experiments and the relative frequency interpretation is still appropriate.

   If 1% of the bits transmitted through a digital communications channel are received in error, then the probability of an error would be assigned 0.01. If we let E denote the event that a bit is received in error, then we would write

$$P(E) = 0.01$$

   Probabilities assigned to events satisfy properties analogous to those in equation 3-1 so that they can be interpreted as relative frequencies. Consequently, (1) the probability assigned to the union of all possible categories is one, (2) $0 \leq P(E) \leq 1$ for any event E, and (3) if E_1, E_2, ..., E_k are mutually exclusive events, then $P(E_1 \cup E_2 \cdots \cup E_k) = P(E_1 \text{ or } E_2 \cdots \text{ or } E_k) = P(E_1) + P(E_2) + \cdots + P(E_k)$. The events E_1, E_2, ..., E_k are mutually exclusive when each refers to a distinct category such as *low*, *medium*, and *high*. If one of the events is a union of categories, they might not be mutually exclusive. As an example of mutually exclusive events, suppose that the probability of *low*, *medium*, and *high* results are 0.1, 0.7, and 0.2, respectively. The probability of a *medium* or *high* result is denoted as $P(medium \text{ or } high)$ and

$$P(medium \text{ or } high) = P(medium) + P(high) = 0.7 + 0.2 = 0.9$$

EXERCISES FOR SECTION 3-3

3-8. State the complement of each of the following sets:
  (a) Engineers with less than 36 months of full-time employment.
  (b) Samples of cement blocks with compressive strength less than 6000 kilograms per square centimeter.
  (c) Measurements of the diameter of forged pistons that do not conform to engineering specifications.
  (d) Cholesterol levels that measure greater than 180 and less than 220.

3-9. If $P(X \in A) = 0.4$, and $P(X \in B) = 0.6$ and the intersection of sets A and B is empty,
  (a) Are sets A and B mutually exclusive?
  (b) Find $P(X \in A')$.
  (c) Find $P(X \in B')$.
  (d) Find $P(X \in A \cup B)$.

3-10. If $P(X \in A) = 0.3$, $P(X \in B) = 0.25$, and $P(X \in C) = 0.60$ and $P(X \in A \cup B) = 0.55$ and $P(X \in B \cup C) = 0.70$, determine the following probabilities.
  (a) $P(X \in A')$
  (b) $P(X \in B')$
  (c) $P(X \in C')$
  (d) Are A and B mutually exclusive?
  (e) Are B and C mutually exclusive?

3-11. Let $P(X \leq 15) = 0.3$, $P(15 < X \leq 24) = 0.6$ and $P(X > 20) = 0.5$.
  (a) Find $P(X > 15)$.
  (b) Find $P(X \leq 24)$.

  (c) Find $P(15 < X \leq 20)$.
  (d) If $P(18 < X \leq 24) = 0.4$ find $P(X \leq 18)$.

3-12. Let $P(X \leq 8) = 0.85$, $P(7 < X \leq 7.5) = 0.25$ and $P(X > 7.5) = 0.35$.
  (a) Find $P(X > 8)$.
  (b) Find $P(X \leq 7.5)$.
  (c) Find $P(7.5 < X \leq 8.0)$.
  (d) If $P(7.25 < X \leq 7.5) = 0.1$, find $P(X \leq 7.25)$.

3-13. Let X denote the life of a semiconductor laser (in hours) with the following probabilities:
$$P(X \leq 5000) = 0.05$$
$$P(X > 7000) = 0.45$$
  (a) What is the probability that the life is less than or equal to 7000 hours?
  (b) What is the probability that the life is greater than 5000 hours?
  (c) What is $P(5000 < X \leq 7000)$?

3-14. Let E_1 denote the event that a structural component fails during a test and E_2 denote the event that the component shows some strain, but does not fail. Given $P(E_1) = 0.15$ and $P(E_2) = 0.30$,
  (a) What is the probability that a structural component does not fail during a test?
  (b) What is the probability that a component either fails or shows strain during a test?
  (c) What is the probability that a component neither fails nor shows strain during a test?

3-4   CONTINUOUS RANDOM VARIABLES

3-4.1   Probability Density Function

The **probability distribution** or simply **distribution** of a random variable X is a description of the set of the probabilities associated with the possible values for X. The probability distribution of a random variable can be specified in more than one way.

Density functions are commonly used in engineering to describe physical systems. For example, consider the density of a loading on a long, thin beam as shown in Fig. 3-5. For any point x along the beam, the density can be described by a function (in grams/cm). Intervals with large loadings correspond to large values for the function. The total loading between points a and b is determined as the integral of the density function from a to b. This integral is the area under the density function over this interval, and it can be loosely interpreted as the sum of all the loadings over this interval.

Similarly, a **probability density function** $f(x)$ can be used to describe the probability distribution of a continuous random variable X. The probability that X is between a and b is determined as the integral of $f(x)$ from a to b. See Fig. 3-6. The notation follows.

The probability density function $f(x)$ of a continuous random variable is used to determine probabilities as follows:

$$P(a < X < b) = \int_a^b f(x)\,dx \qquad (3\text{-}2)$$

A histogram is an approximation to a probability density function. See Fig. 3-7. For each interval of the histogram, the area of the bar equals the relative frequency (proportion) of the measurements in the interval. The relative frequency is an estimate of the probability that a measurement falls in the interval. Similarly, the area under $f(x)$ over any interval equals the true probability that a measurement falls in the interval.

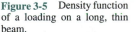

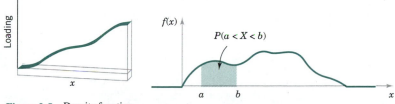

Figure 3-5   Density function of a loading on a long, thin beam.

Figure 3-6   Probability determined from the area under $f(x)$.

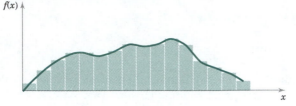

Figure 3-7 Histogram approximates a probability density function. The area of each bar equals the relative frequency of the interval. The area under $f(x)$ over any interval equals the probability of the interval.

A probability density function provides a simple description of the probabilities associated with a random variable. As long as $f(x)$ is nonnegative and $\int_{-\infty}^{\infty} f(x)\, dx = 1$, then $0 \leq P(a < X < b) \leq 1$ so that the probabilities are properly restricted. A probability density function is zero for x values that cannot occur and it is assumed to be zero wherever it is not specifically defined.

The important point is that $f(x)$ **is used to calculate an area** that represents the probability that X assumes a value in $[a, b]$. For the current measurements of Section 3-1, the probability that X results in [14 mA, 15 mA], is the integral of the probability density function of X, $f(x)$, over this interval. The probability that X results in [14.5 mA, 14.6 mA] is the integral of the same function, $f(x)$, over the smaller interval. By appropriate choice of the shape of $f(x)$, we can represent the probabilities associated with any random variable X. The shape of $f(x)$ determines how the probability that X assumes a value in [14.5 mA, 14.6 mA] compares to the probability of any other interval of equal or different length.

For the density function of a loading on a long, thin beam, because every point has zero width, the integral that determines the loading at any point is zero. Similarly, for a continuous random variable X and *any* value x,

$$P(X = x) = 0$$

Based on this result, it might appear that our model of a continuous random variable is useless. However, in practice, when a particular current measurement is observed, such as 14.47 milliamperes, this result can be interpreted as the rounded value of a current measurement that is actually in a range such as $14.465 \leq x \leq 14.475$. Therefore, the probability that the rounded value 14.47 is observed as the value for X is the probability that X assumes a value in the interval [14.465, 14.475], which is not zero. Similarly, our model of a continuous random variable implies the following.

If X is a continuous random variable, then for any x_1 and x_2,

$$P(x_1 \leq X \leq x_2) = P(x_1 < X \leq x_2) = P(x_1 \leq X < x_2) = P(x_1 < X < x_2)$$

EXAMPLE 3-1

Let the continuous random variable X denote the current measured in a thin copper wire in milliamperes. Assume that the range of X is [0, 20 mA], and assume that the probability density function of X is $f(x) = 0.05$ for $0 \leq x \leq 20$. What is the probability that a current measurement is less than 10 milliamperes?

The probability density function is shown in Fig. 3-8. It is assumed that $f(x) = 0$ wherever it is not specifically defined. The probability requested is indicated by the shaded area in Fig. 3-8.

$$P(X < 10) = \int_0^{10} f(x)\, dx = 0.5$$

As another example,

$$P(5 < X < 15) = \int_5^{15} f(x)\, dx = 0.5$$

EXAMPLE 3-2

Let the continuous random variable X denote the diameter of a hole drilled in a sheet metal component. The target diameter is 12.5 millimeters. Most random disturbances to the process result in larger diameters. Historical data show that the distribution of X can be modeled by a probability density function $f(x) = 20e^{-20(x-12.5)}$, $x \geq 12.5$. If a part with a diameter larger than 12.60 millimeters is scrapped, what proportion of parts is scrapped?

The density function and the requested probability are shown in Fig. 3-9. A part is scrapped if $X > 12.60$. Now,

$$P(X > 12.60) = \int_{12.6}^{\infty} f(x)\, dx = \int_{12.6}^{\infty} 20e^{-20(x-12.5)}\, dx = -e^{-20(x-12.5)} \Big|_{12.6}^{\infty} = 0.135$$

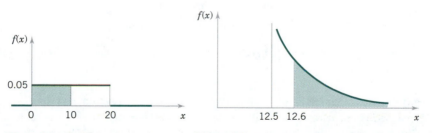

Figure 3-8 Probability density function for Example 3-1.

Figure 3-9 Probability density function for Example 3-2.

What proportion of parts is between 12.5 and 12.6 millimeters? Now,

$$P(12.5 < X < 12.6) = \int_{12.5}^{12.6} f(x)\, dx = \left. -e^{-20(x-12.5)} \right|_{12.5}^{12.6} = 0.865$$

Because the total area under $f(x)$ equals one, we can also calculate $P(12.5 < X < 12.6) = 1 - P(X > 12.6) = 1 - 0.135 = 0.865$.

3-4.2 Cumulative Distribution Function

Another way to describe the probability distribution of a random variable is with a function of a real number x that provides the probability that X is less than or equal to x.

The cumulative distribution function of a continuous random variable X with probability density function $f(x)$ is

$$F(x) = P(X \le x) = \int_{-\infty}^{x} f(u)\, du$$

for $-\infty < x < \infty$.

For a continuous random variable X, the definition can also be $F(x) = P(X < x)$ because $P(X = x) = 0$.

The cumulative distribution function $F(x)$ can be related to the probability density function $f(x)$ and can be used to obtain probabilities as follows.

$$P(a < X < b) = \int_{a}^{b} f(x)\, dx = \int_{-\infty}^{b} f(x)\, dx - \int_{-\infty}^{a} f(x)\, dx = F(b) - F(a)$$

Furthermore, the graph of a cumulative distribution function has specific properties. Because $F(x)$ provides probabilities, it is always nonnegative. Furthermore, as x increases, $F(x)$ is nondecreasing. Finally, as x tends to ∞, $F(x) = P(X \le x)$ tends to 1.

EXAMPLE 3-3

The distance in micrometers from the start of a track on a magnetic disk until the first surface flaw is a random variable with the cumulative distribution function

$$F(x) = 1 - \exp\left(-\frac{x}{2000}\right) \text{ for } x > 0$$

A graph of $F(x)$ is shown in Fig. 3-10. Note that $F(x) = 0$ for $x \le 0$. Also, $F(x)$ increases to 1 as mentioned.

Determine the probability that the distance until the first surface flaw is less than 1000 micrometers. The requested probability is

$$P(X < 1000) = F(1000) = 1 - \exp\left(-\frac{1}{2}\right) = 0.393$$

Determine the probability that the distance until the first surface flaw exceeds 2000 micrometers. Now we use

$$P(2000 < X) = 1 - P(X \le 2000) = 1 - F(2000) = 1 - [1 - \exp(-1)]$$
$$= \exp(-1) = 0.368$$

Determine the probability that the distance is between 1000 and 2000 micrometers. The requested probability is

$$P(1000 < X < 2000) = F(2000) - F(1000) = 1 - \exp(-1) - [1 - \exp(-0.5)]$$
$$= \exp(-0.5) - \exp(-1) = 0.239$$

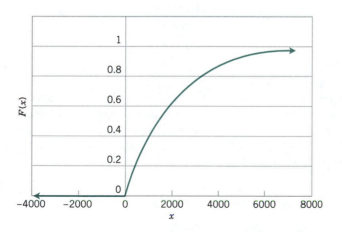

Figure 3-10 Cumulative distribution function for Example 3-3.

The cumulative distribution function is often tabulated to present probabilities. It is convenient to list $F(x)$ for selected values of x. Then additional probabilities can be determined as in the previous example.

3-4.3   Mean and Variance

Just as it is useful to summarize a sample of data by the mean and variance, we can summarize the probability distribution of X by its mean and variance. Recall that for sample data $x_1, x_2, \ldots, x_n$, the sample mean can be written as

$$\bar{x} = \frac{1}{n} x_1 + \frac{1}{n} x_2 + \cdots + \frac{1}{n} x_n$$

That is, $\bar{x}$ uses equal weights of $1/n$ as the multiplier of each measured value x_i. The mean of a random variable X uses the probability model to weight the possible values of X. The mean or expected value of X, denoted as μ or $E(X)$, is

$$\mu = E(X) = \int_{-\infty}^{\infty} x f(x)\, dx$$

The integral in $E(X)$ is analogous to the sum that is used to calculate $\bar{x}$.

Recall that $\bar{x}$ is the balance point when an equal weight is placed at the location of each measurement along a number line. Similarly, if $f(x)$ is the density function of a loading on a long, thin beam, then $E(X)$ is the point at which the beam balances. Consequently, $E(X)$ describes the "center" of the distribution of X in a manner similar to the balance point of a loading.

For sample data $x_1, x_2, \ldots, x_n$, the variance is a summary of the dispersion or scatter in the data. It is

$$s^2 = \frac{1}{n-1}(x_1 - \bar{x})^2 + \frac{1}{n-1}(x_2 - \bar{x})^2 + \cdots + \frac{1}{n-1}(x_n - \bar{x})^2$$

That is, s^2 uses equal weights of $1/(n-1)$ as the multiplier of each squared deviation $(x_i - \bar{x})^2$. As mentioned previously, deviations calculated from $\bar{x}$ tend to be smaller than those calculated from μ, and the weight is adjusted from $1/n$ to $1/(n-1)$ to compensate.

The variance of a random variable X is a measure of dispersion or scatter in the possible values for X. The variance of X, denoted as σ^2 or $V(X)$, is

$$\sigma^2 = V(X) = \int_{-\infty}^{\infty} (x - \mu)^2 f(x)\, dx$$

$V(X)$ uses weight $f(x)$ as the multiplier of each possible squared deviation $(x - \mu)^2$. The integral in $V(X)$ is analogous to the sum that is used to calculate s^2.

Properties of integrals and the definition of μ can be used to show that

$$V(X) = \int_{-\infty}^{\infty} (x - \mu)^2 f(x)\, dx$$

$$= \int_{-\infty}^{\infty} x^2 f(x)\, dx - 2\mu \int_{-\infty}^{\infty} x f(x)\, dx + \int_{-\infty}^{\infty} \mu^2 f(x)\, dx$$

$$= \int_{-\infty}^{\infty} x^2 f(x)\, dx - 2\mu^2 + \mu^2 = \int_{-\infty}^{\infty} x^2 f(x)\, dx - \mu^2$$

so an alternative formula for $V(x)$ can be used.

Definition

Suppose X is a continuous random variable with probability density function $f(x)$. The **mean** or **expected value** of X, denoted as μ or $E(X)$, is

$$\mu = E(X) = \int_{-\infty}^{\infty} x f(x)\, dx \tag{3-3}$$

The **variance** of X, denoted as $V(X)$ or σ^2, is

$$\sigma^2 = V(X) = \int_{-\infty}^{\infty} (x - \mu)^2 f(x)\, dx = \int_{-\infty}^{\infty} x^2 f(x)\, dx - \mu^2$$

The **standard deviation** of X is $\sigma = [V(X)]^{1/2}$.

EXAMPLE 3-4

For the copper current measurement in Example 3-1, the mean of X is

$$E(x) = \int_{-\infty}^{\infty} x f(x)\, dx = \int_{0}^{20} x \left(\frac{1}{20} \right) dx = 0.05 x^2/2 \, \Big|_{0}^{20} = 10$$

The variance of X is

$$V(x) = \int_{-\infty}^{\infty} (x - \mu)^2 f(x)\, dx = \int_{0}^{20} (x - 10)^2 \left(\frac{1}{20} \right) dx = 0.05 (x - 10)^3/3 \, \Big|_{0}^{20} = 33.33$$

EXAMPLE 3-5

For the drilling operation in Example 3-2, the mean of X is

$$E(x) = \int_{-\infty}^{\infty} xf(x)\, dx = \int_{12.5}^{\infty} x\, 20\, e^{-20(x-12.5)}\, dx$$

Integration by parts can be used to show that

$$E(X) = -xe^{-20(x-12.5)} - \left. \frac{e^{-20(x-12.5)}}{20}\right|_{12.5}^{\infty} = 12.5 + 0.05 = 12.55$$

The variance of X is

$$V(X) = \int_{-\infty}^{\infty} (x-\mu)^2 f(x)\, dx = \int_{12.5}^{\infty} (x-12.55)^2\, 20e^{-20(x-12.5)}dx$$

Integration by parts can be used two times to show that

$$V(X) = 0.0025$$

EXERCISES FOR SECTION 3-4

3-15. Show that the following functions are probability density functions for some value of k and determine k. Then, determine the mean and variance of X.
(a) $f(x) = kx^2$ for $0 < x < 4$
(b) $f(x) = k(1 + 2x)$ for $0 < x < 2$
(c) $f(x) = ke^{-x}$ for $0 < x$

3-16. Determine the following probabilities for the probability density function in Exercise 3-15(a).
(a) $P(X > 2)$
(b) $P(1 < X < 3)$
(c) $P(X < 1)$
(d) $P(X < 1) + P(1 < X < 3)$

3-17. Suppose that $f(x) = e^{-(x-6)}$ for $6 < x$ and $f(x) = 0$ for $x \le 6$. Determine the following probabilities.
(a) $P(X > 6)$
(b) $P(6 \le X < 8)$
(c) $P(X < 8)$
(d) $P(X > 8)$
(e) Determine x such that $P(X < x) = 0.95$.

3-18. Suppose that $f(x) = 1.5x^2$ for $-1 < x < 1$ and $f(x) = 0$ otherwise. Determine the following probabilities.
(a) $P(0 < X)$
(b) $P(0.5 < X)$
(c) $P(-0.5 \le X \le 0.5)$
(d) $P(X < -2)$
(e) $P(X < 0 \text{ or } X > -0.5)$
(f) Determine x such that $P(x < X) = 0.05$.

3-19. The probability density function of the time to failure of an electronic component in a copier (in hours) is $f(x) = \exp(-x/3000)/3000$ for $x > 0$ and $f(x) = 0$ for $x \le 0$. Determine the probability that
(a) A component lasts more than 1000 hours before failure.

(b) A component fails in the interval from 1000 to 2000 hours.

(c) A component fails before 3000 hours.

(d) Determine the number of hours at which 10% of all components have failed.

(e) Determine the mean and variance.

3-20. The probability density function of the net weight in pounds of a packaged compound is $f(x) = 4.0$ for $19.875 < x < 20.125$ ounces and $f(x) = 0$ for x elsewhere.

(a) Determine the probability that a package weighs less than 20 ounces.

(b) Suppose that the packaging specifications require that the weight be between 19.9 and 20.1 ounces. What is the probability that a randomly selected package will have a weight within these specifications?

(c) Determine the mean and variance.

3-21. The temperature readings from a thermocouple in a furnace fluctuate according to a cumulative distribution function

$$F(x) = \begin{cases} 0 & x < 800°C \\ 0.1x - 80 & 800°C \le x < 810°C \\ 0 & x > 810°C \end{cases}$$

Determine the following.

(a) $P(X < 805)$

(b) $P(800 < X \le 805)$

(c) $P(X > 808)$

(d) If the specifications for the process require that the furnace temperature be between 802°C and 808°C, what is the probability that the furnace will operate outside of the specifications?

3-22. The probability density function describing the thickness measurement of a wall of plastic tubing is $f(x) = 400$ for $2.0025 < x < 2.0050$ millimeters and $f(x) = 0$ for x elsewhere. Determine the following.

(a) $P(X \le 2.0030)$

(b) $P(X > 2.0045)$

(c) If the specification for the tubing requires that the thickness measurement be between 2.0030 and 2.0040 millimeters, what is the probability that a single measurement will indicate conformance to the specification?

3-23. Suppose that contamination particle size (in micrometers) can be modeled as $f(x) = 2x^{-3}$ for $1 < x$ and $f(x) = 0$ for $x \le 1$.

(a) Confirm that $f(x)$ is a probability density function.

(b) Determine the mean.

(c) What is the probability that the size of a random particle will be less than 5 micrometers?

(d) An optical device is being marketed to detect contamination particles. It is capable of detecting particles exceeding 7 micrometers in size. What proportion of the particles will be detected?

3-24. (Integration by parts is required in this exercise.) The probability density function for the diameter of a drilled hole in millimeters is $10e^{-10(x-5)}$ for $x > 5$ mm and zero for $x \le 5$ mm. Although the target diameter is 5 millimeters, vibrations, tool wear, and other factors can produce diameters larger than 5 millimeters.

(a) Determine the mean and variance of the diameter of the holes.

(b) Determine the probability that a diameter exceeds 5.1 millimeters.

3-25. Suppose the cumulative distribution function of the length (in millimeters) of computer cables is

$$F(x) = \begin{cases} 0 & x \le 1200 \\ 0.1x - 120 & 1200 < x \le 1210 \\ 1 & x > 1210 \end{cases}$$

(a) Determine $P(x < 1208)$.

(b) If the length specifications are $1195 < x < 1205$ millimeters, what is the probability that a randomly selected computer cable will meet the specification requirement?

3-26. The thickness of a conductive coating

in micrometers has a density function of $600x^{-2}$ for 100 μm $< x <$ 120 μm and zero for x elsewhere.
(a) Determine the mean and variance of the coating thickness.
(b) If the coating costs \$0.50 per micrometer of thickness on each part, what is the average cost of the coating per part?

3-27. Suppose that $f(x) = 0.5x - 1$ for $2 < x < 4$ and $f(x) = 0$ for x elsewhere. Determine the following.
(a) $P(X < 2)$
(b) $P(X > 3)$
(c) $P(2.5 < X < 3.5)$
(d) The mean and variance

3-5   NORMAL DISTRIBUTION

Undoubtedly, the most widely used model for the distribution of a random variable is a **normal distribution.** In Chapter 2, several histograms are shown with similar symmetric, bell shapes. A fundamental result, known as the **central limit theorem,** implies that histograms often have this characteristic shape, at least approximately. Whenever a random experiment is replicated, the random variable that equals the average (or total) result over the replicates tends to have a normal distribution as the number of replicates becomes large. De Moivre presented this result in 1733. Unfortunately, his work was lost for some time, and Gauss independently developed a normal distribution nearly 100 years later. Although De Moivre was later credited with the derivation, a normal distribution is also referred to as a **Gaussian** distribution.

When do we average (or total) results? Almost always. In Example 2-1 the average of the eight pull-off force measurements was calculated to be 13.0 foot-pounds. If we assume that each measurement results from a replicate of a random experiment, then the normal distribution can be used to make approximate conclusions about this average. These conclusions are the primary topics in the subsequent chapters of this book.

Furthermore, sometimes the central limit theorem is less obvious. For example, assume that the deviation (or error) in the length of a machined part is the sum of a large number of infinitesimal (small) effects, such as temperature and humidity drifts, vibrations, cutting angle variations, cutting tool wear, bearing wear, rotational speed variations, mounting and fixturing variations, variations in numerous raw material characteristics, and variation in levels of contamination. If the component errors are independent and equally likely to be positive or negative, then the total error can be shown to have an approximate normal distribution. Furthermore, the normal distribution arises in the study of numerous basic physical phenomena. For example, the physicist Maxwell developed a normal distribution from simple assumptions regarding the velocities of molecules.

The theoretical basis of a normal distribution is mentioned to justify the somewhat complex form of the probability density function. Our objective now is to calculate probabilities for a normal random variable. The central limit theorem will be stated more carefully later in this chapter.

Random variables with different means and variances can be modeled by normal probability density functions with appropriate choices of the center and width of the curve.

The value of $E(X) = \mu$ determines the center of the probability density function and the value of $V(X) = \sigma^2$ determines the width. Figure 3-11 illustrates several normal probability density functions with selected values of μ and σ^2. Each has the characteristic symmetric, bell-shaped curve, but the centers and dispersions differ. The following definition provides the formula for normal probability density functions.

Definition

A random variable X with probability density function

$$f(x) = \frac{1}{\sqrt{2\pi}\sigma} e^{\frac{-(x-\mu)^2}{2\sigma^2}} \qquad \text{for} \quad -\infty < x < \infty \qquad (3\text{-}4)$$

has a **normal distribution** (and is called a **normal random variable**) with parameters μ and σ, where $-\infty < \mu < \infty$, and $\sigma > 0$. Also,

$$E(X) = \mu \quad \text{and} \quad V(X) = \sigma^2$$

The notation $N(\mu, \sigma^2)$ is often used to denote a normal distribution with mean μ and variance σ^2.

EXAMPLE 3-6

Assume that the current measurements in a strip of wire follow a normal distribution with a mean of 10 milliamperes and a variance of 4 (milliamperes)^2. What is the probability that a measurement exceeds 13 milliamperes?

Let X denote the current in milliamperes. The requested probability can be represented as $P(X > 13)$. This probability is shown as the shaded area under the normal probability density function in Fig. 3-12. Unfortunately, there is no closed-form expression for the integral of a normal probability density function, and probabilities based on the normal distribution are typically found numerically or from a table (that we will introduce later).

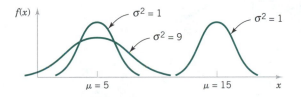

Figure 3-11   Normal probability density functions for selected values of the parameters μ and σ^2.

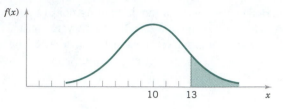

Figure 3-12   Probability that $X > 13$ for a normal random variable with $\mu = 10$ and $\sigma^2 = 4$ in Example 3-6.

Some useful results concerning a normal distribution are summarized in Fig. 3-13. For any normal random variable,

$$P(\mu - \sigma < X < \mu + \sigma) = 0.6827$$
$$P(\mu - 2\sigma < X < \mu + 2\sigma) = 0.9545$$
$$P(\mu - 3\sigma < X < \mu + 3\sigma) = 0.9973$$

From the symmetry of $f(x)$, $P(X > \mu) = P(X < \mu) = 0.5$. Because $f(x)$ is positive for all x, this model assigns some probability to each interval of the real line. However, the probability density function decreases as x moves farther from μ. Consequently, the probability that a measurement falls far from μ is small, and at some distance from μ the probability of an interval can be approximated as zero. The area under a normal probability density function beyond 3σ from the mean is quite small. This fact is convenient for quick, rough sketches of a normal probability density function. The sketches help us determine probabilities. Because more than 0.9973 of the probability of a normal distribution is within the interval $(\mu - 3\sigma, \mu + 3\sigma)$, 6σ *is often referred to as the width of a normal distribution.* Numerical integration can be used to show that the area under the normal probability density function from $-\infty < x < \infty$ is 1.

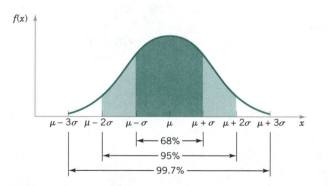

Figure 3-13   Probabilities associated with a normal distribution.

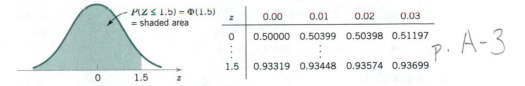

z	0.00	0.01	0.02	0.03
0	0.50000	0.50399	0.50398	0.51197
⋮		⋮		
1.5	0.93319	0.93448	0.93574	0.93699

$P(Z \leq 1.5) = \Phi(1.5)$
= shaded area

p. A-3

Figure 3-14   Standard normal probability density function.

Definition

A normal random variable with $\mu = 0$ and $\sigma^2 = 1$ is called a **standard normal random variable.** A standard normal random variable is denoted as Z.

Appendix A Table I provides cumulative probabilities for a standard normal random variable. The use of Table I is illustrated by the following example.

EXAMPLE 3-7

Assume that Z is a standard normal random variable. Appendix A Table I provides probabilities of the form $P(Z \leq z)$. The use of Table I to find $P(Z \leq 1.5)$ is illustrated in Fig. 3-14. We read down the z column to the row that equals 1.5. The probability is read from the adjacent column, labeled 0.00 to be 0.93319.

The column headings refer to the hundredth's digit of the value of z in $P(Z \leq z)$. For example, $P(Z \leq 1.53)$ is found by reading down the z column to the row 1.5 and then selecting the probability from the column labeled 0.03 to be 0.93699.

Definition

The function

$$\Phi(z) = P(Z \leq z)$$

is used to denote a probability from Appendix A Table I. It is called the cumulative distribution function of a standard normal random variable. A table is required because the probability can't be determined by elementary methods.

Cumulative distribution functions exist for other random variables, and they are widely available in computer software packages. They can be used in the same manner as $\Phi(z)$ to obtain probabilities for these random variables.

Probabilities that are not of the form $P(Z \le z)$ are found by using the basic rules of probability and the symmetry of the normal distribution along with Appendix A Table I. The following examples illustrate the method.

EXAMPLE 3-8

The following calculations are shown pictorially in Fig. 3-15. In practice, a probability is often rounded to one or two significant digits.

(1) $P(Z > 1.26) = 1 - P(Z \le 1.26) = 1 - 0.89616 = 0.10384$

(2) $P(Z < -0.86) = 0.19490$

(3) $P(Z > -1.37) = P(Z < 1.37) = 0.91465$

(4) $P(-1.25 < Z < 0.37)$. This probability can be found from the difference of two areas, $P(Z < 0.37) - P(Z < -1.25)$. Now,

$$P(Z < 0.37) = 0.64431 \quad \text{and} \quad P(Z < -1.25) = 0.10565$$

Therefore,

$$P(-1.25 < Z < 0.37) = 0.64431 - 0.10565 = 0.53866$$

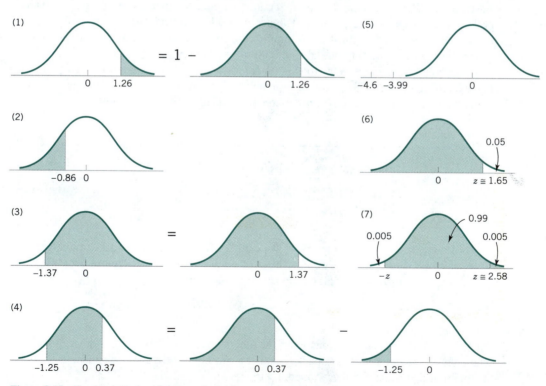

Figure 3-15 Graphical displays for Example 3-8.

(5) $P(Z \leq -4.6)$ cannot be found exactly from Table I. However, the last entry in the table can be used to find that $P(Z \leq -3.99) = 0.00003$. Because $P(Z \leq -4.6) < P(Z \leq -3.99)$, $P(Z \leq -4.6)$ is nearly zero.

(6) Find the value z such that $P(Z > z) = 0.05$. This probability equation can be written as $P(Z \leq z) = 0.95$. Now, Table I is used in reverse. We search through the probabilities to find the value that corresponds to 0.95. The solution is illustrated in Fig. 3-15. We do not find 0.95 exactly; the nearest value is 0.95053, corresponding to $z = 1.65$.

(7) Find the value of z such that $P(-z < Z < z) = 0.99$. Because of the symmetry of the normal distribution, if the area of the shaded region in Fig. 3-15(7) is to equal 0.99, then the area in each tail of the distribution must equal 0.005. Therefore, the value for z corresponds to a probability of 0.995 in Table I. The nearest probability in Table I is 0.99506, when $z = 2.58$.

The preceding examples show how to calculate probabilities for standard normal random variables. To use the same approach for an arbitrary normal random variable would require a separate table for every possible pair of values for μ and σ. Fortunately, all normal probability distributions are related algebraically, and Appendix A Table I can be used to find the probabilities associated with an arbitrary normal random variable by first using a simple transformation.

If X is a normal random variable with $E(X) = \mu$ and $V(X) = \sigma^2$, then the random variable

$$Z = \frac{(X - \mu)}{\sigma}$$

is a normal random variable with $E(Z) = 0$ and $V(Z) = 1$. That is, Z is a standard normal random variable.

Creating a new random variable by this transformation is referred to as **standardizing.** The random variable Z represents the distance of X from its mean in terms of standard deviations. It is the key step in calculating a probability for an arbitrary normal random variable.

EXAMPLE 3-9

Suppose the current measurements in a strip of wire are assumed to follow a normal distribution with a mean of 10 milliamperes and a variance of 4 (milliamperes)^2. What is the probability that a measurement will exceed 13 milliamperes?

Let X denote the current in milliamperes. The requested probability can be represented as $P(X > 13)$. Let $Z = (X - 10)/2$. The relationship between the several values of X and the transformed values of Z are shown in Fig. 3-16. We note that $X > 13$ corresponds to $Z > 1.5$. Therefore, from Table I,

$$P(X > 13) = P(Z > 1.5) = 1 - P(Z \leq 1.5) = 1 - 0.93319 = 0.06681$$

Rather than using Fig. 3-16, the probability can be found from the inequality $X > 13$. That is,

$$P(X > 13) = P\left(\frac{X - 10}{2} > \frac{13 - 10}{2}\right) = P(Z > 1.5) = 0.06681$$

In the preceding example, the value 13 is transformed to 1.5 by standardizing, and 1.5 is often referred to as the **z-value** associated with a probability.

The following box summarizes the calculation of probabilities derived from normal random variables.

Suppose X is a normal random variable with mean μ and variance σ^2. Then,

$$P(X \leq x) = P\left(\frac{X - \mu}{\sigma} \leq \frac{x - \mu}{\sigma}\right) = P(Z \leq z) \qquad (3\text{-}5)$$

where

Z is a **standard normal random variable,** and

$z = (x - \mu)/\sigma$ is the **z-value** obtained by **standardizing** X.

The probability is obtained by entering **Appendix A Table I** with $z = (x - \mu)/\sigma$.

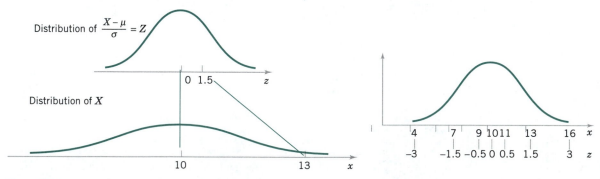

Figure 3-16   Standardizing a normal random variable.

EXAMPLE 3-10

Continuing the previous example, what is the probability that a current measurement is between 9 and 11 milliamperes?

From Fig. 3-16, or by proceeding algebraically, we have

$$P(9 < X < 11) = P\left(\frac{9-10}{2} < \frac{X-10}{2} < \frac{11-10}{2}\right)$$
$$= P(-0.5 < Z < 0.5)$$
$$= P(Z < 0.5) - P(Z < -0.5)$$
$$= 0.69146 - 0.30854$$
$$= 0.38292$$

Determine the value for which the probability that a current measurement is below this value is 0.98. The requested value is shown graphically in Fig. 3-17. We need the value of x such that $P(X < x) = 0.98$. By standardizing, this probability expression can be written as

$$P(X < x) = P\left(\frac{X-10}{2} < \frac{x-10}{2}\right)$$
$$= P\left(Z < \left(\frac{x-10}{2}\right)\right)$$
$$= 0.98$$

Table I is used to find the z value such that $P(Z < z) = 0.98$. The nearest probability from Table I results in

$$P(Z < 2.05) = 0.97982$$

Therefore, $(x - 10)/2 = 2.05$, and the standardizing transformation is used in reverse to solve for x. The result is

$$x = 2(2.05) + 10 = 14.1 \text{ milliamperes}$$

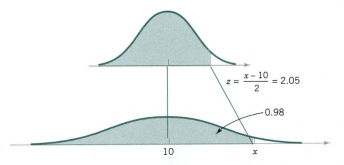

Figure 3-17 Determining the value of x to meet a specified probability.

EXAMPLE 3-11

In the transmission of a digital signal assume that the background noise follows a normal distribution with a mean of 0 volt and standard deviation of 0.45 volt. If the system assumes a digital 1 has been transmitted when the voltage exceeds 0.9, what is the probability of detecting a digital 1 when none was sent?

Let the random variable N denote the voltage of noise. The requested probability is

$$P(N > 0.9) = P\left(\frac{N}{0.45} > \frac{0.9}{0.45}\right) = P(Z > 2) = 1 - 0.97725 = 0.02275$$

This probability can be described as the probability of a false detection.

Determine symmetric bounds about 0 that include 99% of all noise readings. The question requires us to find x such that $P(-x < N < x) = 0.99$. A graph is shown in Fig. 3-18. Now,

$$P(-x < N < x) = P\left(-\frac{x}{0.45} < \frac{N}{0.45} < \frac{x}{0.45}\right)$$

$$= P\left(-\frac{x}{0.45} < Z < \frac{x}{0.45}\right) = 0.99$$

From Table I

$$P(-2.58 < Z < 2.58) = 0.99$$

Therefore,

$$\frac{x}{0.45} = 2.58$$

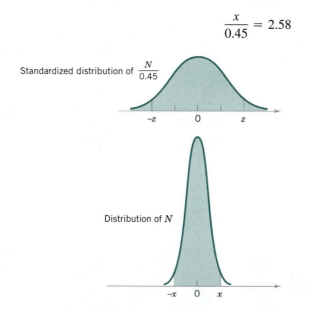

Figure 3-18   Determining the value of x to meet a specified probability.

and

$$x = 2.58(0.45) = 1.16$$

Suppose a digital 1 is represented as a shift in the mean of the noise distribution to 1.8 volts. What is the probability that a digital 1 is not detected? Let the random variable S denote the voltage when a digital 1 is transmitted. Then,

$$P(S < 0.9) = P\left(\frac{S - 1.8}{0.45} < \frac{0.9 - 1.8}{0.45}\right)$$

$$= P(Z < -2)$$

$$= 0.02275$$

This probability can be interpreted as the probability of a missed signal.

EXAMPLE 3-12

The diameter of a shaft in an optical storage drive is normally distributed with mean 0.2508 inch and standard deviation 0.0005 inch. The specifications on the shaft are 0.2500 ± 0.0015 inch. What proportion of shafts conforms to specifications?

Let X denote the shaft diameter in inches. The requested probability is shown in Fig. 3-19 and

$$P(0.2485 < X < 0.2515) = P\left(\frac{0.2485 - 0.2508}{0.0005} < Z < \frac{0.2515 - 0.2508}{0.0005}\right)$$

$$= P(-4.6 < Z < 1.4)$$
$$= P(Z < 1.4) - P(Z < -4.6)$$
$$= 0.91924 - 0.0000$$
$$= 0.91924$$

Most of the nonconforming shafts are too large, because the process mean is located very

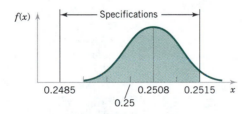

Figure 3-19 Distribution for Example 3-12.

near to the upper specification limit. If the process is centered so that the process mean is equal to the target value of 0.2500, then

$$P(0.2485 < X < 0.2515) = P\left(\frac{0.2485 - 0.2500}{0.0005} < Z < \frac{0.2515 - 0.2500}{0.0005}\right)$$
$$= P(-3 < Z < 3)$$
$$= P(Z < 3) - P(Z < -3)$$
$$= 0.99865 - 0.00135$$
$$= 0.9973$$

By recentering the process, the yield is increased to approximately 99.73%.

EXERCISES FOR SECTION 3-5

3-28. Use Appendix A Table I to determine the following probabilities for the standard normal random variable Z.
(a) $P(-1 < Z < 1)$
(b) $P(-2 < Z < 2)$
(c) $P(-3 < Z < 3)$
(d) $P(Z < -3)$
(e) $P(0 < Z \le 3)$

3-29. Assume that Z has a standard normal distribution. Use Appendix A Table I to determine the value for z that solves each of the following.
(a) $P(Z < z) = 0.50000$
(b) $P(Z < z) = 0.001001$
(c) $P(Z > z) = 0.881000$
(d) $P(Z > z) = 0.866500$
(e) $P(-1.3 < Z < z) = 0.863140$

3-30. Assume that Z has a standard normal distribution. Use Appendix A Table I to determine the value for z that solves each of the following.
(a) $P(-z < Z < z) = 0.95$
(b) $P(-z < Z < z) = 0.99$
(c) $P(-z < Z < z) = 0.68$
(d) $P(-z < Z < z) = 0.9973$

3-31. Assume that X is normally distributed with a mean of 10 and a standard deviation of 2. Determine the following.
(a) $P(X < 14)$
(b) $P(X > 8)$
(c) $P(8 < X < 12)$
(d) $P(4 < X < 16)$
(e) $P(6 < X < 10)$
(f) $P(10 < X < 16)$

3-32. Assume that X is normally distributed with a mean of 10 and a standard deviation of 2. Determine the value for x that solves each of the following.
(a) $P(X > x) = 0.5$
(b) $P(X > x) = 0.95$
(c) $P(x < X < 10) = 0.2$

3-33. Assume that X is normally distributed with a mean of 7 and a standard deviation of 2. Determine the following.
(a) $P(X < 11)$
(b) $P(X > 0)$
(c) $P(3 < X < 7)$
(d) $P(-2 < X < 9)$
(e) $P(2 < X < 8)$

3-34. Assume that X is normally distributed with a mean of 6 and a standard deviation of 3. Determine the value for x that solves each of the following.
(a) $P(X > x) = 0.5$
(b) $P(X > x) = 0.95$
(c) $P(x < X < 9) = 0.2$
(d) $P(3 < X < x) = 0.8$

3-35. The compressive strength of samples of cement can be modeled by a normal distribution with a mean of 6000 kilograms per square centimeter and a standard deviation of 100 kilograms per square centimeter.

(a) What is the probability that a sample's strength is less than 6250 kg/cm^2?

(b) What is the probability that a sample's strength is between 5800 and 5900 kg/cm^2?

(c) What strength is exceeded by 95% of the samples?

3-36. The tensile strength of paper is modeled by a normal distribution with a mean of 35 pounds per square inch and a standard deviation of 2 pounds per square inch.

(a) What is the probability that the strength of a sample is less than 39 lb/in^2?

(b) If the specifications require the tensile strength to exceed 29 lb/in^2, what proportion of the sample is scrapped?

3-37. The line width of a tool used for semiconductor manufacturing is assumed to be normally distributed with a mean of 0.5 micrometer and a standard deviation of 0.05 micrometer.

(a) What is the probability that a line width is greater than 0.62 micrometer?

(b) What is the probability that a line width is between 0.47 and 0.63 micrometer?

(c) The line width of 90% of samples is below what value?

3-38. The fill volume of an automated filling machine used for filling cans of carbonated beverage is normally distributed with a mean of 12.4 fluid ounces and a standard deviation of 0.1 fluid ounce.

(a) What is the probability that a fill volume is less than 12 fluid ounces?

(b) If all cans less than 12.1 or greater than 12.6 ounces are scrapped, what proportion of cans is scrapped?

(c) Determine specifications that are symmetric about the mean that include 99% of all cans.

3-39. Continuation of Exercise 3-38. The mean of the filling operation can be adjusted easily, but the standard deviation remains at 0.1 ounce.

(a) At what value should the mean be set so that 99.9% of all cans exceed 12 ounces?

(b) At what value should the mean be set so that 99.9% of all cans exceed 12 ounces if the standard deviation can be reduced to 0.05 fluid ounce?

3-40. The reaction time of a driver to visual stimulus is normally distributed with a mean of 0.4 second and a standard deviation of 0.05 second.

(a) What is the probability that a reaction requires more than 0.5 second?

(b) What is the probability that a reaction requires between 0.4 and 0.5 second?

(c) What is the reaction time that is exceeded 90% of the time?

3-41. The length of an injected-molded plastic case that holds magnetic tape is normally distributed with a mean length of 90.2 millimeters and a standard deviation of 0.1 millimeter.

(a) What is the probability that a part is longer than 90.3 millimeters or shorter than 89.7 millimeters?

(b) What should the process mean be set at to obtain the greatest number of parts between 89.7 and 90.3 millimeters?

(c) If parts that are not between 89.7 and 90.3 millimeters are scrapped, what is the yield for the process mean that you selected in part (b)?

3-42. Continuation of Exercise 3-41. Assume that the process is centered so that the mean is 90 millimeters and the standard deviation is 0.1 millimeter.

(a) What is the probability that a part is between 89.8 and 90.2 millimeters?

(b) What is the probability that a part is less than 89.8 or greater than 90.2 millimeters?

3-43. The sick-leave time of employees in a firm in a month is normally distributed

with a mean of 60 hours and a standard deviation of 10 hours.
(a) What is the probability that the sick-leave time for next month will be between 50 and 80 hours?
(b) How much time should be budgeted for sick leave if the budgeted amount should be exceeded with a probability of only 10%?

3-44. The life of a semiconductor laser at a constant power is normally distributed with a mean of 7000 hours and a standard deviation of 600 hours.
(a) What is the probability that a laser fails before 5000 hours?
(b) What is the life in hours that 95% of the lasers exceed?

3-45. The diameter of the dot produced by a printer is normally distributed with a mean diameter of 0.002 inch and a standard deviation of 0.0004 inch.
(a) What is the probability that the diameter of a dot exceeds 0.0026 inch?

(b) What is the probability that a diameter is between 0.0014 and 0.0026 inch?
(c) What standard deviation of diameters is needed so that the probability in part (b) is 0.995?

3-46. The weight of a human joint replacement part is normally distributed with a mean of 2 ounces and a standard deviation of 0.05 ounce.
(a) What is the probability that a part weighs more than 2.10 ounces?
(b) What must the standard deviation of weight be for the company to state that 99.9% of its parts are less than 2.10 ounces?
(c) If the standard deviation remains at 0.05 ounce, what must the mean weight be for the company to state that 99.9% of its parts are less than 2.10 ounces?

3-6   PROBABILITY PLOTS

How do we know whether a normal distribution is a reasonable model for data? **Probability plotting** is a graphical method for determining whether sample data conform to a hypothesized distribution based on a subjective visual examination of the data. The general procedure is very simple and can be performed quickly. Probability plotting typically uses special graph paper, known as **probability paper,** that has been designed for the hypothesized distribution. Probability paper is widely available for the normal, lognormal, Weibull, and various chi-square and gamma distributions.

To construct a probability plot, the observations in the sample are first ranked from smallest to largest. That is, the sample $x_1, x_2, \ldots, x_n$ is arranged as $x_{(1)}, x_{(2)}, \ldots, x_{(n)}$, where $x_{(1)}$ is the smallest observation, $x_{(2)}$ is the second smallest observation, and so forth, with $x_{(n)}$ the largest. The ordered observations $x_{(j)}$ are then plotted against their observed cumulative frequency $(j - 0.5)/n$ on the appropriate probability paper. If the hypothesized distribution adequately describes the data, the plotted points will fall approximately along a straight line; if the plotted points deviate significantly and systematically from a straight line, then the hypothesized model is not appropriate. Usually, the determination of whether or not the data plot as a straight line is subjective. The procedure is illustrated in the following example.

EXAMPLE 3-13

Ten observations on the effective service life in minutes of batteries used in a portable personal computer are as follows: 176, 191, 214, 220, 205, 192, 201, 190, 183, 185. We hypothesize that battery life is adequately modeled by a normal distribution. To use probability plotting to investigate this hypothesis, first arrange the observations in ascending order and calculate their cumulative frequencies $(j - 0.5)/10$ as follows.

j	$x_{(j)}$	$(j - 0.5)/10$
1	176	0.05
2	183	0.15
3	185	0.25
4	190	0.35
5	191	0.45
6	192	0.55
7	201	0.65
8	205	0.75
9	214	0.85
10	220	0.95

The pairs of values $x_{(j)}$ and $(j - 0.5)/10$ are now plotted on normal probability paper. This plot is shown in Fig. 3-20. Most normal probability paper plots $100(j - 0.5)/n$ on the left vertical scale and $100[1 - (j - 0.5)/n]$ on the right vertical scale, with the variable value plotted on the horizontal scale. A straight line, chosen subjectively, has been drawn through the plotted points. In drawing the straight line, you should be influenced more by the points near the middle of the plot than by the extreme points. A good rule of thumb is to draw the line approximately between the 25th and 75th percentile points. This is how the line in Fig. 3-20 was determined. In assessing the systematic deviation of the points from the straight line, imagine a "fat pencil" lying along the line. If all the

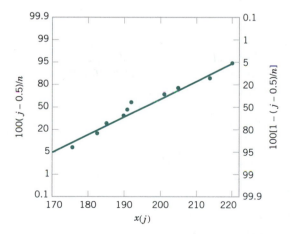

Figure 3-20 Normal probability plot for battery life.

points are covered by this imaginary pencil, then a normal distribution adequately describes the data. Since the points in Fig. 3-20 would pass the "fat pencil" test, we conclude that the normal distribution is an appropriate model.

A normal probability plot can also be constructed on ordinary graph paper by plotting the standardized normal scores z_j against $x_{(j)}$, where the standardized normal scores satisfy

$$\frac{j - 0.5}{n} = P(Z \le z_j) = \Phi(z_j)$$

For example, if $(j - 0.5)/n = 0.05$, then $\Phi(z_j) = 0.05$ implies that $z_j = -1.64$. To illustrate, consider the data from the previous example. In the following table we show the standardized normal scores in the last column.

j	$x_{(j)}$	$(j - 0.5)/10$	z_j
1	176	0.05	-1.64
2	183	0.15	-1.04
3	185	0.25	-0.67
4	190	0.35	-0.39
5	191	0.45	-0.13
6	192	0.55	0.13
7	201	0.65	0.39
8	205	0.75	0.67
9	214	0.85	1.04
10	220	0.95	1.64

Figure 3-21 presents the plot of z_j versus $x_{(j)}$. This normal probability plot is equivalent to the one in Fig. 3-20.

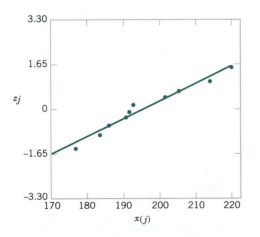

Figure 3-21 Normal probability plot obtained from standardized normal scores.

A very important application of normal probability plotting is in *verification of assumptions* when using statistical inference procedures that require the normality assumption.

Only probability plots for a normal distribution are shown here, but the method is just as easy for other distributions. Computer software packages enable us to generate probability plots to assess many other distributions. A disadvantage is that the method is subjective. Chapter 4 presents a more formal approach to assess a distribution based on a goodness-of-fit test.

EXERCISES FOR SECTION 3-6

3-47. A soft-drink bottler is studying the internal pressure strength of 1-liter glass bottles. A random sample of 16 bottles is tested, and the pressure strengths are obtained. The data are shown next. Plot these data on normal probability paper. Does it seem reasonable to conclude that pressure strength is normally distributed?

226.16 psi	211.14 psi
202.20	203.62
219.54	188.12
193.73	224.39
208.15	221.31
195.45	204.55
193.71	202.21
200.81	201.63

3-48. Samples of 20 parts are selected from two machines, and a critical dimension is measured on each part. The data are shown next. Plot the data on normal probability paper. Does this dimension seem to have a normal distribution? What tentative conclusions can you draw about the two machines?

Machine 1			
99.4	101.5	102.3	96.7
99.1	103.8	100.4	100.9
99.0	99.6	102.5	96.5
98.9	99.4	99.7	103.1
99.6	104.6	101.6	96.8

Machine 2			
90.9	100.7	95.0	98.8
99.6	105.5	92.3	115.5
105.9	104.0	109.5	87.1
91.2	96.5	96.2	109.8
92.8	106.7	97.6	106.5

3-49. After examining the data from the two machines in Exercise 3-48, the process engineer concludes that machine 2 has higher part-to-part variability. She makes some adjustments to the machine that should reduce the variability, and she obtains another sample of 20 parts. The measurements on those parts are shown next. Plot these data on normal probability paper and compare them with the normal probability plot of the data from machine 2 in Exercise 3-48. Is the normal distribution a reasonable mode for the data? Does it appear that the variance has been reduced?

103.4	107.0	107.7	104.5
108.1	101.5	106.2	106.6
103.1	104.1	106.3	105.6
108.2	106.9	107.8	103.7
103.9	103.3	107.4	102.6

3-50. In studying the uniformity of polysilicon thickness on a wafer in semiconductor manufacturing, Lu, Davis, and Gyurcsik (*Journal of the American Statistical Association,* Vol 93, 1998), collect data from 22 independent wafers: 494, 853,

1090, 1058, 517, 882, 732, 1143, 608,
590, 940, 920, 917, 581, 738, 732,
750, 1205, 1194, 1221, 1209, 708.

Is it reasonable to model these data using a normal probability distribution?

3-7   DISCRETE RANDOM VARIABLES

For a discrete random variable, only measurements at discrete points are possible.

EXAMPLE 3-14

A voice communication system for a business contains 48 external lines. At a particular time, the system is observed and some of the lines are being used. Let the random variable X denote the number of lines in use. Then X can assume any of the integer values 0 through 48.

EXAMPLE 3-15

The analysis of the surface of a semiconductor wafer records the number of particles of contamination that exceed a certain size. Define the random variable X to equal the number of particles of contamination.

The possible values of X are integers from 0 up to some large value that represents the maximum number of these particles that can be found on one of the wafers. If this maximum number is very large, it might be convenient to assume that any integer from zero to ∞ is possible.

3-7.1   Probability Mass Function

As mentioned previously, the probability distribution of a random variable X is a description of the probabilities associated with the possible values of X. For a discrete random variable, the distribution is often specified by just a list of the possible values along with the probability of each. In some cases, it is convenient to express the probability in terms of a formula.

EXAMPLE 3-16

There is a chance that a bit transmitted through a digital transmission channel is received in error. Let X equal the number of bits in error in the next four bits transmitted. The possible values for X are $\{0, 1, 2, 3, 4\}$. Based on a model for the errors that is presented in the following section, probabilities for these values will be determined. Suppose that the probabilities are

$$P(X = 0) = 0.6561$$
$$P(X = 1) = 0.2916$$
$$P(X = 2) = 0.0486$$
$$P(X = 3) = 0.0036$$
$$P(X = 4) = 0.0001$$

(handwritten) $x \mid \{0 \ 1 \ 2 \ 3 \ 4\}$ / $f(x) \mid .6561 \quad\quad .0001$

The probability distribution of X is specified by the possible values along with the probability of each. A graphical description of the probability distribution of X is shown in Fig. 3-22.

Suppose a loading on a long, thin beam places mass only at discrete points. See Fig. 3-23. The loading can be described by a function that specifies the mass at each of the discrete points. Similarly, for a discrete random variable X, its distribution can be described by a function that specifies the probability at each of the possible discrete values for X.

Definition

For a discrete random variable X with possible values $x_1, x_2, \ldots, x_n$, the **probability mass function** is

$$f(x_i) = P(X = x_i) \tag{3-6}$$

Because $f(x_i)$ is defined as a probability, $f(x_i) \geq 0$ for all x_i and $\sum_{i=1}^{n} f(x_i) = 1$. The reader should check that the sum of the probabilities in the previous example is 1.

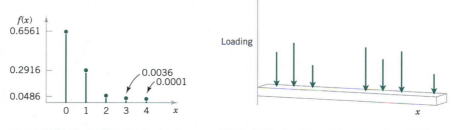

Figure 3-22   Probability distribution for X in Example 3-16.

Figure 3-23   Loadings at discrete points on a long, thin beam.

3-7.2 Cumulative Distribution Function

A cumulative distribution function can also be used to provide the probability distribution of a discrete random variable. The cumulative distribution function at a value x is the sum of the probabilities at all points less than or equal to x.

The **cumulative distribution function** of a discrete random variable X is

$$F(x) = P(X \le x) = \sum_{x_i \le x} f(x_i)$$

EXAMPLE 3-17

In the previous example, the probability mass function of X is

$$P(X = 0) = 0.6561 \qquad P(X = 1) = 0.2916 \qquad P(X = 2) = 0.0486$$
$$P(X = 3) = 0.0036 \qquad P(X = 4) = 0.0001$$

Therefore,

$$F(0) = 0.6561 \qquad F(1) = 0.9477 \qquad F(2) = 0.9963 \qquad F(3) = 0.9999 \qquad F(4) = 1$$

Even if the random variable can assume only integer values, the cumulative distribution function is defined at noninteger values. For example,

$$F(1.5) = P(X \le 1.5) = P(X \le 1) = 0.9477$$

The graph of $F(x)$ is shown in Figure 3-24. Note that the graph has discontinuities (jumps) at the discrete values for X. The size of the jump at a point x equals the probability at x.

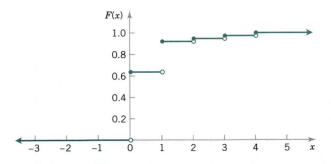

Figure 3-24 Cumulative distribution function for x in Example 3-16.

For example, consider $x = 1$. Here, $F(1) = 0.9477$, but for $0 \le x < 1$, $F(x) = 0.6561$. The change is $P(X = 1) = 0.2916$.

3-7.3   Mean and Variance

The mean and variance of a discrete random variable are defined similarly to a continuous random variable. Summation replaces integration in the definitions.

Definition

Let the possible values of the random variable X be denoted as $x_1, x_2, \ldots, x_n$. The probability mass function of X is $f(x)$, so $f(x_i) = P(X = x_i)$.

The **mean** or **expected value** of the discrete random variable X, denoted as μ or $E(X)$, is

$$\mu = E(X) = \sum_{i=1}^{n} x_i f(x_i) \tag{3-7}$$

The **variance** of X, denoted as σ^2 or $V(X)$, is

$$\sigma^2 = V(X) = E(X - \mu)^2 = \sum_{i=1}^{n} (x_i - \mu)^2 f(x_i) = \sum_{i=1}^{n} x_i^2 f(x_i) - \mu^2$$

The **standard deviation** of X is $\sigma = [V(X)]^{1/2}$.

The mean of X can be interpreted as the center of mass of the range of values of X. That is, if we place mass equal to $f(x_i)$ at each point x_i on the real line, then $E(X)$ is the point at which the real line is balanced. Therefore, the term "probability mass function" can be interpreted by this analogy with mechanics.

EXAMPLE 3-18

For the random variable in the previous example,

$$\begin{aligned} \mu = E(X) &= 0f(0) + 1f(1) + 2f(2) + 3f(3) + 4f(4) \\ &= 0(0.6561) + 1(0.2916) + 2(0.0486) + 3(0.0036) + 4(0.0001) \\ &= 0.4 \end{aligned}$$

Although X never assumes the value 0.4, the weighted average of the possible values is 0.4.

To calculate $V(X)$, a table is convenient.

x	$x - 0.4$	$(x - 0.4)^2$	$f(x)$	$f(x)(x - 0.4)^2$
0	-0.4	0.16	0.6561	0.104976
1	0.6	0.36	0.2916	0.104976
2	1.6	2.56	0.0486	0.124416
3	2.6	6.76	0.0036	0.024336
4	3.6	12.96	0.0001	0.001296

$\Sigma f(x) = 1 \Rightarrow$ prob. mass funct.

$$V(X) = \sigma^2 = \sum_{i=1}^{5} f(x_i)(x_i - 0.4)^2 = 0.36$$

$\mu = \Sigma x \, f(x) = .4000$

EXAMPLE 3-19

Two new product designs are to be compared on the basis of revenue potential. Marketing feels that the revenue from design A can be predicted quite accurately to be \$3 million. The revenue potential of design B is more difficult to assess. Marketing concludes that there is a probability of 0.3 that the revenue from design B will be \$7 million, but there is a 0.7 probability that the revenue will be only \$2 million. Which design would you choose?

Let X denote the revenue from design A. Because there is no uncertainty in the revenue from design A, we can model the distribution of the random variable X as \$3 million with probability one. Therefore, $E(X) =$ \$3 million.

Let Y denote the revenue from design B. The expected value of Y in millions of dollars is

$$E(Y) = \$7(0.3) + \$2(0.7) = \$3.5$$

Because $E(Y)$ exceeds $E(X)$, we might choose design B. However, the variability of the result from design B is larger. That is,

$$\sigma^2 = (7 - 3.5)^2(0.3) + (2 - 3.5)^2(0.7)$$
$$= 5.25 \text{ millions of dollars squared}$$

EXERCISES FOR SECTION 3-7

Verify that the functions in Exercises 3-51 through 3-54 are probability mass functions, and determine the requested values.

(a) $P(X \leq 3)$
(b) $P(3 < X < 5.1)$
(c) $P(X > 4.5)$
(d) Mean and variance
(e) Graph $F(x)$.

3-51.

x	1	2	3	4
$f(x)$	0.326	0.088	0.019	0.251

x	5	6	7
$f(x)$	0.158	0.140	0.018

3-52.

x	0	1	2	3
$f(x)$	0.025	0.041	0.049	0.074

x	4	5	6	7
$f(x)$	0.098	0.205	0.262	0.123

x	8	9
$f(x)$	0.074	0.049

(a) $P(X \le 1)$
(b) $P(2 < X < 7.2)$
(c) $P(X \ge 6)$
(d) Mean and variance
(e) Graph $F(x)$.

3-53. $f(x) = (8/7)(1/2)^x, \quad x = 1, 2, 3$
(a) $P(X \le 1)$
(b) $P(X > 1)$
(c) Mean and variance
(d) Graph $F(x)$.

3-54. $f(x) = (1/2)(x/5), \quad x = 1, 2, 3, 4$
(a) $P(X = 2)$
(b) $P(X \le 2)$
(c) $P(X > 2)$
(d) $P(X \ge 1)$
(e) Mean and variance
(f) Graph $F(x)$.

3-55. Customers purchase a particular make of automobile with a variety of options. The probability mass function of the number of options selected is

x	7	8	9	10
$f(x)$	0.040	0.130	0.190	0.300

x	11	12	13
$f(x)$	0.240	0.050	0.050

(a) What is the probability that a customer will choose fewer than 9 options?
(b) What is the probability that a customer will choose more than 11 options?
(c) What is the probability that a customer will choose between 8 and 12 options, inclusively?
(d) What is the expected number of options chosen? What is the variance?

3-56. Marketing estimates that a new instrument for the analysis of soil samples will be very successful, moderately successful, or unsuccessful, with probabilities 0.3, 0.6, and 0.1, respectively. The yearly revenue associated with a very successful, moderately successful, or unsuccessful product is $10 million, $5 million, and $1 million, respectively. Let the random variable X denote the yearly revenue of the product.
(a) Determine the probability mass function of X.
(b) Determine the expected value and the standard deviation of the yearly revenue.

3-8 BINOMIAL DISTRIBUTION

A widely used discrete random variable is introduced next. Consider the following random experiments and random variables.

1. Flip a coin 10 times. Let $X = $ the number of heads obtained.

2. A worn machine tool produces 1% defective parts. Let $X = $ the number of defective parts in the next 25 parts produced.

3. Each sample of air has a 10% chance of containing a particular rare molecule. Let $X = $ the number of air samples that contain the rare molecule in the next 18 samples analyzed.

4. Of all bits transmitted through a digital transmission channel, 10% are received in error. Let $X = $ the number of bits in error in the next 4 bits transmitted.

5. A multiple choice test contains 10 questions, each with four choices, and you guess at each question. Let X = the number of questions answered correctly.

6. In the next 20 births at a hospital, let X = the number of female births.

7. Of all patients suffering a particular illness, 35% experience improvement from a particular medication. In the next 30 patients administered the medication, let X = the number of patients that experience improvement.

These examples illustrate that a general probability model that includes these experiments as particular cases would be very useful.

Each of these random experiments can be thought of as consisting of a series of repeated, random trials: 10 flips of the coin in experiment (1), the production of 25 parts in experiment (2), and so forth. The random variable in each case is a count of the number of trials that meet a specified criterion. The outcome from each trial either meets the criterion that X counts or it does not; consequently, each trial can be summarized as resulting in either a success or a failure, respectively. For example, in the multiple choice experiment, for each question, only the choice that is correct is considered a **success.** Choosing any one of the three incorrect choices results in the trial being summarized as a failure.

The terms *success* and *failure* are merely labels. We can just as well use "A" and "B" or "0" or "1." Unfortunately, the usual labels can sometimes be misleading. In experiment (2), because X counts defective parts, the production of a defective part is called a success.

A trial with only two possible outcomes is used so frequently as a building block of a random experiment that it is called a **Bernoulli trial.** It is usually assumed that the trials that constitute the random experiment are **independent.** This implies that the outcome from one trial has no effect on the outcome to be obtained from any other trial. Furthermore, it is often reasonable to assume that the **probability of a success on each trial is constant.**

In item 5, the multiple choice experiment, if the test taker has no knowledge of the material and just guesses at each question, we might assume that the probability of a correct answer is 1/4 *for each question.*

To analyze X, recall the relative frequency interpretation of probability. The proportion of times that Question 1 is expected to be correct is 1/4 and the proportion of times that Question 2 is expected to be correct is 1/4. For simple guesses, the proportion of times both questions are correct is expected to be

$$(1/4)(1/4) = 1/16$$

Furthermore, if one merely guesses, then the proportion of times Question 1 is correct and Question 2 is incorrect is expected to be

$$(1/4)(3/4) = 3/16$$

Similarly, if one merely guesses, then the proportion of times Question 1 is incorrect and Question 2 is correct is expected to be

$$(3/4)(1/4) = 3/16$$

Finally, if one merely guesses, then the proportion of times Question 1 is incorrect and Question 2 is incorrect is expected to be

$$(3/4)(3/4) = 9/16$$

We have accounted for all of the possible correct and incorrect combinations for these two questions, and the four probabilities associated with these possibilities sum to one:

$$1/16 + 3/16 + 3/16 + 9/16 = 1$$

This approach is used to derive the binomial distribution in the following example.

EXAMPLE 3-20

In Example 3-16, assume that the chance that a bit transmitted through a digital transmission channel is received in error is 0.1. Also, assume that the transmission trials are independent. Let X = the number of bits in error in the next four bits transmitted. Determine $P(X = 2)$.

Let the letter E denote a bit in error, and let the letter O denote that the bit is okay—that is, received without error. We can represent the outcomes of this experiment as a list of four letters that indicate the bits that are in error and okay. For example, the outcome *OEOE* indicates that the second and fourth bits are in error and the other two bits are okay. The corresponding values for x are

Outcome	x	Outcome	x
OOOO	0	*EOOO*	1
OOOE	1	*EOOE*	2
OOEO	1	*EOEO*	2
OOEE	2	*EOEE*	3
OEOO	1	*EEOO*	2
OEOE	2	*EEOE*	3
OEEO	2	*EEEO*	3
OEEE	3	*EEEE*	4

The event that $X = 2$ consists of the six outcomes

$$\{EEOO, EOEO, EOOE, OEEO, OEOE, OOEE\}$$

Using the assumption that the trials are independent, the probability of $\{EEOO\}$ is

$$P(EEOO) = P(E)P(E)P(O)P(O) = (0.1)^2(0.9)^2 = 0.0081$$

Also, any one of the six mutually exclusive outcomes for which $X = 2$ has the same probability of occurring. Therefore,

$$P(X = 2) = 6(0.0081) = 0.0486$$

In general,

$$P(X = x) = \text{(number of outcomes that result in } x \text{ errors)} \times (0.1)^x(0.9)^{4-x}$$

To complete a general probability formula, only an expression for the number of outcomes that contain x errors is needed. An outcome that contains x errors can be constructed by partitioning the four trials (letters) in the outcome into two groups. One group is of size x and contains the errors, and the other group is of size $n - x$ and consists of those trials that are okay. The number of ways of partitioning four objects into two groups, one of which is of size x, is $\binom{4}{x} = 4!/[x!(4-x)!]$. Therefore, in this example

$$P(X = x) = \binom{4}{x}(0.1)^x(0.9)^{4-x}$$

Note that $\binom{4}{2} = 4!/[2!\,2!] = 6$, as was found previously. The probability mass function of X was shown in Fig. 3-22.

The previous example motivates the following result.

Definition

A random experiment consisting of n repeated trials such that

(1) the trials are independent,

(2) each trial results in only two possible outcomes, labeled as "success" and "failure," and

(3) the probability of a success on each trial, denoted as p, remains constant

is called a *binomial experiment.*

The random variable X that equals the number of trials that result in a success has a **binomial distribution** with parameters p and n where $0 < p < 1$ and $n = \{1, 2, 3, \ldots\}$.

The probability mass function of X is

$$f(x) = \binom{n}{x}p^x(1 - p)^{n-x}, \qquad x = 0, 1, \ldots, n \qquad (3\text{-}8)$$

As before, $\binom{n}{x}$ equals the number of different sequences of trials that contain x successes and $n - x$ failures. The number of different sequences that contain x successes and $n - x$ failures times the probability of each sequence equals $P(X = x)$.

It can be shown (by using the binomial expansion formula) that the sum of the probabilities for a binomial random variable is 1. Furthermore, because each trial in the experiment is classified into two outcomes, {success, failure}, the distribution is called a "bi"-nomial. A more general distribution, which includes the binomial as a special case, is the multinomial distribution.

Examples of binomial distributions are shown in Fig. 3-25. For a fixed n, the distribution becomes more symmetric as p increases from 0 to 0.5 or decreases from 1 to 0.5. For a fixed p, the distribution becomes more symmetric as n increases.

EXAMPLE 3-21

Several examples using the binomial coefficient $\binom{n}{x}$ follow.

$$\binom{10}{3} = 10!/[3!\ 7!] = (10 \cdot 9 \cdot 8)/(3 \cdot 2) = 120$$

$$\binom{15}{10} = 15!/[10!\ 5!] = (15 \cdot 14 \cdot 13 \cdot 12 \cdot 11)/(5 \cdot 4 \cdot 3 \cdot 2) = 3003$$

$$\binom{100}{4} = 100!/[4!\ 96!] = (100 \cdot 99 \cdot 98 \cdot 97)/(4 \cdot 3 \cdot 2) = 3{,}921{,}225$$

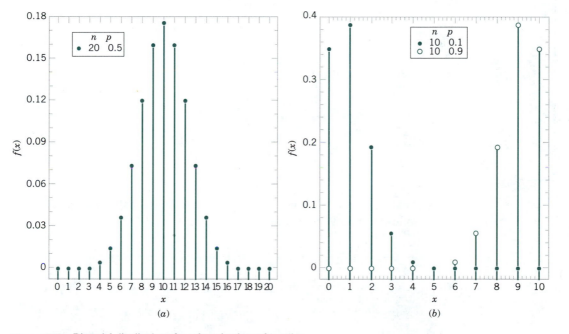

Figure 3-25 Binomial distributions for selected values of n and p.

EXAMPLE 3-22

Each sample of air has a 10% chance of containing a particular rare molecule. Assume that the samples are independent with regard to the presence of the rare molecule. Find the probability that in the next 18 samples, exactly 2 contain the rare molecule.

Let X = the number of air samples that contain the rare molecule in the next 18 samples analyzed. Then X is a binomial random variable with $p = 0.1$ and $n = 18$. Therefore,

$$P(X = 2) = \binom{18}{2}(0.1)^2(0.9)^{16}$$

Now $\binom{18}{2} = (18!/[2! \; 16!]) = 18(17)/2 = 153$. Therefore,

$$P(X = 2) = 153(0.1)^2(0.9)^{16} = 0.284$$

Find the probability that at least four samples contain the rare molecule. The requested probability is

$$P(X \geq 4) = \sum_{x=4}^{18} \binom{18}{x}(0.1)^x(0.9)^{18-x}$$

However, it is easier to use the complementary event,

$$P(X \geq 4) = 1 - P(X < 4)$$
$$= 1 - \sum_{x=0}^{3} \binom{18}{x}(0.1)^x(0.9)^{18-x}$$
$$= 1 - [0.150 + 0.300 + 0.284 + 0.168]$$
$$= 0.098$$

Furthermore, the probability that $3 \leq X < 7$ is

$$P(3 \leq X < 7) = \sum_{x=3}^{6} \binom{18}{x}(0.1)^x(0.9)^{18-x}$$
$$= 0.168 + 0.070 + 0.022 + 0.005$$
$$= 0.265$$

The mean and variance of a binomial random variable depend only on the parameters p and n. The following result can be shown.

If X is a binomial random variable with parameters p and n, then

$$\mu = E(X) = np \qquad \text{and} \qquad \sigma^2 = V(X) = np(1 - p) \qquad (3\text{-}9)$$

EXAMPLE 3-23

For the number of transmitted bits received in error in Example 3-20, $n = 4$ and $p = 0.1$ so

$$E(X) = 4(0.1) = 0.4$$

The variance of the number of defective bits is

$$V(X) = 4(0.1)(0.9) = 0.36$$

These results match those that were calculated directly from the probabilities in Example 3-18.

EXERCISES FOR SECTION 3-8

3-57. For each scenario described, state whether or not the binomial distribution is a reasonable model for the random variable and why. State any assumptions you make.

(a) A production process produces thousands of temperature transducers. Let X denote the number of nonconforming transducers in a sample of size 30 selected at random from the process.

(b) From a batch of 50 temperature transducers, a sample of size 30 is selected without replacement. Let X denote the number of nonconforming transducers in the sample.

(c) Four identical electronic components are wired to a controller. Let X denote the number of components that have failed after a specified period of operation.

(d) Let X denote the number of express mail packages received by the post office in a 24-hour period.

(e) Let X denote the number of correct answers by a student taking a multiple choice exam in which a student can eliminate some of the choices as being incorrect in some questions and all of the incorrect choices in other questions.

(f) Forty randomly selected semiconductor chips are tested. Let X denote the number of chips in which the test finds at least one contamination particle.

(g) Let X denote the number of contamination particles found on forty randomly selected semiconductor chips.

(h) A filling operation attempts to fill detergent packages to the advertised weight. Let X denote the number of detergent packages that are underfilled.

(i) Errors in a digital communication channel occur in bursts that affect several consecutive bits. Let X denote the number of bits in error in a transmission of 100,000 bits.

(j) Let X denote the number of surface flaws in a large coil of galvanized steel.

3-58. The random variable X has a binomial distribution with $n = 10$ and $p = 0.5$. Sketch the probability mass function of X.

(a) What value of X is most likely?

(b) What value(s) of X is (are) least likely?

3-59. The random variable X has a binomial distribution with $n = 10$ and $p = 0.5$.

Determine the following probabilities.
(a) $P(X = 5)$
(b) $P(X \le 2)$
(c) $P(X \ge 9)$
(d) $P(3 \le X < 5)$
(e) Sketch the cumulative distribution function.

3-60. Sketch the probability mass function of a binomial distribution with $n = 10$ and $p = 0.01$.
(a) What value of X is most likely?
(b) What value of X is least likely?

3-61. The random variable X has a binomial distribution with $n = 10$ and $p = 0.1$. Determine the following probabilities.
(a) $P(X = 5)$
(b) $P(X \le 2)$
(c) $P(X \ge 9)$
(d) $P(3 \le X < 5)$

3-62. An electronic product contains 40 integrated circuits. The probability that any integrated circuit is defective is 0.01, and the integrated circuits are independent. The product operates only if there are no defective integrated circuits. What is the probability that the product operates?

3-63. A hip joint replacement part is being stress-tested in a laboratory. The probability of successfully completing the test is 0.80. Seven randomly and independently chosen parts are tested. What is the probability that exactly 2 of the 7 parts successfully complete the test?

3-64. The phone lines to an airline reservation system are occupied 40% of the time. Assume that the events that the lines are occupied on successive calls are independent. Assume that 10 calls are placed to the airline.
(a) What is the probability that for exactly 3 calls the lines are occupied?
(b) What is the probability that for at least 1 call the lines are occupied?
(c) What is the expected number of calls in which the lines are occupied?

3-65. Batches that consist of 50 coil springs from a production process are checked for conformance to customer requirements. The mean number of nonconforming coil springs in a batch is 5. Assume that the number of nonconforming springs in a batch, denoted as X, is a binomial random variable.
(a) What are n and p?
(b) What is $P(X \le 2)$?
(c) What is $P(X \ge 49)$?

3-66. A statistical process control chart example. Samples of 20 parts from a metal punching process are selected every hour. Typically, 1% of the parts require rework. Let X denote the number of parts in the sample of 20 that require rework. A process problem is suspected if X exceeds its mean by more than three standard deviations.
(a) If the percentage of parts that require rework remains at 1%, what is the probability that X exceeds its mean by more than three standard deviations?
(b) If the rework percentage increases to 4%, what is the probability that X exceeds one?
(c) If the rework percentage increases to 4%, what is the probability that X exceeds one in at least 1 of the next 5 hours of samples?

3-67. Because not all airline passengers show up for their reserved seat, an airline sells 125 tickets for a flight that holds only 120 passengers. The probability that a passenger does not show up is 0.10, and the passengers behave independently.
(a) What is the probability that every passenger who shows up can take the flight?
(b) What is the probability that the flight departs with empty seats?
(c) What are the mean and standard deviation of the number of passengers who show up?

3-68. This exercise illustrates that poor quality

can impact schedules and costs. A manufacturing process has 100 customer orders to fill. Each order requires one component part that is purchased from a supplier. However, typically, 2% of the components are identified as defective, and the components can be assumed to be independent.

(a) If the manufacturer stocks 100 components, what is the probability that the 100 orders can be filled without reordering components?

(b) If the manufacturer stocks 102 components, what is the probability that the 100 orders can be filled without reordering components?

(c) If the manufacturer stocks 105 components, what is the probability that the 100 orders can be filled without reordering components?

3-69. The probability of successfully landing a plane using a flight simulator is given as 0.70. Six randomly and independently chosen student pilots are asked to try to fly the plane using the simulator. What is the probability that two of the six students successfully land the plane using the simulator?

3-70. Traffic engineers install 15 street lights with new bulbs. The probability that any one bulb fails within 200 hours of operation is 0.15. Assume that each of the bulbs fail independently.

(a) What is the probability that fewer than two of the original bulbs fail within 200 hours of operation?

(b) What is the probability that no bulbs will have to be replaced within 200 hours of operation?

(c) What is the probability that more than four of the original bulbs will need replacing within 200 hours

3-9 POISSON PROCESS

Consider e-mail messages that arrive at a mail server on a computer network. This is an example of events (such as message arrivals) that occur randomly in an interval (such as time). The number of events over an interval (such as the number of messages that arrive in one hour) is a discrete random variable that is often modeled by a Poisson distribution. The length of the interval between events (such as the time between messages) is often modeled by an exponential distribution. These distributions are related; they provide probabilities for different random variables in the same random experiment.

3-9.1 Poisson Distribution

We introduce the Poisson distribution with an example.

EXAMPLE 3-24

Consider the transmission of n bits over a digital communication channel. Let the random variable X equal the number of bits in error. When the probability that a bit is in error is constant and the transmissions are independent, X has a binomial distribution. Let p denote the probability that a bit is in error. Then, $E(X) = pn$. Now, suppose that the number of bits transmitted increases and the probability of an error decreases exactly enough that pn remains equal to a constant—say, λ. That is, n increases and p decreases accordingly, such that $E(X)$ remains constant. Then,

$$P(X = x) = \binom{n}{x} p^x (1 - p)^{n-x}$$

$$= \frac{n(n - 1)(n - 2) \cdots (n - x + 1)}{n^x \, x!} (np)^x (1 - p)^n (1 - p)^{-x}$$

With some work, it can be shown that

$$\lim_{n \to \infty} P(X = x) = \frac{e^{-\lambda} \lambda^x}{x!}, \qquad x = 0, 1, 2, \ldots$$

Also, because the number of bits transmitted tends to infinity, the number of errors can equal any nonnegative integer. Therefore, the possible values for X are the integers from zero to infinity.

The distribution obtained as the limit in the previous example is more useful than the derivation implies. The following example illustrates the broader applicability.

EXAMPLE 3-25

Flaws occur at random along the length of a thin copper wire. Let X denote the random variable that counts the number of flaws in a length of L millimeters of wire and suppose that the average number of flaws in L millimeters is λ.

The probability distribution of X can be found by reasoning in a manner similar to Example 3-24. Partition the length of wire into n subintervals of small length—say, 1 micrometer each. If the subinterval chosen is small enough, the probability that more than one flaw occurs in the subinterval is negligible. Furthermore, we can interpret the assumption that flaws occur at random to imply that every subinterval has the same probability of containing a flaw—say, p. Finally, if we assume that the probability that a subinterval contains a flaw is independent of other subintervals, then we can model the distribution of X as approximately a binomial random variable. Because

$$E(X) = \lambda = np$$

we obtain

$$p = \lambda/n$$

That is, the probability that a subinterval contains a flaw is λ/n. With small enough subintervals, n is very large and p is very small. Therefore, the distribution of X is obtained as in the previous example.

Clearly, Example 3-25 can be generalized to include a broad array of random experiments. The interval that was partitioned in Example 3-25 was a length of wire. However, the same reasoning can be applied to any interval, including an interval of time, an area, or a volume. For example, counts of (1) particles of contamination in semiconductor manufacturing, (2) flaws in rolls of textiles, (3) calls to a telephone exchange, (4) power outages, and (5) atomic particles emitted from a specimen have all been successfully modeled by the probability mass function in the following definition.

Definition

Assume that events occur at random throughout the interval. If the interval can be partitioned into subintervals of small enough length such that

   (1)   the probability of more than one count in a subinterval is zero,

   (2)   the probability of one count in a subinterval is the same for all subintervals and proportional to the length of the subinterval, and

   (3)   the count in each subinterval is independent of other subintervals,

then the random experiment is called a *Poisson process.*

If the mean number of counts in the interval is $\lambda > 0$, the random variable X that equals the number of counts in the interval has a **Poisson distribution** with parameter λ, and the probability mass function of X is

$$f(x) = \frac{e^{-\lambda}\lambda^x}{x!}, \qquad x = 0, 1, 2, \ldots \tag{3-10}$$

The mean and variance of X are

$$E(X) = \lambda \qquad \text{and} \qquad V(X) = \lambda \tag{3-11}$$

Historically, the term *process* has been used to suggest the observation of a system over time. In our example with the copper wire, we showed that the Poisson distribution could also apply to intervals such as lengths. Figure 3-26 shows graphs of selected Poisson distributions. Figure 3-27 provides a general description of a Poisson process.

It is important to **use consistent units** in the calculation of probabilities, means, and variances involving Poisson random variables. The following example illustrates unit conversions. For example, if the

mean number of flaws per millimeter of wire is 3.4, then the

mean number of flaws in 10 millimeters of wire is 34, and the

mean number of flaws in 100 millimeters of wire is 340.

If a Poisson random variable represents the number of counts in some interval, then the mean of the random variable must equal the expected number of counts in the same length of interval.

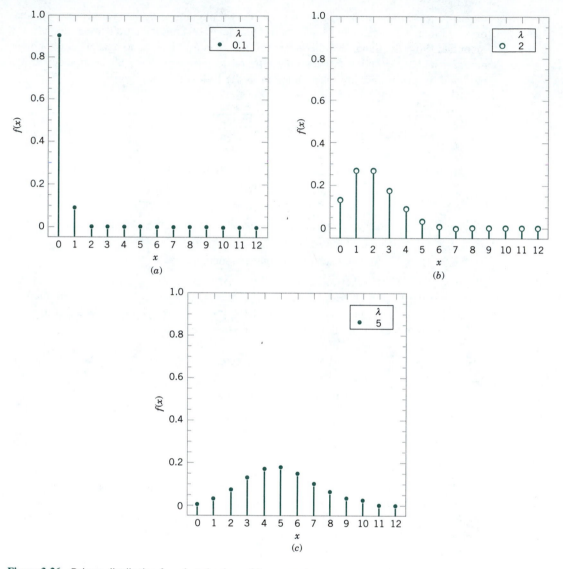

Figure 3-26   Poisson distribution for selected values of the parameters.

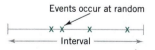

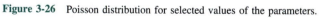

Figure 3-27   In a Poisson process, events occur at random in an interval.

EXAMPLE 3-26

For the case of the thin copper wire, suppose that the number of flaws follows a Poisson distribution with a mean of 2.3 flaws per millimeter. Determine the probability of exactly 2 flaws in 1 millimeter of wire.

Let X denote the number of flaws in 1 millimeter of wire. Then $E(X) = 2.3$ flaws and

$$P(X = 2) = \frac{e^{-2.3}2.3^2}{2!} = 0.265$$

Determine the probability of 10 flaws in 5 millimeters of wire. Let X denote the number of flaws in 5 millimeters of wire. Then X has a Poisson distribution with

$$E(X) = 5 \text{ mm} \times 2.3 \text{ flaws/mm} = 11.5 \text{ flaws}$$

Therefore,

$$P(X = 10) = e^{-11.5} \, 11.5^{10}/10! = 0.113$$

Determine the probability of at least one flaw in 2 millimeters of wire. Let X denote the number of flaws in 2 millimeters of wire. Then X has a Poisson distribution with

$$E(X) = 2 \text{ mm} \times 2.3 \text{ flaws/mm} = 4.6 \text{ flaws}$$

Therefore,

$$\begin{aligned} P(X \geq 1) &= 1 - P(X = 0) \\ &= 1 - e^{-4.6} \\ &= 0.9899 \end{aligned}$$

The next example uses a computer program to sum Poisson probabilities.

EXAMPLE 3-27

Contamination is a problem in the manufacture of optical storage disks. The number of particles of contamination that occur on an optical disk has a Poisson distribution, and the average number of particles per centimeter squared of media surface is 0.1. The area of a disk under study is 100 squared centimeters. Find the probability that 12 particles occur in the area of a disk under study.

Let X denote the number of particles in the area of a disk under study. Because the mean number of particles is 0.1 particles per cm^2,

$$\begin{aligned} E(X) &= 100 \text{ cm}^2 \times 0.1 \text{ particles/cm}^2 \\ &= 10 \text{ particles} \end{aligned}$$

Therefore,

$$P(X = 12) = \frac{e^{-10}10^{12}}{12!}$$
$$= 0.095$$

Find the probability that zero particles occur in the area of the disk under study. Now, $P(X = 0) = e^{-10} = 4.54 \times 10^{-5}$.

Find the probability that 12 or fewer particles occur in the area of a disk under study. This probability is

$$P(X \le 12) = P(X = 0) + P(X = 1) + \cdots + P(X = 12)$$
$$= \sum_{i=0}^{12} \frac{e^{-10}10^i}{i!}$$

Because this sum is tedious to compute, many computer programs calculate cumulative Poisson probabilities. From one such program, $P(X \le 12) = 0.791$.

The variance of a Poisson random variable can be shown to equal its mean. For example, if particle counts follow a Poisson distribution with a mean of 25 particles per square centimeter, then the standard deviation of the counts is 5 per square centimeter. Consequently, information on the variability is very easily obtained. Conversely, if the variance of count data is much greater than the mean of the same data, then the Poisson distribution is not a good model for the distribution of the random variable.

EXERCISES FOR SECTION 3-9.1

3-71. Suppose X has a Poisson distribution with a mean of 0.3. Determine the following probabilities.
  (a) $P(X = 0)$
  (b) $P(X \le 3)$
  (c) $P(X = 6)$
  (d) $P(X = 2)$

3-72. Suppose X has a Poisson distribution with a mean of 5. Determine the following probabilities.
  (a) $P(X = 0)$
  (b) $P(X \le 3)$
  (c) $P(X = 6)$
  (d) $P(X = 9)$

3-73. Suppose that the number of customers who enter a post office in a 30-minute period is a Poisson random variable and that $P(X = 0) = 0.018$. Determine the mean and variance of X.

3-74. Suppose that the number of customers that enter a bank in an hour is a Poisson random variable and that $P(X = 0) = 0.05$. Determine the mean and variance of X.

3-75. The number of telephone calls that arrive at a phone exchange is often modeled as a Poisson random variable. Assume that on the average there are 20 calls per hour.
  (a) What is the probability that there are exactly 18 calls in 1 hour?
  (b) What is the probability that there

are 3 or fewer calls in 30 minutes?

(c) What is the probability that there are exactly 30 calls in 2 hours?

(d) What is the probability that there are exactly 10 calls in 30 minutes?

3-76. The number of earthquake tremors in a 12-month period appears to be distributed as a Poisson random variable with a mean of 8. Assume the number of tremors from one 12-month period is independent of the number in the next 12-month period.

(a) What is the probability that there are 12 tremors in 1 year?

(b) What is the probability that there are 20 tremors in 2 years?

(c) What is the probability that there are no tremors in a 1-month period?

(d) What is the probability that there are more than 5 tremors in a 6-month period?

3-77. The number of cracks in a section of interstate highway that are significant enough to require repair is assumed to follow a Poisson distribution with a mean of two cracks per mile.

(a) What is the probability that there are no cracks that require repair in 5 miles of highway?

(b) What is the probability that at least one crack requires repair in one-half mile of highway?

(c) If the number of cracks is related to the vehicle load on the highway and some sections of the highway have a heavy load of vehicles while other sections carry a light load, how do you feel about the assumption of a Poisson distribution for the number of cracks that require repair for all sections?

3-78. The number of surface flaws in plastic panels used in the interior of automobiles has a Poisson distribution with a mean of 0.05 flaw per square foot of plastic panel. Assume an automobile in-

terior contains 10 square feet of plastic panel.

(a) What is the probability that there are no surface flaws in an auto's interior?

(b) If 10 cars are sold to a rental company, what is the probability that none of the 10 cars has any surface flaws?

(c) If 10 cars are sold to a rental company, what is the probability that at most one car has any surface flaws?

3-79. The number of failures of a testing instrument from contamination particles on the product is a Poisson random variable with a mean of 0.04 failure per hour.

(a) What is the probability that the instrument does not fail in an 8-hour shift?

(b) What is the probability of at least three failures in a 24-hour day?

3-80. Assume that the number of errors along a magnetic recording surface is a Poisson random variable with a mean of one error every 10^5 bits. A sector of data consists of 4096 eight-bit bytes.

(a) What is the probability of more than one error in a sector?

(b) What is the probability of observing fewer than two errors in a sector?

3-81. A telecommunication station is designed to receive a maximum of 10 calls per half-second. If the number of calls to the station is modeled as a Poisson random variable with a mean of 9 calls per half-second, what is the probability that the number of calls will exceed the maximum design constraint of the station?

3-82. Flaws occur in the interior of plastic used for office furniture according to a Poisson distribution with a mean of 0.02 flaw per panel.

(a) If 50 panels are inspected, what is the probability that there are no flaws?

(b) What is the probability that a randomly selected panel has no flaws?

(c) If 50 panels are inspected, what is the probability that the number of panels that have two or more flaws is less than or equal to two?

3-83. Messages arrive to a computer server according to a Poisson distribution with a mean rate of 10 per hour.

(a) What is the probability that three messages will arrive in 1 hour?

(b) What is the probability that six messages arrive in 30 minutes?

3-9.2 Exponential Distribution

The discussion of the Poisson distribution defined a random variable to be the number of flaws along a length of copper wire. The distance between flaws is another random variable that is often of interest. Let the random variable X denote the length from any starting point on the wire until a flaw is detected.

As you might expect, the distribution of X can be obtained from knowledge of the distribution of the number of flaws. The key to the relationship is the following concept. The distance to the first flaw exceeds 3 millimeters if and only if there are no flaws within a length of 3 millimeters—simple, but sufficient for an analysis of the distribution of X.

In general, let the random variable N denote the number of flaws in x millimeters of wire. If the mean number of flaws is λ per millimeter, then N has a Poisson distribution with mean λx. We assume that the wire is longer than the value of x. Now,

$$P(X > x) = P(N = 0) = \frac{e^{-\lambda x}(\lambda x)^0}{0!} = e^{-\lambda x}$$

and

$$P(X \le x) = 1 - e^{-\lambda x}$$

for $x \ge 0$.

If $f(x)$ is the probability density function of X, then the cumulative distribution function is

$$F(x) = P(X \le x) = \int_{-\infty}^{x} f(u)\, du$$

From the fundamental theorem of calculus, the derivative of $F(x)$ (with respect to x) is $f(x)$. Therefore, the probability density function of X is

$$f(x) = \frac{d}{dx}(1 - e^{-\lambda x}) = \lambda e^{-\lambda x} \qquad \text{for } x \ge 0$$

The distribution of X depends only on the assumption that the flaws in the wire follow a Poisson process. Also, the starting point for measuring X doesn't matter because the probability of the number of flaws in an interval of a Poisson process depends only on the length of the interval, not on the location. For any Poisson process, the following general result applies.

Definition

The random variable X that equals the distance between successive counts of a Poisson process with mean $\lambda > 0$ has an **exponential distribution** with parameter λ. The probability density function of X is

$$f(x) = \lambda e^{-\lambda x}, \qquad \text{for } 0 \leq x < \infty \tag{3-12}$$

The mean and variance of X are

$$E(X) = \frac{1}{\lambda} \quad \text{and} \quad V(X) = \frac{1}{\lambda^2} \tag{3-13}$$

The exponential distribution obtains its name from the exponential function in the probability density function. Plots of the exponential distribution for selected values of λ are shown in Fig. 3-28. For any value of λ, the exponential distribution is quite skewed. The mean and variance results are easily obtained and are left as an exercise.

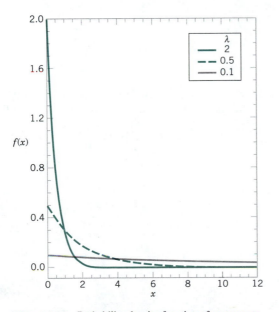

Figure 3-28   Probability density function of an exponential random variable for selected values of λ.

It is important to **use consistent units** in the calculation of probabilities, means, and variances involving exponential random variables. The following example illustrates unit conversions.

EXAMPLE 3-28

In a large corporate computer network, user log-ons to the system can be modeled as a Poisson process with a mean of 25 log-ons per hour. What is the probability that there are no log-ons in an interval of 6 minutes?

Let X denote the time in hours from the start of the interval until the first log-on. Then X has an exponential distribution with $\lambda = 25$ log-ons per hour. We are interested in the probability that X exceeds 6 minutes. Because λ is given in log-ons per hour, we express all time units in hours; that is, 6 minutes = 0.1 hour. The probability requested is shown as the shaded area under the probability density function in Fig. 3-29. Therefore,

$$P(X > 0.1) = \int\limits_{0.1}^{\infty} 25e^{-25x}\, dx$$

$$= e^{-25(0.1)} = 0.082$$

An identical answer is obtained by expressing the mean number of log-ons as 0.417 log-ons per minute and computing the probability that the time until the next log-on exceeds 6 minutes. Try it!

What is the probability that the time until the next log-on is between 2 and 3 minutes? On converting all units to hours,

$$P(0.033 < X < 0.05) = \int\limits_{0.033}^{0.05} 25e^{-25x}\, dx = -e^{-25x}\Big|_{0.033}^{0.05} = 0.152$$

Determine the interval of time such that the probability that no log-on occurs in the interval is 0.90. The question asks for the length of time x such that $P(X > x) = 0.90$. Now,

$$P(X > x) = e^{-25x} = 0.90$$

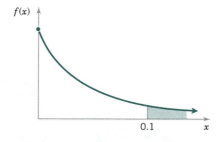

Figure 3-29   Probability for the exponential distribution in Example 3-28.

Therefore, on taking logarithms of both sides,

$$x = 0.00421 \text{ hour} = 0.25 \text{ minute}$$

Furthermore, the mean time until the next log-on is

$$E(X) = 1/25 = 0.04 \text{ hour} = 2.4 \text{ minutes}$$

The standard deviation of the time until the next log-on is

$$\sigma_X = 1/25 \text{ hours} = 2.4 \text{ minutes}$$

In the previous example, the probability that there are no log-ons in a 6-minute interval is 0.082 regardless of the starting time of the interval. A Poisson process assumes that events occur uniformly throughout the interval of observation; that is, there is no clustering of events. If the log-ons are well modeled by a Poisson process, the probability that the first log-on after noon occurs after 12:06 P.M. is the same as the probability that the first log-on after 3:00 P.M. occurs after 3:06 P.M. If someone logs on at 2:22 P.M., the probability that the next log-on occurs after 2:28 P.M. is still 0.082.

Our starting point for observing the system does not matter. However, if there are high-use periods during the day, such as right after 8:00 A.M., followed by a period of low use, a Poisson process is not an appropriate model for log-ons and the distribution is not appropriate for computing probabilities. It might be reasonable to model each of the high- and low-use periods by a separate Poisson process, employing a larger value for λ during the high-use periods and a smaller value otherwise. Then, an exponential distribution with the corresponding value of λ can be used to calculate log-on probabilities for the high- and low-use periods.

An even more interesting property of an exponential random variable is the **lack of memory property.** Suppose that there are no log-ons from 12:00 to 12:15; the probability that there are no log-ons from 12:15 to 12:21 is still 0.082. Because we have already been waiting for 15 minutes, we feel that we are "due." That is, the probability of a log-on in the next 6 minutes should be greater than 0.082. However, for an exponential distribution this is not true.

The lack of memory property is not that surprising when you consider the development of a Poisson process. In that development, we assumed that an interval could be partitioned into small intervals that were independent. The presence or absence of events in subintervals are similar to independent, Bernoulli trials that comprise a binomial process; knowledge of previous results does not affect the probabilities of events in future subintervals.

The exponential distribution is often used in reliability studies as the model for the time until failure of a device. For example, the lifetime of a semiconductor chip might be modeled as an exponential random variable with a mean of 40,000 hours. The lack of memory property of the exponential distribution implies that the device does not wear out. That is, regardless of how long the device has been operating, the probability of a failure in the next 1000 hours is the same as the probability of a failure in the first 1000 hours of operation. The lifetime of a device with failures caused by random shocks might be appropriately modeled as an exponential random variable. However, the lifetime of a device that suffers slow mechanical wear, such as bearing wear, is better modeled by a distribution that does not lack memory.

EXERCISES FOR SECTION 3-9.2

3-84. Suppose X has an exponential distribution with $\lambda = 3$. Determine the following.
(a) $P(X \le 0)$
(b) $P(X \ge 3)$
(c) $P(X \le 2)$
(d) $P(2 < X < 3)$
(e) Find the value of x such that $P(X < x) = 0.05$.

3-85. Suppose X has an exponential distribution with mean equal to 5. Determine the following.
(a) $P(X > 5)$
(b) $P(X > 15)$
(c) $P(X > 20)$
(d) Find the value of x such that $P(X < x) = 0.95$.

3-86. Suppose the counts recorded by a geiger counter follow a Poisson process with an average of two counts per minute.
(a) What is the probability that there are no counts in a 30-second interval?
(b) What is the probability that the first count occurs in less than 10 seconds?
(c) What is the probability that the first count occurs between 1 and 2 minutes after start-up?

3-87. Continuation of Exercise 3-86.
(a) What is the mean time between counts?
(b) What is the standard deviation of the time between counts?
(c) Determine x, such that the probability that at least one count occurs before time x minutes is 0.95.

3-88. The time between calls to a plumbing supply business is exponentially distributed with a mean time between calls of 15 minutes.

(a) What is the probability that there are no calls within a 30-minute interval?
(b) What is the probability that at least one call arrives within a 10-minute interval?
(c) What is the probability that the first call arrives within 5 and 10 minutes after opening?
(d) Determine the length of an interval of time such that the probability of at least one call in the interval is 0.90.

3-89. A robot completes a weld operation in an automobile at an average rate of 12 per hour. The time to complete a weld operation is defined from the time of the start of the weld procedure to the time until the start of the next weld procedure. The random variable X represents the time to complete a weld operation and is assumed to be modeled by the exponential distribution.
(a) What is the probability that a completed weld operation requires more than 6 minutes to complete?
(b) What is the probability that a weld operation is completed in less than 8 minutes?

3-90. Continuation of Exercise 3-89.
(a) What is the expected number of completed weld operations performed by a robot in a 10-minute interval?
(b) What is the probability that the number of completed weld operations will equal 1 in a 10-minute interval?

3-91. The distance between major cracks in a highway follows an exponential distribution with a mean of 5 miles.
(a) What is the probability that there

are no major cracks in a 10-mile stretch of the highway?

(b) What is the probability that there are two major cracks in a 10-mile stretch of the highway?

(c) What is the standard deviation of the distance between major cracks?

3-92. Continuation of Exercise 3-91.

(a) What is the probability that the first major crack occurs between 12 and 15 miles of the start of inspection?

(b) What is the probability that there are no major cracks in two separate 5-mile stretches of the highway?

(c) Given that there are no cracks in the first 5 miles inspected, what is the probability that there are no major cracks in the next 10 miles inspected?

3-93. The time between the arrival of electronic messages at your computer is exponentially distributed with a mean of 2 hours.

(a) What is the probability that you do not receive a message during a 2-hour period?

(b) If you have not had a message in the last 4 hours, what is the probability that you do not receive a message in the next 2 hours?

(c) What is the expected time between your fifth and sixth message?

3-94. The time-to-failure of a certain type of electrical component is assumed to follow an exponential distribution with a mean of 4 years. The manufacturer replaces for free all components that fail while under guarantee.

(a) What percentage of the components will fail in 1 year?

(b) What is the probability that a component will fail in 2 years?

(c) What is the probability that a component will fail in 4 years?

3-95. Continuation of Exercise 3-94.

(a) If the manufacturer wants to replace a maximum of 3% of the components, for how long should the manufacturer's stated guarantee on the component be?

(b) By redesigning the component, the manufacturer could increase the life. What does the mean time-to-failure have to be so that the manufacturer can offer a 1-year guarantee, yet still replace at most 3% of the components?

3-96. The time between calls to a corporate office is exponentially distributed with a mean of 10 minutes.

(a) What is the probability that there are more than three calls in one-half hour?

(b) What is the probability that there are no calls within one-half hour?

(c) Determine x such that the probability that there are no calls within x hours is 0.01.

(d) What is the probability that there are no calls within a 2-hour interval?

(e) If four nonoverlapping one-half hour intervals are selected, what is the probability that none of these intervals contains any call?

3-10 NORMAL APPROXIMATION TO THE BINOMIAL AND POISSON DISTRIBUTIONS

Because a binomial random variable is a count from repeated independent trials, the central limit theorem can be applied. Consequently, it should not be surprising to use the normal distribution to approximate binomial probabilities for cases in which n is large. The following example illustrates that for many physical systems the binomial model is appropriate with an extremely large value for n. In these cases, it is difficult to calculate probabilities by using the binomial distribution. Fortunately, the normal approximation is most effective in these cases. An illustration is provided in Fig. 3-30.

EXAMPLE 3-29

In a digital communication channel, assume that the number of bits received in error can be modeled by a binomial random variable, and assume that the probability that a bit is received in error is 1×10^{-5}. If 16 million bits are transmitted, what is the probability that more than 150 errors occur?

Let the random variable X denote the number of errors. Then X is a binomial random variable and

$$P(X > 150) = 1 - P(X \le 150)$$

$$= 1 - \sum_{x=0}^{150} \binom{16,000,000}{x} (10^{-5})^x (1 - 10^{-5})^{16,000,000 - x}$$

Clearly, the probability in the previous example is difficult to compute. Fortunately, the normal distribution can be used to provide an excellent approximation in this example.

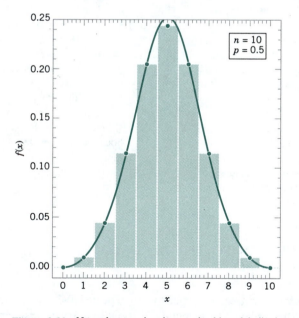

Figure 3-30   Normal approximation to the binomial distribution.

If X is a binomial random variable, then

$$Z = \frac{X - np}{\sqrt{np\,(1-p)}} \qquad (3\text{-}14)$$

is approximately a standard normal random variable. Consequently, probabilities computed from Z can be used to approximate probabilities for X.

Recall that for a binomial variable X, $E(X) = np$ and $V(X) = np(1 - p)$. Consequently, the normal approximation is nothing more than the formula for standardizing the random variable X. Probabilities involving X can be approximated by using a standard normal random variable. The normal approximation to the binomial distribution is good *if n is large enough relative to p;* in particular, whenever

$$np > 5 \quad \text{and} \quad n(1 - p) > 5$$

The digital communication problem is solved as follows:

$$P(X > 150) = P\left(\frac{X - 160}{\sqrt{160(1 - 10^{-5})}} > \frac{150 - 160}{\sqrt{160(1 - 10^{-5})}}\right)$$

$$= P(Z > -0.79) = P(Z < 0.79) = 0.785$$

EXAMPLE 3-30

Again consider the transmission of bits in the previous example. To judge how well the normal approximation works, assume that only $n = 50$ bits are to be transmitted and that the probability of an error is $p = 0.1$. The exact probability that 2 or fewer errors occur is

$$P(X \le 2) = \binom{50}{0}0.9^{50} + \binom{50}{1}0.1(0.9^{49}) + \binom{50}{2}0.1^2(0.9^{48})$$

$$= 0.11$$

Based on the normal approximation,

$$P(X \le 2) = P\left(\frac{X - 5}{2.12} < \frac{2 - 5}{2.12}\right) = P(Z < -1.415) = 0.08$$

For a sample as small as 50 bits, with $np = 5$, the normal approximation is reasonable.

A correction factor can be used that will further improve the approximation. (This is discussed in the exercises at the end of this section.) However, if np or $n(1 - p)$ is small, the binomial distribution is quite skewed and the symmetric normal distribution is not a good approximation. Two cases are illustrated in Fig. 3-31.

Recall that the Poisson distribution was developed as the limit of a binomial distribution as the number of trials increased to infinity. Consequently, the normal distribution can also be used to approximate probabilities of a Poisson random variable. The approximation is good for

$$\lambda > 5$$

If X is a Poisson random variable with $E(X) = \lambda$ and $V(X) = \lambda$, then

$$Z = \frac{X - \lambda}{\sqrt{\lambda}} \qquad (3\text{-}15)$$

is approximately a standard normal random variable.

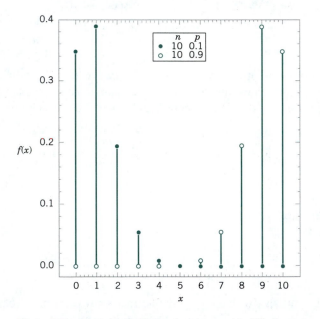

Figure 3-31   Binomial distribution is not symmetrical if p is near 0 or 1.

EXAMPLE 3-31

Assume that the number of asbestos particles in a squared centimeter of dust follows a Poisson distribution with a mean of 1000. If a squared centimeter of dust is analyzed, what is the probability that fewer than 950 particles are found?

This probability can be expressed exactly as

$$P(X \le 950) = \sum_{x=0}^{950} \frac{e^{-1000} 1000^x}{x!}$$

The computational difficulty is clear. The probability can be approximated as

$$P(X \le x) = P\left(Z \le \frac{950 - 1000}{\sqrt{1000}}\right) = P(Z \le -1.58) = 0.057$$

EXERCISES FOR SECTION 3-10

3-97. Suppose that X is a binomial random variable with $n = 200$ and $p = 0.4$.
(a) Approximate the probability that X is less than or equal to 70.
(b) Approximate the probability that X is greater than 70 and less than 90.

3-98. Suppose that X is a binomial random variable with $n = 100$ and $p = 0.1$.
(a) Compute the exact probability that X is less than 4.
(b) Approximate the probability that X is less than 4 and compare to the result in part (a).
(c) Approximate the probability that $8 < X < 12$.

3-99. The manufacturing of semiconductor chips produces 2% defective chips. Assume that the chips are independent and that a lot contains 1000 chips.
(a) Approximate the probability that more than 25 chips are defective.
(b) Approximate the probability that between 20 and 30 chips are defective.

3-100. A particular vendor produces parts with a defect rate of 8%. Incoming inspection to a manufacturing plant samples 100 delivered parts from this vendor and rejects the delivery if he discovers 6 defective parts.
(a) Compute the exact probability that the inspector accepts delivery.
(b) Approximate the probability of acceptance and compare the result to part (a).

3-101. A large electronic office product contains 2000 electronic components. Assume that the probability that each component operates without failure during the useful life of the product is 0.995, and assume that the components fail independently. Approximate the probability that 5 or more of the original 2000 components fail during the useful life of the product.

3-102. Suppose that the number of asbestos particles in a sample of 1 squared centimeter of dust is a Poisson random variable with a mean of 1000. Approximate the probability that 10 squared centimeters of dust contains more than 10,000 particles.

3-103. Continuity correction. The normal approximation of a binomial probability is sometimes modified by a correction factor of 0.5 that improves the approximation. Suppose that X is binomial with $n = 50$ and $p = 0.1$. Because X is a discrete random variable, $P(X \leq 2) = P(X \leq 2.5)$. However, the normal approximation to $P(X \leq 2)$ can be improved by applying the approximation to $P(X \leq 2.5)$.

(a) Approximate $P(X \leq 2)$ by computing the z-value corresponding to $x = 2.5$.

(b) Approximate $P(X \leq 2)$ by computing the z-value corresponding to $x = 2$.

(c) Compare the results in parts (a) and (b) to the exact value of $P(X \leq 2)$ to evaluate the effectiveness of the continuity correction.

(d) Use the continuity correction to approximate $P(X \leq 10)$.

(e) Use the continuity correction to approximate $P(X < 10)$.

3-104.   **Continuity correction.** Suppose that X is binomial with $n = 50$ and $p = 0.1$. Because X is a discrete random

variable, $P(X \geq 2) = P(X \geq 1.5)$. However, the normal approximation to $P(X \geq 2)$ can be improved by applying the approximation to $P(X \geq 1.5)$. The continuity correction of 0.5 is either added or subtracted. The easy rule to remember is that the continuity correction is always applied to make the approximating normal probability greatest.

(a) Approximate $P(X \geq 2)$ by computing the z-value corresponding to 1.5.

(b) Approximate $P(X \geq 2)$ by computing the z-value corresponding to 2.

(c) Compare the results in parts (a) and (b) to the exact value of $P(X \geq 2)$ to evaluate the effectiveness of the continuity correction.

(d) Use the continuity correction to approximate $P(X \geq 6)$.

(e) Use the continuity correction to approximate $P(X > 6)$.

3-11   MORE THAN ONE RANDOM VARIABLE AND INDEPENDENCE

3-11.1   Joint Distributions

In many experiments, more than one variable is measured. For example, suppose both the diameter and thickness of an injection-molded disk are measured and denoted by X and Y, respectively. These two random variables are often related. If a pressure in the mold increases, there might be an increase in the fill of the cavity that results in larger values for both X and Y. Similarly, a pressure decrease might result in smaller values for both X and Y. Suppose that diameter and thickness measurements from many parts are plotted in an X–Y plane (scatter diagram). As shown in Fig. 3-32, the relationship between X and Y implies that some regions of the X–Y plane are more likely to contain measurements than others.

This tendency can be modeled by a probability density function [denoted as $f(x, y)$] over the X–Y plane as shown in Fig. 3-33. The analogies that related a probability density function to the loading on a long, thin beam can be applied to relate this two-dimensional probability density function to the density of a loading over a large, flat surface. The probability that the random experiment (part production) generates measurements in a region of the X–Y plane is determined from the integral of $f(x, y)$ over the region

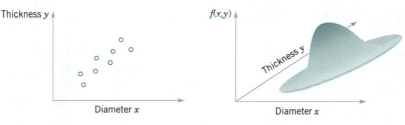

Figure 3-32 Scatter diagram of diameter and thickness measurements.

Figure 3-33 Joint probability density function of x and y.

as shown in Fig. 3-34. This is the volume over the region enclosed by $f(x, y)$. Because $f(x, y)$ determines probabilities for two random variables, it is referred to as a **joint probability density function.** From Fig. 3-34, the probability that a part is produced in the region shown is

$$P(a < X < b, c < Y < d) = \int_a^b \int_c^d f(x, y)\, dx\, dy$$

Similar concepts can be applied to discrete random variables. For example, suppose the quality of each bit received through a digital communications channel is categorized into one of four classes, "excellent," "good," "fair," and "poor," denoted by E, G, F, and P, respectively. Let the random variables X, Y, W, and Z denote the numbers of bits that are E, G, F, and P, respectively, in a transmission of 20 bits. In this example, we are interested in the joint probability distribution of four random variables. To simplify, we only consider X and Y. The joint probability distribution of X and Y can be specified by a **joint probability mass function** $f(x, y) = P(X = x, Y = y)$. Because each of the 20 bits is categorized into one of the four classes, $X + Y + W + Z = 20$, so that only integers such that $X + Y \leq 20$ have positive probability in the joint probability mass function of X and Y. The joint probability mass function is zero elsewhere. For a general discussion of joint distributions, which we do not present here, we refer the interested reader to Montgomery and Runger (1999). Instead, we focus here on the important special case of independent random variables.

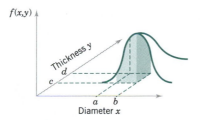

Figure 3-34 Probability of a region is the volume enclosed by $f(x, y)$ over the region.

3-11.2   Independence

If we make some assumptions regarding our probability models, a probability involving more than one random variable can often be simplified. In Example 3-12, the probability that a diameter meets specifications was determined to be 0.919. What can we say about 10 such diameters? What is the probability that they all meet specifications? This is the type of question of interest to a customer of optical drives.

Such questions lead to an important concept and definition. To accommodate more than just the two random variables X and Y, we adopt the notation $X_1, X_2, \ldots, X_n$ to represent n random variables.

Definition

The random variables $X_1, X_2, \ldots, X_n$ are **independent** if

$$P(X_1 \in E_1, X_2 \in E_2, \ldots, X_n \in E_n) = P(X_1 \in E_1)P(X_2 \in E_2) \cdots P(X_n \in E_n)$$

for *any* sets $E_1, E_2, \ldots, E_n$.

The importance of independence is illustrated in the following example.

EXAMPLE 3-32

In Example 3-12, the probability that a diameter meets specifications was determined to be 0.919. What is the probability that 10 diameters all meet specifications, assuming that the diameters are independent?

Denote the diameter of the first shaft as X_1, the diameter of the second shaft as X_2, and so forth, so that the diameter of the tenth shaft is denoted as X_{10}. The probability that all shafts meet specifications can be written as

$$P(0.2485 < X_1 < 0.2515, 0.2485 < X_2 < 0.2515, \ldots, 0.2485 < X_{10} < 0.2515)$$

In this example, the only set of interest is

$$E_1 = (0.2485, 0.2515)$$

With respect to the notation used in the definition of independence,

$$E_1 = E_2 = \cdots = E_{10}$$

Recall the relative frequency interpretation of probability. The proportion of times that shaft 1 is expected to meet the specifications is 0.919, the proportion of times that shaft 2 is expected to meet the specifications is 0.919, and so forth. If the random variables are independent, then the proportion of times in which we measure 10 shafts that we expect all to meet the specifications is

$$P(0.2485 < X_1 < 0.2515, 0.2485 < X_2 < 0.2515, \ldots, 0.2485 < X_{10} < 0.2515)$$
$$= P(0.2485 < X_1 < 0.2515)P(0.2485 < X_2 < 0.2515) \cdots P(0.2485 < X_{10} < 0.2515)$$
$$= 0.919^{10} = 0.430$$

Independent random variables are fundamental to the analyses in the remainder of the book. We often assume that random variables that record the replicates of a random experiment are independent, as in the previous example. Really what we assume is that the disturbances in the model

$$X = \mu + \epsilon$$

are unrelated because it is the disturbances that generate the randomness and the probabilities associated with the measurements.

Note that independence implies that the probabilities can be multiplied for *any* sets $E_1, E_2, \ldots, E_n$. Therefore, it should not be surprising to learn that an equivalent definition of independence is that the joint probability density function of the random variables equals the product of the probability density function of each random variable. This definition also holds for the joint probability mass function if the random variables are discrete.

EXAMPLE 3-33

Suppose X_1, X_2, and X_3 represent the thickness in micrometers of a substrate, an active layer, and a coating layer of a chemical product, respectively. Assume that X_1, X_2, and X_3 are independent and normally distributed with $\mu_1 = 10000$, $\mu_2 = 1000$, $\mu_3 = 80$, $\sigma_1 = 250$, $\sigma_2 = 20$, and $\sigma_3 = 4$. The specifications for the thickness of the substrate, active layer, and coating layer are $9200 < x_1 < 10800$, $950 < x_2 < 1050$, and $75 < x_3 < 85$, respectively. What proportion of chemical products meets all thickness specifications? Which one of the three thicknesses has the least probability of meeting specifications?

The requested probability is $P(9200 < X_1 < 10800, 950 < X_2 < 1050, 75 < X_3 < 85)$. Using the notation in the definition of independence, $E_1 = (9200, 10,800)$, $E_2 = (950, 1050)$, and $E_3 = (75, 85)$ in this example. Because the random variables are independent,

$$P(9200 < X_1 < 10800, 950 < X_2 < 1050, 75 < X_3 < 85)$$
$$= P(9200 < X_1 < 10800)P(950 < X_2 < 1050)P(75 < X_3 < 85)$$

After standardizing, the above equals

$$P(-3.2 < Z < 3.2)P(-2.5 < Z < 2.5)P(-1.25 < Z < 1.25)$$

where Z is a standard normal random variable. From the table of the standard normal distribution, the above equals

$$(0.99862)(0.98758)(0.78870) = 0.7778$$

The thickness of the coating layer has the least probability of meeting specifications. Consequently, a priority should be to reduce variability in this part of the process.

The concept of independence can also be applied to experiments that classify results. We used this concept to derive the binomial distribution. Recall that a test taker who just guesses from four multiple choices has probability 1/4 that any question is answered correctly. If it is assumed that the correct or incorrect outcome from one question is independent of others, then the probability that, say five questions are answered correctly can be determined by multiplication to equal

$$(1/4)^5 = 0.00098$$

Some additional applications of independence frequently occur in the area of system analysis. Consider a system that consists of devices that are either functional or failed. It is assumed that the devices are independent.

EXAMPLE 3-34

The system shown here operates only if there is a path of functional components from left to right. The probability that each component functions is shown in the diagram. Assume that the components function or fail independently. What is the probability that

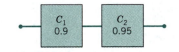

the system operates?

Let C_1 and C_2 denote the events that components 1 and 2 are functional, respectively. For the system to operate, both components must be functional. The probability that the system operates is

$$P(C_1, C_2) = P(C_1)P(C_2) = (0.9)(0.95) = 0.855$$

Note that the probability that the system operates is smaller than the probability that any component operates. This system fails whenever *any* component fails. A system of this type is called a **series system.**

EXAMPLE 3-35

The system shown here operates only if there is a path of functional components from left to right. The probability that each component functions is shown. Assume that the components function or fail independently. What is the probability that the system operates?

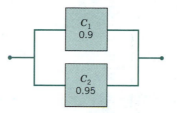

Let C_1 and C_2 denote the events that components 1 and 2 are functional, respectively. Also, C_1' and C_2' denote the events that components 1 and 2 fail, respectively, with associated probabilities $P(C_1') = 1 - 0.9 = 0.1$ and $P(C_2') = 1 - 0.95 = 0.05$. The system will operate if either component is functional. The probability that the system operates is one minus the probability that the system fails, and this occurs whenever both independent components fail. Therefore, the requested probability is

$$P(C_1 \text{ or } C_2) = 1 - P(C_1', C_2') = 1 - P(C_1')P(C_2') = 1 - (0.1)(0.05) = 0.995$$

Note that the probability that the system operates is greater than the probability that any component operates. This is a useful design strategy to decrease system failures. This system fails whenever *all* components fail. A system of this type is called a **parallel system.**

More general results can be obtained. The probability that a component does not fail over the time of its mission is called its **reliability.** Suppose that r_i denotes the reliability of component i in a system that consists of k components and that r denotes the probability that the system does not fail over the time of the mission. That is, r can be called the system reliability. The previous examples can be extended to obtain the following. For a series system

$$r = r_1 r_2 \cdots r_k$$

and for a parallel system

$$r = 1 - (1 - r_1)(1 - r_2) \cdots (1 - r_k)$$

The analysis of a complex system can be accomplished by a partition into sybsystems, which are sometimes called blocks.

EXAMPLE 3-36

The system shown here operates only if there is a path of functional components from left to right. The probability that each component functions is shown. Assume that the components function or fail independently. What is the probability that the system operates?

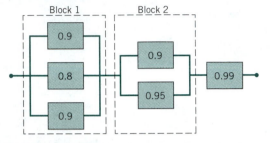

The system can be partitioned into blocks that are exclusively parallel subsystems. The result for a parallel system can be applied to each block, and the block results can be combined by the analysis for a series system. For block 1, the reliability is obtained from the result for a parallel system to be

$$1 - (0.1)(0.2)(0.1) = 0.998$$

Similarly, for block 2, the reliability is

$$1 - (0.1)(0.05) = 0.995$$

The system reliability is determined from the result for a series system to be

$$(0.998)(0.995)(0.99) = 0.983$$

EXERCISES FOR SECTION 3-11

3-105.   Let X be a normal random variable with $\mu = 10$ and $\sigma = 1.5$ and Y be a normal random variable with $\mu = 2$ and $\sigma = 0.25$. Assume X and Y are independent. Find the following probabilities.
(a) $P(X < 9, Y < 2.5)$
(b) $P(X > 8, Y < 2.25)$
(c) $P(8.5 \leq X \leq 11.5, Y > 1.75)$
(d) $P(X < 13, 1.5 \leq Y \leq 1.8)$

3-106.   Let X be a normal random variable with $\mu = 15.0$ and $\sigma = 3$ and Y be a normal random variable with $\mu = 20$ and $\sigma = 1$. Assume X and Y are independent. Find the following probabilities.
(a) $P(X < 12, Y < 19)$
(b) $P(X > 16, Y < 18)$
(c) $P(14 \leq X < 16, Y > 22)$
(d) $P(11 \leq X \leq 20, 17.5 \leq Y \leq 21)$

3-107.   Let X be a Poisson random variable with $\lambda = 2$ and Y be a Poisson random variable with $\lambda = 4$. Assume X and Y are independent. Find the following probabilities.
(a) $P(X < 4, Y < 4)$

(b) $P(X > 2, Y < 4)$

(c) $P(2 \le X < 4, Y \ge 3)$

(d) $P(X < 5, 1 \le Y \le 4)$

3-108. Let X be an exponential random variable with mean equal to 5 and Y be an exponential random variable with mean equal to 8. Assume X and Y are independent. Find the following probabilities.

(a) $P(X \le 5, Y \le 8)$

(b) $P(X > 5, Y \le 6)$

(c) $P(3 < X \le 7, Y > 7)$

(d) $P(X > 7, 5 < Y \le 7)$

3-109. Two independent vendors supply cement to a highway contractor. Through previous experience it is known that the compressive strength of samples of cement can be modeled by a normal distribution, with $\mu_1 = 6000$ kilograms per square centimeter and $\sigma_1 = 100$ kilograms per square centimeter for vendor 1, and $\mu_2 = 5825$ and $\sigma_2 = 90$ for vendor 2. What is the probability that both vendors supply a sample with compressive strength

(a) Less than 6100 kg/cm^2?

(b) Between 5800 and 6050?

(c) In excess of 6200?

3-110. The time between surface finish problems in a galvanizing process is exponentially distributed with a mean of 40 hours. A single plant operates three galvanizing lines that are assumed to operate independently.

(a) What is the probability that none of the lines experiences a surface finish problem in 40 hours of operation?

(b) What is the probability that all three lines experience a surface finish problem between 20 and 40 hours of operation?

3-111. The weights of adobe bricks used for construction are normally distributed with a mean of 3 pounds and a standard deviation of 0.25 pound. Assume that the weights of the bricks are independent and that a random sample of 20 bricks is selected.

(a) What is the probability that all the bricks in the sample exceed 2.75 pounds?

(b) What is the probability that none of the bricks exceed 3.75 pounds?

3-112. The yield in pounds from a day's production is normally distributed with a mean of 1500 pounds and a variance of 10,000 pounds squared. Assume that the yield on different days are independent random variables.

(a) What is the probability that the production yield exceeds 1400 pounds on each of 5 days?

(b) What is the probability that the production yield exceeds 1400 pounds on none of the next 5 days?

3-113. Consider the series system described in Example 3-34. Suppose that the probability that component C_1 functions is 0.95 and that the probability that component C_2 functions is 0.92.

(a) What is the probability that the system operates?

(b) What is the probability that the system does not operate?

3-114. Suppose a series system has three components C_1, C_2, and C_3 with the probability that each component functions equal to 0.90, 0.99, and 0.95, respectively.

(a) What is the probability that the system operates?

(b) What is the probability that the system does not operate?

3-115. Consider the parallel system described in Example 3-35. Suppose the probability that component C_1 functions is 0.85 and the probability that component C_2 functions is 0.92.

(a) Determine the probability that component C_1 fails.

(b) Determine the probability that component C_2 fails.

(c) What is the probability that the system operates?

(d) What is the probability that the system does not operate?

3-116. Suppose a parallel system has three components C_1, C_2, and C_3, in parallel, with the probability that each component functions equal to 0.90, 0.99, and 0.95, respectively. What is the probability that the system operates?

3-12 RANDOM SAMPLES, STATISTICS, AND THE CENTRAL LIMIT THEOREM

Previously in this chapter it was mentioned that data are the measured values of random variables obtained from replicates of a random experiment. Let the random variables that represent the measurements from the n replicates be denoted by $X_1, X_2, \ldots, X_n$. Because the replicates are identical, each random variable has the same distribution. Furthermore, the random variables are often assumed to be independent. That is, the results from some replicates do not affect the results from others. Throughout the remainder of the book, a common model is that data are measurements from independent random variables with the same distribution. That is, data are measurements from independent replicates of a random experiment. This model is used so frequently that we provide a definition.

Definition

Independent random variables $X_1, X_2, \ldots, X_n$ with the same distribution are called a **random sample.**

The term random sample stems from the historical use of statistical methods. Suppose that from a large population of objects, a sample of n objects is selected randomly. Here, randomly means that each subset of size n is equally likely to be selected. If the number of objects in the population is much larger than n, then the random variables $X_1, X_2, \ldots, X_n$ that represent the measurements from the sample can be shown to be approximately independent random variables with the same distribution. Consequently, independent random variables with the same distribution are referred to as a random sample.

EXAMPLE 3-37

In Example 2-1 in Chapter 2, the average pull-off force of eight connectors was 13.0 pounds. Two obvious questions are the following: What can we conclude about the average pull-off force of the future connectors? How wrong might we be if we concluded that the average pull-off force of this future population of connectors is 13.0?

There are two important issues to be considered in the answer to these questions.

1. First, because a conclusion is needed for a future population, this is an example of an analytic study. Certainly, we need to assume that the current prototypes are representative of the connectors that will be produced. This is related to the issue of stability in analytic studies that we discussed in Chapter 1. The usual

approach is to assume that these connectors are a random sample from the population. Suppose that the mean of this future population is denoted as μ. The objective is to estimate μ.

2. Second, even if we assume that these connectors are a random sample from future production, the average of these eight items might not equal the average of future production. However, the error that can occur can be quantified.

The key concept is the following: The average is a function of the individual pull-off forces of the eight connectors. That is, the average is a function of a random sample. Consequently, the average is a random variable with its own distribution. Recall that the distribution of an individual random variable can be used to determine the probability that a measurement is more than one, two, or three standard deviations from the mean of the distribution. In the same manner, the distribution of an average provides the probability that the average is more than a specified distance from μ. Consequently, if we conclude that μ is 13.0, the error is determined by the distribution of the average. We discuss this distribution in the remainder of the section.

Example 3-37 illustrates that a typical summary of data, such as an average, can be thought of as a function of a random sample.

Definition

A **statistic** is a function of the random variables in a random sample.

Given data, we calculate statistics all the time. All of the numerical summaries in Chapter 1 such as the sample mean $\overline{X}$, the sample variance S^2, and the sample standard deviation S are statistics. Although the definition of a statistic might seem overly complex, this is because we do not usually consider the distribution of a statistic. However, once we ask how wrong we might be, we are forced to think of a statistic as a function of random variables. Consequently, each statistic has a distribution. It is the distribution of a statistic that determines how well it estimates a quantity such as μ. Often, the probability distribution of a statistic can be determined from the probability distribution of the random sample and the sample size. Another definition is in order.

Definition

The probability distribution of a statistic is called its **sampling distribution.**

EXAMPLE 3-38

Suppose that the eight pull-off force measurements are assumed to be a random sample from a normal distribution with $\mu = 14.0$ and $\sigma = 0.5$. The probability density function of this distribution is illustrated in Fig. 3-35a. The sampling distributions of $\overline{X}$ and S^2 can be determined and will be discussed later in the book. The probability density functions of $\overline{X}$ and S^2 are illustrated in Figs. 3-35b and c. Note that the sampling distribution of a statistic can be very different from the random variables from which it is derived. Given the data, the values of these statistics were calculated in Chapter 1 to be $\overline{x} = 13.0$ and $s^2 = 0.2286$.

An important special case of a sampling distribution relates the sampling distribution of a linear function to the distribution of the random variables.

If $X_1, X_2, \ldots, X_n$ are random variables with $E(X_i) = \mu_i$ and $V(X_i) = \sigma_i^2$ and the random variable Y is

$$Y = c_1 X_1 + c_2 X_2 + \cdots + c_n X_n \tag{3-16}$$

then

$$E(Y) = c_1 \mu_1 + c_2 \mu_2 + \cdots + c_n \mu_n$$

and if $X_1, X_2, \ldots, X_n$ are *independent* random variables, then

$$V(Y) = c_1^2 \sigma_1^2 + c_2^2 \sigma_2^2 + \cdots + c_n^2 \sigma_n^2$$

Furthermore, if $X_1, X_2, \ldots, X_n$ are *independent, normal* random variables, then Y is a normal random variable.

EXAMPLE 3-39

Suppose that the random variables X_1 and X_2 denote the length and width, respectively, of a manufactured part. Assume that X_1 is normal with $E(X_1) = 2$ centimeters and standard deviation 0.1 centimeter and that X_2 is normal with $E(X_2) = 5$ centimeters and standard deviation 0.2 centimeter. Also, assume that X_1 and X_2 are independent. Determine the probability that the perimeter exceeds 14.5 centimeters.

Then, $Y = 2X_1 + 2X_2$ is a random variable that represents the perimeter of the part. Furthermore, Y is normally distributed, $E(Y) = 14$ centimeters, and $V(Y) = 2^2(0.1^2) + 2^2(0.2^2) = 0.2$. Also, $\sigma_Y = \sqrt{0.2} = 0.447$.

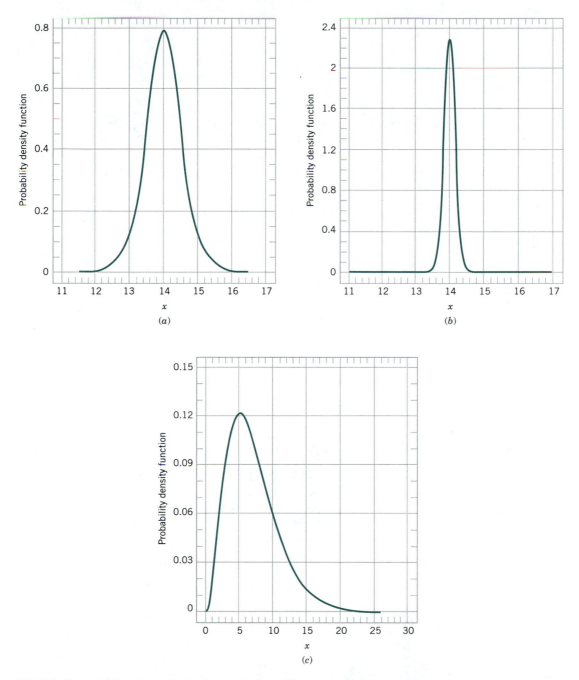

Figure 3-35 Probability distribution for Example 3-38. (a) Probability density function of a pull-off force measurement. (b) Probability density function of the average of eight pull-off force measurements. (c) Probability density function of the sample variance of eight pull-off force measurements.

Now,

$$P(Y > 14.5) = P\left(\frac{Y - \mu_Y}{\sigma_Y} > \frac{14.5 - 14}{0.447}\right)$$

$$= P(Z > 1.12) = 0.13$$

Consider the sampling distribution of the sample mean $\overline{X}$. Suppose that a random sample of size n is taken from a normal population with mean μ and variance σ^2. Now each random variable in this sample—say, $X_1, X_2, \ldots, X_n$—is a normally and independently distributed random variable with mean μ and variance σ^2. Then from (3-16), we conclude that the sample mean

$$\overline{X} = \frac{X_1 + X_2 + \cdots + X_n}{n}$$

has a normal distribution with mean

$$E(\overline{X}) = \frac{\mu + \mu + \cdots + \mu}{n} = \mu \tag{3-17}$$

and variance

$$V(\overline{X}) = \frac{\sigma^2 + \sigma^2 + \cdots + \sigma^2}{n^2} = \frac{\sigma^2}{n} \tag{3-18}$$

The mean and variance of $\overline{X}$ are also denoted as $\mu_{\overline{X}}$ and $\sigma_{\overline{X}}$, respectively.

EXAMPLE 3-40

Soft-drink cans are filled by an automated filling machine. The mean fill volume is 12.1 fluid ounces, and the standard deviation is 0.05 fluid ounce. Assume that the fill volumes of the cans are independent, normal random variables. What is the probability that the average volume of 10 cans selected from this process is less than 12 fluid ounces?

Let $X_1, X_2, \ldots, X_{10}$ denote the fill volumes of the 10 cans. The average fill volume (denoted as $\overline{X}$) is a normal random variable with

$$E(\overline{X}) = 12.1 \quad \text{and} \quad V(\overline{X}) = \frac{0.05^2}{10} = 0.00025$$

Consequently, $\sigma_{\overline{X}} = \sqrt{0.00025} = 0.0158$ and

$$P(\overline{X} < 12) = P\left(\frac{\overline{X} - \mu_{\overline{X}}}{\sigma_{\overline{X}}} < \frac{12 - 12.1}{0.0158}\right)$$

$$= P(Z < -6.32) = 0$$

If we are sampling from a population that has an unknown probability distribution, the sampling distribution of the sample mean will still be approximately normal with mean μ and variance σ^2/n, if the sample size n is large. This is one of the most useful theorems in statistics. It is called the **central limit theorem.** The statement is as follows:

Central Limit Theorem

If $X_1, X_2, \ldots, X_n$ is a random sample of size n taken from a population with mean μ and variance σ^2, and if $\overline{X}$ is the sample mean, then the limiting form of the distribution of

$$ Z = \frac{\overline{X} - \mu}{\sigma/\sqrt{n}} \qquad (3\text{-}19) $$

as $n \to \infty$, is the standard normal distribution.

The normal approximation for $\overline{X}$ depends on the sample size n. Figure 3-36a shows the distribution obtained for throws of a single, six-sided true die. The probabilities are equal (1/6) for all the values obtained, 1, 2, 3, 4, 5, or 6. Figure 3-36b shows the distribution of the average score obtained when tossing 2 dice, and Figs. 3-36c, 3-36d, and 3-36e show the distributions of average scores obtained when tossing 3, 5, and 10 dice, respectively. Notice that, while the distribution of 1 die is relatively far from normal, the distribution of averages is approximated reasonably well by the normal distribution for sample sizes as small as 5. (The dice throw distributions are discrete, however, while the normal is continuous.) Although the central limit theorem will work well for small samples ($n = 4, 5$) in most cases—particularly where the population is continuous, unimodal, and symmetric—larger samples will be required in other situations, depending on the shape of the population. In many cases of practical interest, if $n \geq 30$, the normal approximation will be satisfactory regardless of the shape of the population. If $n < 30$, the central limit theorem will work if the distribution of the population is not severely nonnormal.

EXAMPLE 3-41

An electronics company manufactures resistors that have a mean resistance of 100 Ω and a standard deviation of 10 Ω. Find the probability that a random sample of $n = 25$ resistors will have an average resistance less than 95 Ω.

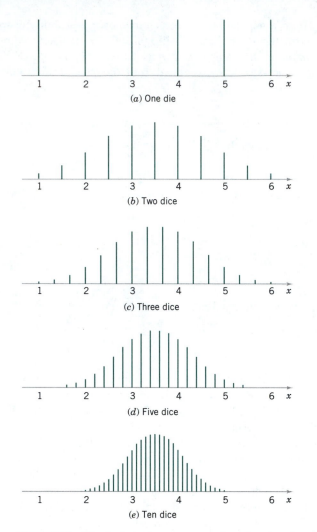

Figure 3-36   Distributions of average scores from throwing dice.
[Adapted with permission from Box, Hunter, and Hunter (1978).]

Note that the sampling distribution of $\overline{X}$ is approximately normal, with mean $\mu_{\overline{X}}$ = 100 Ω and a standard deviation of

$$\sigma_{\overline{X}} = \frac{\sigma}{\sqrt{n}} = \frac{10}{\sqrt{25}} = 2$$

Therefore, the desired probability corresponds to the shaded area in Fig. 3-37. Standardizing the point $\overline{X}$ = 95 in Fig. 3-37, we find that

$$z = \frac{95 - 100}{2} = -2.5$$

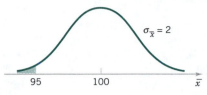

Figure 3-37   Probability density function of resistance.

and therefore,

$$P(\overline{X} < 95) = P(Z < -2.5)$$
$$= 0.0062$$

EXERCISES FOR SECTION 3-12

3-117. If X and Y are independent, normal random variables with $E(X) = 5$, $V(X) = 1$, $E(Y) = 12$, and $V(Y) = 4$, determine the following.
(a) $E(2X + 3Y)$
(b) $V(2X + 3Y)$
(c) $P(2X + 3Y < 35)$
(d) $P(2X + 3Y < 40)$

3-118. If W, X, and Y are independent, normal random variables with $E(X) = 5$, $V(X) = 1$, $E(Y) = 16$, $V(Y) = 16$, $E(W) = 20$, and $V(W) = 4$, determine the following.
(a) $E(W + 2X + 3Y)$
(b) $V(W + 2X + 3Y)$
(c) $P(W + 2X + 3Y > 60)$
(d) $P(W + 2X + 3Y \leq 30)$

> **3-119.** A plastic casing for a magnetic disk is composed of two halves. The thickness of each half is normally distributed with a mean of 1.5 millimeters and a standard deviation of 0.1 millimeter and the halves are independent.
(a) Determine the mean and standard deviation of the total thickness of the two halves.
(b) What is the probability that the total thickness exceeds 3.3 millimeters?

3-120. The width of a casing for a door is normally distributed with a mean of 24 inches and a standard deviation of 1/8 inch. The width of a door is normally distributed with a mean of 23 and 7/8 inches and a standard deviation of 1/16 inch. Assume independence.
(a) Determine the mean and standard deviation of the difference between the width of the casing and the width of the door.
(b) What is the probability that the width of the casing minus the width of the door exceeds 1/4 inch?
(c) What is the probability that the door does not fit in the casing?

> **3-121.** A U-shaped assembly is to be formed from the three parts A, B, and C. The picture is shown in Fig. 3-38. The

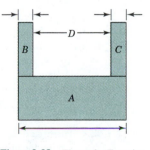

Figure 3-38   Figure for Exercise 3-121.

length of A is normally distributed with a mean of 10 millimeters and a standard deviation of 0.1 millimeter. The thickness of parts B and C is normally distributed with a mean of 2 millimeters and a standard deviation of 0.05 millimeter. Assume that all dimensions are independent.

(a) Determine the mean and standard deviation of the length of the gap D.

(b) What is the probability that the gap D is less than 5.9 millimeters?

3-122. Intravenous fluid bags are filled by an automated filling machine. Assume that the fill volumes of the bags are independent, normal random variables with a standard deviation of 0.08 fluid ounces.

(a) What is the standard deviation of the average fill volume of 20 bags?

(b) If the mean fill volume of the machine is 6.16 ounces, what is the probability that the average fill volume of 20 bags is below 5.95 ounces?

(c) What should the mean fill volume equal in order that the probability that the average of 20 bags is below 6 ounces is 0.001?

3-123. The photoresist thickness in semiconductor manufacturing has a mean of 10 micrometers and a standard deviation of 1 micrometer. Assume that the thickness is normally distributed and that the thicknesses of different wafers are independent.

(a) Determine the probability that the average thickness of 10 wafers is either greater than 11 or less than 9 micrometers.

(b) Determine the number of wafers that need to be measured such that the probability that the average thickness exceeds 11 micrometers is 0.01.

3-124. The time to complete a manual task in a manufacturing operation is consid-

ered a normally distributed random variable with mean of 0.50 minute and a standard deviation of 0.05 minute. Find the probability that the average time to complete the manual task, after 49 repetitions, is less than 0.465 minute.

3-125. A synthetic fiber used in manufacturing carpet has tensile strength that is normally distributed with mean 75.5 psi and standard deviation 3.5 psi. Find the probability that a random sample of $n = 6$ fiber specimens will have sample mean tensile strength that exceeds 75.75 psi.

3-126. The compressive strength of concrete has a mean of 2500 psi and a standard deviation of 50 psi. Find the probability that a random sample of $n = 5$ specimens will have a sample mean diameter that falls in the interval from 2499 psi to 2510 psi.

3-127. The amount of time that a customer spends waiting at an airport check-in counter is a random variable with mean 8.2 minutes and standard deviation 1.5 minutes. Suppose that a random sample of $n = 49$ customers is observed. Find the probability that the average time waiting in line for these customers is

(a) Less than 8 minutes

(b) Between 8 and 9 minutes

(c) Less than 7.5 minutes

3-128. Suppose that X has the following discrete distribution

$$f(x) = \begin{cases} \frac{1}{3}, & x = 1, 2, 3 \\ 0, & \text{otherwise} \end{cases}$$

A random sample of $n = 36$ is selected from this population. Approximate the probability that the sample mean is greater than 2.1 but less than 2.5.

3-129. The viscosity of a fluid can be measured in an experiment by dropping a small ball into a calibrated tube containing the fluid and observing the random variable X, the time it takes for the ball to drop the measured distance. Assume that X is normally distributed with a mean of 20 seconds and a standard deviation of 0.5 second for a particular type of liquid.

(a) What is the standard deviation of the average time of 40 experiments?

(b) What is the probability that the average time of 40 experiments will exceed 20.1 seconds?

(c) Suppose the experiment is repeated only 20 times. What is the probability that the average value of X will exceed 20.1 seconds?

(d) Is the probability computed in part (b) greater than or less than the probability computed in part (c)? Explain why this inequality occurs.

3-130. A random sample of $n = 9$ structural elements is tested for compressive strength. We know that the true mean compressive strength $\mu = 5500$ psi and the standard deviation is $\sigma = 100$ psi. Find the probability that the sample mean compressive strength exceeds 4985 psi.

SUPPLEMENTAL EXERCISES

3-131. Suppose that $f(x) = e^{-x}$ for $0 < x$ and $f(x) = 0$ for $x < 0$. Determine the following probabilities.

(a) $P(X \leq 1.5)$

(b) $P(X < 1.5)$

(c) $P(1.5 < X < 3)$

(d) $P(X = 3)$

(e) $P(X > 3)$

3-132 Suppose that $f(x) = e^{-x/2}$ for $0 < x$ and $f(x) = 0$ for $x < 0$.

(a) Determine x such that $P(x < X) = 0.20$.

(b) Determine x such that $P(X \leq x) = 0.75$.

3-133. The random variable X has the following probability distribution.

x	2	3	5	8
probability	0.2	0.4	0.3	0.1

Determine the following.

(a) $P(X \leq 3)$

(b) $P(X > 2.5)$

(c) $P(2.7 < X < 5.1)$

(d) $E(X)$

(e) $V(X)$

3-134. A driveshaft will suffer fatigue failure with a mean time-to-failure of 40,000 hours of use. If it is known that the probability of failure before 36,000 hours is 0.04 and that the distribution governing time-to-failure is a normal distribution, what is the standard deviation of the time-to-failure distribution?

3-135. A standard fluorescent tube has a life length that is normally distributed with a mean of 7000 hours and a standard deviation of 1000 hours. A competitor has developed a compact fluorescent lighting system that will fit into incandescent sockets. It claims that a new compact tube has a normally distributed life length with a mean of 7500 hours and a standard deviation of 1200 hours. Which fluorescent tube is more likely to have a life length greater than 9000 hours? Justify your answer.

3-136. The average life of a certain type of compressor is 10 years with a standard deviation of 1 year. The manufacturer replaces free all compressors that fail while under guarantee. The manufacturer is willing to replace 3% of all compressors sold. For how many years should the guarantee be in effect? Assume a normal distribution.

3-137. The probability that a call to an emergency help line is answered in less than 15 seconds is 0.85. Assume that all calls are independent.

(a) What is the probability that exactly 7 of 10 calls are answered within 15 seconds?

(b) What is the probability that at least 16 of 20 calls are answered in less than 15 seconds?

(c) For 50 calls, what is the mean number of calls that are answered in less than 15 seconds?

(d) Repeat parts (a)–(c) using the normal approximation.

3-138. The number of messages sent to a computer bulletin board is a Poisson random variable with a mean of five messages per hour.

(a) What is the probability that five messages are received in 1 hour?

(b) What is the probability that 10 messages are received in 1.5 hours?

(c) What is the probability that fewer than 2 messages are received in one-half hour?

3-139. Continuation of Exercise 3-138. Let Y be the random variable defined as the time between messages arriving to the computer bulletin board.

(a) What is the distribution of Y? What is the mean of Y?

(b) What is the probability that the time between messages exceeds 15 minutes?

(c) What is the probability that the time between messages is less than 5 minutes?

(d) Given that 10 minutes has passed without a message arriving, what is the probability that there will not be a message in the next 10 minutes?

3-140. The number of errors in a textbook follows a Poisson distribution with mean of 0.01 error per page.

(a) What is the probability that there are three or fewer errors in 100 pages?

(b) What is the probability that there are four or more errors in 100 pages?

(c) What is the probability that there are three or fewer errors in 200 pages?

3-141. Continuation of Exercise 3-140. Let Y be the random variable defined as the number of pages between errors.

(a) What is the distribution of Y? What is the mean of Y?

(b) What is the probability that there are fewer than 100 pages between errors?

(c) What is the probability that there are no errors in 200 consecutive pages?

(d) Given that there are 100 consecutive pages without errors, what is the probability that there will not be an error in the next 50 pages?

3-142. Polyelectrolytes are typically used to separate oil and water in industrial applications. The separation process is dependent on controlling the pH. Fifteen pH readings of wastewater following these processes were recorded. Is it reasonable to model these data using a normal distribution?

6.2 6.5 7.6 7.7 7.1 7.1 7.9 8.4
7.0 7.3 6.8 7.6 8.0 7.1 7.0

3-143. The lifetimes of six major components in a copier are independent exponential random variables with means of 8000 hours, 10,000 hours, 10,000 hours, 20,000 hours, 20,000 hours, and 25,000 hours, respectively.

(a) What is the probability that the lifetimes of all the components exceed 5000 hours?

(b) What is the probability that none of the components have a lifetime that exceeds 5000 hours?

(c) What is the probability that the lifetimes of all the components are less than 3000 hours?

3-144. A random sample of 36 observations has been drawn. Find the probability that the sample mean is in the interval $47 < X < 53$, for each of the following population distributions and population parameter values.

(a) Normal with mean 50 and standard deviation 12.

(b) Exponential with mean 50.

(c) Poisson with mean 50.

(d) Compare the probabilities obtained in parts (a)–(c) and explain why the probabilities differ.

3-145. From contractual commitments and extensive past laboratory testing, we know that compressive strength measurements are normally distributed with the true mean compressive strength $\mu = 5500$ psi and standard deviation $\sigma = 100$ psi. A random sample of structural elements is tested for compressive strength at the customer's receiving location.

(a) What is the standard deviation of the sampling distribution of the sample mean for this problem if $n = 9$?

(b) What is the standard deviation of the sampling distribution of the sample mean for this problem if $n = 20$?

(c) Compare your results of parts (a) and (b), and comment on why they are the same or different.

3-146. The weight of adobe bricks for construction is normally distributed with a mean of 3 pounds and a standard deviation of 0.25 pound. Assume that the weights of the bricks are independent and that a random sample of 25 bricks is chosen. What is the probability that the mean weight of the sample is less than 2.95 pounds?

3-147. A disk drive assembly consists of one hard disk and spacers on each side, as shown in Fig. 3-39. The height of the top spacer, W, is normally distributed with mean 120 mm and standard deviation 0.5 mm, and the height of the disk, X, is normally distributed with mean 20 mm and standard deviation 0.1 mm,

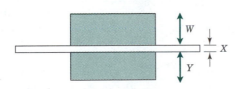

Figure 3-39 Figure for Exercise 3-147.

and the height of the bottom spacer, Y, is normally distributed with mean 100 mm and standard deviation 0.4 mm.

(a) What are the distribution, the mean, and the variance of the height of the stack?

(b) Assume that the stack must fit into a space with a height of 242 mm. What is the probability that the stack height will exceed the space height?

3-148. The time for an automated system in a warehouse to locate a part is normally distributed with a mean of 45 seconds and a standard deviation of 30 seconds. Suppose that independent requests are made for 10 parts.

(a) What is the probability that the average time to locate 10 parts exceeds 60 seconds?

(b) What is the probability that the total time to locate 10 parts exceeds 600 seconds?

3-149. A mechanical assembly used in an automobile engine contains four major components. The weights of the components are independent and normally distributed with the following means and standard deviations (in ounces).

Component	Mean	Standard Deviation
left case	4	0.4
right case	5.5	0.5
bearing assembly	10	0.2
bolt assembly	8	0.5

(a) What is the probability that the weight of an assembly exceeds 29.5 ounces?

(b) What is the probability that the mean weight of eight independent assemblies exceeds 29 ounces?

3-150. A bearing assembly contains 10 bearings. The bearing diameters are assumed to be independent and normally distributed with a mean of 1.5 millimeters and a standard deviation of 0.025

millimeter. What is the probability that the maximum diameter bearing in the assembly exceeds 1.6 millimeters?

3-151. A process is said to be of **six-sigma quality** if the process mean is at least six standard deviations from the nearest specification. Assume a normally distributed measurement.

(a) If a process mean is centered between the upper and lower specifications at a distance of six standard deviations from each, what is the probability that a product does not meet specifications? Using the result that 0.000001 equals one part per million, express the answer in parts per million.

(b) Because it is difficult to maintain a process mean centered between the specifications, the probability of a product not meeting specifications is often calculated after assuming the process shifts. If the process mean positioned as in part (a) shifts upward by 1.5 standard deviations, what is the probability that a product does not meet specifications? Express the answer in parts per million.

3-152. Continuation of Exercise 3-47. Recall that it was determined that a normal distribution adequately fit the internal pressure strength data. Use this distribution and suppose that the sample mean of 206.04 and standard deviation of 11.57 are used to estimate the population parameters. Estimate the following probabilities.

(a) What is the probability that the internal pressure strength measurement will be between 210 and 220 psi?

(b) What is the probability that the internal pressure strength measurement will exceed 228 psi?

(c) Find x such that $P(X \geq x) = 0.02$, where X is the internal pressure strength random variable.

3-153. Continuation of Exercise 3-48. Recall that it was determined that a normal distribution adequately fit the dimensional measurements for parts from two different machines. Using this distribution, suppose that $\bar{x}_1 = 100.27$ and $s_1 = 2.28$ and $\bar{x}_2 = 100.11$ and $s_2 = 7.58$ are used to estimate the population parameters. Estimate the following probabilities. Assume that the engineering specifications indicate that acceptable parts measure between 96 and 104.

(a) What is the probability that machine 1 produces acceptable parts?

(b) What is the probability that machine 2 produces acceptable parts?

(c) Use your answers from parts (a) and (b) to determine which machine is preferable.

(d) Recall that the data reported in Exercise 3-49 were a result of a process engineer making adjustments to machine 2. Use the new sample mean 105.39 and sample standard deviation 2.08 to estimate the population parameters. What is the probability that the newly adjusted machine 2 will produce acceptable parts? Did adjusting machine 2 improve its overall performance?

3-154. Continuation of Exercise 2-1.

(a) Plot the data on normal probability paper. Does this dimension appear to have a normal distribution?

(b) Suppose it has been determined that the largest observation, 74.015, was an outlier due to a machining problem. Consequently, it can be removed from the data set. Does this improve the fit of the normal distribution to the data?

3-155. Continuation of Exercise 2-2.

(a) Plot the data on normal probability paper. Do these data appear to have a normal distribution?

(b) Remove the largest observation from the data set. Does this improve the fit of the normal distribution to the data?

3-156. Consider Exercise 3-35. What is the probability that four random, independent samples of cement will all have compressive strength greater than 6250 kg/cm^2?

3.157. Consider Exercise 3-36. What is the probability that six random, independent samples of paper will all have a tensile strength less than 30 lb/in^2?

3-158. Consider Exercise 3-37. What is the probability that two randomly selected tools will have a line width exceeding 0.62 micrometers?

3-159. Consider Exercise 3-40. Suppose that the driver's reaction time is measured for 10 independent visual stimuli. What is the probability that the reaction time will always be less than 0.35 second?

3-160. Consider the following system made up of functional components in parallel and series. The probability that each component functions is shown in Fig. 3-40.

(a) What is the probability that the system operates?

(b) What is the probability that the system fails due to the components in series? Assume parallel components do not fail.

(c) What is the probability that the system fails due to the components in parallel? Assume series components do not fail.

(d) Compute the probability that the system fails using the following formula:

$[1 - P(C_1) \cdot P(C_4)] \cdot [1 - P(C_2') \cdot P(C_4')] + P(C_1) \cdot P(C_4) \cdot P(C_2') \cdot P(C_3') + [1 - P(C_1)P(C_4)] \cdot P(C_2') \cdot P(C_3')$.

(e) Describe in words the meaning of each of the terms in the formula in part (d).

(f) Use part (a) to compute the probability that the system fails.

3-161. Consider Exercise 3-160.

(a) Improve the probability that component C_1 functions to a value of 0.95 and recompute parts (a), (b), (c), and (f).

(b) Alternatively, do not change the original probability associated with C_1; rather, increase the probability that component C_2 functions to a value of 0.95 and recompute parts (a), (b), (c), and (f).

(c) Based on your answers in parts (a) and (b) of this exercise, comment on whether you would recommend increasing the reliability of a series component or a parallel component to increase overall system reliability.

3-162. **Lognormal Example.** In some data sets, a transformation by some mathematical function applied to the original data, such as log y, can result in data that are simpler to analyze statistically. When the transformation log y results in normally distributed data, then we call the original data "lognormal." To illustrate the effect of a log transformation, consider the following data, which represent cycles to failure for a yarn product:

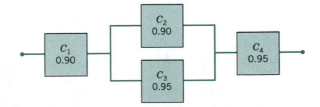

Figure 3-40 Figure for Exercise 3-160.

675, 3650, 175, 1150, 290, 2000, 100, 375.

(a) Plot the data on normal probability paper and comment on the adequacy of the fit.

(b) Transform the data using logarithms, that is, let y^* (new value) = log y (old value). Plot the transformed data on normal probability paper and comment on the adequacy of the fit.

(c) Engineering has specified that acceptable yarn strength should exceed 200 cycles prior to failure. Use your results from part (b) to estimate the proportion of acceptable yarn. (*Hint:* Be sure to transform the lower specification limit, 200, prior to computing the proportion. Suppose that the sample mean and sample standard deviation are used to estimate the population parameters in your calculations.)

3-163. Lognormal Example. Consider the following data, which represent the number of hours of operation of a surveillance camera until failure:

246,785	183,424	1060
22,310	921	35,659
127,015	10,649	13,859
53,731	10,763	1456
189,880	2414	21,414
411,884	29,644	1473

(a) Plot the data on normal probability paper and comment on the adequacy of the fit.

(b) Transform the data using logarithms, that is, let y^* (new value) = log y (old value). Plot the transformed data on normal probability paper and comment on the adequacy of the fit.

(c) The manufacturer of the cameras is interested in defining a warranty limit such that not more than 2% of the cameras will need to be replaced. Use your fitted model from part (b) to propose a warranty limit on the time to failure of the number of hours of a random surveillance camera. (*Hint:* Be sure to give the warranty limit in the original units of hours. Suppose that the sample mean and sample standard deviation are used to estimate the population parameters in your calculations.)

3-164. Weibull Example. The Weibull distribution is often used to model the time until failure of many different physical systems. If the random variable Y has a Weibull distribution, then

$$P(Y \le y) = 1 - e^{-(y/\delta)^\beta}$$

We call $\delta > 0$ the scale parameter and $\beta > 0$ the shape parameter. The two parameters of this model provide a great deal of flexibility to model systems in which the number of failures increases with time (bearing wear), decreases with time (some semiconductors), or remains constant with time (failures caused by external shocks to the system).

(a) Show, by substitution, that the exponential distribution is a special case of the Weibull, when the shape parameter is set to 1.

(b) A typical approach to fitting data to a Weibull distribution is to plot the sample data on Weibull probability plotting paper. Plot the following data, which represent the life of roller bearings (in hours), on Weibull probability paper and determine the adequacy of the fit.

7203	3917	7476	5410
7891	10,033	4484	12,539
2933	16,710	10,702	16,122
13,295	12,653	5610	6466
5263	2,504	9098	7759

(c) Using the estimated shape parameter = 2.2 and the estimated scale parameter 9525, estimate the probability that a bearing lasts at least 7500 hours.

(d) If 5 bearings are in use and failures occur independently, what is the probability that all 5 bearings last at least 7500 hours?

3-165. **Weibull Example.** Consider the following data that represent the life of packaged magnetic disks exposed to corrosive gases (in hours):

4	86	335	746	80
1510	195	562	137	1574
7600	4394	4	98	
1196	15	934	11	

(a) Plot the data on Weibull probability paper and determine the adequacy of the fit.

(b) Using the estimated shape parameter $= 0.53$ and the estimated scale parameter 604, estimate the probability that a disk fails before 150 hours.

(c) If a warranty is planned to cover no more than 10% of the manufactured disks, at what value should the warranty level be set?

3-166. **Non-uniqueness of Probability Models.** It is possible to fit more than one model to a set of data. Consider the life data given in Exercise 3-165.

(a) Transform the data using logarithms; that is, let y^* (new value) $= \log y$ (old value). Plot the transformed data on normal probability paper and comment on the adequacy of the fit.

(b) Use the fitted normal distribution from part (c) to estimate the probability that the disk fails before 150 hours. Compare your results with your answer in part (b).

Team Exercises

3-167. Using the data set that you found or collected in the first team exercise of

Chapter 2, or another data set of interest, answer the following questions:

(a) Is a continuous or discrete distribution model more appropriate to model your data? Explain.

(b) You have studied the normal, exponential, Poisson, and binomial distributions in this chapter. Based on your recommendation in part (a), attempt to fit at least one model to your data set. Report on your results.

3-168. Computer software can be used to simulate data from a normal distribution. Use a package such as Minitab to simulate dimensions for parts A, B, and C in Fig. 3-38 of Exercise 3-121.

(a) Simulate 500 assemblies from simulated data for parts A, B, and C and calculate the length of gap D for each.

(b) Summarize the data for gap D with a histogram and relevant summary statistics.

(c) Compare your simulated results with those obtained in Exercise 3-121.

(d) Describe a problem for which simulation is a good method of analysis.

3-169. Consider the data on weekly waste (percent) as reported for five suppliers of the Levi-Strauss clothing plant in Albuquerque and reported on the Web site http://lib.stat.cmu.edu/DASL/Stories/wasterunup.html. Test each of the data sets for conformance to a normal probability model using a normal probability plot. For those that do not pass the test for normality, delete any outliers (these can be identified using a box plot) and re-plot the data. Summarize your findings.

4 Decision Making for a Single Sample

CHAPTER OUTLINE

4-1   STATISTICAL INFERENCE

The field of statistical inference consists of those methods used to make decisions or to draw conclusions about a **population.** These methods utilize the information contained in a **sample** from the population in drawing conclusions. Figure 4-1 illustrates the relationship between a population and a sample. This chapter begins our study of the statistical methods used for inference and decision making.

Statistical inference may be divided into two major areas: **parameter estimation** and **hypothesis testing.** As an example of a parameter estimation problem, suppose that a structural engineer is analyzing the tensile strength of a component used in an automobile chassis. Since variability in tensile strength is naturally present between the individual components because of differences in raw material batches, manufacturing processes, and measurement procedures (for example), the engineer is interested in estimating the mean tensile strength of the components. Knowledge of the statistical sampling properties of the estimator used would enable the engineer to establish the precision of the estimate.

Now consider a situation in which two different reaction temperatures can be used in a chemical process—say, t_1 and t_2. The engineer conjectures that t_1 results in higher yields than does t_2. Statistical hypothesis testing is a framework for solving problems of this type. In this case, the hypothesis would be that the mean yield using temperature t_1 is greater than the mean yield using temperature t_2. Note that there is no emphasis on estimating yields; instead, the focus is on drawing conclusions about a stated hypothesis.

This chapter begins by discussing methods for estimating parameters. Then we introduce the basic principles of hypothesis testing. Once these statistical fundamentals have been presented, we will apply them to several situations that arise frequently in engineering practice. These include inference on the mean of a population, the variance of a population, and a population proportion.

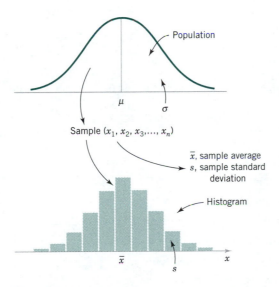

Figure 4-1   Relationship between a population and a sample.

4-2 POINT ESTIMATION

One very important application of statisitics is in obtaining **point estimates** of parameters such as the population mean and the population variance. When discussing inference problems, it is convenient to have a general symbol to represent the parameter of interest. We will use the Greek symbol θ (theta) to represent the parameter. The objective of point estimation is to select a single number, based on sample data, that is the most plausible value for θ. A numerical value of a sample statistic will be used as the point estimate.

For example, suppose that the random variable X is normally distributed with unknown mean μ. The sample mean is a point estimator of the unknown population mean μ; that is, $\hat{\mu} = \overline{X}$. After the sample has been selected, the numerical value $\overline{x}$ is the point estimate of μ. Thus, if $x_1 = 25$, $x_2 = 30$, $x_3 = 29$, and $x_4 = 31$, then the point estimate of μ is

$$\overline{x} = \frac{25 + 30 + 29 + 31}{4} = 28.75$$

Similarly, if the population variance σ^2 is also unknown, a point estimator for σ^2 is the sample variance S^2, and the numerical value $s^2 = 6.9$ calculated from the sample data is called the point estimate of σ^2.

In general, if X is a random variable with probability distribution $f(x)$, characterized by the unknown parameter θ, and if $X_1, X_2, \ldots, X_n$ is a random sample of size n from $f(x)$, then the statistic $\hat{\Theta} = h(X_1, X_2, \ldots, X_n)$ is called a **point estimator** of θ. Here h is just a function of observations in the random sample. Note that $\hat{\Theta}$ is a random variable, because it is a function of random variables. After the sample has been selected, $\hat{\Theta}$ takes on a particular numerical value $\hat{\theta}$ called the **point estimate** of θ.

Definition

A **point estimate** of some population parameter θ is a single numerical value $\hat{\theta}$ of a statistic $\hat{\Theta}$.

Estimation problems occur frequently in engineering. We often need to estimate

- The mean μ of a single population
- The variance σ^2 (or standard deviation σ) of a single population
- The proportion p of items in a population that belong to a class of interest
- The difference in means of two populations, $\mu_1 - \mu_2$
- The difference in two population proportions, $p_1 - p_2$

Reasonable point estimates of these parameters are as follows:

- For μ, the estimate is $\hat{\mu} = \bar{x}$, the sample mean.
- For σ^2, the estimate is $\hat{\sigma}^2 = s^2$, the sample variance.
- For p, the estimate is $\hat{p} = x/n$, the sample proportion, where x is the number of items in a random sample of size n that belong to the class of interest.
- For $\mu_1 - \mu_2$, the estimate is $\hat{\mu}_1 - \hat{\mu}_2 = \bar{x}_1 - \bar{x}_2$, the difference between the sample means of two independent random samples.
- For $p_1 - p_2$, the estimate is $\hat{p}_1 - \hat{p}_2$, the difference between two sample proportions computed from two independent random samples.

The following display summarizes the relationship between the unknown parameters and their typical associated statistics and point estimates.

Unknown Parameter θ	Statistic $\hat{\Theta}$	Point Estimate $\hat{\theta}$
μ	$\bar{X} = \dfrac{\sum X_i}{n}$	$\bar{x}$
σ^2	$S^2 = \dfrac{\sum (X_i - \bar{X})^2}{n - 1}$	s^2
p	$\hat{P} = \dfrac{X}{n}$	$\hat{p}$
$\mu_1 - \mu_2$	$\bar{X}_1 - \bar{X}_2 = \dfrac{\sum X_{1i}}{n_1} - \dfrac{\sum X_{2i}}{n_2}$	$\bar{x}_1 - \bar{x}_2$
$p_1 - p_2$	$\hat{P}_1 - \hat{P}_2 = \dfrac{X_1}{n_1} - \dfrac{X_2}{n_2}$	$\hat{p}_1 - \hat{p}_2$

We may have several different choices for the point estimator of a parameter. For example, if we wish to estimate the mean of a population, we might consider the sample mean, the sample median, or perhaps the average of the smallest and largest observations in the sample as point estimators. In order to decide which point estimator of a particular parameter is the best one to use, we need to examine their statistical properties and develop some criteria for comparing estimators.

An estimator should be "close" in some sense to the true value of the unknown parameter. Formally, we say that $\hat{\Theta}$ is an unbiased estimator of θ if the expected value of $\hat{\Theta}$ is equal to θ. This is equivalent to saying that the mean of the probability distribution of $\hat{\Theta}$ (or the mean of the sampling distribution of $\hat{\Theta}$) is equal to θ.

Definition

The point estimator $\hat{\Theta}$ is an **unbiased estimator** for the parameter θ if

$$E(\hat{\Theta}) = \theta \qquad (4\text{-}1)$$

If the estimator is not unbiased, then the difference

$$E(\hat{\Theta}) - \theta \qquad (4\text{-}2)$$

is called the **bias** of the estimator $\hat{\Theta}$.

When an estimator is unbiased, the $E(\hat{\Theta}) - \theta = 0$; that is, the bias is zero.

EXAMPLE 4-1

Suppose that X is a random variable with mean μ and variance σ^2. Let $X_1, X_2, \ldots, X_n$ be a random sample of size n from the population represented by X. Show that the sample mean $\overline{X}$ and sample variance S^2 are unbiased estimators of μ and σ^2, respectively.

First consider the sample mean. In Chapter 3, we indicated that $E(\overline{X}) = \mu$. Therefore, the sample mean $\overline{X}$ is an unbiased estimator of the population mean μ.

Now consider the sample variance. We have

$$E(S^2) = E\left[\frac{\sum\limits_{i=1}^{n}(X_i - \overline{X})^2}{n-1}\right] = \frac{1}{n-1}E\sum\limits_{i=1}^{n}(X_i - \overline{X})^2$$

$$= \frac{1}{n-1}E\sum\limits_{i=1}^{n}(X_i^2 + \overline{X}^2 - 2\overline{X}X_i) = \frac{1}{n-1}E\left(\sum\limits_{i=1}^{n}X_i^2 - n\overline{X}^2\right)$$

$$= \frac{1}{n-1}\left[\sum\limits_{i=1}^{n}E(X_i^2) - nE(\overline{X}^2)\right]$$

The last equality follows from equation 3-21 in Chapter 3. However, since $E(X_i^2) = \mu^2 + \sigma^2$ and $E(\overline{X}^2) = \mu^2 + \sigma^2/n$, we have

$$E(S^2) = \frac{1}{n-1}\left[\sum\limits_{i=1}^{n}(\mu^2 + \sigma^2) - n\left(\mu^2 + \frac{\sigma^2}{n}\right)\right]$$

$$= \frac{1}{n-1}(n\mu^2 + n\sigma^2 - n\mu^2 - \sigma^2)$$

$$= \sigma^2$$

Therefore, the sample variance S^2 is an unbiased estimator of the population variance σ^2. However, we can show that the sample standard deviation S is a biased estimator of the population standard deviation. For large samples this bias is negligible.

Sometimes there are several unbiased estimators of the sample population parameter. For example, suppose we take a random sample of size $n = 10$ from a normal population and obtain the data $x_1 = 12.8$, $x_2 = 9.4$, $x_3 = 8.7$, $x_4 = 11.6$, $x_5 = 13.1$, $x_6 = 9.8$, $x_7 = 14.1$, $x_8 = 8.5$, $x_9 = 12.1$, $x_{10} = 10.3$. Now the sample mean is

$$\bar{x} = \frac{12.8 + 9.4 + 8.7 + 11.6 + 13.1 + 9.8 + 14.1 + 8.5 + 12.1 + 10.3}{10}$$
$$= 11.04$$

the sample median is

$$\tilde{x} = \frac{10.3 + 11.6}{2} = 10.95$$

and a single observation from this normal population, say, $x_1 = 12.8$.

We can show that all of these values result from unbiased estimators of μ. Since there is not a unique unbiased estimator, we cannot rely on the property of unbiasedness alone to select our estimator. We need a method to select among unbiased estimators.

Suppose that $\hat{\Theta}_1$ and $\hat{\Theta}_2$ are unbiased estimators of θ. This indicates that the distribution of each estimator is centered at the true value of θ. However, the variances of these distributions may be different. Figure 4-2 illustrates the situation. Since $\hat{\Theta}_1$ has a smaller variance than $\hat{\Theta}_2$, the estimator $\hat{\Theta}_1$ is more likely to produce an estimate close to the true value θ. A logical principle of estimation, when selecting among several estimators, is to choose the estimator that has minimum variance.

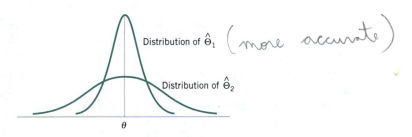

Figure 4-2   The sampling distributions of two unbiased estimators $\hat{\Theta}_1$ and $\hat{\Theta}_2$.

Definition

If we consider all unbiased estimators of θ, the one with the smallest variance is called the **minimum variance unbiased estimator** (MVUE).

The concepts of an unbiased estimator and an estimator with minimum variance are extremely important. There are methods for formally deriving estimates of the parameters of a probability distribution. One of these methods, the **method of maximum likelihood,** produces point estimators that are approximately unbiased and very close to the minimum variance estimator. For further information on the method of maximum likelihood, see Montgomery and Runger (1999).

In practice, one must occasionally use a biased estimator (such as S for σ). In such cases, the mean square error of the estimator can be important. The **mean square error** of an estimator $\hat{\Theta}$ is the expected squared difference between $\hat{\Theta}$ and θ.

Definition

The **mean square error** of an estimator $\hat{\Theta}$ of the parameter θ is defined as

$$\text{MSE}(\hat{\Theta}) = E(\hat{\Theta} - \theta)^2 \tag{4-3}$$

The mean square error can be rewritten as follows:

$$\text{MSE}(\hat{\Theta}) = E[\hat{\Theta} - E(\hat{\Theta})]^2 + [\theta - E(\hat{\Theta})]^2$$
$$= V(\hat{\Theta}) + (\text{bias})^2$$

That is, the mean square error of $\hat{\Theta}$ is equal to the variance of the estimator plus the squared bias. If $\hat{\Theta}$ is an unbiased estimator of θ, the mean square error of $\hat{\Theta}$ is equal to the variance of $\hat{\Theta}$.

The mean square error is an important criterion for comparing two estimators. Let $\hat{\Theta}_1$ and $\hat{\Theta}_2$ be two estimators of the parameter θ, and let $\text{MSE}(\hat{\Theta}_1)$ and $\text{MSE}(\hat{\Theta}_2)$ be the mean square errors of $\hat{\Theta}_1$ and $\hat{\Theta}_2$. Then the **relative efficiency** of $\hat{\Theta}_2$ to $\hat{\Theta}_1$ is defined as

$$\frac{\text{MSE}(\hat{\Theta}_1)}{\text{MSE}(\hat{\Theta}_2)} \tag{4-4}$$

If this relative efficiency is less than one, we would conclude that $\dot{\Theta}_1$ is a more efficient estimator of θ than $\hat{\Theta}_2$, in the sense that it has smaller mean square error.

Previously, we suggested several estimators of μ: the sample average, the sample median, and a single observation. Because the variance of the sample median is somewhat awkward to work with, we consider only the sample mean $\hat{\Theta}_1 = \overline{X}$ and $\hat{\Theta}_2 = X_i$. Note that both $\overline{X}$ and X_i are unbiased estimators of μ; consequently, the mean square error of both estimators is simply the variance. For the sample mean, we have $\text{MSE}(\overline{X}) = V(\overline{X}) = \sigma^2/n$ from equation 3-23. Therefore, the **relative efficiency** of X_i to $\overline{X}$ is

$$\frac{\text{MSE}(\hat{\Theta}_1)}{\text{MSE}(\hat{\Theta}_2)} = \frac{\sigma^2/n}{\sigma^2} = \frac{1}{n}$$

Since $(1/n) < 1$ for sample sizes $n \geq 2$, we would conclude that the sample mean is a better estimator of μ than a single observation X_i. This is an important point, because it illustrates why, in general, large samples are preferable to small ones for many kinds of statistics problems.

The variance of an estimator, $V(\hat{\Theta})$, can be thought of as the variance of the sampling distribution of $\hat{\Theta}$. The square root of this quantity, $\sqrt{V(\hat{\Theta})}$, is usually called the standard error of the estimator.

Definition

The **standard error** of a statistic is the standard deviation of its sampling distribution. If the standard error involves unknown parameters whose values can be estimated, substitution of these estimates into the standard error results in an **estimated standard error**.

The standard error gives some idea about the **precision of estimation.** For example, if the sample mean $\overline{X}$ is used as a point estimator of the population mean μ, the standard error of $\overline{X}$ measures how precisely $\overline{X}$ estimates μ.

Suppose we are sampling from a normal distribution with mean μ and variance σ^2. Now the distribution of $\overline{X}$ is normal with mean μ and variance σ^2/n, and so the standard error of $\overline{X}$ is

$$\sigma_{\overline{X}} = \frac{\sigma}{\sqrt{n}}$$

If we did not know σ but substituted the sample standard deviation S into the above equation, then the estimated standard error of $\overline{X}$ is

$$\hat{\sigma}_{\overline{X}} = \frac{S}{\sqrt{n}}$$

To illustrate this definition, an article in the *Journal of Heat Transfer* (*Trans. ASME, Ses. C, 96*, 1974, p. 59) described a new method of measuring the thermal conductivity of Armco iron. Using a temperature of 100°F and a power input of 550 W, the following 10 measurements of thermal conductivity (in Btu/hr-ft-°F) were obtained:

$$41.60, 41.48, 42.34, 41.95, 41.86,$$
$$42.18, 41.72, 42.26, 41.81, 42.04$$

A point estimate of the mean thermal conductivity at 100°F and 550 W is the sample mean or

$$\bar{x} = 41.924 \text{ Btu/hr-ft-°F}$$

The standard error of the sample mean is $\sigma_{\bar{X}} = \sigma/\sqrt{n}$, and since σ is unknown, we may replace it by the sample standard deviation $s = 0.284$ to obtain the estimated standard error of $\bar{X}$ as

$$\hat{\sigma}_{\bar{X}} = \frac{s}{\sqrt{n}} = \frac{0.284}{\sqrt{10}} = 0.0898$$

Note that the standard error is about 0.2% of the sample mean, implying that we have obtained a relatively precise point estimate of thermal conductivity.

EXERCISES FOR SECTION 4-2

4-1.   Suppose we have a random sample of size $2n$ from a population denoted by X, and $E(X) = \mu$ and $V(X) = \sigma^2$. Let

$$\bar{X}_1 = \frac{1}{2n} \sum_{i=1}^{2n} X_i \quad \text{and} \quad \bar{X}_2 = \frac{1}{n} \sum_{i=1}^{n} X_i$$

be two estimators of μ. Which is the better estimator of μ? Explain your choice.

4-2.   Let $X_1, X_2, \ldots, X_9$ denote a random sample from a population having mean μ and variance σ^2. Consider the following estimators of μ:

$$\hat{\Theta}_1 = \frac{X_1 + X_2 + \cdots + X_9}{9}$$

$$\hat{\Theta}_2 = \frac{3X_1 - X_6 + 2X_4}{2}$$

(a) Is either estimator unbiased?

(b) Which estimator is "best"? In what sense is it best?

4-3.   Suppose that $\hat{\Theta}_1$ and $\hat{\Theta}_2$ are unbiased estimators of the parameter θ. We know that $V(\hat{\Theta}_1) = 2$ and $V(\hat{\Theta}_2) = 4$. Which estimator is better and in what sense is it better?

4-4.   Calculate the relative efficiency of the two estimators in Exercise 4-2.

4-5.   Calculate the relative efficiency of the two estimators in Exercise 4-3.

4-6.   Suppose that $\hat{\Theta}_1$ and $\hat{\Theta}_2$ are estimators of the parameter θ. We know that $E(\hat{\Theta}_1) = \theta$, $E(\hat{\Theta}_2) = \theta/2$, $V(\hat{\Theta}_1) = 10$, $V(\hat{\Theta}_2) = 4$. Which estimator is "best"? In what sense is it best?

4-7.   Suppose that $\hat{\Theta}_1$, $\hat{\Theta}_2$, and $\hat{\Theta}_3$ are estimators of θ. We know that $E(\hat{\Theta}_1) = E(\hat{\Theta}_2) = \theta$, $E(\hat{\Theta}_3) \neq \theta$, $V(\hat{\Theta}_1) = 16$, $V(\hat{\Theta}_2) = 11$, and $E(\hat{\Theta}_3 - \theta)^2 = 6$. Compare these three estimators. Which do you prefer? Why?

4-8.   Let three random samples of sizes $n_1 = 20$, $n_2 = 10$, and $n_3 = 8$ be taken from a population with mean μ and variance σ^2. Let S_1^2, S_2^2, and S_3^2 be the sample variances. Show that $S^2 = (20S_1^2 + 10S_2^2 + 8S_3^2)/38$ is an unbiased estimator of σ^2.

4-9. (a) Show that $\sum_{i=1}^{n} (X_i - \overline{X})^2/n$ is a biased estimator of σ^2.
(b) Find the amount of bias in the estimator.
(c) What happens to the bias as the sample size n increases?

4-10. Let $X_1, X_2, \ldots, X_n$ be a random sample of size n.
(a) Show that $\overline{X}^2$ is a biased estimator for μ^2.
(b) Find the amount of bias in this estimator.
(c) What happens to the bias as the sample size n increases?

4-3 HYPOTHESIS TESTING

4-3.1 Statistical Hypotheses

In the previous section we illustrated how a parameter can be estimated from sample data. However, many problems in engineering require that we decide whether to accept or reject a statement about some parameter. The statement is called a **hypothesis,** and the decision-making procedure about the hypothesis is called **hypothesis testing.** This is one of the most useful aspects of statistical inference, since many types of decision-making problems, tests, or experiments in the engineering world can be formulated as hypothesis testing problems. We like to think of statistical hypothesis testing as the data analysis stage of a **comparative experiment,** in which the engineer is interested, for example, in comparing the mean of a population to a specified value. These simple comparative experiments are frequently encountered in practice and provide a good foundation for the more complex experimental design problems that we will discuss in Chapter 7. In this chapter we discuss comparative experiments involving one population, and our focus is on testing hypotheses concerning the parameters of the population.

We now give a formal definition of a statistical hypothesis.

Definition

A **statistical hypothesis** is a statement about the parameters of one or more populations.

Since we use probability distributions to represent populations, a statistical hypothesis may also be thought of as a statement about the probability distribution of a random variable. The hypothesis will usually involve one or more parameters of this distribution.

For example, suppose that we are interested in the burning rate of a solid propellant used to power aircrew escape systems; burning rate is a random variable that can be described by a probability distribution. Suppose that our interest focuses on the mean burning rate (a parameter of this distribution). Specifically, we are interested in deciding whether or not the mean burning rate is 50 cm/s. We may express this formally as

$$H_0: \mu = 50 \text{ cm/s}$$
$$H_1: \mu \neq 50 \text{ cm/s} \tag{4-5}$$

The statement $H_0: \mu = 50$ cm/s in equation 4-5 is called the **null hypothesis,** and the statement $H_1: \mu \neq 50$ cm/s is called the **alternative hypothesis.** Since the alternative hypothesis specifies values of μ that could be either greater or less than 50 cm/s, it is called a **two-sided alternative hypothesis.** In some situations, we may wish to formulate a **one-sided alternative hypothesis,** as in

$$H_0: \mu = 50 \text{ cm/s} \qquad H_0: \mu = 50 \text{ cm/s}$$
$$\text{or} \tag{4-6}$$
$$H_1: \mu < 50 \text{ cm/s} \qquad H_1: \mu > 50 \text{ cm/s}$$

It is important to remember that hypotheses are always statements about the population or distribution under study, not statements about the sample. The value of the population parameter specified in the null hypothesis (50 cm/s in the above example) is usually determined in one of three ways. First, it may result from past experience or knowledge of the process, or even from previous tests or experiments. The objective of hypothesis testing then is usually to determine whether the parameter value has changed. Second, this value may be determined from some theory or model regarding the process under study. Here the objective of hypothesis testing is to verify the theory or model. A third situation arises when the value of the population parameter results from external considerations, such as design or engineering specifications, or from contractual obligations. In this situation, the usual objective of hypothesis testing is conformance testing.

A procedure leading to a decision about a particular hypothesis is called a **test of a hypothesis.** Hypothesis testing procedures rely on using the information in a random sample from the population of interest. If this information is consistent with the hypothesis, then we will conclude that the hypothesis is true; however, if this information is inconsistent with the hypothesis, we will conclude that the hypothesis is false. We emphasize that the truth or falsity of a particular hypothesis can never be known with certainty unless we can examine the entire population. This is usually impossible in most practical situations. Therefore, a hypothesis testing procedure should be developed with the probability of reaching a wrong conclusion in mind.

The structure of hypothesis testing problems is identical in all the applications that we will consider. The null hypothesis is the hypothesis we wish to test. Rejection of the null hypothesis always leads to accepting the alternative hypothesis. In our treatment of hypothesis testing, the null hypothesis will always be stated so that it specifies an exact value of the parameter (as in the statement $H_0: \mu = 50$ cm/s in equation 4-5). The alternative hypothesis will allow the parameter to take on several values (as in the statement H_1: $\mu \neq 50$ cm/s in equation 4-5). Testing the hypothesis involves taking a random sample, computing a **test statistic** from the sample data, and then using the test statistic to make a decision about the null hypothesis.

4-3.2   Testing Statistical Hypotheses

To illustrate the general concepts, consider the propellant burning rate problem introduced earlier. The null hypothesis is that the mean burning rate is 50 cm/s, and the alternative is that it is not equal to 50 cm/s. That is, we wish to test

$$H_0: \mu = 50 \text{ cm/s}$$

$$H_1: \mu \neq 50 \text{ cm/s}$$

Suppose that a sample of $n = 10$ specimens is tested and that the sample mean burning rate $\bar{x}$ is observed. The sample mean is an estimate of the true population mean μ. A value of the sample mean $\bar{x}$ that falls close to the hypothesized value of $\mu = 50$ cm/s is evidence that the true mean μ is really 50 cm/s; that is, such evidence supports the null hypothesis H_0. On the other hand, a sample mean that is considerably different from 50 cm/s is evidence in support of the alternative hypothesis H_1. Thus, the sample mean is the test statistic in this case.

The sample mean can take on many different values. Suppose that if $48.5 \leq \bar{x} \leq 51.5$, we will not reject the null hypothesis H_0: $\mu = 50$, and if either $\bar{x} < 48.5$ or $\bar{x} > 51.5$, we will reject the null hypothesis in favor of the alternative hypothesis H_1: $\mu \neq 50$. This situation is illustrated in Fig. 4-3. The values of $\bar{x}$ that are less than 48.5 and greater than 51.5 constitute the **critical region** for the test, whereas all values that are in the interval $48.5 \leq \bar{x} \leq 51.5$ form a region for which we will fail to reject the null hypothesis. By convention, this is usually called the **acceptance region.** The boundaries between the critical regions and acceptance region are called the **critical values.** In our example the critical values are 48.5 and 51.5. It is customary to state conclusions relative to the null hypothesis H_0. Therefore, we reject H_0 in favor of H_1 if the test statistic falls in the critical region and fail to reject H_0 otherwise.

This decision procedure can lead to either of two wrong conclusions. For example, the true mean burning rate of the propellant could be equal to 50 cm/s. However, for the

Reject H_0	Fail to Reject H_0	Reject H_0
$\mu \neq 50$ cm/s	$\mu = 50$ cm/s	$\mu \neq 50$ cm/s

Figure 4-3   Decision criteria for testing H_0: $\mu = 50$ cm/s versus H_1: $\mu \neq 50$ cm/s.

randomly selected propellant specimens that are tested, we could observe a value of the test statistic $\bar{x}$ that falls into the critical region. We would then reject the null hypothesis H_0 in favor of the alternative H_1 when, in fact, H_0 is really true. This type of wrong conclusion is called a **type I error.**

ALPHA
ERROR

α

> **Definition**
>
> Rejecting the null hypothesis H_0 when it is true is defined as a **type I error.**

Now suppose that the true mean burning rate is different from 50 cm/s, yet the sample mean $\bar{x}$ falls in the acceptance region. In this case we would fail to reject H_0 when it is false. This type of wrong conclusion is called a **type II error.**

BETA

error

β

> **Definition**
>
> Failing to reject the null hypothesis when it is false is defined as a **type II error.**

Thus, in testing any statistical hypothesis, four different situations determine whether the final decision is correct or in error. These situations are presented in Table 4-1.

Because our decision is based on random variables, probabilities can be associated with the type I and type II errors in Table 4-1. The probability of making a type I error is denoted by the Greek letter α (alpha). That is,

$$\alpha = P(\text{type I error}) = P(\text{reject } H_0 \text{ when } H_0 \text{ is true}) \tag{4-7}$$

Sometimes the type I error probability is called the **significance level** or **size** of the test. In the propellant burning rate example, a type I error will occur when either $\bar{x} > 51.5$ or $\bar{x} < 48.5$ when the true mean burning rate is $\mu = 50$ cm/s. Suppose that the standard deviation of burning rate is $\sigma = 2.5$ cm/s and that the burning rate has a distribution for which the conditions of the central limit theorem apply, so that the distribution of the sample mean is approximately normal with mean $\mu = 50$ and standard deviation

Table 4-1   Decisions in Hypothesis Testing

Decision	H_0 Is True	H_0 Is False
Fail to reject H_0	no error	type II error
Reject H_0	type I error	no error

$\sigma/\sqrt{n} = 2.5/\sqrt{10} = 0.79$. The probability of making a type I error (or the significance level of our test) is equal to the sum of the areas that have been shaded in the tails of the normal distribution in Fig. 4-4. We may find this probability as

$$\alpha = P(\overline{X} < 48.5 \text{ when } \mu = 50) + P(\overline{X} > 51.5 \text{ when } \mu = 50)$$

The z-values that correspond to the critical values 48.5 and 51.5 are

$$z_1 = \frac{48.5 - 50}{0.79} = -1.90$$

and

$$z_2 = \frac{51.5 - 50}{0.79} = 1.90$$

Therefore,

$$\alpha = P(Z < -1.90) + P(Z > 1.90)$$
$$= 0.0288 + 0.0288$$
$$= 0.0576$$

This implies that 5.76% of all random samples would lead to rejection of the hypothesis H_0: $\mu = 50$ cm/s when the true mean burning rate is really 50 cm/s.

From inspection of Fig. 4-4, note that we can reduce α by widening the acceptance region. For example, if we make the critical values 48 and 52, the value of α is

$$\alpha = P\left(Z < \frac{48 - 50}{0.79}\right) + P\left(Z > \frac{52 - 50}{0.79}\right)$$
$$= P(Z < -2.53) + P(Z > 2.53)$$
$$= 0.0057 + 0.0057$$
$$= 0.0114$$

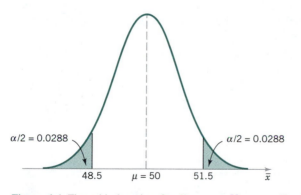

Figure 4-4 The critical region for H_0: $\mu = 50$ versus H_1: $\mu \neq 50$ and $n = 10$.

We could also reduce α by increasing the sample size. If $n = 16$, then $\sigma/\sqrt{n} = 2.5/\sqrt{16} = 0.625$, and using the original critical region in Fig. 4-3, we find

$$z_1 = \frac{48.5 - 50}{0.625} = -2.40$$

and

$$z_2 = \frac{51.5 - 50}{0.625} = 2.40$$

Therefore,

$$\alpha = P(Z < -2.40) + P(Z > 2.40)$$
$$= 0.0082 + 0.0082$$
$$= 0.0164$$

In evaluating a hypothesis testing procedure, it is also important to examine the probability of a type II error, which we will denote by β (beta). That is,

$$\beta = P(\text{type II error}) = P(\text{fail to reject } H_0 \text{ when } H_0 \text{ is false}) \tag{4-8}$$

To calculate β, we must have a specific alternative hypothesis; that is, we must have a particular value of μ. For example, suppose that it is important to reject the null hypothesis H_0: $\mu = 50$ whenever the mean burning rate μ is greater than 52 cm/s or less than 48 cm/s. We could calculate the probability of a type II error β for the values $\mu = 52$ and $\mu = 48$ and use this result to tell us something about how the test procedure would perform. Specifically, how will the test procedure work if we wish to detect—that is, reject H_0—for a mean value of $\mu = 52$ or $\mu = 48$? Because of symmetry, it is only necessary to evaluate one of the two cases—say, find the probability of accepting the null hypothesis H_0: $\mu = 50$ cm/s when the true mean is $\mu = 52$ cm/s.

Figure 4-5 will help us calculate the probability of type II error β. The normal distribution on the left in Fig. 4-5 is the distribution of the test statistic $\overline{X}$ when the null hypothesis H_0: $\mu = 50$ is true (this is what is meant by the expression "under H_0: $\mu = 50$"), and the normal distribution on the right is the distribution of $\overline{X}$ when the alternative hypothesis is true and the value of the mean is 52 (or "under H_1: $\mu = 52$"). Now a type II error will be committed if the sample mean $\overline{x}$ falls between 48.5 and 51.5 (the critical region boundaries) when $\mu = 52$. As seen in Fig. 4-5, this is just the probability that $48.5 \leq \overline{X} \leq 51.5$ when the true mean is $\mu = 52$, or the shaded area under the normal distribution on the right. Therefore, referring to Fig. 4-5, we find that

$$\beta = P(48.5 \leq \overline{X} \leq 51.5 \text{ when } \mu = 52)$$

The z-values corresponding to 48.5 and 51.5 when $\mu = 52$ are

$$z_1 = \frac{48.5 - 52}{0.79} = -4.43$$

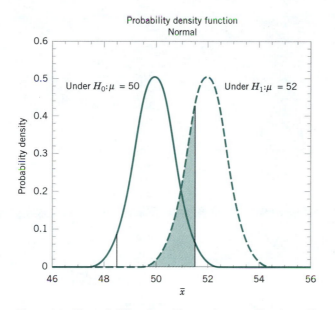

Probability density function
Normal

Under $H_0: \mu = 50$

Under $H_1: \mu = 52$

Figure 4-5 The probability of type II error when $\mu = 52$ and $n = 10$.

and

$$z_2 = \frac{51.5 - 52}{0.79} = -0.63$$

Therefore,

$$
\begin{aligned}
\beta &= P(-4.43 \le Z \le -0.63) \\
&= P(Z \le -0.63) - P(Z \le -4.43) \\
&= 0.2643 - 0.000 \\
&= 0.2643
\end{aligned}
$$

Thus, if we are testing $H_0: \mu = 50$ against $H_1: \mu \ne 50$ with $n = 10$, and the true value of the mean is $\mu = 52$, the probability that we will fail to reject the false null hypothesis is 0.2643. By symmetry, if the true value of the mean is $\mu = 48$, the value of β will also be 0.2643.

The probability of making a type II error β increases rapidly as the true value of μ approaches the hypothesized value. For example, see Fig. 4-6, where the true value of the mean is $\mu = 50.5$ and the hypothesized value is $H_0: \mu = 50$. The true value of μ is very close to 50, and the value for β is

$$\beta = P(48.5 \le \overline{X} \le 51.5 \text{ when } \mu = 50.5)$$

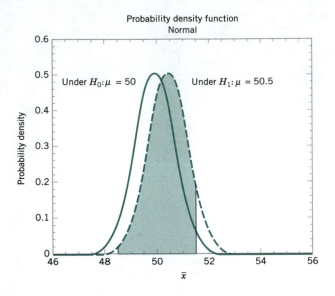

Figure 4-6   The probability of type II error when $\mu = 50.5$ and $n = 10$.

As shown in Fig. 4-6, the z-values corresponding to 48.5 and 51.5 when $\mu = 50.5$ are

$$z_1 = \frac{48.5 - 50.5}{0.79} = -2.53$$

and

$$z_2 = \frac{51.5 - 50.5}{0.79} = 1.27$$

Therefore,

$$
\begin{aligned}
\beta &= P(-2.53 \leq Z \leq 1.27) \\
&= P(Z \leq 1.27) - P(Z \leq -2.53) \\
&= 0.8980 - 0.0057 \\
&= 0.8923
\end{aligned}
$$

Thus, the type II error probability is much higher for the case in which the true mean is 50.5 cm/s than for the case in which the mean is 52 cm/s. Of course, in many practical situations we would not be as concerned with making a type II error if the mean were "close" to the hypothesized value. We would be much more interested in detecting large differences between the true mean and the value specified in the null hypothesis.

The type II error probability also depends on the sample size n. Suppose that the null hypothesis is H_0: $\mu = 50$ cm/s and that the true value of the mean is $\mu = 52$. If the sample size is increased from $n = 10$ to $n = 16$, the situation of Fig. 4-7 results. The normal distribution on the left is the distribution of $\overline{X}$ when the mean $\mu = 50$, and the normal distribution on the right is the distribution of $\overline{X}$ when $\mu = 52$. As shown in Fig. 4-7, the type II error probability is

$$\beta = P(48.5 \leq \overline{X} \leq 51.5 \text{ when } \mu = 52)$$

When $n = 16$, the standard deviation of $\overline{X}$ is $\sigma/\sqrt{n} = 2.5/\sqrt{16} = 0.625$, and the z-values corresponding to 48.5 and 51.5 when $\mu = 52$ are

$$z_1 = \frac{48.5 - 52}{0.625} = -5.60 \quad \text{and} \quad z_2 = \frac{51.5 - 52}{0.625} = -0.80$$

Therefore,

$$\begin{aligned} \beta &= P(-5.60 \leq Z \leq -0.80) \\ &= P(Z \leq -0.80) - P(Z \leq -5.60) \\ &= 0.2119 - 0.000 \\ &= 0.2119 \end{aligned}$$

Recall that when $n = 10$ and $\mu = 52$ we found that $\beta = 0.2643$; therefore, increasing the sample size results in a decrease in the probability of type II error.

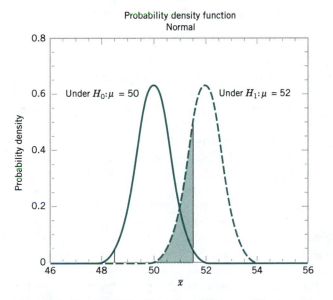

Figure 4-7   The probability of type II error when $\mu = 52$ and $n = 16$.

The results from this section and a few other similar calculations are summarized next:

acceptance region	sample size	α	β at $\mu = 52$	β at $\mu = 50.5$
$48.5 < \bar{x} < 51.5$	10	0.0576	0.2643	0.8923
$48 < \bar{x} < 52$	10	0.0114	0.5000	0.9705
$48.5 < \bar{x} < 51.5$	16	0.0164	0.2119	0.9445
$48 < \bar{x} < 52$	16	0.0014	0.5000	0.9918

The results in boxes were not calculated in the text but can be easily verified by the reader. This display and the preceding discussion reveal four important points:

1. The size of the critical region, and consequently the probability of a type I error α, can always be reduced by appropriate selection of the critical values.

2. Type I and type II errors are related. A decrease in the probability of one type of error always results in an increase in the probability of the other, provided that the sample size n does not change.

3. An increase in sample size will generally reduce both α and β, provided that the critical values are held constant.

4. When the null hypothesis is false, β increases as the true value of the parameter approaches the value hypothesized in the null hypothesis. The value of β decreases as the difference between the true mean and the hypothesized value increases.

Generally, the analyst controls the type I error probability α when he or she selects the critical values. Thus, it is usually easy for the analyst to set the type I error probability at (or near) any desired value. Since the analyst can directly control the probability of wrongly rejecting H_0, we always think of rejection of the null hypothesis H_0 as a **strong conclusion.**

On the other hand, the probability of type II error β is not a constant, but depends on the true value of the parameter. It also depends on the sample size that we have selected. Because the type II error probability β is a function of both the sample size and the extent to which the null hypothesis H_0 is false, it is customary to think of the decision to accept H_0 as a **weak conclusion,** unless we know that β is acceptably small. Therefore, rather than saying we "accept H_0," we prefer the terminology "fail to reject H_0." Failing to reject H_0 implies that we have not found sufficient evidence to reject H_0—that is, to make a strong statement. Failing to reject H_0 does not necessarily mean there is a high probability that H_0 is true. It may simply mean that more data are required to reach a strong conclusion. This can have important implications for the formulation of hypotheses.

An important concept that we will make use of is the **power** of a statistical test.

Definition

The **power** of a statistical test is the probability of rejecting the null hypothesis H_0 when the alternative hypothesis is true.

The power is computed as $1 - \beta$, and power can be interpreted as *the probability of correctly rejecting a false null hypothesis.* We often compare statistical tests by comparing their power properties. For example, consider the propellant burning rate problem when we are testing H_0: $\mu = 50$ cm/s against H_1: $\mu \neq 50$ cm/s. Suppose that the true value of the mean is $\mu = 52$. When $n = 10$, we found that $\beta = 0.2643$, so the power of this test is $1 - \beta = 1 - 0.2643 = 0.7357$ when $\mu = 52$.

Power is a very descriptive and concise measure of the **sensitivity** of a statistical test, where by sensitivity we mean the ability of the test to detect differences. In this case, the sensitivity of the test for detecting the difference between a mean burning rate of 50 cm/s and 52 cm/s is 0.7357. That is, if the true mean is really 52 cm/s, this test will correctly reject H_0: $\mu = 50$ and "detect" this difference 73.57% of the time. If this value of power is judged to be too low, the analyst can increase either α or the sample size n.

4-3.3   One-Sided and Two-Sided Hypotheses

A test of any hypothesis such as

$$H_0: \mu = \mu_0$$

$$H_1: \mu \neq \mu_0$$

is called a **two-sided** test, because it is important to detect differences from the hypothesized value of the mean μ_0 that lie on either side of μ_0. In such a test, the critical region is split into two parts, with (usually) equal probability placed in each tail of the distribution of the test statistic.

Many hypothesis testing problems naturally involve a **one-sided** alternative hypothesis, such as

$$H_0: \mu = \mu_0$$

$$H_1: \mu > \mu_0$$

or

$$H_0: \mu = \mu_0$$

$$H_1: \mu < \mu_0$$

If the alternative hypothesis is $H_1: \mu > \mu_0$, the critical region should lie in the upper tail of the distribution of the test statistic, whereas if the alternative hypothesis is $H_1: \mu < \mu_0$, the critical region should lie in the lower tail of the distribution. Consequently, these tests are sometimes called **one-tailed** tests. The location of the critical region for one-sided tests is usually easy to determine. Simply visualize the behavior of the test statistic if the null hypothesis is true and place the critical region in the appropriate end or tail of the distribution. Generally, the inequality in the alternative hypothesis "points" in the direction of the critical region.

In constructing hypotheses, we will always state the null hypothesis as an equality, so that the probability of type I error α can be controlled at a specific value. The alternative hypothesis might be either one-sided or two-sided, depending on the conclusion to be drawn if H_0 is rejected. If the objective is to make a claim involving statements such as "greater than," "less than," "superior to," "exceeds," "at least," and so forth, then a one-sided alternative is appropriate. If no direction is implied by the claim or if the claim "not equal to" is to be made, then a two-sided alternative should be used.

EXAMPLE 4-2

Consider the propellant burning rate problem. Suppose that if the burning rate is less than 50 cm/s, we wish to show this with a strong conclusion. The hypotheses should be stated as

$$H_0: \mu = 50 \text{ cm/s}$$
$$H_1: \mu < 50 \text{ cm/s}$$

Here the critical region lies in the lower tail of the distribution of $\overline{X}$. Since the rejection of H_0 is always a strong conclusion, this statement of the hypotheses will produce the desired outcome if H_0 is rejected. Note that, although the null hypothesis is stated with an equal sign, it is understood to include any value of μ not specified by the alternative hypothesis. Therefore, failing to reject H_0 does not mean that $\mu = 50$ cm/s exactly, but only that we do not have strong evidence in support of H_1.

In some real-world problems where one-sided test procedures are indicated, it is occasionally difficult to choose an appropriate formulation of the alternative hypothesis. For example, suppose that a soft-drink beverage bottler purchases 10-ounce bottles from a glass company. The bottler wants to be sure that the bottles meet the specification on mean internal pressure or bursting strength, which for 10-ounce bottles is a minimum strength of 200 psi. The bottler has decided to formulate the decision procedure for a specific lot of bottles as a hypothesis problem. There are two possible formulations for this problem, either

$$H_0: \mu = 200 \text{ psi}$$
$$H_1: \mu > 200 \text{ psi} \tag{4-9}$$

or

$$H_0: \mu = 200 \text{ psi}$$

$$H_1: \mu < 200 \text{ psi} \qquad (4\text{-}10)$$

Consider the formulation in equation 4-9. If the null hypothesis is rejected, the bottles will be judged satisfactory, whereas if H_0 is not rejected, the implication is that the bottles do not conform to specifications and should not be used. Because rejecting H_0 is a strong conclusion, this formulation forces the bottle manufacturer to "demonstrate" that the mean bursting strength of the bottles exceeds the specification. Now consider the formulation in equation 4-10. In this situation, the bottles will be judged satisfactory unless H_0 is rejected. That is, we conclude that the bottles are satisfactory unless there is strong evidence to the contrary.

Which formulation is correct, the one of equation 4-9 or of equation 4-10? The answer is, "it depends." For equation 4-9, there is some probability that H_0 will not be rejected (i.e., we would decide that the bottles are not satisfactory), even though the true mean is slightly greater than 200 psi. This formulation implies that we want the bottle manufacturer to demonstrate that the product meets or exceeds our specifications. Such a formulation could be appropriate if the manufacturer has experienced difficulty in meeting specifications in the past, or if product safety considerations force us to hold tightly to the 200 psi specification. On the other hand, for the formulation of equation 4-10 there is some probability that H_0 will be accepted and the bottles judged satisfactory, even though the true mean is slightly less than 200 psi. We would conclude that the bottles are unsatisfactory only when there is strong evidence that the mean does not exceed 200 psi—that is, when $H_0: \mu = 200 \text{ psi}$ is rejected. This formulation assumes that we are relatively happy with the bottle manufacturer's past performance and that small deviations from the specification of $\mu \geq 200 \text{ psi}$ are not harmful.

In formulating one-sided alternative hypotheses, we should remember that rejecting H_0 is always a strong conclusion. Consequently, **we should put the statement about which it is important to make a strong conclusion in the alternative hypothesis.** In real-world problems, this will often depend on our point of view and experience with the situation.

4-3.4   General Procedure for Hypothesis Testing

This chapter develops hypothesis testing procedures for many practical problems. Use of the following sequence of steps in applying hypothesis testing methodology is recommended.

1. From the problem context, identify the parameter of interest.
2. State the null hypothesis, H_0.
3. Specify an appropriate alternative hypothesis, H_1.
4. Choose a significance level α.
5. State an appropriate test statistic.

6. State the rejection region for the statistic.

7. Compute any necessary sample quantities, substitute these into the equation for the test statistic, and compute that value.

8. Decide whether or not H_0 should be rejected and report that in the problem context.

Steps 1–4 should be completed prior to examination of the sample data. This sequence of steps will be illustrated in subsequent sections.

EXERCISES FOR SECTION 4-3

4-11. A textile fiber manufacturer is investigating a new drapery yarn, which the company claims has a mean thread elongation force of 14 kg with a standard deviation of 0.3 kg. The company wishes to test the hypothesis H_0: $\mu = 14$ against H_1: $\mu < 14$, using a random sample of five specimens.

(a) What is the type I error probability if the critical region is defined as $\bar{x} < 13.7$ kg?

(b) Find β for the case where the true mean elongation force is 13.5 kg.

4-12. Repeat Exercise 4-11 using a sample size of $n = 16$ and the same critical region.

4-13. In Exercise 4-11, find the boundary of the critical region if the type I error probability is specified to be $\alpha = 0.01$.

4-14. In Exercise 4-12, find the boundary of the critical region if the type I error probability is specified to be 0.05.

4-15. The heat evolved in calories per gram of a cement mixture is approximately normally distributed. The mean is thought to be 100 and the standard deviation is 2. We wish to test H_0: $\mu = 100$ versus H_1: $\mu \neq 100$ with a sample of $n = 9$ specimens.

(a) If the acceptance region is defined as $98.5 \leq \bar{x} \leq 101.5$, find the type I error probability α.

(b) Find β for the case where the true mean heat evolved is 103.

(c) Find β for the case where the true mean heat evolved is 105. This

value of β is smaller than the one found in part (b). Why?

4-16. Repeat Exercise 4-15 using a sample size of $n = 5$ and the same acceptance region.

4-17. A consumer products company is formulating a new shampoo and is interested in foam height (in mm). Foam height is approximately normally distributed and has a standard deviation of 20 mm. The company wishes to test H_0: $\mu = 175$ mm versus H_1: $\mu > 175$ mm, using the results of $n = 10$ samples.

(a) Find the type I error probability α if the critical region is $\bar{x} > 185$.

(b) What is the probability of type II error if the true mean foam height is 195 mm?

4-18. In Exercise 4-17, suppose that the sample data result in $\bar{x} = 190$ mm.

(a) What conclusion would you reach?

(b) How "unusual" is the sample value $\bar{x} = 190$ mm if the true mean is really 175 mm? That is, what is the probability that you would observe a sample average as large as 190 mm (or larger), if the true mean foam height was really 175 mm?

4-19. Repeat Exercise 4-17 assuming that the sample size is $n = 16$ and the boundary of the critical region is the same.

4-20. Consider Exercise 4-17, and suppose that the sample size is increased to $n = 16$.

(a) Where would the boundary of the

critical region be placed if the type I error probability were to remain equal to the value that it took on when $n = 10$?

(b) Using $n = 16$ and the new critical region found in part (a), find the type II error probability β if the true mean foam height is 195 mm.

(c) Compare the value of β obtained in part (b) with the value from Exercise 4-17 (b). What conclusions can you draw?

4-21.   A manufacturer is interested in the output voltage of a power supply used in a PC. Output voltage is assumed to be normally distributed, with standard deviation 0.25 V, and the manufacturer wishes to test H_0: $\mu = 9$ V against H_1: $\mu \neq 9$ V, using $n = 10$ units.

(a) The acceptance region is $8.85 \leq \bar{x} \leq 9.15$. Find the value of α.

(b) Find the power of the test for detecting a true mean output voltage of 9.1 V.

4-22.   Rework Exercise 4-21 when the sample size is 16 and the boundaries of the acceptance region do not change.

4-23.   Consider Exercise 4-21, and suppose that the manufacturer wants the type I error probability for the test to be $\alpha = 0.05$. Where should the acceptance region be located?

4-4   INFERENCE ON THE MEAN OF A POPULATION, VARIANCE KNOWN

In this section, we consider making inferences about the mean μ of a single population where the variance of the population σ^2 is known.

Assumptions

1. $X_1, X_2, \ldots, X_n$ is a random sample of size n from a population.
2. The population is normal, or if it is not normal, the conditions of the central limit theorem apply.

Based on our previous discussion in Section 4.2, the sample mean $\bar{X}$ is an **unbiased point estimator** of μ. With these assumptions, the distribution of $\bar{X}$ is approximately normal with mean μ and variance σ^2/n.

Under the previous assumptions, the quantity

$$Z = \frac{\bar{X} - \mu}{\sigma/\sqrt{n}} \tag{4-11}$$

has a standard normal distribution, $N(0,1)$.

4-4.1   Hypothesis Testing on the Mean

Suppose that we wish to test the hypotheses

$$H_0: \mu = \mu_0$$

$$H_1: \mu \neq \mu_0$$

where μ_0 is a specified constant. We have a random sample $X_1, X_2, \ldots, X_n$ from the population. Since $\overline{X}$ has an approximate normal distribution (i.e., the **sampling distribution** of $\overline{X}$ is approximately normal) with mean μ_0 and standard deviation $\sigma/\sqrt{n}$, if the null hypothesis is true, we could construct a critical region for the computed value of the sample mean $\overline{x}$, as in Section 4-3.1.

It is usually more convenient to *standardize* the sample mean and use a test statistic based on the standard normal distribution. That is, the test procedure for $H_0: \mu = \mu_0$ uses the **test statistic.**

$$Z_0 = \frac{\overline{X} - \mu_0}{\sigma/\sqrt{n}} \tag{4-12}$$

If the null hypothesis $H_0: \mu = \mu_0$ is true, then $E(\overline{X}) = \mu_0$, and it follows that the distribution of Z_0 is the standard normal distribution [denoted $N(0, 1)$]. Consequently, if $H_0: \mu = \mu_0$ is true, the probability is $1 - \alpha$ that the test statistic Z_0 falls between $-z_{\alpha/2}$ and $z_{\alpha/2}$, where $z_{\alpha/2}$ is the $100\alpha/2$ percentage point of the standard normal distribution. The regions associated with $z_{\alpha/2}$ and $-z_{\alpha/2}$ are illustrated in Fig. 4-8. Note that the probability is α that the test statistic Z_0 will fall in the region $Z_0 > z_{\alpha/2}$ or $Z_0 < -z_{\alpha/2}$ when $H_0: \mu = \mu_0$ is true. Clearly, a sample producing a value of the test statistic that falls in the tails of the distribution of Z_0 would be unusual if $H_0: \mu = \mu_0$ is true; therefore, it is an indication that H_0 is false. Thus, we should reject H_0 if either

$$z_0 > z_{\alpha/2} \tag{4-13}$$

or

$$z_0 < -z_{\alpha/2} \tag{4-14}$$

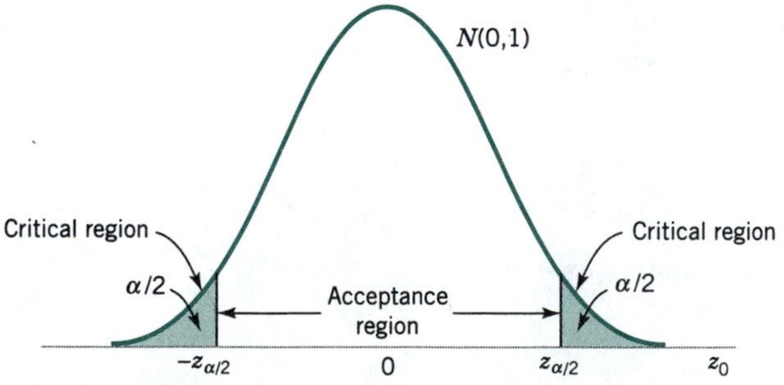

Figure 4-8   The distribution of Z_0 when $H_0: \mu = \mu_0$ is true, with critical region for $H_1: \mu \neq \mu_0$.

and we should fail to reject H_0 if

$$-z_{\alpha/2} \le z_0 \le z_{\alpha/2} \tag{4-15}$$

Equation 4-15 defines the **acceptance region** for H_0, and equations 4-13 and 4-14 define the **critical region** or **rejection region.** The type I error probability for this test procedure is α.

It is easier to understand the critical region and the test procedure, in general, when the test statistic is Z_0 rather than $\overline{X}$. However, the same critical region can always be written for the computed value of the sample mean $\overline{x}$. A procedure identical to the above is as follows:

$$\text{Reject } H_0: \mu = \mu_0 \text{ if either } \overline{x} > a \text{ or } \overline{x} < b$$

where

$$a = \mu_0 + z_{\alpha/2}\, \sigma/\sqrt{n} \tag{4-16}$$

$$b = \mu_0 - z_{\alpha/2}\, \sigma/\sqrt{n} \tag{4-17}$$

EXAMPLE 4-3

Aircrew escape systems are powered by a solid propellant. The burning rate of this propellant is an important product characteristic. Specifications require that the mean burning rate must be 50 cm/s. We know that the standard deviation of burning rate is $\sigma = 2$ cm/s. The experimenter decides to specify a type I error probability or significance level of $\alpha = 0.05$. He selects a random sample of $n = 25$ and obtains a sample average burning rate of $\overline{x} = 51.3$ cm/s. What conclusions should he draw?

We may solve this problem by following the eight-step procedure outlined in Section 4-3.4. This results in the following.

1. The parameter of interest is μ, the mean burning rate.

2. $H_0: \mu = 50$ cm/s

3. $H_1: \mu \ne 50$ cm/s

4. $\alpha = 0.05$

5. The test statistic is

$$z_0 = \frac{\overline{x} - \mu_0}{\sigma/\sqrt{n}}$$

6. Reject H_0 if $z_0 > 1.96$ or if $z_0 < -1.96$. Note that this results from step 4, where we specified $\alpha = 0.05$, and so the boundaries of the critical region are at $z_{0.025} = 1.96$ and $-z_{0.025} = -1.96$.

7. Computation: Since $\overline{x} = 51.3$ and $\sigma = 2$,

$$z_0 = \frac{51.3 - 50}{2/\sqrt{25}} = 3.25$$

8. Conclusion: Since $z_0 = 3.25 > 1.96$, we reject H_0: $\mu = 50$ at the 0.05 level of significance. Stated more completely, we conclude that the mean burning rate differs from 50 cm/s, based on a sample of 25 measurements. In fact, there is strong evidence that the mean burning rate exceeds 50 cm/s.

We may also develop procedures for testing hypotheses on the mean μ where the alternative hypothesis is one-sided. Suppose that we specify the hypotheses as

$$H_0: \mu = \mu_0$$

$$H_1: \mu > \mu_0 \qquad (4\text{-}18)$$

In defining the critical region for this test, we observe that a negative value of the test statistic Z_0 would never lead us to conclude that H_0: $\mu = \mu_0$ is false. Therefore, we would place the critical region in the upper tail of the standard normal distribution and reject H_0 if the computed value of z_0 is too large. That is, we would reject H_0 if

$$z_0 > z_\alpha \qquad (4\text{-}19)$$

Similarly, to test

$$H_0: \mu = \mu_0$$

$$H_1: \mu < \mu_0 \qquad (4\text{-}20)$$

we would calculate the test statistic Z_0 and reject H_0 if the value of Z_0 is too small. That is, the critical region is in the lower tail of the standard normal distribution and we reject H_0 if

$$z_0 < -z_\alpha \qquad (4\text{-}21)$$

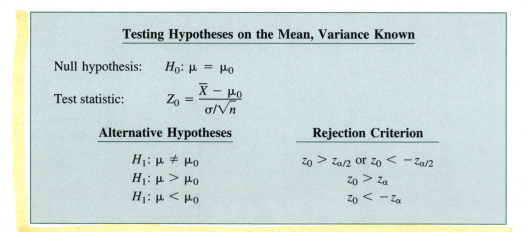

Testing Hypotheses on the Mean, Variance Known

Null hypothesis: H_0: $\mu = \mu_0$

Test statistic: $Z_0 = \dfrac{\overline{X} - \mu_0}{\sigma/\sqrt{n}}$

Alternative Hypotheses	Rejection Criterion
H_1: $\mu \neq \mu_0$	$z_0 > z_{\alpha/2}$ or $z_0 < -z_{\alpha/2}$
H_1: $\mu > \mu_0$	$z_0 > z_\alpha$
H_1: $\mu < \mu_0$	$z_0 < -z_\alpha$

4-4.2 *P*-Values in Hypothesis Testing

One way to report the results of a hypothesis test is to state that the null hypothesis was or was not rejected at a specified α-value or level of significance. For example, in the preceding propellant problem, we can say that $H_0: \mu = 50$ was rejected at the 0.05 level of significance. This statement of conclusions is often inadequate, because it gives the decision maker no idea about whether the computed value of the test statistic was just barely in the rejection region or whether it was very far into this region. Furthermore, stating the results this way imposes the predefined level of significance on other users of the information. This approach may be unsatisfactory, as some decision makers might be uncomfortable with the risks implied by $\alpha = 0.05$.

To avoid these difficulties the **P-value approach** has been adopted widely in practice. The *P*-value is the probability that the test statistic will take on a value that is at least as extreme as the observed value of the statistic when the null hypothesis H_0 is true. Thus, a *P*-value conveys much information about the weight of evidence against H_0, and so a decision maker can draw a conclusion at *any* specified level of significance. We now give a formal definition of a *P*-value.

Definition

The **P-value** is the smallest level of significance that would lead to rejection of the null hypothesis H_0.

It is customary to call the test statistic (and the data) significant when the null hypothesis H_0 is rejected; therefore, we may think of the *P*-value as the smallest level α at which the data are significant. Once the *P*-value is known, the decision maker can determine for himself or herself how significant the data are without the data analyst formally imposing a preselected level of significance.

For the foregoing normal distribution tests it is relatively easy to compute the *P*-value. If z_0 is the computed value of the test statistic, then the *P*-value is

$$P = \begin{cases} 2[1 - \Phi(|z_0|)] & \text{for a two-tailed test: } H_0: \mu = \mu_0, \quad H_1: \mu \neq \mu_0 \\ 1 - \Phi(z_0) & \text{for an upper-tailed test: } H_0: \mu = \mu_0, \quad H_1: \mu > \mu_0 \\ \Phi(z_0) & \text{for a lower-tailed test: } H_0: \mu = \mu_0, \quad H_1: \mu < \mu_0 \end{cases} \quad (4\text{-}22)$$

Here, $\Phi(z)$ is the standard normal cumulative distribution function defined in Chapter 3. Recall that $\Phi(z) = P(Z \leq z)$, where Z is $N(0, 1)$. To illustrate this, consider the propellant problem in Example 4-3. The computed value of the test statistic is $z_0 = 3.25$ and since the alternative hypothesis is two-tailed, the *P*-value is

$$P\text{-value} = 2[1 - \Phi(3.25)] = 0.0012$$

Thus, H_0: $\mu = 50$ would be rejected at any level of significance $\alpha \geq P\text{-value} = 0.0012$. For example, H_0 would be rejected if $\alpha = 0.01$, but it would not be rejected if $\alpha = 0.001$. The P-value is illustrated in Fig. 4-9.

It is not always easy to compute the exact P-value for a test. However, most modern computer programs for statistical analysis report P-values, and they can be obtained on some hand-held calculators. We will also show how to approximate the P-value. Finally, if the P-value approach is used, then step 6 of the hypothesis testing procedure can be modified. Specifically, it is not necessary to state explicitly the critical region.

4-4.3   Type II Error and Choice of Sample Size

In testing hypotheses, the analyst directly selects the type I error probability. However, the probability of type II error β depends on the choice of sample size. In this section, we will show how to calculate the probability of type II error β. We will also show how to select the sample size in order to obtain a specified value of β.

Finding the Probability of Type II Error β

Consider the two-sided hypothesis

$$H_0: \mu = \mu_0$$
$$H_1: \mu \neq \mu_0$$

Suppose that the null hypothesis is false and that the true value of the mean is $\mu = \mu_0 + \delta$—say, where $\delta > 0$. The test statistic Z_0 is

$$Z_0 = \frac{\overline{X} - \mu_0}{\sigma/\sqrt{n}}$$

$$= \frac{\overline{X} - (\mu_0 + \delta)}{\sigma/\sqrt{n}} + \frac{\delta}{\sigma/\sqrt{n}}$$

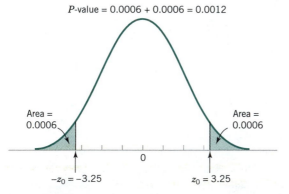

P-value = 0.0006 + 0.0006 = 0.0012

Area = 0.0006

Area = 0.0006

$-z_0 = -3.25$ $z_0 = 3.25$

Figure 4-9   Illustration of the P-value for the two-tailed test in the propellant example.

Therefore, the distribution of Z_0 when H_1 is true is

$$Z_0 \sim N\left(\frac{\delta}{\sigma/\sqrt{n}}, 1\right) \tag{4-23}$$

Here the notation "$\sim$" means "is distributed as." The distribution of the test statistic Z_0 under both the null hypothesis H_0 and the alternative hypothesis H_1 is shown in Fig. 4-10. From examining this figure, we note that if H_1 is true, a type II error will be made only if $-z_{\alpha/2} \le Z_0 \le z_{\alpha/2}$ where $Z_0 \sim N(\delta\sqrt{n}/\sigma, 1)$. That is, the probability of the type II error β is the probability that Z_0 falls between $-z_{\alpha/2}$ and $z_{\alpha/2}$ given that H_1 is true. This probability is shown as the shaded portion of Fig. 4-10 and is expressed mathematically in the following equation.

> The probability of a type II error for the two-sided alternative hypothesis on the mean, variance known, is
>
> $$\beta = \Phi\left(z_{\alpha/2} - \frac{\delta\sqrt{n}}{\sigma}\right) - \Phi\left(-z_{\alpha/2} - \frac{\delta\sqrt{n}}{\sigma}\right) \tag{4-24}$$

where $\Phi(z)$ denotes the probability to the left of z in the standard normal distribution. Note that equation 4-24 was obtained by evaluating the probability that Z_0 falls in the interval $[-z_{\alpha/2}, z_{\alpha/2}]$ when H_1 is true. Furthermore, note that equation 4-24 also holds if $\delta < 0$, due to the symmetry of the normal distribution. It is also possible to derive an equation similar to equation 4-24 for a one-sided alternative hypothesis.

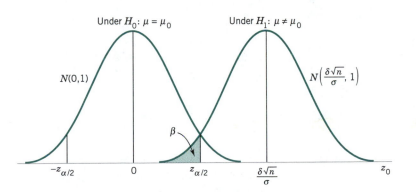

Figure 4-10   The distribution of Z_0 under H_0 and H_1.

Sample Size Formulas

One may easily obtain formulas that determine the appropriate sample size to obtain a particular value of β for a given δ and α. For the two-sided alternative hypothesis, we know from equation 4-24 that

$$\beta = \Phi\left(z_{\alpha/2} - \frac{\delta\sqrt{n}}{\sigma}\right) - \Phi\left(-z_{\alpha/2} - \frac{\delta\sqrt{n}}{\sigma}\right)$$

or if $\delta > 0$,

$$\beta \simeq \Phi\left(z_{\alpha/2} - \frac{\delta\sqrt{n}}{\sigma}\right) \qquad (4\text{-}25)$$

since $\Phi(-z_{\alpha/2} - \delta\sqrt{n}/\sigma) \simeq 0$ when δ is positive. Let z_β be the 100β upper percentile of the standard normal distribution. Then, $\beta = \Phi(-z_\beta)$. From equation 4-25

$$-z_\beta \simeq z_{\alpha/2} - \frac{\delta\sqrt{n}}{\sigma}$$

which leads to the following equation.

for $H_1 : \mu \ne \mu_0$

Sample Size for Two-Sided Alternative Hypothesis on the Mean, Variance Known

For the two-sided alternative hypothesis with significance level α, the sample size required to detect a difference between the true and hypothesized mean of δ with power at least $1 - \beta$ is

$$n \simeq \frac{(z_{\alpha/2} + z_\beta)^2 \sigma^2}{\delta^2} \qquad (4\text{-}26)$$

where

$$\delta = \mu - \mu_0$$

This approximation is good when $\Phi(-z_{\alpha/2} - \delta\sqrt{n}/\sigma)$ is small compared to β. For either of the one-sided alternative hypotheses the sample size required to produce a specified type II error with probability β given δ and α is as follows:

for $H_1 \; \mu < \mu_0$
$H_1 \; \mu > \mu_0$
4-4 INFERENCE ON THE MEAN OF A POPULATION, VARIANCE KNOWN   165

Sample Size for One-Sided Alternative Hypothesis on the Mean, Variance Known

For the one-sided alternative hypothesis with significance level α, the sample size required to detect a difference between the true and hypothesized mean of δ with power at least $1 - \beta$ is

$$n = \frac{(z_\alpha + z_\beta)^2 \sigma^2}{\delta^2} \qquad (4\text{-}27)$$

where

$$\delta = \mu - \mu_0$$

EXAMPLE 4-4

Consider the rocket propellant problem of Example 4-3. Suppose that the analyst wishes to design the test so that if the true mean burning rate differs from 50 cm/s by as much as 1 cm/s, the test will detect this (i.e., reject H_0: $\mu = 50$) with a high probability—say, 0.90. Now, we note that $\sigma = 2$, $\delta = 51 - 50 = 1$, $\alpha = 0.05$, and $\beta = 0.10$. Since $z_{\alpha/2} = z_{0.025} = 1.96$ and $z_\beta = z_{0.10} = 1.28$, the sample size required to detect this departure from H_0: $\mu = 50$ is found by equation 4-26 as

$$n \simeq \frac{(z_{\alpha/2} + z_\beta)^2 \sigma^2}{\delta^2} = \frac{(1.96 + 1.28)^2 2^2}{(1)^2} \simeq 42$$

The approximation is good here, since $\Phi(-z_{\alpha/2} - \delta\sqrt{n}/\sigma) = \Phi(-1.96 - (1)\sqrt{42}/2) = \Phi(-5.20) \simeq 0$, which is small relative to β.

4-4.4 Large-Sample Test

Although we have developed the test procedure for the null hypothesis H_0: $\mu = \mu_0$ assuming that σ^2 is known, in many if not most practical situations σ^2 will be unknown. In general, if $n \geq 30$, then the sample variance s^2 will be close to σ^2 for most samples, and so s can be substituted for σ in the test procedures with little harmful effect. Thus, although we have given a test for known σ^2, it can be easily converted into a *large-sample test procedure for unknown* σ^2. Exact treatment of the case where σ^2 is unknown and n is small involves use of the t distribution and will be deferred until Section 4-5.

4-4.5   Some Practical Comments on Hypothesis Testing

The Eight-Step Procedure

In Section 4-3.4 we described an eight-step procedure for statistical hypothesis testing. This procedure was illustrated in Example 4-3 and will be encountered many times in this chapter. In practice, such a formal and (seemingly) rigid procedure is not always necessary. Generally, once the experimenter (or decision maker) has decided on the question of interest and has determined the *design of the experiment* (that is, how the data are to be collected, how the measurements are to be made, and how many observations are required), then only three steps are really required:

1. Specify the test statistic to be used (such as z_0).

2. Specify the location of the critical region (two-tailed, upper-tailed, or lower-tailed).

3. Specify the criteria for rejection (typically, the value of α, or the *P*-value at which rejection should occur).

These steps are often completed almost simultaneously in solving real-world problems, although we emphasize that it is important to think carefully about each step. That is why we present and use the eight-step process: it seems to reinforce the essentials of the correct approach. Although you may not use it every time in solving real problems, it is a helpful framework when you are first learning about hypothesis testing.

Statistical Versus Practical Significance

We noted previously that reporting the results of a hypothesis test in terms of a *P*-value is very useful, because it conveys more information than just the simple statement "reject H_0" or "fail to reject H_0." That is, rejection of H_0 at the 0.05 level of significance is much more meaningful if the value of the test statistic is well into the critical region, greatly exceeding the 5% critical value, than if it barely exceeds that value.

Even a very small *P*-value can be difficult to interpret from a practical viewpoint when we are making decisions; although a small *P*-value indicates **statistical significance** in the sense that H_0 should be rejected in favor of H_1, the actual departure from H_0 that has been detected may have little (if any) **practical significance** (engineers like to say "engineering significance"). This is particularly true when the sample size n is large.

For example, consider the propellant burning rate problem of Example 4-3 where we are testing H_0: $\mu = 50$ cm/s versus H_1: $\mu \neq 50$ cm/s with $\sigma = 2$. If we suppose that the mean rate is really 50.5 cm/s, then this is not a serious departure from H_0: $\mu = 50$ cm/s in the sense that if the mean really is 50.5 cm/s there is no practical observable effect on the performance of the aircrew escape system. In other words, concluding that $\mu = 50$ cm/s when it is really 50.5 cm/s is an inexpensive error and has no practical significance. For a reasonably large sample size, a true value of $\mu = 50.5$ will lead to a sample $\bar{x}$ that is close to 50.5 cm/s, and we would not want this value of $\bar{x}$ from the sample to result in rejection of H_0. The accompanying display shows the P-value for testing H_0: $\mu = 50$ when we observe $\bar{x} = 50.5$ cm/s and the power of the test at $\alpha = 0.05$ when the true mean is 50.5 for various sample sizes n.

Sample Size n	P-value When $\bar{x} = 50.5$	Power (at $\alpha = 0.05$) When $\mu = 50.5$
10	0.4295	0.1241
25	0.2113	0.2396
50	0.0767	0.4239
100	0.0124	0.7054
400	5.73×10^{-7}	0.9988
1000	2.57×10^{-15}	1.0000

The P-value column in this display indicates that for large sample sizes the observed sample value of $\bar{x} = 50.5$ would strongly suggest that H_0: $\mu = 50$ should be rejected, even though the observed sample results imply that from a practical viewpoint the true mean does not differ much at all from the hypothesized value $\mu_0 = 50$. The power column indicates that if we test a hypothesis at a fixed significance level α and even if there is little practical difference between the true mean and the hypothesized value, a large sample size will almost always lead to rejection of H_0. The moral of this demonstration is clear: **Be careful when interpreting the results from hypothesis testing when the sample size is large, because any small departure from the hypothesized value μ_0 will probably be detected, even when the difference is of little or no practical significance.**

4-4.6   Confidence Interval on the Mean

In many situations, a point estimate does not provide enough information about a parameter. For example, in the rocket propellant problem we have rejected the null hypothesis H_0: $\mu = 50$, and our point estimate of the mean burning rate is $\bar{x} = 51.3$ cm/s. However, the engineer would prefer to have an **interval** in which we would expect to find the true mean burning rate, since it is unlikely that $\mu = 51.3$. One way to accomplish this is with an interval estimate called a **confidence interval.**

An interval estimate of the unknown parameter μ is an interval of the form $l \leq \mu \leq u$, where the end-points l and u depend on the numerical value of the sample mean $\bar{X}$ for a particular sample. Since different samples will produce different values of $\bar{x}$ and, consequently, different values of the end-points l and u, these end-points are values of random variables—say, L and U, respectively. From the sampling distribution of the sample mean $\bar{X}$ we will be able to determine values of L and U such that the following probability statement is true:

$$P(L \leq \mu \leq U) = 1 - \alpha \tag{4-28}$$

where $0 < \alpha < 1$. Thus, we have a probability of $1 - \alpha$ of selecting a sample that will produce an interval containing the true value of μ.

The resulting interval

$$l \le \mu \le u \qquad (4\text{-}29)$$

is called a $100(1 - \alpha)$ percent **confidence interval** for the parameter μ. The quantities l and u are called the **lower- and upper-confidence limits,** respectively, and $1 - \alpha$ is called the **confidence coefficient.** The interpretation of a confidence interval is that, if an infinite number of random samples are collected and a $100(1 - \alpha)$ percent confidence interval for μ is computed from each sample, then $100(1 - \alpha)$ percent of these intervals will contain the true value of μ.

The situation is illustrated in Fig. 4-11, which shows several $100(1 - \alpha)$ percent confidence intervals for the mean μ of a distribution. The dots at the center of each interval indicate the point estimate of μ (that is, $\bar{x}$). Note that one of the 15 intervals fails to contain the true value of μ. If this were a 95 percent confidence interval, in the long run only 5 percent of the intervals would fail to contain μ.

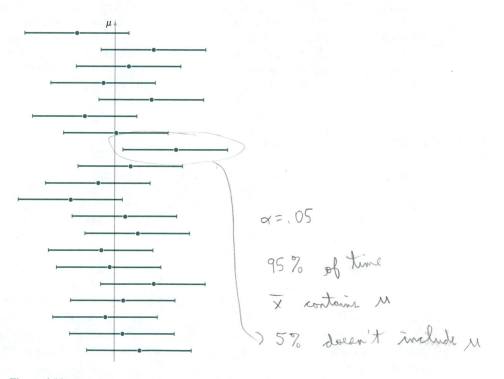

$\alpha = .05$

95% of time

$\bar{x}$ contains μ

5% doesn't include μ

Figure 4-11 Repeated construction of a confidence interval for μ.

Now in practice, we obtain only one random sample and calculate one confidence interval. Since this interval either will or will not contain the true value of μ, it is not reasonable to attach a probability level to this specific event. The appropriate statement is that the observed interval $[l, u]$ brackets the true value of μ with confidence $100(1 - \alpha)$. This statement has a frequency interpretation; that is, we don't know whether the statement is true for this specific sample, but the method used to obtain the interval $[l, u]$ yields correct statements $100(1 - \alpha)$ percent of the time.

The confidence interval in equation 4-29 is more properly called a **two-sided confidence interval,** as it specifies both a lower and an upper limit on μ. Occasionally, a **one-sided confidence interval** might be more appropriate. A one-sided $100(1 - \alpha)\%$ lower-confidence interval on μ is given by the interval

$$l \leq \mu \tag{4-30}$$

where the lower-confidence limit l is chosen so that

$$P(L \leq \mu) = 1 - \alpha \tag{4-31}$$

Similarly, a one-sided $100(1 - \alpha)\%$ upper-confidence interval on μ is given by the interval

$$\mu \leq u \tag{4-32}$$

where the upper-confidence limit u is chosen so that

$$P(\mu \leq U) = 1 - \alpha \tag{4-33}$$

The length $u - l$ of the observed confidence interval is an important measure of the quality of the information obtained from the sample. The half-interval length $\mu - l$ or $u - \mu$ is called the **precision** of the estimator. The longer the confidence interval, the more confident we are that the interval actually contains the true value of μ. On the other hand, the longer the interval is, the less information we have about the true value of μ. In an ideal situation, we obtain a relatively short interval with high confidence.

It is very easy to find the quantities L and U that define the confidence interval for μ. We know that the sampling distribution of $\overline{X}$ is normal with mean μ and variance σ^2/n. Therefore, the distribution of the statistic

$$Z = \frac{\overline{X} - \mu}{\sigma/\sqrt{n}}$$

is a standard normal distribution.

The distribution of $Z = (\overline{X} - \mu)/(\sigma/\sqrt{n})$ is shown in Fig. 4-12. From an examination of this figure we see that

$$P\{-z_{\alpha/2} \leq Z \leq z_{\alpha/2}\} = 1 - \alpha$$

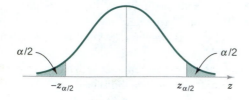

Figure 4-12   The distribution of Z.

so that

$$P\left\{ -z_{\alpha/2} \le \frac{\overline{X} - \mu}{\sigma/\sqrt{n}} \le z_{\alpha/2} \right\} = 1 - \alpha$$

This can be rearranged as

$$P\left\{ \overline{X} - \frac{z_{\alpha/2}\sigma}{\sqrt{n}} \le \mu \le \overline{X} + \frac{z_{\alpha/2}\sigma}{\sqrt{n}} \right\} = 1 - \alpha \qquad (4\text{-}34)$$

From consideration of equation 4-28, the lower and upper limits of the inequalities in equation 4-34 are the lower- and upper-confidence limits L and U, respectively. This leads to the following definition.

Confidence Interval on the Mean, Variance Known

If $\overline{x}$ is the sample mean of a random sample of size n from a population with known variance σ^2, a $100(1 - \alpha)\%$ confidence interval on μ is given by

$$\overline{x} - \frac{z_{\alpha/2}\sigma}{\sqrt{n}} \le \mu \le \overline{x} + \frac{z_{\alpha/2}\sigma}{\sqrt{n}} \qquad (4\text{-}35)$$

where $z_{\alpha/2}$ is the upper $100\alpha/2$ percentage point and $-z_{\alpha/2}$ is the lower $100\,\alpha/2$ percentage point of the standard normal distribution in Appendix A Table I.

For samples from a normal population or for samples of size $n \ge 30$ regardless of the shape of the population, the confidence interval in equation 4-35 will provide good results. However, for small samples from nonnormal populations we cannot expect the confidence level $1 - \alpha$ to be exact.

EXAMPLE 4-5

Consider the rocket propellant problem in Example 4-3. Suppose that we want to find a 95% confidence interval on the mean burning rate. We can use equation 4-35 to construct the confidence interval. A 95% interval implies that $1 - \alpha = 0.95$, so $\alpha = 0.05$ and from Table I in the Appendix $z_{\alpha/2} = z_{0.05/2} = z_{0.025} = 1.96$.

The lower confidence limit is

$$
\begin{aligned}
l &= \bar{x} - z_{\alpha/2}\sigma/\sqrt{n} \\
&= 51.3 - 1.96(2)/\sqrt{25} \\
&= 51.3 - 0.78 \\
&= 50.52
\end{aligned}
$$

and the upper confidence limit is

$$
\begin{aligned}
u &= \bar{x} + z_{\alpha/2}\sigma/\sqrt{n} \\
&= 51.3 + 1.96(2)/\sqrt{25} \\
&= 51.3 + 0.78 \\
&= 52.08
\end{aligned}
$$

Thus, the 95% two-sided confidence interval is

$$50.52 \le \mu \le 52.08$$

which is our interval of reasonable values for mean burning rate at 95% confidence.

Relationship between Tests of Hypotheses and Confidence Intervals

There is a close relationship between the test of a hypothesis about any parameter—say, θ—and the confidence interval for θ. If $[l, u]$ is a $100(1 - \alpha)\%$ confidence interval for the parameter θ, then the test of significance level α of the hypothesis

$$H_0: \theta = \theta_0$$

$$H_1: \theta \ne \theta_0$$

will lead to rejection of H_0 if and only if θ_0 is not in the $100(1 - \alpha)\%$ confidence interval $[l, u]$. As an illustration, consider the escape system propellant problem discussed above. The null hypothesis $H_0: \mu = 50$ was rejected, using $\alpha = 0.05$. The 95% two-sided confidence interval on μ is $50.52 \le \mu \le 52.08$. That is, the interval $[l, u]$ is [50.52, 52.08], and since $\mu_0 = 50$ is not included in this interval, the null hypothesis $H_0: \mu = 50$ is rejected.

Confidence Level and Precision of Estimation

Note in the previous example that our choice of the 95% level of confidence was essentially arbitrary. What would have happened if we had chosen a higher level of confidence— say, 99%? In fact, doesn't it seem reasonable that we would want the higher level of confidence? At $\alpha = 0.01$, we find $z_{\alpha/2} = z_{0.01/2} = z_{0.005} = 2.58$, whereas for $\alpha = 0.05$, $z_{0.025} = 1.96$. Thus, the length of the 95% confidence interval is

$$2(1.96 \; \sigma/\sqrt{n}) = 3.92 \; \sigma/\sqrt{n}$$

whereas the length of the 99% confidence interval is

$$2(2.58 \; \sigma/\sqrt{n}) = 5.16 \; \sigma/\sqrt{n}$$

The 99% confidence interval is longer than the 95% confidence interval, which is why we have a higher level of confidence in the 99% confidence interval. Generally, for a fixed sample size n and standard deviation σ, the higher the confidence level is, the longer is the resulting confidence interval.

Since the half-length of the confidence interval measures the precision of estimation, we see that precision is inversely related to the confidence level. As noted earlier, it is desirable to obtain a confidence interval that is short enough for decision-making purposes and that also has adequate confidence. One way to achieve this is by choosing the sample size n to be large enough to give a confidence interval of specified length with prescribed confidence.

Choice of Sample Size

The precision of the confidence interval in equation 4-35 is $z_{\alpha/2}\sigma/\sqrt{n}$. This means that in using $\bar{x}$ to estimate μ, the error $E = |\bar{x} - \mu|$ is less than or equal to $z_{\alpha/2}\sigma/\sqrt{n}$ with confidence $100(1 - \alpha)$. This is shown graphically in Fig. 4-13. In situations where the sample size can be controlled, we can choose n so that we are $100(1 - \alpha)\%$ confident that the error in estimating μ is less than a specified error E. The appropriate sample size is found by choosing n such that $z_{\alpha/2}\sigma/\sqrt{n} = E$. Solving this equation gives the following formula for n.

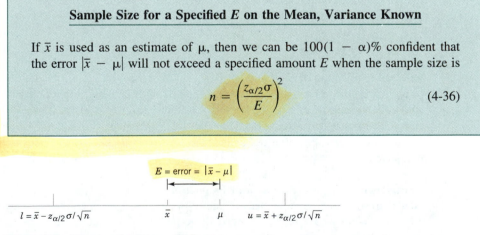

Sample Size for a Specified E on the Mean, Variance Known

If $\bar{x}$ is used as an estimate of μ, then we can be $100(1 - \alpha)\%$ confident that the error $|\bar{x} - \mu|$ will not exceed a specified amount E when the sample size is

$$n = \left(\frac{z_{\alpha/2}\sigma}{E}\right)^2 \qquad (4\text{-}36)$$

$E = \text{error} = |\bar{x} - \mu|$

$l = \bar{x} - z_{\alpha/2}\sigma/\sqrt{n} \qquad \bar{x} \qquad \mu \qquad u = \bar{x} + z_{\alpha/2}\sigma/\sqrt{n}$

Figure 4-13 Error in estimating μ with $\bar{x}$.

If the right-hand side of equation 4-36 is not an integer, it must be rounded up, which will ensure that the level of confidence does not fall below $100(1 - \alpha)\%$. Note that $2E$ is the length of the resulting confidence interval.

EXAMPLE 4-6

To illustrate the use of this procedure, suppose that we wanted the error in estimating the mean burning rate of the rocket propellant to be less than 1.5 cm/s, with 95% confidence. Since $\sigma = 2$ and $z_{0.025} = 1.96$, we may find the required sample size from equation 4-36 as

$$n = \left(\frac{z_{\alpha/2}\sigma}{E}\right)^2 = \left[\frac{(1.96)2}{1.5}\right]^2 = 6.83 \cong 7$$

Note the general relationship between sample size, desired length of the confidence interval $2E$, confidence level $100(1 - \alpha)\%$, and standard deviation σ:

- As the desired length of the interval $2E$ decreases, the required sample size n increases for a fixed value of σ and specified confidence.

- As σ increases, the required sample size n increases for a fixed desired length $2E$ and specified confidence.

- As the level of confidence increases, the required sample size n increases for fixed desired length $2E$ and standard deviation σ.

One-Sided Confidence Intervals

It is also possible to obtain one-sided confidence intervals for μ by setting either $l = -\infty$ or $u = \infty$ and replacing $z_{\alpha/2}$ by z_α.

One-Sided Confidence Intervals on the Mean, Variance Known

The $100(1 - \alpha)\%$ **upper-confidence interval** for μ is

$$\mu \le u = \bar{x} + z_\alpha\sigma/\sqrt{n} \tag{4-37}$$

and the $100(1 - \alpha)\%$ **lower-confidence interval** for μ is

$$\bar{x} - z_\alpha\sigma/\sqrt{n} = l \le \mu \tag{4-38}$$

4-4.7   General Method for Deriving a Confidence Interval

It is easy to give a general method for finding a confidence interval for an unknown parameter θ. Let $X_1, X_2, \ldots, X_n$ be a random sample of n observations. Suppose we can find a statistic $g(X_1, X_2, \ldots, X_n; \theta)$ with the following properties:

1. $g(X_1, X_2, \ldots, X_n; \theta)$ depends on both the sample and θ, and
2. the probability distribution of $g(X_1, X_2, \ldots, X_n; \theta)$ does not depend on θ or any other unknown parameter.

In the case considered in this section, the parameter $\theta = \mu$. The random variable $g(X_1, X_2, \ldots, X_n; \mu) = (\bar{X} - \mu)/(\sigma/\sqrt{n})$, and satisfies both these conditions; it depends on the sample and on μ, and it has a standard normal distribution since σ is known. Now one must find constants C_L and C_U so that

$$P[C_L \le g(X_1, X_2, \ldots, X_n; \theta) \le C_U] = 1 - \alpha$$

Because of property 2, C_L and C_U do not depend on θ. In our example, $C_L = -z_{\alpha/2}$ and $C_U = z_{\alpha/z}$. Finally, you must manipulate the inequalities in the probability statement so that

$$P[L(X_1, X_2, \ldots, X_n) \le \theta \le U(X_1, X_2, \ldots, X_n)] = 1 - \alpha$$

This gives $L(X_1, X_2, \ldots, X_n)$ and $U(X_1, X_2, \ldots, X_n)$ as the lower and upper confidence limits defining the $100(1 - \alpha)\%$ confidence interval for θ. In our example, we found $L(X_1, X_2, \ldots, X_n) = \bar{X} - z_{\alpha/2}\sigma/\sqrt{n}$ and $U(X_1, X_2, \ldots, X_n) = \bar{X} + z_{\alpha/2}\sigma/\sqrt{n}$.

EXERCISES FOR SECTION 4-4

4-24. The mean breaking strength of yarn used in manufacturing drapery material is required to be at least 100 psi. Past experience has indicated that the standard deviation of breaking strength is 2 psi. A random sample of nine specimens is tested, and the average breaking strength is found to be 98.03 psi.
(a) Should the fiber be judged acceptable with $\alpha = 0.05$?
(b) What is the P-value for this test?
(c) What is the probability of accepting the null hypothesis at $\alpha = 0.05$ if the fiber has a true breaking strength of 104 psi?
(d) Find a 95% two-sided confidence interval on the true mean breaking strength.

4-25. The yield of a chemical process is being studied. From previous experience with this process the standard deviation of yield is known to be 3. The past 5 days of plant operation have resulted in the following yields: 91.6%, 88.75%, 90.8%, 89.95%, and 91.3%. Use $\alpha = 0.05$.
(a) Is there evidence that the mean yield is not 90%?
(b) What is the P-value for this test?
(c) What sample size would be required to detect a true mean yield of 85% with probability 0.95?
(d) What is the type II error probability if the true mean yield is 92%?
(e) Find a 95% two-sided confidence interval on the true mean yield.

4-26. The diameter of holes for cable harness is known to have a standard deviation of 0.02 in. A random sample of size 10 yields the following data: 1.76, 1.69, 1.74, 1.73, 1.76, 1.77, 1.75, 1.78, 1.75, and 1.76. Use $\alpha = 0.01$.
(a) Test the hypothesis that the true mean hole diameter equals 1.75 in.
(b) What is the P-value for this test?
(c) What size sample would be necessary to detect a true mean hole diameter of 1.755 in. with probability at least 0.90?
(d) What is the β-value if the true mean hole diameter is 1.755 in.?
(e) Find a 99% two-sided confidence interval on the mean hole diameter. Do the results of this calculation seem intuitive based on the answer to parts (a) and (b)? Please discuss.

4-27. A manufacturer produces piston rings for an automobile engine. It is known that ring diameter is approximately normally distributed and has standard deviation $\sigma = 0.001$ mm. A random sample of 15 rings has a mean diameter of $\bar{x} = 74.036$ mm.
(a) Test the hypothesis that the mean piston ring diameter is 74.035 mm. Use $\alpha = 0.01$.
(b) What is the P-value for this test?
(c) Construct a 99% two-sided confidence interval on the mean piston ring diameter.
(d) Construct a 95% lower-confidence limit on the mean piston ring diameter.

4-28. The life in hours of a thermocouple used in a furnace is known to be approximately normally distributed, with standard deviation $\sigma = 20$ hours. A random sample of 15 thermocouples resulted in the following data: 553, 552, 567, 579, 550, 541, 537, 553, 552, 546, 538, 553, 581, 539, 529.
(a) Is there evidence to support the claim that mean life exceeds 540 hours? Use $\alpha = 0.05$.
(b) What is the P-value for the test in part (a)?
(c) What is the β-value for the test in part (a) if the true mean life is 560 hours?
(d) What sample size would be required to ensure that β does not exceed 0.10 if the true mean life is 560 hours?
(e) Construct a 95% two-sided confidence interval on the mean life.
(f) Construct a 95% lower confidence interval on the mean life.

4-29. A civil engineer is analyzing the compressive strength of concrete. Compressive strength is approximately normally distributed with variance $\sigma^2 = 1000(\text{psi})^2$. A random sample of 12 specimens has a mean compressive strength of $\bar{x} = 3255.42$ psi.
(a) Test the hypothesis that mean compressive strength is 3500 psi. Use $\alpha = 0.01$.
(b) What is the smallest level of significance at which you would be willing to reject the null hypothesis?
(c) Construct a 95% two-sided confidence interval on mean compressive strength.
(d) Construct a 99% two-sided confidence interval on mean compressive strength. Compare the width of this confidence interval with the width of the one found in part (c).

4-30. Suppose that in Exercise 4-28 we wanted to be 95% confident that the error in estimating the mean life is less than 5 hours. What sample size should we use?

4-31. Suppose that in Exercise 4-27 we wanted to be 95% confident that the error in estimating the mean life is less than 0.0005 mm. What sample size should we use?

4-32. Suppose that in Exercise 4-29 it is desired to estimate the compressive strength with an error that is less than 15 psi at 99% confidence. What sample size is required?

4-5   INFERENCE ON THE MEAN OF A POPULATION, VARIANCE UNKNOWN

When we are testing hypotheses or constructing confidence intervals on the mean μ of a population when σ^2 is unknown, we can use the test procedures in Section 4-4, provided that the sample size is large ($n \geq 30$, say). These procedures are approximately valid (because of the central limit theorem) regardless of whether or not the underlying population is normal. However, when the sample is small and σ^2 is unknown, we must make an assumption about the form of the underlying distribution in order to obtain a test procedure. A reasonable assumption in many cases is that the underlying distribution is normal.

Many populations encountered in practice are well approximated by the normal distribution, so this assumption will lead to inference procedures of wide applicability. In fact, moderate departure from normality will have little effect on validity. When the assumption is unreasonable, an alternative is to use nonparametric procedures that are valid for any underlying distribution. See Montgomery and Runger (1999) for an introduction to these techniques.

4-5.1   Hypothesis Testing on the Mean

Suppose that the population of interest has a normal distribution with unknown mean μ and variance σ^2. We wish to test the hypothesis that μ equals a constant μ_0. Note that this situation is similar to that in Section 4-4, except that now both μ and σ^2 are unknown. Assume that a random sample of size n—say, $X_1, X_2, \ldots, X_n$—is available, and let $\overline{X}$ and S^2 be the sample mean and variance, respectively.

We wish to test the two-sided alternative hypothesis

$$H_0: \mu = \mu_0$$

$$H_1: \mu \neq \mu_0$$

If the variance σ^2 is known, the test statistic is equation 4-12:

$$Z_0 = \frac{\overline{X} - \mu_0}{\sigma/\sqrt{n}}$$

When σ^2 is unknown, a reasonable procedure is to replace σ in the above expression with the sample standard deviation S. The test statistic is now

$$T_0 = \frac{\overline{X} - \mu_0}{S/\sqrt{n}} \tag{4-39}$$

A logical question is what effect does replacing σ by S have on the distribution of the statistic T_0? If n is large, the answer to this question is "very little," and we can proceed to use the test procedure based on the normal distribution from Section 4-4. However, n is usually small in most engineering problems, and in this situation a different distribution must be employed.

Let $X_1, X_2, \ldots, X_n$ be a random sample for a normal distribution with unknown mean μ and unknown variance σ^2. The quantity

$$T = \frac{\overline{X} - \mu}{S/\sqrt{n}}$$

has a t distribution with $n - 1$ degrees of freedom.

The t probability density function is

$$f(x) = \frac{\Gamma[(k + 1)/2]}{\sqrt{\pi k}\,\Gamma(k/2)} \cdot \frac{1}{[(x^2/k) + 1]^{(k+1)/2}} \qquad -\infty < x < \infty \qquad (4\text{-}40)$$

where k is the number of degrees of freedom. The mean and variance of the t distribution are zero and $k/(k - 2)$ (for $k > 2$), respectively. The function $\Gamma(m) = \int_0^\infty e^{-x} x^{m-1}\, dx$ is the gamma function. It is used commonly in engineering analyses. Although it is defined for $m \geq 0$, in the special case that m is an integer, it can be shown that $\Gamma(m) = (m - 1)!$ Also, $\Gamma(1) = \Gamma(0) = 1$.

Several t distributions are shown in Fig. 4-14. The general appearance of the t distribution is similar to the standard normal distribution, in that both distributions are symmetric and unimodal, and the maximum ordinate value is reached when the mean

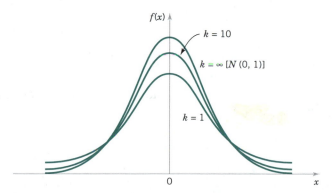

Figure 4-14 Probability density functions of several t distributions.

$\mu = 0$. However, the t distribution has heavier tails than the normal; that is, it has more probability in the tails than the normal distribution. As the number of degrees of freedom $k \to \infty$, the limiting form of the t distribution is the standard normal distribution. In visualizing the t distribution, it is sometimes useful to know that the ordinate of the density at the mean $\mu = 0$ is approximately four to five times larger than the ordinate at the 5th and 95th percentiles. For example, with 10 degrees of freedom for t this ratio is 4.8, with 20 degrees of freedom it is 4.3, and with 30 degrees of freedom, 4.1. By comparison, this factor is 3.9 for the normal distribution.

Appendix A Table II provides **percentage points** of the t distribution. We will let $t_{\alpha,k}$ be the value of the random variable T with k degrees of freedom above which we find an area (or probability) α. Thus, $t_{\alpha,k}$ is an upper-tail 100α percentage point of the t distribution with k degrees of freedom. This percentage point is shown in Fig. 4-15. In the Appendix Table II the α values are the column headings, and the degrees of freedom are listed in the left column. To illustrate the use of the table, note that the t-value with 10 degrees of freedom having an area of 0.05 to the right is $t_{0.05,10} = 1.812$. That is,

$$P(T_{10} > t_{0.05,10}) = P(T_{10} > 1.812) = 0.05$$

Since the t distribution is symmetric about zero, we have $t_{1-\alpha} = -t_{\alpha}$; that is, the t-value having an area of $1 - \alpha$ to the right (and therefore an area of α to the left) is equal to the negative of the t-value that has area α in the right tail of the distribution. Therefore, $t_{0.95,10} = -t_{0.05,10} = -1.812$.

It is now straightforward to see that the distribution of the test statistic in equation 4-39 is t with $n - 1$ degrees of freedom if the null hypothesis $H_0: \mu = \mu_0$ is true. To test $H_0: \mu = \mu_0$, the value of the test statistic t_0 in equation 4-39 is calculated, and H_0 is rejected if either

$$t_0 > t_{\alpha/2,n-1} \tag{4-41a}$$

or

$$t_0 < -t_{\alpha/2,n-1} \tag{4-41b}$$

where $t_{\alpha/2,n-1}$ and $-t_{\alpha/2,n-1}$ are the upper and lower $100\alpha/2$ percentage points of the t distribution with $n - 1$ degrees of freedom defined previously.

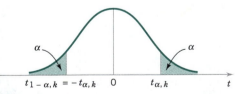

Figure 4-15   Percentage points of the t distribution.

For the one-sided alternative hypothesis

$$H_0: \mu = \mu_0$$

$$H_1: \mu > \mu_0 \tag{4-42}$$

we calculate the test statistic t_0 from equation 4-39 and reject H_0 if

$$t_0 > t_{\alpha, n-1} \tag{4-43}$$

For the other one-sided alternative

$$H_0: \mu = \mu_0$$

$$H_1: \mu < \mu_0 \tag{4-44}$$

we reject H_0 if

$$t_0 < -t_{\alpha, n-1} \tag{4-45}$$

Testing Hypotheses on the Mean of a Normal Distribution, Variance Unknown

Null hypothesis: $H_0: \mu = \mu_0$

Test statistic: $T_0 = \dfrac{\bar{X} - \mu_0}{S/\sqrt{n}}$

Alternative Hypotheses	Rejection Criterion
$H_1: \mu \neq \mu_0$	$t_0 > t_{\alpha/2, n-1}$ or $t_0 < -t_{\alpha/2, n-1}$
$H_1: \mu > \mu_0$	$t_0 > t_{\alpha, n-1}$
$H_1: \mu < \mu_0$	$t_0 < -t_{\alpha, n-1}$

EXAMPLE 4-7

An article in the journal *Materials Engineering* (1989, Vol. II, No. 4, pp. 275–281) describes the results of tensile adhesion tests on 22 U-700 alloy specimens. The load at specimen failure is as follows (in MPa):

19.8	18.5	17.6	16.7	15.8
15.4	14.1	13.6	11.9	11.4
11.4	8.8	7.5	15.4	15.4
19.5	14.9	12.7	11.9	11.4
10.1	7.9			

The sample mean is $\bar{x} = 13.71$, and the sample standard deviation is $s = 3.55$. Do the data suggest that the mean load at failure exceeds 10 MPa? Assume that load at failure has a normal distribution, and use $\alpha = 0.05$.

The solution using the eight-step procedure for hypothesis testing is as follows:

1. The parameter of interest is the mean load at failure, μ.

2. H_0: $\mu = 10$

3. H_1: $\mu > 10$. We want to reject H_0 if the mean load at failure exceeds 10 MPa.

4. $\alpha = 0.05$

5. The test statistic is

$$t_0 = \frac{\bar{x} - \mu_0}{s/\sqrt{n}}$$

6. Reject H_0 if $t_0 > t_{0.05,21} = 1.721$.

7. Computation: Since $\bar{x} = 13.71$, $s = 3.55$, $\mu_0 = 10$, and $n = 22$, we have

$$t_0 = \frac{13.71 - 10}{3.55/\sqrt{22}} = 4.90$$

8. Conclusion: Since $t_0 = 4.90 > 1.721$, we reject H_0 and conclude at the 0.05 level of significance that the mean load at failure exceeds 10 MPa.

As noted previously, the t-test assumes that the observations are a random sample from a normal population. It is always a good idea to investigate the **validity of the assumptions** when applying any statistical procedure. Figure 4-16 presents a box plot of the 22 observations on load at failure from Example 4-7. The impression from looking at this plot is that the sample comes from a symmetric population, and there is no immediate reason to question the normality assumption.

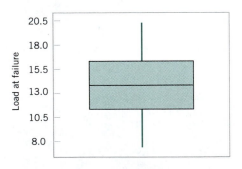

Figure 4-16   Box and whisker plot for the failure load data in Example 4-7.

Another excellent way to check the normality assumption in the t-test is to examine a normal probability plot of the sample data. Figure 4-17 is a normal probability plot of the load at failure data. The observations lie very close to the straight line, and so we conclude that the assumption of normality is reasonable.

4-5.2 *P*-Value for a *t*-Test

The P-value for a t-test is simply the smallest level of significance at which the null hypothesis would be rejected. That is, it is the tail area beyond the value of the test statistic t_0 for a one-sided test, or twice this area for a two-sided test. Because the t-table in Appendix A Table II contains only 10 critical values for each t distribution, computation of the exact P-value directly from the table is usually impossible. However, it is easy to find upper and lower bounds on the P-value from this table.

To illustrate, consider the t-test based on 21 degrees of freedom in Example 4-7. The relevant critical values from Appendix A Table II are as follows:

Critical Value: 0.257 0.686 1.323 1.721 2.080 2.518 2.831 3.135 3.527 3.819

Tail Area: 0.40   0.25   0.10   0.05   0.025 0.01   0.005 0.0025 0.001 0.0005

Note that, since $t_0 = 4.90$ in Example 4-7, we have $P(T_{21} > 4.90) < 0.0005$, because $t_0 = 4.90$ is greater than 3.819. Therefore, the P-value must be less than 0.0005 for this test, and we could say that an upper bound for the P-value is 0.0005.

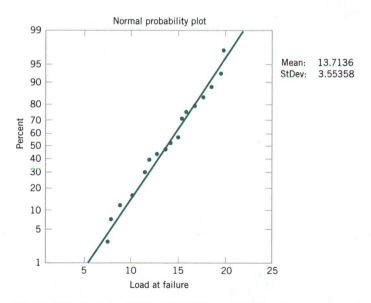

Figure 4-17 Normal probability plot of the load at failure data from Example 4-7.

Suppose that the computed value of the test statistic had been 2.75. Now this computed value is between the two critical values 2.518 (corresponding to $\alpha = 0.01$) and 2.831 (corresponding to 0.005). Therefore, we know that the P-value is less than 0.01 but must be greater than 0.005. This gives the lower and upper bounds on the P-value as $0.005 < P < 0.01$. This is illustrated in Fig. 4-18.

Example 4-7 is an upper-tailed test. If the test is lower-tailed, just change the sign of t_0 and proceed as above in the upper-tailed test. For example, if $t_0 = -2.75$ for a lower-tailed test with 21 degrees of freedom, we would find the lower and upper bounds to be $0.005 < P < 0.01$, exactly as above. Remember that for a two-tailed test the level of significance associated with a particular critical value is twice the corresponding tail area in the column heading. This consideration must be taken into account when we compute the bound on the P-value. For example, suppose that $t_0 = 2.75$ for a two-tailed alternative based on 21 degrees of freedom. The value $t_0 > 2.518$ (corresponding to $\alpha = 0.02$) and $t_0 < 2.831$ (corresponding to $\alpha = 0.01$), so the lower and upper bounds on the P-value would be $0.01 < P < 0.02$ for this case.

Finally, most computer programs report P-values along with the computed value of the test statistic. Some hand-held calculators also have this capability. From one such calculator, we obtain the P-value for the value $t_0 = 4.90$ in Example 4-7 as 0.000038.

4-5.3   Computer Solution

There are many widely available statistical software packages, and most of these packages have some capability for statistical hypothesis testing. The output from Minitab applied to the data in Example 4-7 is shown in Table 4-2. Note that the output includes some summary statistics about the sample, as well as a 95% confidence interval on the mean. (The confidence level may be chosen by the analyst.) The program also tests the hypothesis

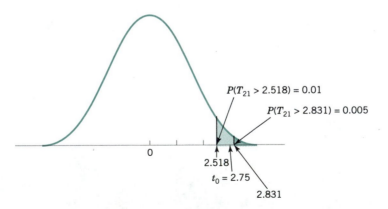

Figure 4-18 P-value for $t_0 = 2.75$ and an upper-tailed test is shown to be between 0.005 and 0.01.

Table 4-2 One-Sample Analysis for Example 4-7

Confidence Intervals					
Variable	N	Mean	StDev	SE Mean	95.0%CI
Load at	22	13.714	3.554	0.758	(12.138, 15.289)

T-Test of the Mean						

Test of mu = 10.000 vs mu > 10.000

Variable	N	Mean	StDev	SE Mean	T	P
Load at	22	13.714	3.554	0.758	4.90	0.0000

of interest, allowing the analyst to specify μ_0 (in this case 10) and the nature of the alternative hypothesis (H_1: $\mu > 10$).

The output includes the computed value of t_0 and the P-value. The P-value is rounded to the nearest 0.0001, which results in 0.0000 in this example. Note that H_0: $\mu = 10$ should be rejected in favor of H_1: $\mu > 10$, which is the identical conclusion we reached in Example 4-7.

4-5.4   Type II Error and Choice of Sample Size

The type II error probability for tests on the mean of a normal distribution with unknown variance depends on the distribution of the test statistic in equation 4-39 when the null hypothesis H_0: $\mu = \mu_0$ is false. When the true value of the mean is $\mu = \mu_0 + \delta$, the distribution for T_0 is called the **noncentral t distribution** with $n - 1$ degrees of freedom and noncentrality parameter $\delta\sqrt{n}/\sigma$. Note that if $\delta = 0$, then the noncentral t distribution reduces to the usual **central t distribution.** Therefore, the type II error of the two-sided alternative (for example) would be

$$\beta = P\{-t_{\alpha/2,n-1} \leq T_0 \leq t_{\alpha/2,n-1} \text{ when } \delta \neq 0\}$$
$$= P\{-t_{\alpha/2,n-1} \leq T_0' \leq t_{\alpha/2,n-1}\}$$

where T_0' denotes the noncentral t random variable. Finding the type II error probability β for the t-test involves finding the probability contained between two points on the noncentral t distribution. Because the noncentral t-random variable has a messy density function, this integration must be done numerically.

Fortunately, this unpleasant task has already been done, and the results are summarized in a series of graphs in Appendix A Charts Va, Vb, Vc, and Vd that plot β for the t-test against a parameter d for various sample sizes n. These graphics are called **operating characteristic** (or **OC**) **curves.** Curves are provided for two-sided alternatives on Charts Va and Vb. The abscissa scale factor d on these charts is defined as

$$d = \frac{|\mu - \mu_0|}{\sigma} = \frac{|\delta|}{\sigma} \qquad (4-46)$$

For the one-sided alternative $\mu > \mu_0$ as in equation 4-42, we use Charts Vc and Vd with

$$d = \frac{\mu - \mu_0}{\sigma} = \frac{\delta}{\sigma} \tag{4-47}$$

whereas if $\mu < \mu_0$, as in equation 4-44,

$$d = \frac{\mu_0 - \mu}{\sigma} = \frac{\delta}{\sigma} \tag{4-48}$$

We note that d depends on the unknown parameter σ^2. We can avoid this difficulty in several ways. In some cases, we may use the results of a previous experiment or prior information to make a rough initial estimate of σ^2. If we are interested in evaluating test performance after the data have been collected, we could use the sample variance s^2 to estimate σ^2. If there is no previous experience on which to draw in estimating σ^2, we then define the difference in the mean d that we wish to detect relative to σ. For example, if we wish to detect a small difference in the mean, we might use a value of $d = |\delta|/\sigma \le 1$ (for example), whereas if we are interested in detecting only moderately large differences in the mean, we might select $d = |\delta|/\sigma = 2$ (for example). That is, it is the value of the ratio $|\delta|/\sigma$ that is important in determining sample size, and if it is possible to specify the relative size of the difference in means that we are interested in detecting, then a proper value of d usually can be selected.

EXAMPLE 4-8

Consider the tensile adhesion testing problem from Example 4-7. If the mean load at failure differs from 10 MPa by as much as 1 MPa, is the sample size $n = 22$ adequate to ensure that H_0: $\mu = 10$ will be rejected with probability at least 0.8?

To solve this problem, we will use the sample standard deviation $s = 3.55$ to estimate σ. Then $d = |\delta|/\sigma = 1.0/3.55 = 0.28$. By referring to the operating characteristic curves in Appendix A Chart Vc (for $\alpha = 0.05$) with $d = 1/3.55 = 0.28$ and $n = 22$, we find that $\beta = 0.68$, approximately. Thus, the probability of rejecting H_0: $\mu = 10$ if the true mean exceeds this by 1.0 MPa is approximately $1 - \beta = 1 - 0.68 = 0.32$, and we conclude that a sample size of $n = 22$ is not adequate to provide the desired sensitivity. To find the sample size required to give the desired degree of sensitivity, enter the operating characteristic curves in Chart Vc with $d = 0.28$ and $\beta = 0.2$, and read the corresponding sample size as $n = 75$.

4-5.5 Confidence Interval on the Mean

It is easy to find a $100(1 - \alpha)$ percent confidence interval on the mean of a normal distribution with unknown variance by proceeding as we did in Section 4-4.6. In general, the distribution of $T = (\overline{X} - \mu)/(S/\sqrt{n})$ is t with $n - 1$ degrees of freedom. Letting

$t_{\alpha/2,n-1}$ be the upper $100\alpha/2$ percentage point of the t distribution with $n-1$ degrees of freedom, we may write:

$$P(-t_{\alpha/2,n-1} \leq T \leq t_{\alpha/2,n-1}) = 1 - \alpha$$

or

$$P\left(-t_{\alpha/2,n-1} \leq \frac{\overline{X} - \mu}{S/\sqrt{n}} \leq t_{\alpha/2,n-1}\right) = 1 - \alpha$$

Rearranging this last equation yields

$$P(\overline{X} - t_{\alpha/2,n-1}S/\sqrt{n} \leq \mu \leq \overline{X} + t_{\alpha/2,n-1}S/\sqrt{n}) = 1 - \alpha \qquad (4\text{-}49)$$

This leads to the following definition of the $100(1-\alpha)\%$ two-sided confidence interval on μ.

Confidence Interval on the Mean of a Normal Distribution, Variance Unknown

If $\overline{x}$ and s are the mean and standard deviation of a random sample from a normal distribution with unknown variance σ^2, then a $100(1-\alpha)\%$ confidence interval on μ is given by

$$\overline{x} - t_{\alpha/2,n-1}s/\sqrt{n} \leq \mu \leq \overline{x} + t_{\alpha/2,n-1}s/\sqrt{n} \qquad (4\text{-}50)$$

where $t_{\alpha/2,n-1}$ is the upper $100\alpha/2$ percentage point of the t distribution with $n-1$ degrees of freedom.

One-Sided Confidence Interval

To find a $100(1-\alpha)\%$ lower confidence interval on μ, with unknown σ^2, simply replace $-t_{\alpha/2,n-1}$ with $-t_{\alpha,n-1}$ in the lower bound of equation 4-50 and set the upper bound to ∞. Similarly, to find a $100(1-\alpha)\%$ upper confidence interval on μ, with unknown σ^2, replace $t_{\alpha/2,n-1}$ with $t_{\alpha,n-1}$ in the upper bound and set the lower bound to $-\infty$. These formulas are given in the table on the inside front cover.

EXAMPLE 4-9

Reconsider the tensile adhesion problem in Example 4-7. We know that $n = 22$, $\overline{X} = 13.71$, and $s = 3.55$. We will find a 95% confidence interval on μ. From equation 4-50 we find ($t_{\alpha/2,n-1} = t_{0.025,21} = 2.080$):

$$\overline{X} - t_{\alpha/2,n-1}S/\sqrt{n} \leq \mu \leq \overline{X} + t_{\alpha/2,n-1}S/\sqrt{n}$$
$$13.71 - 2.080(3.55)/\sqrt{22} \leq \mu \leq 13.71 + 2.080(3.55)/\sqrt{22}$$
$$13.71 - 1.57 \leq \mu \leq 13.71 + 1.57$$
$$12.14 \leq \mu \leq 15.28$$

In Example 4-7, we tested a one-sided alternative hypothesis on μ. Some engineers might be interested in a one-sided confidence interval. The 95% lower confidence interval on mean load at failure is

$$\bar{X} - t_{.05, n-1}S/\sqrt{n} \leq \mu$$
$$13.71 - 1.721(3.25)/\sqrt{22} \leq \mu$$
$$12.41 \leq \mu$$

Thus we can state with 95% confidence that the mean load at failure exceeds 12.41 MPa.

EXERCISES FOR SECTION 4-5

4-33. A research engineer for a tire manufacturer is investigating tire life for a new rubber compound. He has built 10 tires and tested them to end-of-life in a road test. The sample mean and standard deviation are 61,492 and 3035 km, respectively.

(a) The engineer would like to demonstrate that the mean life of this new tire is in excess of 60,000 km. Formulate and test appropriate hypotheses, being sure to state (test, if possible) assumptions and draw conclusions using $\alpha = 0.05$.

(b) Suppose that if the mean life is as long as 61,000 km, the engineer would like to detect this difference with probability at least 0.90. Was the sample size $n = 10$ used in part (a) adequate? Use the sample standard deviation s as an estimate of σ in reaching your decision.

(c) Find a 95% confidence interval on mean tire life.

4-34. An Izod impact test was performed on 20 specimens of PVC pipe. The ASTM standard for this material requires that Izod impact strength must be greater than 1.0 ft-lb/in. The sample average and standard deviation obtained were $\bar{x} = 1.121$ and $s = 0.328$, respectively. Test H_0: $\mu = 1.0$ versus H_1: $\mu > 1.0$ using $\alpha = 0.01$ and draw conclusions.

State any necessary assumptions about the underlying distribution of the data.

4-35. The life in hours of a biomedical device under development in the laboratory is known to be approximately normally distributed. A random sample of 15 devices is selected and found to have an average life of 5625.1 hours and a sample standard deviation of 226.1 hours.

(a) At the $\alpha = 0.05$ level of significance, test the hypothesis H_0: $\mu = 5500$ versus H_1: $\mu > 5500$. On completing the hypothesis test, do you believe that the true mean life of a biomedical device is greater than 5500? Clearly state your answer.

(b) Find the p-value of the test statistic.

(c) Construct a 95% lower confidence interval on the mean and describe how this interval can be used to test the alternative hypothesis of part (a).

4-36. A particular brand of diet margarine was analyzed to determine the level of polyunsaturated fatty acid (in percent). A sample of six packages resulted in the following data: 16.8, 17.2, 17.4, 16.9, 16.5, 17.1.

(a) Test the hypothesis H_0: $\mu = 17.0$ versus H_1: $\mu \neq 17.0$ using $\alpha = 0.01$. What are your conclusions? Use a normal probability plot to test the normality assumption.

(b) Find the P-value for the test in part (a).

(c) Suppose that if the mean polyunsaturated fatty acid content is really $\mu = 17.5$, it is important to detect this with probability at least 0.90. Is the sample size $n = 6$ adequate? Use the sample standard deviation to estimate the population standard deviation σ. Use $\alpha = 0.01$.

(d) Find a 99% confidence interval on the mean μ. Provide a practical interpretation of this interval.

4-37. The compressive strength of concrete is being tested by a civil engineer, who tests 12 specimens and obtains the following data.

2256	2257	2243	2199
2227	2230	2238	2248
2332	2230	2264	2243

(a) Check the normality assumption for these compressive strength data.

(b) Test the hypotheses H_0: $\mu = 2250$ psi versus H_1: $\mu \neq 2250$ psi, using $\alpha = 0.05$. Draw conclusions based on the outcome of this test.

(c) Construct a 95% two-sided confidence interval on the mean strength.

(d) Construct a 95% lower confidence interval on the mean strength.

4-38. A machine produces metal rods used in an automobile suspension system. A random sample of 15 rods is selected, and the diameter is measured. The resulting data are shown here.

8.24 mm	8.23 mm	8.20 mm
8.21	8.20	8.28
8.23	8.26	8.24
8.25	8.19	8.25
8.26	8.23	8.24

(a) Check the assumption of normality for rod diameter.

(b) Is there strong evidence to indicate that mean rod diameter exceeds 8.20 mm using $\alpha = 0.05$?

(c) Find the P-value for the statistical test you performed in part (b).

(d) Find a 95% two-sided confidence interval on mean rod diameter.

4-39. The wall thickness of 25 glass 2-liter bottles was measured by a quality-control engineer. The sample mean was $\bar{x} = 4.058$ mm, and the sample standard deviation was $s = 0.081$ mm.

(a) Suppose that it is important to demonstrate that the wall thickness exceeds 4.0 mm. Formulate and test appropriate hypotheses using these data. Draw conclusions at $\alpha = 0.05$. Calculate the P-value for this test.

(b) Find a 95% confidence interval for mean wall thickness. Interpret the interval you obtain.

4-40. Measurements on the percent enrichment of 12 fuel rods used in a nuclear reactor were reported as follows:

3.11	2.88	3.08	3.01
2.84	2.86	3.04	3.09
3.08	2.89	3.12	2.98

(a) Use a normal probability plot to check the normality assumption.

(b) Test the hypothesis H_0: $\mu = 2.95$ versus H_1: $\mu \neq 2.95$, using $\alpha = 0.05$, and draw appropriate conclusions. Calculate the P-value for this test.

(c) Find a 99% two-sided confidence interval on the mean percent enrichment. Are you comfortable with the statement that the mean percent enrichment is 2.95%? Why?

4-41. A post-mix beverage machine is adjusted to release a certain amount of syrup into a chamber where it is mixed with carbonated water. A random sample of 25 beverages was found to have a mean syrup content of $\bar{x} = 1.098$ fluid ounces and a standard deviation of $s = 0.016$ fluid ounce.

(a) Do the data presented in this exercise support the claim that the mean amount of syrup dispensed is not 1.0

fluid ounce? Test this claim using
α = 0.05.

(b) Do the data support the claim that
the mean amount of syrup dispensed
exceeds 1.0 fluid ounce? Test this
claim using α = 0.05.

(c) Consider the hypothesis test in part
(a). If the mean amount of syrup
dispensed differs from μ = 1.0 by
as much as 0.05, it is important to
detect this with a high probability
(at least 0.90, say). Using s as an
estimate of σ, what can you say
about the adequacy of the sample
size n = 25 used by the experimenters?

(d) Find a 95% confidence interval
on the mean amount of syrup
dispensed.

4-42.   An article in the *Journal of Composite
Materials* (December 1989, Vol 23,
p. 1200) describes the effect of delami-
nation on the natural frequency of beams
made from composite laminates. Five
such delaminated beams were subjected
to loads, and the resulting frequencies
were as follows (in Hz):

230.66, 233.05, 232.58, 229.48, 232.58

Find a 90% two-sided confidence interval on mean natural frequency. Do the
results of your calculations support the
claim that mean natural frequency is 235
Hz? Please discuss your findings and
state any necessary assumptions.

4-6   INFERENCE ON THE VARIANCE OF A NORMAL POPULATION

Sometimes hypothesis tests and confidence intervals on the population variance or standard
deviation are needed. If we have a random sample $X_1, X_2, \ldots, X_n$, the sample variance
S^2 is an unbiased point estimator of σ^2. When the population is modeled by a normal
distribution, the tests and intervals described in this section are applicable.

4-6.1   Hypothesis Testing on the Variance of a Normal Population

Suppose that we wish to test the hypothesis that the variance of a normal population σ^2
equals a specified value—say, σ_0^2. Let $X_1, X_2, \ldots, X_n$ be a random sample of n observations
from this population. To test

$$H_0: \sigma^2 = \sigma_0^2$$

$$H_1: \sigma^2 \neq \sigma_0^2 \tag{4-51}$$

we will use the following test statistic:

$$X_0^2 = \frac{(n-1)S^2}{\sigma_0^2} \tag{4-52}$$

To define the test procedure, we will need to know the distribution of the test statistic X_0^2 in equation 4-52 when the null hypothesis is true.

Let $X_1, X_2, \ldots, X_n$ be a random sample from a normal distribution with unknown mean μ and unknown variance σ^2. The quantity

$$X^2 = \frac{(n-1)S^2}{\sigma^2} \qquad (4\text{-}53)$$

has a chi-square distribution with $n - 1$ degrees of freedom, abbreviated as χ_{n-1}^2. In general, the probability density function of a chi-square random variable is

$$f(x) = \frac{1}{2^{k/2}\Gamma(k/2)} x^{(k/2)-1} e^{-x/2} \qquad x > 0 \qquad (4\text{-}54)$$

where k is the number of degrees of freedom and $\Gamma(k/2)$ was defined in Section 4-5.1.

The mean and variance of the χ^2 distribution are

$$\mu = k$$

and $\qquad (4\text{-}55)$

$$\sigma^2 = 2k$$

Several chi-square distributions are shown in Fig. 4-19. Note that the chi-square random variable is nonnegative and that the probability distribution is skewed to the right. However, as k increases, the distribution becomes more symmetric. As $k \to \infty$, the limiting form of the chi-square distribution is the normal distribution.

The **percentage points** of the χ^2 distribution are given in Table III of Appendix A. Define $\chi_{\alpha,k}^2$ as the percentage point or value of the chi-square random variable with k degrees of freedom such that the probability that X^2 exceeds this value is α. That is,

$$P(X^2 > \chi_{\alpha,k}^2) = \int_{\chi_{\alpha,k}^2}^{\infty} f(u) \, du = \alpha$$

This probability is shown as the shaded area in Fig. 4-20. To illustrate the use of Table III, note that the areas α are the column headings and the degrees of freedom k are given in the left column, labeled v. Therefore, the value with 10 degrees of freedom

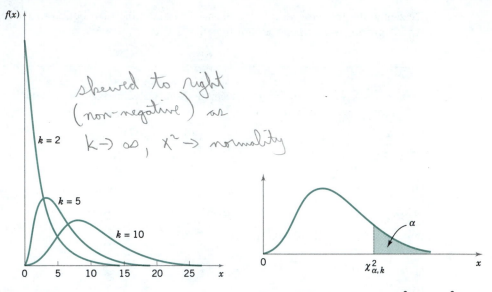

*skewed to right
(non-negative) as
$k \to \infty$, $\chi^2 \to$ normality*

Figure 4-19 Probability density functions of several χ^2 distributions.

Figure 4-20 Percentage point $\chi^2_{\alpha,k}$ of the χ^2 distribution.

related to gamma dist.

having an area (probability) of 0.05 to the right is $\chi^2_{0.05,10} = 18.31$. This value is often called an upper 5% point of chi-square with 10 degrees of freedom. We may write this as a probability statement as follows:

$$P(X^2 > \chi^2_{0.05,10}) = P(X^2 > 18.31) = 0.05$$

It is relatively easy to construct a test for the hypothesis in equation 4-51. If the null hypothesis $H_0: \sigma^2 = \sigma_0^2$ is true, then the test statistic X_0^2 defined in equation 4-52 follows the chi-square distribution with $n - 1$ degrees of freedom. Therefore, we calculate the value of the test statistic χ_0^2, and the hypothesis $H_0: \sigma^2 = \sigma_0^2$ would be rejected if

$$\chi_0^2 > \chi^2_{\alpha/2, n-1} \tag{4-56a}$$

or if

$$\chi_0^2 < \chi^2_{1-\alpha/2, n-1} \tag{4-56b}$$

where $\chi^2_{\alpha/2, n-1}$ and $\chi^2_{1-\alpha/2, n-1}$ are the upper and lower $100\alpha/2$ percentage points of the chi-square distribution with $n - 1$ degrees of freedom, respectively.

The same test statistic is used for one-sided alternative hypotheses. For the one-sided hypothesis

$$H_0: \sigma^2 = \sigma_0^2$$
$$H_1: \sigma^2 > \sigma_0^2 \tag{4-57}$$

we would reject H_0 if

$$\chi_0^2 > \chi_{\alpha,n-1}^2 \qquad (4\text{-}58)$$

For the other one-sided hypothesis

$$H_0: \sigma^2 = \sigma_0^2$$

$$H_1: \sigma^2 < \sigma_0^2 \qquad (4\text{-}59)$$

we would reject H_0 if

$$\chi_0^2 < \chi_{1-\alpha,n-1}^2 \qquad (4\text{-}60)$$

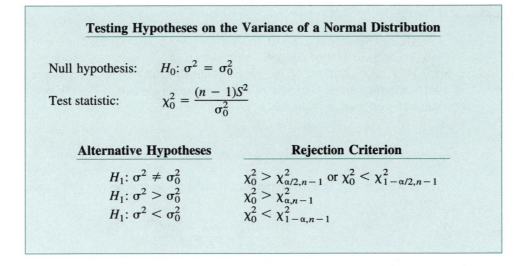

Testing Hypotheses on the Variance of a Normal Distribution

Null hypothesis: $H_0: \sigma^2 = \sigma_0^2$

Test statistic: $\chi_0^2 = \dfrac{(n-1)S^2}{\sigma_0^2}$

Alternative Hypotheses	Rejection Criterion
$H_1: \sigma^2 \neq \sigma_0^2$	$\chi_0^2 > \chi_{\alpha/2,n-1}^2$ or $\chi_0^2 < \chi_{1-\alpha/2,n-1}^2$
$H_1: \sigma^2 > \sigma_0^2$	$\chi_0^2 > \chi_{\alpha,n-1}^2$
$H_1: \sigma^2 < \sigma_0^2$	$\chi_0^2 < \chi_{1-\alpha,n-1}^2$

EXAMPLE 4-10

An automatic filling machine is used to fill bottles with liquid detergent. A random sample of 20 bottles results in a sample variance of fill volume of $s^2 = 0.0153$ (fluid ounces)^2. If the variance of fill volume exceeds 0.01 (fluid ounces)^2, an unacceptable proportion of bottles will be underfilled and overfilled. Is there evidence in the sample data to suggest that the manufacturer has a problem with underfilled and overfilled bottles? Use $\alpha = 0.05$, and assume that fill volume has a normal distribution.

Using the eight-step procedure results in the following:

1. The parameter of interest is the population variance σ^2.
2. $H_0: \sigma^2 = 0.01$
3. $H_1: \sigma^2 > 0.01$
4. $\alpha = 0.05$

5. The test statistic is

$$\chi_0^2 = \frac{(n-1)s^2}{\sigma_0^2}$$

6. Reject H_0 if $\chi_0^2 > \chi_{0.05,19}^2 = 30.14$.

7. Computation:

$$\chi_0^2 = \frac{19(0.0153)}{0.01} = 29.07$$

8. Conclusion: Since $\chi_0^2 = 29.07 < \chi_{0.05,19}^2 = 30.14$, we conclude that there is no strong evidence that the variance of fill volume exceeds 0.01 (fluid ounces)^2.

Using Appendix A Table III, it is easy to place bounds on the P-value of a chi-square test. From inspection of the table, we find that $\chi_{0.10,19}^2 = 27.20$ and $\chi_{0.05,19}^2 = 30.14$. Since $27.20 < 29.07 < 30.14$, we conclude that the P-value for the test in Example 4-10 is in the interval $0.05 < P < 0.10$. The actual P-value is $P = 0.0649$. (This value was obtained using a calculator.)

4-6.2   Confidence Interval on the Variance of a Normal Population

It was noted in the previous section that if the population is normal, the sampling distribution of

$$X^2 = \frac{(n-1)S^2}{\sigma^2}$$

is chi-square with $n-1$ degrees of freedom. To develop the confidence interval, we first write

$$P(\chi_{1-\alpha/2,n-1}^2 \le X^2 \le \chi_{\alpha/2,n-1}^2) = 1 - \alpha$$

so that

$$P\left(\chi_{1-\alpha/2,n-1}^2 \le \frac{(n-1)S^2}{\sigma^2} \le \chi_{\alpha/2,n-1}^2\right) = 1 - \alpha$$

This last equation can be rearranged as

$$P\left(\frac{(n-1)S^2}{\chi_{\alpha/2,n-1}^2} \le \sigma^2 \le \frac{(n-1)S^2}{\chi_{1-\alpha/2,n-1}^2}\right) = 1 - \alpha \tag{4-61}$$

This leads to the following definition of the confidence interval for σ^2.

Confidence Interval on the Variance of a Normal Distribution

If s^2 is the sample variance from a random sample of n observations from a normal distribution with unknown variance σ^2, then a $100(1 - \alpha)\%$ confidence interval on σ^2 is

$$\frac{(n-1)s^2}{\chi^2_{\alpha/2,n-1}} \leq \sigma^2 \leq \frac{(n-1)s^2}{\chi^2_{1-\alpha/2,n-1}} \tag{4-62}$$

where $\chi^2_{\alpha/2,n-1}$ and $\chi^2_{1-\alpha/2,n-1}$ are the upper and lower $100\alpha/2$ percentage points of the chi-square distribution with $n - 1$ degrees of freedom, respectively.

One-Sided Confidence Intervals
To find a $100(1 - \alpha)\%$ lower confidence interval on σ^2, set the upper confidence limit in equation 4-62 equal to ∞ and replace $\chi^2_{\alpha/2,n-1}$ with $\chi^2_{\alpha,n-1}$. The $100(1 - \alpha)\%$ upper confidence interval is found by setting the lower confidence limit in equation 4-62 equal to zero and replacing $\chi^2_{1-\alpha/2,n-1}$ with $\chi^2_{1-\alpha,n-1}$. For your convenience, these equations for constructing the one-sided upper and lower confidence intervals are given in the table on the inside front cover of this text.

EXAMPLE 4-11
Reconsider the bottle filling machine from Example 4-10. We will continue to assume that fill volume is approximately normally distributed. A random sample of 20 bottles results in a sample variance of $s^2 = 0.0153$ (fluid ounce)^2. A 95% upper confidence interval is found from equation 4-62 as follows:

$$\sigma^2 \leq \frac{(n-1)s^2}{\chi^2_{0.95,19}}$$

or

$$\sigma^2 \leq \frac{(19)0.0153}{10.12} = 0.0287 \text{ (fluid ounce)}^2$$

This last statement may be converted into a confidence interval on the standard deviation σ by taking the square root of both sides, resulting in

$$\sigma \leq 0.17$$

Therefore, at the 95% level of confidence, the data indicate that the process standard deviation could be as large as 0.17 fluid ounce.

EXERCISES FOR SECTION 4-6

4-43. A rivet is to be inserted into a hole. If the standard deviation of hole diameter exceeds 0.02 mm, there is an unacceptably high probability that the rivet will not fit. A random sample of $n = 15$ parts is selected, and the hole diameter is measured. The sample standard deviation of the hole diameter measurements is $s = 0.016$ mm.

(a) Is there strong evidence to indicate that the standard deviation of hole diameter exceeds 0.02 mm? Use $\alpha = 0.05$. State any necessary assumptions about the underlying distribution of the data.

(b) Find the P-value for this test.

(c) Construct a 95% lower confidence interval for σ. Explain how this confidence interval can be used to test the hypothesis in part (a).

4-44. The sugar content of the syrup in canned peaches is normally distributed, and the variance is thought to be $\sigma^2 = 18$ (mg)^2.

(a) Test the hypothesis $H_0: \sigma^2 = 18$ versus $H_1: \sigma^2 \neq 18$ if a random sample of $n = 10$ cans yields a sample standard deviation of $s = 4$ mg, using $\alpha = 0.05$. State any necessary assumptions about the underlying distribution of the data.

(b) What is the P-value for this test?

(c) Find a 95% two-sided confidence interval for σ.

4-45. Consider the tire life data in Exercise 4-33.

(a) Can you conclude, using $\alpha = 0.05$, that the standard deviation of tire life exceeds 4000 km? State any necessary assumptions about the underlying distribution of the data.

(b) Find the P-value for this test.

(c) Find a 95% lower confidence interval for σ.

4-46. Consider the Izod impact test data in Exercise 4-34.

(a) Test the hypothesis that $\sigma = 0.10$ against an alternative specifying that $\sigma \neq 0.10$, using $\alpha = 0.01$, and draw a conclusion. State any necessary assumptions about the underlying distribution of the data.

(b) What is the P-value for this test?

(c) Find a 99% two-sided confidence interval for σ^2.

4-47. The percentage of titanium in an alloy used in aerospace castings is measured in 51 randomly selected parts. The sample standard deviation is $s = 0.37$.

(a) Test the hypothesis $H_0: \sigma = 0.25$ versus $H_1: \sigma \neq 0.25$ using $\alpha = 0.05$. State any necessary assumptions about the underlying distribution of the data.

(b) Construct a 95% two-sided confidence interval for σ.

4-7   INFERENCE ON A POPULATION PROPORTION

It is often necessary to test hypotheses and construct confidence intervals on a population proportion. For example, suppose that a random sample of size n has been taken from a large (possibly infinite) population and that $X(\leq n)$ observations in this sample belong to a class of interest. Then $\hat{P} = X/n$ is a point estimator of the proportion of the population p that belongs to this class. Note that n and p are the parameters of a binomial distribution. Furthermore, from Chapter 3 we know that the sampling distribution of $\hat{P}$ is approximately normal with mean p and variance $p(1 - p)/n$, if p is not too close to either 0 or 1 and if n is relatively large. Typically, to apply this approximation we require that np and

$n(1 - p)$ be greater than or equal to 5. We will make use of the normal approximation in this section.

4-7.1 Hypothesis Testing on a Binomial Proportion

In many engineering problems, we are concerned with a random variable that follows the binomial distribution. For example, consider a production process that manufactures items classified as either acceptable or defective. It is usually reasonable to model the occurrence of defectives with the binomial distribution, where the binomial parameter p represents the proportion of defective items produced. Consequently, many engineering decision problems include hypothesis testing about p.

We will consider testing

$$H_0: p = p_0$$

$$H_1: p \neq p_0 \tag{4-63}$$

An approximate test based on the normal approximation to the binomial will be given. As noted above, this approximate procedure will be valid as long as p is not extremely close to zero or one, and if the sample size is relatively large. The following result will be used to perform hypothesis testing and to construct confidence intervals on p.

Let X be the number of observations in a random sample of size n that belongs to the class associated with p. Then, the quantity

$$Z = \frac{X - np}{\sqrt{np(1 - p)}} \tag{4-64}$$

has approximately a standard normal distribution, $N(0, 1)$.

Then, if the null hypothesis $H_0: p = p_0$ is true, we have $X \sim N(np_0, np_0(1 - p_0))$, approximately. To test $H_0: p = p_0$, calculate the **test statistic**

$$Z_0 = \frac{X - np_0}{\sqrt{np_0(1 - p_0)}}$$

and reject $H_0: p = p_0$ if

$$z_0 > z_{\alpha/2} \quad \text{or} \quad z_0 < -z_{\alpha/2}$$

Critical regions for the one-sided alternative hypotheses would be constructed in the usual manner.

Testing Hypotheses on a Binomial Proportion

Null hypotheses: $H_0: p = p_0$

Test statistic: $Z_0 = \dfrac{X - np_0}{\sqrt{np_0(1 - p_0)}}$

Alternative Hypotheses	Rejection Criterion
$H_1: p \neq p_0$	$z_0 > z_{\alpha/2}$ or $z_0 < -z_{\alpha/2}$
$H_1: p > p_0$	$z_0 > z_\alpha$
$H_1: p < p_0$	$z_0 < -z_\alpha$

EXAMPLE 4-12

A semiconductor manufacturer produces controllers used in automobile engine applications. The customer requires that the process fallout or fraction defective at a critical manufacturing step not exceed 0.05 and that the manufacturer demonstrate process capability at this level of quality using $\alpha = 0.05$. The semiconductor manufacturer takes a random sample of 200 devices and finds that four of them are defective. Can the manufacturer demonstrate process capability for the customer?

We may solve this problem using the eight-step hypothesis testing procedure as follows:

1. The parameter of interest is the process fraction defective p.
2. $H_0: p = 0.05$
3. $H_1: p < 0.05$
    This formulation of the problem will allow the manufacturer to make a strong claim about process capability if the null hypothesis $H_0: p = 0.05$ is rejected.
4. $\alpha = 0.05$
5. The test statistic is (from equation 4-64)

$$z_0 = \frac{x - np_0}{\sqrt{np_0(1 - p_0)}}$$

where $x = 4$, $n = 200$, and $p_0 = 0.05$.
6. Reject $H_0: p = 0.05$ if $z_0 < -z_{0.05} = -1.645$

7. Computation: The test statistic is

$$z_0 = \frac{4 - 200(0.05)}{\sqrt{200(0.05)(0.95)}} = -1.95$$

8. Conclusion: Since $z_0 = -1.95 < -z_{0.05} = -1.645$, we reject H_0 and conclude that the process fraction defective p is less than 0.05. The P-value for this value of the test statistic z_0 is $P = 0.0256$, which is less than $\alpha = 0.05$. We conclude that the process is capable.

We occasionally encounter another form of the test statistic Z_0 in equation 4-64. Note that if X is the number of observations in a random sample of size n that belongs to a class of interest, then $\hat{P} = X/n$ is the sample proportion that belongs to that class. Now divide both numerator and denominator of Z_0 in equation 4-64 by n, giving

$$Z_0 = \frac{X/n - p_0}{\sqrt{p_0(1 - p_0)/n}}$$

or

$$Z_0 = \frac{\hat{P} - p_0}{\sqrt{p_0(1 - p_0)/n}} \tag{4-65}$$

This presents the test statistic in terms of the sample proportion instead of the number of items X in the sample that belongs to the class of interest.

4-7.2 Type II Error and Choice of Sample Size

It is possible to obtain closed-form equations for the approximate β-error for the tests in Section 4-7.1. Suppose that p is the true value of the population proportion.

The approximate β-error for the two-sided alternative H_1: $p \neq p_0$ is

$$\beta = \Phi\left(\frac{p_0 - p + z_{\alpha/2}\sqrt{p_0(1 - p_0)/n}}{\sqrt{p(1 - p)/n}}\right)$$

$$- \Phi\left(\frac{p_0 - p - z_{\alpha/2}\sqrt{p_0(1 - p_0)/n}}{\sqrt{p(1 - p)/n}}\right) \tag{4-66}$$

If the alternative is $H_1: p < p_0$, then

$$\beta = 1 - \Phi\left(\frac{p_0 - p - z_\alpha\sqrt{p_0(1 - p_0)/n}}{\sqrt{p(1 - p)/n}}\right) \qquad (4\text{-}67)$$

whereas if the alternative is $H_1: p > p_0$, then

$$\beta = \Phi\left(\frac{p_0 - p + z_\alpha\sqrt{p_0(1 - p_0)/n}}{\sqrt{p(1 - p)/n}}\right) \qquad (4\text{-}68)$$

These equations can be solved to find the approximate sample size n that gives a test of level α that has a specified β risk.

controlling alpha & beta

one sided

z_α

Sample Size for a Two-Sided Hypothesis Test on a Binomial Proportion

$$n = \left(\frac{z_{\alpha/2}\sqrt{p_0(1 - p_0)} + z_\beta\sqrt{p(1 - p)}}{p - p_0}\right)^2 \qquad (4\text{-}69)$$

For a one-sided alternative, replace $z_{\alpha/2}$ in equation 4-69 by z_α.

EXAMPLE 4-13

Consider the semiconductor manufacturer from Example 4-12. Suppose that the process fallout is really $p = 0.03$. What is the β-error for his test of process capability, which uses $n = 200$ and $\alpha = 0.05$?

The β-error can be computed using equation 4-67 as follows:

$$\beta = 1 - \Phi\left(\frac{0.05 - 0.03 - (1.645)\sqrt{0.05(0.95)/200}}{\sqrt{0.03(1 - 0.03)/200}}\right) = 1 - \Phi(-0.44) = 0.67$$

Thus, the probability is about 0.7 that the semiconductor manufacturer will fail to conclude that the process is capable if the true process fraction defective is $p = 0.03$ (3%). This appears to be a large β-error, but the difference between $p = 0.05$ and $p = 0.03$ is fairly small, and the sample size $n = 200$ is not particularly large.

Suppose that the semiconductor manufacturer was willing to accept a β-error as large as 0.10 if the true value of the process fraction defective was $p = 0.03$. If the manufacturer continues to use $\alpha = 0.05$, what sample size would be required?

The required sample size can be computed from equation 4-69 as follows:

$$n = \left(\frac{1.645\sqrt{0.05(0.95)} + 1.28\sqrt{0.03(0.97)}}{0.03 - 0.05} \right)^2$$

$$\approx 832$$

where we have used $p = 0.03$ in equation 4-69 and $z_{\alpha/2}$ is replaced by z_α for the one-sided alternative. Note that $n = 832$ is a very large sample size. However, we are trying to detect a fairly small deviation from the null value $p_0 = 0.05$.

4-7.3 Confidence Interval on a Binomial Proportion

It is straightforward to find an approximate $100(1 - \alpha)\%$ confidence interval on a binomial proportion using the normal approximation. Recall that the sampling distribution of $\hat{P}$ is approximately normal with mean p and variance $p(1 - p)/n$, if p is not too close to either 0 or 1 and if n is relatively large. Then the distribution of

$$Z = \frac{X - np}{\sqrt{np(1 - p)}} = \frac{\hat{P} - p}{\sqrt{\dfrac{p(1 - p)}{n}}} \tag{4-70}$$

is approximately standard normal.

To construct the confidence interval on p, note that

$$P(-z_{\alpha/2} \le Z \le z_{\alpha/2}) \approx 1 - \alpha$$

so that

$$P\left(-z_{\alpha/2} \le \frac{\hat{P} - p}{\sqrt{\dfrac{p(1 - p)}{n}}} \le z_{\alpha/2} \right) \approx 1 - \alpha$$

This may be rearranged as

$$P\left(\hat{P} - z_{\alpha/2}\sqrt{\frac{p(1 - p)}{n}} \le p \le \hat{P} + z_{\alpha/2}\sqrt{\frac{p(1 - p)}{n}} \right) \approx 1 - \alpha \tag{4-71}$$

The quantity $\sqrt{p(1 - p)/n}$ in equation 4-71 is called the **standard error of the point estimator** $\hat{P}$. Unfortunately, the upper and lower limits of the confidence interval obtained from equation 4-71 contain the unknown parameter p. However, a satisfactory solution is to replace p by $\hat{P}$ in the standard error, which results in

$$P\left(\hat{P} - z_{\alpha/2}\sqrt{\frac{\hat{P}(1 - \hat{P})}{n}} \le p \le \hat{P} + z_{\alpha/2}\sqrt{\frac{\hat{P}(1 - \hat{P})}{n}} \right) \approx 1 - \alpha \tag{4-72}$$

Equation 4-72 leads to the approximate $100(1 - \alpha)\%$ confidence interval on p.

Confidence Interval on a Binomial Proportion

If $\hat{p}$ is the proportion of observations in a random sample of size n that belong to a class of interest, then an approximate $100(1 - \alpha)\%$ confidence interval on the proportion p of the population that belongs to this class is

$$\hat{p} - z_{\alpha/2} \sqrt{\frac{\hat{p}(1 - \hat{p})}{n}} \leq p \leq \hat{p} + z_{\alpha/2} \sqrt{\frac{\hat{p}(1 - \hat{p})}{n}} \qquad (4\text{-}73)$$

where $z_{\alpha/2}$ is the upper $100 \, \alpha/2$ percentage point of the standard normal distribution.

This procedure depends on the adequacy of the normal approximation to the binomial. To be reasonably conservative, this requires that np and $n(1 - p)$ be greater than or equal to 5. In situations where this approximation is inappropriate, particularly in cases where n is small, other methods must be used. Tables of the binomial distribution could be used to obtain a confidence interval for p. However, we prefer to use numerical methods based on the binomial probability mass function that are implemented in computer programs.

EXAMPLE 4-14

In a random sample of 85 automobile engine crankshaft bearings, 10 have a surface finish that is rougher than the specifications allow. Therefore, a point estimate of the proportion of bearings in the population that exceed the roughness specification is $\hat{p} = x/n = 10/85 = 0.12$. A 95% two-sided confidence interval for p is computed from equation 4-73 as

$$\hat{p} - z_{0.025} \sqrt{\frac{\hat{p}(1 - \hat{p})}{n}} \leq p \leq \hat{p} + z_{0.025} \sqrt{\frac{\hat{p}(1 - \hat{p})}{n}}$$

or

$$0.12 - 1.96 \sqrt{\frac{0.12(0.88)}{85}} \leq p \leq 0.12 + 1.96 \sqrt{\frac{0.12(0.88)}{85}}$$

which simplifies to

$$0.05 \leq p \leq 0.19$$

Choice of Sample Size

Since $\hat{P}$ is the point estimator of p, we can define the error in estimating p by $\hat{P}$ as $E = |\hat{P} - p|$. Note that we are approximately $100(1 - \alpha)\%$ confident that this error is less than $z_{\alpha/2}\sqrt{p(1-p)/n}$. For instance, in Example 4-14, we are 95% confident that the sample proportion $\hat{p} = 0.12$ differs from the true proportion p by an amount not exceeding 0.07.

In situations where the sample size can be selected, we may choose n to be 100 $(1 - \alpha)\%$ confident that the error is less than some specified value E. If we set $E = z_{\alpha/2}\sqrt{p(1-p)/n}$ and solve for n, we obtain the following formula.

Sample Size for a Specified E on a Binomial Proportion

If $\hat{P}$ is used as an estimate of p, we can be $100(1 - \alpha)\%$ confident that the error $|\hat{P} - p|$ will not exceed a specified amount E when the sample size is

$$n = \left(\frac{z_{\alpha/2}}{E}\right)^2 p(1 - p) \tag{4-74}$$

$$\frac{\sqrt{n}}{\sqrt{p(1-p)}} = \frac{z_{\alpha/2}}{E}$$

$$z_{\frac{\alpha}{2}} = \frac{E}{\sqrt{\frac{p(1-p)}{n}}}$$

An estimate of p is required to use equation 4-74. If an estimate $\hat{p}$ from a previous sample is available, it can be substituted for p in equation 4-74, or perhaps a subjective estimate can be made. If these alternatives are unsatisfactory, a preliminary sample can be taken, $\hat{p}$ computed, and then equation 4-74 used to determine how many additional observations are required to estimate p with the desired accuracy. Another approach to choosing n uses the fact that the sample size from equation 4-74 will always be a maximum for $p = 0.5$ [that is, $p(1 - p) \le 0.25$ with equality for $p = 0.5$], and this can be used to obtain an upper bound on n. In other words, we are at least $100(1 - \alpha)\%$ confident that the error in estimating p by $\hat{p}$ is less than E if the sample size is selected as follows.

For a specified error E, an upper bound on the sample size for estimating p is

$$n = \left(\frac{z_{\alpha/2}}{E}\right)^2 \frac{1}{4} \tag{4-75}$$

EXAMPLE 4-15

Consider the situation in Example 4-14. How large a sample is required if we want to be 95% confident that the error in using $\hat{p}$ to estimate p is less than 0.05? Using $\hat{p} = 0.12$ as an initial estimate of p, we find from equation 4-74 that the required sample size is

$$n = \left(\frac{z_{0.025}}{E}\right)^2 \hat{p}(1 - \hat{p}) = \left(\frac{1.96}{0.05}\right)^2 0.12(0.88) \cong 163$$

If we wanted to be *at least* 95% confident that our estimate $\hat{p}$ of the true proportion p was within 0.05 regardless of the value of p, then we would use equation 4-75 to find the sample size

$$n = \left(\frac{z_{0.025}}{E}\right)^2 (0.25) = \left(\frac{1.96}{0.05}\right)^2 (0.25) \cong 385$$

Note that if we have information concerning the value of p, either from a preliminary sample or from past experience, we could use a smaller sample while maintaining both the desired precision of estimation and the level of confidence.

One-Sided Confidence Intervals

To find an approximate $100(1 - \alpha)\%$ lower confidence interval on p, simply replace $-z_{\alpha/2}$ with $-z_{\alpha}$ in the lower bound of equation 4-73 and set the upper bound to ∞. Similarly, to find an approximate $100(1 - \alpha)\%$ upper confidence interval on p, replace $z_{\alpha/2}$ with z_{α} in the upper bound of equation 4-73 and set the lower bound to $-\infty$. These formulas are given in the table on the inside front cover.

EXERCISES FOR SECTION 4-7

4-48. Of 1000 randomly selected cases of lung cancer, 823 resulted in death.
(a) Test the hypotheses $H_0: p = 0.85$ versus $H_1: p \neq 0.85$ with $\alpha = 0.05$.
(b) Construct a 95% two-sided confidence interval on the death rate from lung cancer.

4-49. How large a sample would be required in Exercise 4-48 to be at least 95% confident that the error in estimating the death rate from lung cancer is less than 0.03?

4-50. A random sample of 50 suspension helmets used by motorcycle riders and automobile race-car drivers was subjected to an impact test, and on 18 of these helmets some damage was observed.
(a) Test the hypotheses $H_0: p = 0.3$ versus $H_1: p \neq 0.3$ with $\alpha = 0.05$.
(b) Find a 95% two-sided confidence interval on the true proportion of helmets of this type that would show damage from this test. Explain how this confidence interval can be used to test the hypothesis in part (a).
(c) Using the point estimate of p obtained from the preliminary sample of 50 helmets, how many helmets must be tested to be 95% confident that the error in estimating the true value of p is less than 0.02?
(d) How large must the sample be if we wish to be at least 95% confident that the error in estimating p is less than 0.02, regardless of the true value of p?

4-51. The Arizona Department of Transportation wishes to survey state residents to determine what proportion of the population would be in favor of building a citywide light-rail system. How many residents do they need to survey if they want to be at least 99% confident that the sample proportion is within 0.05 of the true proportion?

4-52. A manufacturer of electronic calculators is interested in estimating the fraction of defective units produced. A random sample of 800 calculators contains 10 defectives.
(a) Formulate and test an appropriate hypothesis to determine if the frac-

tion defective exceeds 0.01 at the 0.05 level of significance.

(b) Compute a 99% upper confidence interval on the fraction defective.

4-53. A study is to be conducted of the percentage of homeowners who have a high-speed Internet connection. How large a sample is required if we wish to be 95% confident that the error in estimating this quantity is less than 0.02?

4-54. The fraction of defective integrated circuits produced in a photolithography process is being studied. A random sample of 300 circuits is tested, revealing 18 defectives.

(a) Use the data to test $H_0: p = 0.04$ versus $H_1: p \neq 0.04$. Use $\alpha = 0.05$.

(b) Find the P-value for the test.

4-55. Consider the defective circuit data and hypotheses in Exercise 4-54.

(a) Suppose that the fraction defective is actually $p = 0.05$. What is the β-error for this test?

(b) Suppose that the manufacturer is willing to accept a β-error of 0.10 if the true value of p is 0.05. With $\alpha = 0.05$, what sample size would be required?

4-56. An article in *Fortune* (September 21, 1992) claimed that about one-half of all engineers continue academic studies beyond the B.S. degree, ultimately receiving either an M.S. or a Ph.D. degree. Data from an article in *Engineering Horizons* (Spring 1990) indicated that 117 of 484 new engineering graduates were planning graduate study.

(a) Are the data from *Engineering Horizons* consistent with the claim reported by *Fortune*? Use $\alpha = 0.05$ in reaching your conclusions.

(b) Find the P-value for this test.

4-57. A manufacturer of interocular lenses is qualifying a new grinding machine. He will qualify the machine if the percentage of polished lenses that contain surface defects does not exceed 4%. A ran-

dom sample of 300 lenses contains 14 defective lenses.

(a) Formulate and test an appropriate set of hypotheses to determine whether the machine can be qualified. Use $\alpha = 0.05$.

(b) Find the P-value for the test in part (a).

(c) Suppose that the fraction of defective lenses is actually $p = 0.02$. What is the β-error for this test?

(d) Suppose that a β-error of 0.05 is acceptable if the true value of $p = 0.02$. With $\alpha = 0.05$, what is the required sample size?

4-58. A researcher claims that at least 10% of all football helmets have manufacturing flaws that could potentially cause injury to the wearer. A sample of 200 helmets revealed that 24 helmets contained such defects.

(a) Does this finding support the researcher's claim? Use $\alpha = 0.01$.

(b) Find the P-value for this test.

4-59. A random sample of 500 registered voters in Phoenix is asked whether they favor the use of oxygenated fuels year-round to reduce air pollution. If more than 315 voters respond positively, we will conclude that at least 60% of the voters favor the use of these fuels.

(a) Find the probability of type I error if exactly 60% of the voters favor the use of these fuels.

(b) What is the type II error probability β if 75% of the voters favor this action?

4-60. The warranty for batteries for mobile phones is set at 400 operating hours, with proper charging procedures. A study of 1000 batteries is carried out and four stop operating prior to 400 hours. Do these experimental results support the claim that less than 0.2% of the company's batteries will fail during the warranty period, with proper charging procedures? Use a hypothesis testing procedure with $\alpha = 0.01$.

4-8   SUMMARY TABLES OF INFERENCE PROCEDURES FOR A SINGLE SAMPLE

The tables on the inside front cover present a summary of all the single-sample inference procedures from this chapter. The tables contain the null hypothesis statement, the test statistic, the various alternative hypotheses and the criteria for rejecting H_0, and the formulas for constructing the $100(1 - \alpha)\%$ confidence intervals.

4-9   TESTING FOR GOODNESS OF FIT

The hypothesis testing procedures that we have discussed in previous sections are designed for problems in which the population or probability distribution is known and the hypotheses involve the parameters of the distribution. Another kind of hypothesis is often encountered: we do not know the underlying distribution of the population, and we wish to test the hypothesis that a particular distribution will be satisfactory as a population model. For example, we might wish to test the hypothesis that the population is normal.

In Chapter 3, we discussed a very useful graphical technique for this problem called **probability plotting** and illustrated how it was applied in the case of a normal distribution. In this section, we describe a formal goodness-of-fit test procedure based on the chi-square distribution.

The test procedure requires a random sample of size n from the population whose probability distribution is unknown. These n observations are arranged in a frequency histogram, having k bins or class intervals. Let O_i be the observed frequency in the ith class interval. From the hypothesized probability distribution, we compute the expected frequency in the ith class interval, denoted E_i. The test statistic is

$$X_0^2 = \sum_{i=1}^{k} \frac{(O_i - E_i)^2}{E_i} \qquad (4\text{-}76)$$

It can be shown that, if the population follows the hypothesized distribution, X_0^2 has, approximately, a chi-square distribution with $k - p - 1$ degrees of freedom, where p represents the number of parameters of the hypothesized distribution estimated by sample statistics. This approximation improves as n increases. We would reject the hypothesis that the distribution of the population is the hypothesized distribution if the calculated value of the test statistic $\chi_0^2 > \chi_{\alpha,\, k-p-1}^2$.

One point to be noted in the application of this test procedure concerns the magnitude of the expected frequencies. If these expected frequencies are too small, then the test statistic X_0^2 will not reflect the departure of observed from expected, but only the small magnitude of the expected frequencies. There is no general agreement regarding the minimum value of expected frequencies, but values of 3, 4, and 5 are widely used as minimal. Some writers suggest that an expected frequency could be as small as 1 or 2, so long as most of them exceed 5. Should an expected frequency be too small, it can be combined with the expected frequency in an adjacent class interval. The corresponding observed frequencies would then also be combined, and k would be reduced by one. Class intervals are not required to be of equal width.

We now give an example of the test procedure.

EXAMPLE 4-16

A Poisson Distribution

The number of defects in printed circuit boards is hypothesized to follow a Poisson distribution. A random sample of $n = 60$ printed boards has been collected, and the number of defects per printed circuit board observed. The following data result:

Number of Defects	Observed Frequency
0	32
1	15
2	9
3	4

The mean of the assumed Poisson distribution in this example is unknown and must be estimated from the sample data. The estimate of the mean number of defects per board is the sample average—that is, $(32 \cdot 0 + 15 \cdot 1 + 9 \cdot 2 + 4 \cdot 3)/60 = 0.75$. From the Poisson distribution with parameter 0.75, we may compute p_i, the theoretical, hypothesized probability associated with the ith class interval. Since each class interval corresponds to a particular number of defects, we may find the p_i as follows:

$$p_1 = P(X = 0) = \frac{e^{-0.75}(0.75)^0}{0!} = 0.472$$

$$p_2 = P(X = 1) = \frac{e^{-0.75}(0.75)^1}{1!} = 0.354$$

$$p_3 = P(X = 2) = \frac{e^{-0.75}(0.75)^2}{2!} = 0.133$$

$$p_4 = P(X \geq 3) = 1 - (p_1 + p_2 + p_3) = 0.041$$

The expected frequencies are computed by multiplying the sample size $n = 60$ by the probabilities p_i; that is, $E_i = np_i$. The expected frequencies are shown next.

Number of Defects	Probability	Expected Frequency
0	0.472	28.32
1	0.354	21.24
2	0.133	7.98
3 (or more)	0.041	2.46

Since the expected frequency in the last cell is less than 3, we combine the last two cells:

Number of Defects	Observed Frequency	Expected Frequency
0	32	28.32
1	15	21.24
2 (or more)	13	10.44

The chi-square test statistic in equation 4-76 will have $k - p - 1 = 3 - 1 - 1 = 1$ degree of freedom, because the mean of the Poisson distribution was estimated from the data.

The eight-step hypothesis testing procedure may now be applied, using $\alpha = 0.05$, as follows:

1. The variable of interest is the form of the distribution of defects in printed circuit boards.

2. H_0: The form of the distribution of defects is Poisson.

3. H_1: The form of the distribution of defects is not Poisson.

4. $\alpha = 0.05$

5. The test statistic is

$$\chi_0^2 = \sum_{i=1}^{k} \frac{(o_i - E_i)^2}{E_i}$$

6. Reject H_0 if $\chi_0^2 > \chi_{0.05,1}^2 = 3.84$.

7. Computation:

$$\chi_0^2 = \frac{(32 - 28.32)^2}{28.32} + \frac{(15 - 21.24)^2}{21.24} + \frac{(13 - 10.44)^2}{10.44} = 2.94$$

8. Conclusion: Since $\chi_0^2 = 2.94 < \chi_{0.05,1}^2 = 3.84$, we are unable to reject the null hypothesis that the distribution of defects in printed circuit boards is Poisson. The P-value for the test is $P = 0.0864$. (This value was computed using a calculator.)

EXERCISES FOR SECTION 4-9

4-61. Consider the following frequency table of observations on the random variable X.

Values	0	1	2	3	4	5
Observed Frequency	8	25	23	21	16	7

(a) Based on these 100 observations, is a Poisson distribution with a mean of 2.4 an appropriate model? Perform a goodness-of-fit procedure with $\alpha = 0.05$.

(b) Calculate the P-value for this test.

4-62. Let X denote the number of flaws observed on a large coil of galvanized steel. Seventy-five coils are inspected and the following data were observed for the values of X.

Values	1	2	3	4	5	6	7	8
Observed Frequency	1	11	8	13	11	12	10	9

(a) Does the assumption of Poisson distribution with a mean of 6.0 seem appropriate as a probability model for these data? Use $\alpha = 0.01$.

(b) Calculate the P-value for this test.

4-63. The number of calls arriving to a switchboard from noon to 1 P.M. during the business days Monday through Friday is monitored for 6 weeks (i.e., 30 days). Let X be defined as the number of calls during that 1-hour period. The observed frequency of calls was recorded and reported as follows:

Value	5	6	8	9	10
Observed Frequency	2	3	3	3	6

Value	11	12	13	14	15
Observed Frequency	4	3	3	1	2

(a) Does the assumption of a Poisson distribution seem appropriate as a probability model for these data? Use $\alpha = 0.05$.

(b) Calculate the P-value for this test.

4-64. The number of cars passing eastbound through the intersection of Mill Avenue and University Avenue has been tabulated by a group of civil engineering students. They have obtained the following data:

Vehicles per Minute	Observed Frequency	Vehicles per Minute	Observed Frequency
40	14	53	102
41	24	54	96
42	57	55	90
43	111	56	81
44	194	57	73
45	256	58	64
46	296	59	61
47	378	60	59
48	250	61	50
49	185	62	42
50	171	63	29
51	150	64	18
52	110	65	15

(a) Does the assumption of a Poisson distribution seem appropriate as a probability model for this process? Use $\alpha = 0.05$.

(b) Calculate the P-value for this test.

4-65. Consider the following frequency table of observations on the random variable X.

Values	0	1	2	3	4
Frequency	4	21	10	13	2

(a) Based on these 50 observations, is a binomial distribution with $n = 6$ and $p = 0.25$ an appropriate model? Perform a goodness-of-fit procedure with $\alpha = 0.05$.

(b) Calculate the P-value for this test.

4-66. Define X as the number of underfilled bottles in a filling operation in a carton of 12 bottles. Eighty cartons are inspected and the following observations on X are recorded.

Values	0	1	2	3	4
Frequency	23	39	12	5	1

(a) Based on these 80 observations, is a binomial distribution an appropriate model? Perform a goodness-of-fit procedure with $\alpha = 0.10$.

(b) Calculate the P-value for this test.

SUPPLEMENTAL EXERCISES

4-67. If we plot the probability of accepting H_0: $\mu = \mu_0$ versus various values of μ and connect the points with a smooth curve, we obtain the **operating characteristic curve** (or the **OC curve**) of the test procedure. These curves are used extensively in industrial applications of hypothesis testing to display the sensitivity and relative performance of the test. When the true mean is really equal to μ_0, the probability of accepting H_0 is $1 - \alpha$. Construct an OC curve for Exercise 4-17, using values of the true mean μ of 178, 181, 184, 187, 190, 193, 196, and 199.

4-68. Convert the OC curve in the previous problem into a plot of the **power function** of the test.

4-69. Consider the confidence interval for μ with known standard deviation σ:

$$\bar{x} - z_{\alpha_1}\sigma/\sqrt{n} \le \mu \le \bar{x} + z_{\alpha_2}\sigma/\sqrt{n}$$

where $\alpha_1 + \alpha_2 = \alpha$. Let $\alpha = 0.05$ and find the interval for $\alpha_1 = \alpha_2 = \alpha/2 = 0.025$. Now find the interval for the case $\alpha_1 = 0.01$ and $\alpha_2 = 0.04$. Which interval is shorter? Is there any advantage to a "symmetric" confidence interval?

4-70. Formulate the appropriate null and alternative hypotheses to test the following claims.

(a) A plastics production engineer claims that 99.95% of the plastic tube manufactured by his company meets the engineering specifications requiring the length to exceed 6.5 inches.

(b) A chemical and process engineering

team claims that the mean temperature of a resin bath is greater than 45°C.

(c) The proportion of start-up software companies that successfully market their product within 3 years of company formation is less than 0.05.

(d) A chocolate bar manufacturer claims that, at the time of purchase by a consumer, the mean life of its product is less than 90 days.

(e) The designer of a computer laboratory at a major university claims that the standard deviation of time of a student on the network is less than 10 minutes.

(f) A manufacturer of traffic signals advertises that its signals will have a mean operating life in excess of 2160 hours.

4-71. A normal population has known mean $\mu = 50$ and variance $\sigma^2 = 5$. What is the approximate probability that the sample variance is greater than or equal to 7.44? less than or equal to 2.56?

(a) For a random sample of $n = 16$.

(b) For a random sample of $n = 30$.

(c) For a random sample of $n = 71$.

(d) Compare your answers to parts (a)–(c) for the approximate probability that the sample variance is greater than or equal to 7.44. Explain why this tail probability is increasing or decreasing with increased sample size.

(e) Compare your answers to parts (a)–(c) for the approximate probability that the sample variance is less than or equal to 2.56. Explain why this tail probability is increasing or decreasing with increased sample size.

4-72. An article in the *Journal of Sports Science* (1987, Vol. 5, pp. 261–271) presents the results of an investigation of the hemoglobin level of Canadian Olympic ice hockey players. The data reported are as follows (in g/dl):

15.3	16.0	14.4	16.2	16.2
14.9	15.7	15.3	14.6	15.7
16.0	15.0	15.7	16.2	14.7
14.8	14.6	15.6	14.5	15.2

(a) Given the probability plot of the data in Fig. 4-21, what is a logical assumption about the underlying distribution of the data?

(b) Explain why this check of the distribution underlying the sample data is important if we want to construct a confidence interval on the mean.

(c) Based on these sample data, a 95% confidence interval for the mean is [15.04, 15.62]. Is it reasonable to infer that the true mean could be 14.5? Explain your answer.

(d) Explain why this check of the distribution underlying the sample data is important if we want to construct a confidence interval on the variance.

(e) Based on these sample data, a 95% confidence interval for the variance is [0.22, 0.82]. Is it reasonable to infer that the true variance could be 0.35? Explain your answer.

(f) Is it reasonable to use these confidence intervals to draw an inference about the mean and variance of hemoglobin levels
  (i) Of Canadian doctors? Explain your answer.
  (ii) Of Canadian children ages 6–12? Explain your answer.

4-73. The article "Mix Design for Optimal Strength Development of Fly Ash Concrete" (*Cement and Concrete Research,* 1989, Vol. 19, No. 4, pp. 634–640) investigates the compressive strength of concrete when mixed with fly ash (a mixture of silica, alumina, iron, magnesium oxide, and other ingredients). The compressive strength for nine samples in dry conditions on the twenty-eighth day are as follows (in Mpa):

40.2	30.4	28.9	30.5	22.4
25.8	18.4	14.2	15.3	

(a) Given the probability plot of the data in Fig. 4-22, what is a logical

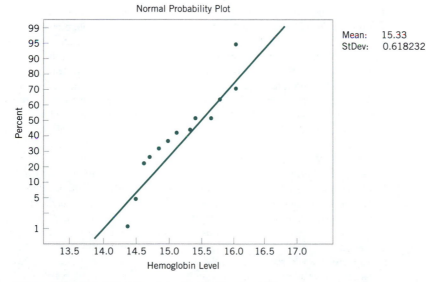

Mean: 15.33
StDev: 0.618232

Figure 4-21 Probability plot of the data for Exercise 4-72.

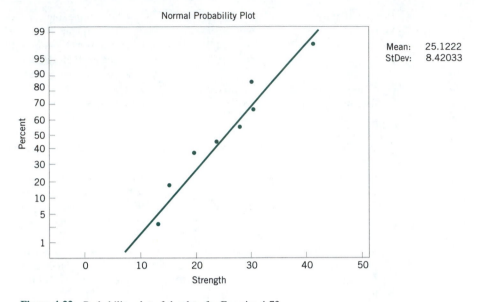

Normal Probability Plot

Mean: 25.1222
StDev: 8.42033

Figure 4-22   Probability plot of the data for Exercise 4-73.

assumption about the underlying distribution of the data?

(b) Find a 99% one-sided lower confidence interval on mean compressive strength. Provide a practical interpretation of this interval.

(c) Find a 98% two-sided confidence interval on mean compressive strength. Provide a practical interpretation of this interval and explain why the lower end-point of the interval is or is not the same as in part (b).

(d) Find a 99% one-sided upper confidence interval on the variance of compressive strength. Provide a practical interpretation of this interval.

(e) Find a 98% two-sided confidence interval on the variance of compression strength. Provide a practical interpretation of this interval and explain why the upper end-point of the interval is or is not the same as in part (d).

(f) Suppose it was discovered that the largest observation 40.2 was misrecorded and should actually be 20.4.

Now the sample mean $\bar{x} = 22.9$ and the sample variance $s^2 = 39.83$. Use these new values and repeat parts (c) and (e). Compare the original computed intervals and the newly computed intervals with the corrected observation value. How does this mistake affect the values of the sample mean, sample variance, and the width of the two-sided confidence intervals?

(g) Suppose, instead, it was discovered that the largest observation 40.2 is correct, but that the observation 25.8 is incorrect and should actually be 24.8. Now the sample mean $\bar{x} = 25.0$ and the sample variance $s^2 = 70.84$. Use these new values and repeat parts (c) and (e). Compare the original computed intervals and the newly computed intervals with the corrected observation value. How does this mistake affect the values of the sample mean, sample variance, and the width of the two-sided confidence intervals?

(h) Use the results from parts (f) and (g)

to explain the effect of mistakenly recorded values on sample estimates. Comment on the effect when the mistaken values are near the sample mean and when they are not.

4-74. An operating system for a personal computer has been studied extensively, and it is known that the standard deviation of the response time following a particular command is $\sigma = 8$ milliseconds. A new version of the operating system is installed, and we wish to estimate the mean response time for the new system to ensure that a 95% confidence interval for μ has length at most 5 milliseconds.

(a) If we can assume that response time is normally distributed and that $\sigma = 8$ for the new system, what sample size would you recommend?

(b) Suppose we are told by the vendor that the standard deviation of the response time of the new system is smaller—say, $\sigma = 6$; give the sample size that you recommend and comment on the effect the smaller standard deviation has on this calculation.

(c) Suppose you cannot assume that the response time of the new system is normally distributed but think that it may follow a Weibull distribution. What is the minimum sample size you would recommend to construct any confidence interval on the true mean response time?

4-75. A manufacturer of semiconductor devices takes a random sample of size n of chips and tests them, classifying each chip as defective or nondefective. Let $X_i = 0$ if the chip is nondefective and $X_i = 1$ if the chip is defective. The sample fraction defective is

$$\hat{p}_i = \frac{X_1 + X_2 + \cdots + X_n}{n}$$

What are the sampling distribution, the sample mean, and sample variance estimates of $\hat{p}$ when

(a) The sample size is $n = 60$?
(b) The sample size is $n = 70$?
(c) The sample size is $n = 100$?
(d) Compare your answers to parts (a)–(c) and comment on the effect of sample size on the variance of the sampling distribution.

4-76. Consider the description of Exercise 4-75. After collecting a sample, we are interested in computing the error in estimating the true value p. For each of the sample sizes and estimates of p, compute the error at the 95% confidence level.

(a) $n = 60$ and $\hat{p} = 0.10$
(b) $n = 70$ and $\hat{p} = 0.10$
(c) $n = 100$ and $\hat{p} = 0.10$
(d) Compare your results from parts (a)–(c) and comment on the effect of sample size on the error in estimating the true value of p and the 95% confidence level.
(e) Repeat parts (a)–(d), this time using a 99% confidence level.
(f) Examine your results when the 95% confidence level and then the 99% confidence level are used to compute the error and explain what happens to the magnitude of the error as the percentage confidence increases.

4-77. A quality control inspector of flow metering devices, used to administer fluid intravenously, will perform a hypothesis test to determine whether the mean flow rate is different from the flow rate setting of 200 ml/h. Based on prior information the standard deviation of the flow rate is assumed to be known and equal to 12 ml/h. For each of the following sample sizes and a fixed $\alpha = 0.05$, find the probability of a type II error if the true mean is 205 ml/h.

(a) $n = 25$
(b) $n = 60$
(c) $n = 100$
(d) Does the probability of a type II error increase or decrease as the sample size increases? Explain your answer.

4-78. Suppose that in Exercise 4-77, the experimenter had believed that $\sigma = 14$. For each of the following sample sizes and a fixed $\alpha = 0.05$, find the probability of a type II error if the true mean is 205 ml/h.

(a) $n = 20$

(b) $n = 50$

(c) $n = 100$

(d) Comparing your answers to those in Exercise 4-77, does the probability of a type II error increase or decrease with the increase in standard deviation? Explain your answer.

4-79. The life in hours of a heating element used in a furnace is known to be approximately normally distributed. A random sample of 15 heating elements is selected and found to have an average life of 598.14 hours and a sample standard deviation of 16.93 hours.

(a) At the $\alpha = 0.05$ level of significance, use the appropriate eight-step procedure to test the hypotheses $H_0: \mu = 550$ versus $H_1: \mu > 550$. On completing the hypothesis test, do you believe that the true mean life of a heating element is greater than 550 hours? Clearly state your answer.

(b) Find the p-value of the test statistic.

(c) Construct a 95% lower confidence interval on the mean and describe how this interval can be used to test the alternative hypothesis of part (a).

(d) Construct a two-sided 95% confidence interval for the underlying variance.

4-80. Suppose we wish to test the hypothesis $H_0: \mu = 85$ versus the alternative $H_1: \mu > 85$ where $\sigma = 16$. Suppose that the true mean is $\mu = 86$ and that in the practical context of the problem this is not a departure from $\mu_0 = 85$ that has practical significance.

(a) For a test with $\alpha = 0.01$, compute β for the sample sizes $n = 25, 100, 400$, and 2500 assuming that $\mu = 86$.

(b) Suppose the sample average is $\bar{x} = 86$. Find the P-value for the test statistic for the different sample sizes specified in part (a). Would the data be statistically significant at $\alpha = 0.01$?

(c) Comment on the use of a large sample size in this exercise.

4-81. The cooling system in a nuclear submarine consists of an assembly of welded pipes through which a coolant is circulated. Specifications require that weld strength must meet or exceed 150 psi.

(a) Suppose that the design engineers decide to test the hypothesis $H_0: \mu = 150$ versus $H_1: \mu > 150$. Explain why this choice of alternative hypothesis is better than $H_1: \mu < 150$.

(b) A random sample of 20 welds results in $\bar{x} = 157.65$ psi and $s = 12.39$ psi. What conclusions can you draw about the hypothesis in part (a)? State any necessary assumptions about the underlying distribution of the data. Assume $\alpha = 0.05$.

4-82. Suppose we are testing $H_0: p = 0.5$ versus $H_1: p \neq 0.5$. Suppose that p is the true value of the population proportion.

(a) Using $\alpha = 0.05$, find the power of the test for $n = 100, 150$, and 300 assuming that $p = 0.6$. Comment on the effect of sample size on the power of the test.

(b) Using $\alpha = 0.01$, find the power of the test for $n = 100, 150$, and 300 assuming that $p = 0.6$. Compare your answers to those from part (a) and comment on the effect of α on the power of the test for different sample sizes.

(c) Using $\alpha = 0.05$, find the power of the test for $n = 100$, assuming $p = 0.08$. Compare your answer to part (a) and comment on the effect of the true value of p on the power of the test for the same sample size and α level.

(d) Using $\alpha = 0.01$, what sample size is required if $p = 0.6$ and we want $\beta = 0.05$? What sample is required if $p = 0.8$ and we want $\beta = 0.05$? Compare the two sample sizes and comment on the effect of the true value of p on sample size required when β is held approximately constant.

4-83. Consider the biomedical device experiment described in Exercise 4-35.
  (a) For this sample size $n = 15$, do the data support the claim that the standard deviation of life is less than 280 hours?
  (b) Suppose that instead of $n = 15$, the sample size was 51. Repeat the analysis performed in part (a) using $n = 51$.
  (c) Compare your answers and comment on how sample size affects your conclusions drawn in parts (a) and (b).

4-84. Consider the fatty acid measurements for the diet margarine described in Exercise 4-36.
  (a) For this sample size $n = 6$, using a two-sided alternative hypothesis and $\alpha = 0.01$, test $H_0: \sigma^2 = 1.0$.
  (b) Suppose that instead of $n = 6$, the sample size was $n = 51$. Repeat the analysis performed in part (a) using $n = 51$.
  (c) Compare your answers and comment on how sample size affects your conclusions drawn in parts (a) and (b).

4-85. A manufacturer of precision measuring instruments claims that the standard deviation in the use of the instruments is at most 0.00002 mm. An analyst, who is unaware of the claim, uses the instrument eight times and obtains a sample standard deviation of 0.00001 mm.
  (a) Confirm using a test procedure and an α level of 0.01 that there is insufficient evidence to support the claim that the standard deviation of the instruments is at most 0.00002. State any necessary assumptions about the underlying distribution of the data.
  (b) Explain why the sample standard deviation, $s = 0.00001$, is less than 0.00002, yet the statistical test procedure results do not support the claim.

4-86. A biotechnology company produces a therapeutic drug whose concentration has a standard deviation of 4 g/l. A new method of producing this drug has been proposed, although some additional cost is involved. Management will authorize a change in production technique only if the standard deviation of the concentration in the new process is less than 4 g/l. The researchers chose $n = 10$ and obtained the following data. Perform the necessary analysis to determine whether a change in production technique should be implemented.

16.628 g/l	16.630 g/l
16.622	16.631
16.627	16.624
16.623	16.622
16.618	16.626

4-87. A manufacturer of electronic calculators claims that less than 1% of his production output is defective. A random sample of 1200 calculators contains 8 defective units.
  (a) Confirm using a test procedure and an α level of 0.01 that there is insufficient evidence to support the claim that the percentage defective is less than 1%.
  (b) Explain why the sample percentage is less than 1%, yet the statistical test procedure results do not support the claim.

4-88. An article in *The Engineer* ("Redesign for Suspect Wiring," June 1990) reported the results of an investigation into wiring errors on commercial transport aircraft that may produce faulty information to the flight crew. Such a wiring error may have been responsible for the crash of a British Midland Airways aircraft in January 1989 by causing the pilot to shut down the wrong engine. Of 1600 randomly selected aircraft, eight were found to have wiring errors that could display incorrect information to the flight crew.

(a) Find a 99% confidence interval on the proportion of aircraft that have such wiring errors.

(b) Suppose we use the information in this example to provide a preliminary estimate of p. How large a sample would be required to produce an estimate of p that we are 99% confident differs from the true value by at most 0.008?

(c) Suppose we did not have a preliminary estimate of p. How large a sample would be required if we wanted to be at least 99% confident that the sample proportion differs from the true proportion by at most 0.008 regardless of the true value of p?

(d) Comment on the usefulness of preliminary information in computing the needed sample size.

4-89. A standardized test for graduating high school seniors is designed to be completed by 75% of the students within 40 minutes. A random sample of 100 graduates showed that 64 completed the test within 40 minutes.

(a) Find a 90% confidence interval on the proportion of such graduates completing the test within 40 minutes.

(b) Find a 95% confidence interval on the proportion of such graduates completing the test within 40 minutes.

(c) Compare your answers to parts (a) and (b) and explain why they are the same or different.

(d) Could you use either of these confidence intervals to determine whether the proportion is significantly different from 0.75? Explain your answer.

[*Hint:* Use the normal approximation to the binomial.]

4-90. The proportion of adults that live in Tempe, Arizona, who are college graduates is estimated to be $p = 0.4$. To test this hypothesis, a random sample of 15 Tempe adults is selected. If the number of college graduates is between 4 and 8, the hypothesis will be accepted; otherwise, we will conclude that $p \neq 0.4$.

(a) Find the type I error probability for this procedure, assuming that $p = 0.4$.

(b) Find the probability of committing a type II error if the true proportion is really $p = 0.2$.

4-91. The proportion of residents in Phoenix favoring the building of toll roads to complete the freeway system is believed to be $p = 0.3$. If a random sample of 20 residents shows that 2 or fewer favor this proposal, we will conclude that $p < 0.3$.

(a) Find the probability of type I error if the true proportion is $p = 0.3$.

(b) Find the probability of committing a type II error with this procedure if $p = 0.2$.

(c) What is the power of this procedure if the true proportion is $p = 0.2$?

4-92. Consider the 40 observations collected on the number of nonconforming coil springs in production batches of size 50 given in Exercise 2-41 of Chapter 2.

(a) Based on the description of the random variable and these 40 observations, is a binomial distribution an appropriate model? Perform a goodness-of-fit procedure with $\alpha = 0.05$.

(b) Calculate the P-value for this test.

4-93. Consider the 20 observations collected on the number of errors in a string of 1000 bits of a communication channel given in Exercise 2-42 of Chapter 2.

(a) Based on the description of the random variable and these 20 observations, is a binomial distribution an appropriate model? Perform a goodness-of-fit procedure with $\alpha = 0.05$.

(b) Calculate the P-value for this test.

4-94. State the null and the alternative hypotheses, and indicate the type of critical region (either two-tailed, lower-tailed, or upper-tailed) to test the following claims.

(a) a manufacturer of lightbulbs has a new type of lightbulb that is advertised to have a mean burning lifetime in excess of 5000 hours.

(b) a chemical engineering firm claims that their new material can be used to make automobile tires with a mean life of more than 60,000 miles.

(c) The standard deviation of breaking strength of fiber used in making drapery material does not exceed 2 psi.

(d) A safety engineer claims that more than 60% of all drivers wear safety belts for automobile trips of less than 2 miles.

(e) A biomedical device is claimed to have a mean time to failure greater than 42,000 hours.

(f) Producers of 1-inch diameter plastic pipe claim that the standard deviation of the inside diameter is less than 0.02 inch.

(g) Lightweight, hand-held, laser range finders used by civil engineers are claimed to have a variance smaller than 0.05 meter^2.

Team Exercises

4-95. Identify an example in which a standard is specified or claim is made about a population. For example, "This type of car gets an average of 30 miles per gallon in urban driving." The standard or claim may be expressed as a mean (average), variance, standard deviation, or proportion. Collect an appropriate random sample of data and perform a hypothesis test to assess the standard or claim. Report on your results. Be sure to include in your report the claim expressed as a hypothesis test, a description of the data collected, the analysis performed, and the conclusion reached.

4-96. Consider the experimental data collected to verify that the "true" speed of light is 710.5 (299,710.5 km/sec) in 1879 and in 1882 by the physicist A. A. Michelson. Read the story associated with the data and reported on the Web site http:/lib.stat.cmu.edu/DASL/Stories/SpeedofLight.html. Use the data file to duplicate the analysis, and write a brief report summarizing your findings.

5 Decision Making for Two Samples

5-1 INTRODUCTION

The previous chapter presented hypothesis tests and confidence intervals for a single population parameter (the mean μ, the variance σ^2, or a proportion p). This chapter extends those results to the case of two independent populations.

The general situation is shown in Figure 5-1. Population 1 has mean μ_1 and variance σ_1^2, while population 2 has mean μ_2 and variance σ_2^2. Inferences will be based on two random samples of sizes n_1 and n_2, respectively. That is, $X_{11}, X_{12}, \ldots, X_{1n_1}$ is a random sample of n_1 observations from population 1, and $X_{21}, X_{22}, \ldots, X_{2n_2}$ is a random sample of n_2 observations from population 2.

5-2 INFERENCE ON THE MEANS OF TWO POPULATIONS, VARIANCES KNOWN

In this section we consider statistical inferences on the difference in means $\mu_1 - \mu_2$ of the populations shown in Figure 5-1, where the variances σ_1^2 and σ_2^2 are known. The assumptions for this section are summarized next.

Assumptions

1. $X_{11}, X_{12}, \ldots, X_{1n_1}$ is a random sample of size n_1 from population 1.
2. $X_{21}, X_{22}, \ldots, X_{2n_2}$ is a random sample of size n_2 from population 2.
3. The two populations represented by X_1 and X_2 are independent.
4. Both populations are normal, or if they are not normal, the conditions of the central limit theorem apply.

A logical point estimator of $\mu_1 - \mu_2$ is the difference in sample means $\overline{X}_1 - \overline{X}_2$. Based on the properties of expected values in Chapter 3, we have

$$E(\overline{X}_1 - \overline{X}_2) = E(\overline{X}_1) - E(\overline{X}_2) = \mu_1 - \mu_2$$

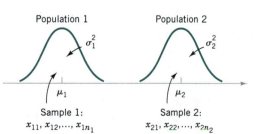

Figure 5-1 Two independent populations.

and the variance of $\overline{X}_1 - \overline{X}_2$ is

$$V(\overline{X}_1 - \overline{X}_2) = V(\overline{X}_1) + V(\overline{X}_2) = \frac{\sigma_1^2}{n_1} + \frac{\sigma_2^2}{n_2}$$

Based on the assumptions and the preceding results, we may state the following.

Under the previous assumptions, the quantity

$$Z = \frac{\overline{X}_1 - \overline{X}_2 - (\mu_1 - \mu_2)}{\sqrt{\dfrac{\sigma_1^2}{n_1} + \dfrac{\sigma_2^2}{n_2}}} \tag{5-1}$$

has a standard normal distribution, $N(0, 1)$.

This result will be used to form tests of hypotheses and confidence intervals on $\mu_1 - \mu_2$. Essentially, we may think of $\mu_1 - \mu_2$ as a parameter θ, and its estimator is $\hat{\Theta} = \overline{X}_1 - \overline{X}_2$ with variance $\sigma_{\hat{\theta}}^2 = \sigma_1^2/n_1 + \sigma_2^2/n_2$. If θ_0 is the null hypothesis value specified for θ, then the test statistic will be $(\hat{\Theta} - \theta_0)/\sigma_{\hat{\theta}}$. Note how similar this is to the test statistic for a single mean used in Chapter 4.

5-2.1   Hypothesis Testing on the Difference in Means, Variances Known

We now consider hypothesis testing on the difference in the means $\mu_1 - \mu_2$ of the two populations in Figure 5-1. Suppose that we are interested in testing that the difference in means $\mu_1 - \mu_2$ is equal to a specified value Δ_0. Thus, the null hypothesis will be stated as H_0: $\mu_1 - \mu_2 = \Delta_0$. Obviously, in many cases, we will specify $\Delta_0 = 0$ so that we are testing the equality of two means (i.e., H_0: $\mu_1 = \mu_2$). The appropriate test statistic would be found by replacing $\mu_1 - \mu_2$ in equation 5-1 by Δ_0, and this test statistic would have a standard normal distribution under H_0. Suppose that the alternative hypothesis is H_1: $\mu_1 - \mu_2 \neq \Delta_0$. Now, a sample value of $\overline{x}_1 - \overline{x}_2$ that is considerably different from Δ_0 is evidence that H_1 is true. Because Z_0 has the $N(0, 1)$ distribution when H_0 is true, we would take $-z_{\alpha/2}$ and $z_{\alpha/2}$ as the boundaries of the critical region just as we did in the single-sample hypothesis testing problem of Section 4-4.1. This would give a test with level of significance α. Critical regions for the one-sided alternatives would be located similarly. Formally, we summarize these results next.

Testing Hypotheses on the Difference in Means, Variances Known

Null hypothesis: $H_0: \mu_1 - \mu_2 = \Delta_0$

Test statistic: $Z_0 = \dfrac{\overline{X}_1 - \overline{X}_2 - \Delta_0}{\sqrt{\dfrac{\sigma_1^2}{n_1} + \dfrac{\sigma_2^2}{n_2}}}$

Alternative Hypotheses	Rejection Criterion
$H_1: \mu_1 - \mu_2 \neq \Delta_0$	$z_0 > z_{\alpha/2}$ or $z_0 < -z_{\alpha/2}$
$H_1: \mu_1 - \mu_2 > \Delta_0$	$z_0 > z_\alpha$
$H_1: \mu_1 - \mu_2 < \Delta_0$	$z_0 < -z_\alpha$

EXAMPLE 5-1

A product developer is interested in reducing the drying time of a primer paint. Two formulations of the paint are tested; formulation 1 is the standard chemistry, and formulation 2 has a new drying ingredient that should reduce the drying time. From experience, it is known that the standard deviation of drying time is 8 minutes, and this inherent variability should be unaffected by the addition of the new ingredient. Ten specimens are painted with formulation 1, and another 10 specimens are painted with formulation 2; the 20 specimens are painted in random order. The two sample average drying times are $\overline{x}_1 = 121$ min and $\overline{x}_2 = 112$ min, respectively. What conclusions can the product developer draw about the effectiveness of the new ingredient, using $\alpha = 0.05$?

We apply the eight-step procedure to this problem as follows:

1. The quantity of interest is the difference in mean drying times, $\mu_1 - \mu_2$, and $\Delta_0 = 0$.

2. $H_0: \mu_1 - \mu_2 = 0$, or $H_0: \mu_1 = \mu_2$.

3. $H_1: \mu_1 > \mu_2$. We want to reject H_0 if the new ingredient reduces mean drying time.

4. $\alpha = 0.05$

5. The test statistic is

$$z_0 = \frac{\overline{x}_1 - \overline{x}_2 - 0}{\sqrt{\dfrac{\sigma_1^2}{n_1} + \dfrac{\sigma_2^2}{n_2}}}$$

where $\sigma_1^2 = \sigma_2^2 = (8)^2 = 64$ and $n_1 = n_2 = 10$.

6. Reject H_0: $\mu_1 = \mu_2$ if $z_0 > 1.645 = z_{0.05}$.

7. Computation: Since $\bar{x}_1 = 121$ min and $\bar{x}_2 = 112$ min, the test statistic is

$$z_0 = \frac{121 - 112}{\sqrt{\dfrac{(8)^2}{10} + \dfrac{(8)^2}{10}}} = 2.52$$

8. Conclusion: Since $z_0 = 2.52 > 1.645$, we reject H_0: $\mu_1 = \mu_2$ at the $\alpha = 0.05$ level and conclude that adding the new ingredient to the paint significantly reduces the drying time. Alternatively, we can find the *P*-value for this test as

$$P\text{-value} = 1 - \Phi(2.52) = 0.0059$$

Therefore, H_0: $\mu_1 = \mu_2$ would be rejected at any significance level $\alpha \geq 0.0059$.

5-2.2   Type II Error and Choice of Sample Size

Suppose that the null hypothesis H_0: $\mu_1 - \mu_2 = \Delta_0$ is false and that the true difference in means is $\mu_1 - \mu_2 = \Delta$, where $\Delta > \Delta_0$. We may find formulas for the sample size required to obtain a specific value of the type II error probability β for a given difference in means Δ and level of significance α.

Sample Size for Two-Sided Alternative Hypothesis on the Difference in Means, Variances Known, when $n_1 = n_2$

For the two-sided alternative hypothesis with significance level α, the sample size $n_1 = n_2 = n$ required to detect a true difference in means of Δ with power at least $1 - \beta$ is

$$n \simeq \frac{(z_{\alpha/2} + z_\beta)^2(\sigma_1^2 + \sigma_2^2)}{(\Delta - \Delta_0)^2} \tag{5-2}$$

This approximation is valid when $\Phi(-z_{\alpha/2} - (\Delta - \Delta_0)\sqrt{n}/\sqrt{\sigma_1^2 + \sigma_2^2})$ is small compared to β.

Sample Size for One-Sided Alternative Hypothesis on the Difference in Means, Variances Known, when $n_1 = n_2$

For a one-sided alternative hypothesis with significance level α, the sample size $n_1 = n_2 = n$ required to detect a true difference in means of $\Delta(\neq \Delta_0)$ with power at least $1 - \beta$ is

$$n = \frac{(z_\alpha + z_\beta)^2(\sigma_1^2 + \sigma_2^2)}{(\Delta - \Delta_0)^2} \qquad (5\text{-}3)$$

The derivation of equations 5-2 and 5-3 closely follows the single-sample case in Section 4-4.3. For example, to obtain equation 5-2, we first write the expression for the β-error for the two-sided alternative, which is

$$\beta = \Phi\left(z_{\alpha/2} - \frac{\Delta - \Delta_0}{\sqrt{\dfrac{\sigma_1^2}{n_1} + \dfrac{\sigma_2^2}{n_2}}}\right) - \Phi\left(-z_{\alpha/2} - \frac{\Delta - \Delta_0}{\sqrt{\dfrac{\sigma_1^2}{n_1} + \dfrac{\sigma_2^2}{n_2}}}\right)$$

where Δ is the true difference in means of interest and Δ_0 is specified in the null hypothesis. Then by following a procedure similar to that used to obtain equation 4-26, the expression for β can be obtained for the case where $n = n_1 = n_2$.

EXAMPLE 5-2

To illustrate the use of these sample size equations, consider the situation described in Example 5-1, and suppose that if the true difference in drying times is as much as 10 minutes, we want to detect this with probability at least 0.90. Under the null hypothesis, $\Delta_0 = 0$. We have a one-sided alternative hypothesis with $\Delta = 10$, $\alpha = 0.05$ (so $z_\alpha = z_{0.05} = 1.645$), and since the power is 0.9, $\beta = 0.10$ (so $z_\beta = z_{0.10} = 1.28$). Therefore, we may find the required sample size from equation 5-3 as follows:

$$n = \frac{(z_\alpha + z_\beta)^2(\sigma_1^2 + \sigma_2^2)}{(\Delta - \Delta_0)^2} = \frac{(1.645 + 1.28)^2[(8)^2 + (8)^2]}{(10 - 0)^2}$$

$$= 11$$

5-2.3 Confidence Interval on the Difference in Means, Variances Known

The $100(1 - \alpha)\%$ confidence interval on the difference in two means $\mu_1 - \mu_2$ when the variances are known can be found directly from results given previously in this section. Recall that $X_{11}, X_{12}, \ldots, X_{1n_1}$ is a random sample of n_1 observations from the first population and $X_{21}, X_{22}, \ldots, X_{2n_2}$ is a random sample of n_2 observations from the second population. The difference in sample means $\overline{X}_1 - \overline{X}_2$ is a point estimator of $\mu_1 - \mu_2$, and

$$Z = \frac{\overline{X}_1 - \overline{X}_2 - (\mu_1 - \mu_2)}{\sqrt{\dfrac{\sigma_1^2}{n_1} + \dfrac{\sigma_2^2}{n_2}}}$$

has a standard normal distribution if the two populations are normal or is approximately standard normal if the conditions of the central limit theorem apply, respectively. This implies that

$$P(-z_{\alpha/2} \leq Z \leq z_{\alpha/2}) = 1 - \alpha$$

or

$$P\left(-z_{\alpha/2} \leq \frac{\overline{X}_1 - \overline{X}_2 - (\mu_1 - \mu_2)}{\sqrt{\dfrac{\sigma_1^2}{n_1} + \dfrac{\sigma_2^2}{n_2}}} \leq z_{\alpha/2}\right) = 1 - \alpha$$

This can be rearranged as

$$P\left(\overline{X}_1 - \overline{X}_2 - z_{\alpha/2}\sqrt{\frac{\sigma_1^2}{n_1} + \frac{\sigma_2^2}{n_2}} \leq \mu_1 - \mu_2 \leq \overline{X}_1 - \overline{X}_2 + z_{\alpha/2}\sqrt{\frac{\sigma_1^2}{n_1} + \frac{\sigma_2^2}{n_2}}\right) = 1 - \alpha$$

Therefore, the $100(1 - \alpha)\%$ confidence interval for $\mu_1 - \mu_2$ is defined as follows.

Confidence Interval on the Difference in Means, Variances Known

If $\overline{x}_1$ and $\overline{x}_2$ are the means of independent random samples of sizes n_1 and n_2 from populations with known variances σ_1^2 and σ_2^2, respectively, then a $100(1 - \alpha)\%$ confidence interval for $\mu_1 - \mu_2$ is

$$\overline{x}_1 - \overline{x}_2 - z_{\alpha/2}\sqrt{\frac{\sigma_1^2}{n_1} + \frac{\sigma_2^2}{n_2}} \leq \mu_1 - \mu_2 \leq \overline{x}_1 - \overline{x}_2 + z_{\alpha/2}\sqrt{\frac{\sigma_1^2}{n_1} + \frac{\sigma_2^2}{n_2}} \qquad (5\text{-}4)$$

where $z_{\alpha/2}$ is the upper $100\,\alpha/2$ percentage point and $-z_{\alpha/2}$ is the lower $100\,\alpha/2$ percentage point of the standard normal distribution in Appendix A Table I.

The confidence level $1 - \alpha$ is exact when the populations are normal. For nonnormal populations, the confidence level is approximately valid for large sample sizes.

EXAMPLE 5-3

Tensile strength tests were performed on two different grades of aluminum spars used in manufacturing the wing of a commercial transport aircraft. From past experience with the spar manufacturing process and the testing procedure, the standard deviations of tensile strengths are assumed to be known. The data obtained are shown in Table 5-1. If μ_1 and μ_2 denote the true mean tensile strengths for the two grades of spars, then we may find a 90% confidence interval on the difference in mean strength $\mu_1 - \mu_2$ as follows:

$$l = \bar{x}_1 - \bar{x}_2 - z_{\alpha/2} \sqrt{\frac{\sigma_1^2}{n_1} + \frac{\sigma_2^2}{n_2}}$$

$$= 87.6 - 74.5 - 1.645 \sqrt{\frac{(1.0)^2}{10} + \frac{(1.5)^2}{12}}$$

$$= 13.1 - 0.88$$

$$= 12.22 \text{ kg/mm}^2$$

$$u = \bar{x}_1 - \bar{x}_2 + z_{\alpha/2} \sqrt{\frac{\sigma_1^2}{n_1} + \frac{\sigma_2^2}{n_2}}$$

$$= 87.6 - 74.5 + 1.645 \sqrt{\frac{(1.0)^2}{10} + \frac{(1.5)^2}{12}}$$

$$= 13.1 + 0.88$$

$$= 13.98 \text{ kg/mm}^2$$

Therefore, the 90% confidence interval on the difference in mean tensile strength is

$$12.22 \text{ kg/mm}^2 \leq \mu_1 - \mu_2 \leq 13.98 \text{ kg/mm}^2$$

Note that the confidence interval does not include zero, which implies that the mean strength of aluminum grade 1 (μ_1) exceeds the mean strength of aluminum grade 2 (μ_2). In fact, we can state that we are 90% confident that the mean tensile strength of aluminum grade 1 exceeds that of aluminum grade 2 by between 12.22 and 13.98 kg/mm^2.

Table 5-1 Tensile Strength Test Result for Aluminum Spars

Spar Grade	Sample Size	Sample Mean Tensile Strength (kg/mm^2)	Standard Deviation (kg/mm^2)
1	$n_1 = 10$	$\bar{x}_1 = 87.6$	$\sigma_1 = 1.0$
2	$n_2 = 12$	$\bar{x}_2 = 74.5$	$\sigma_2 = 1.5$

Choice of Sample Size

If the standard deviations σ_1 and σ_2 are known (at least approximately) and the two sample sizes n_1 and n_2 are equal ($n_1 = n_2 = n$, say), then we can determine the sample size required so that the error in estimating $\mu_1 - \mu_2$ by $\bar{x}_1 - \bar{x}_2$ will be less than E at $100(1 - \alpha)\%$ confidence. The required sample size from each population is as follows.

Sample Size for a Specified E on the Difference in Means, and Variances Known when $n_1 = n_2$

If $\bar{x}_1$ and $\bar{x}_2$ are used as estimates of μ_1 and μ_2, respectively, then we can be $100(1 - \alpha)\%$ confident that the error $|(\bar{x}_1 - \bar{x}_2) - (\mu_1 - \mu_2)|$ will not exceed a specified amount E when the sample size $n_1 = n_2 = n$ is

$$n = \left(\frac{z_{\alpha/2}}{E}\right)^2 (\sigma_1^2 + \sigma_2^2) \tag{5-5}$$

Remember to round up if n is not an integer. This will ensure that the level of confidence does not drop below $100(1 - \alpha)\%$.

EXERCISES FOR SECTION 5-2

5-1. Two machines are used for filling plastic bottles with a net volume of 16.0 ounces. The fill volume can be assumed normal, with standard deviation $\sigma_1 = 0.020$ and $\sigma_2 = 0.025$ ounces. A member of the quality engineering staff suspects that both machines fill to the same mean net volume, whether or not this volume is 16.0 ounces. A random sample of 10 bottles is taken from the output of each machine.

Machine 1		Machine 2	
16.03	16.01	16.02	16.03
16.04	15.96	15.97	16.04
16.05	15.98	15.96	16.02
16.05	16.02	16.01	16.01
16.02	15.99	15.99	16.00

(a) Do you think the engineer is correct? Use $\alpha = 0.05$.

(b) What is the P-value for this test?

(c) What is the power of the test in (a) for a true difference in means of 0.04?

(d) Find a 95% confidence interval on the difference in means. Provide a practical interpretation of this interval.

(e) Assuming equal sample sizes, what sample size should be used to ensure that $\beta = 0.01$ if the true difference in means is 0.04? Assume that $\alpha = 0.05$.

5-2. Two types of plastic are suitable for use by an electronics component manufacturer. The breaking strength of this plastic is important. It is known that $\sigma_1 = \sigma_2 = 1.0$ psi. From a random sample of size $n_1 = 10$ and $n_2 = 12$, we obtain $\bar{x}_1 = 162.7$ and $\bar{x}_2 = 155.4$. The company will not adopt plastic 1 unless its mean breaking strength exceeds that of plastic 2 by at least 10 psi. Based on the sample information, should they use plastic 1? Use $\alpha = 0.05$ in reaching a decision.

5-3. The burning rates of two different solid-fuel propellants used in aircrew escape systems are being studied. It is known that both propellants have approximately the same standard deviation of burning rate; that is, $\sigma_1 = \sigma_2 = 3$ cm/s. Two random samples of $n_1 = 20$ and $n_2 = 20$ specimens are tested; the sample mean burning rates are $\bar{x}_1 = 18.02$ cm/s and $\bar{x}_2 = 24.37$ cm/s.

(a) Test the hypothesis that both propellants have the same mean burning rate. Use $\alpha = 0.05$.

(b) What is the P-value of the test in part (a)?

(c) What is the β-error of the test in part (a) if the true difference in mean burning rate is 2.5 cm/s?

(d) Construct a 95% confidence interval on the difference in means $\mu_1 - \mu_2$. What is the practical meaning of this interval?

5-4. Two machines are used to fill plastic bottles with dishwashing detergent. The standard deviations of fill volume are known to be $\sigma_1 = 0.10$ fluid ounces and $\sigma_2 = 0.15$ fluid ounces for the two machines, respectively. Two random samples of $n_1 = 12$ bottles from machine 1 and $n_2 = 10$ bottles from machine 2 are selected, and the sample mean fill volumes are $\bar{x}_1 = 30.61$ fluid ounces and $\bar{x}_2 = 30.34$ fluid ounces. Assume normality.

(a) Construct a 90% two-sided confidence interval on the mean difference in fill volume. Interpret this interval.

(b) Construct a 95% two-sided confidence interval on the mean difference in fill volume. Compare and comment on the width of this interval to the width of the interval in part (a).

(c) Construct a 95% upper-confidence interval on the mean difference in fill volume. Interpret this interval.

5-5. Reconsider the situation described in Exercise 5-4.

(a) Test the hypothesis that both machines fill to the same mean volume. Use $\alpha = 0.05$.

(b) What is the P-value of the test in part (a)?

(c) If the β-error of the test when the true difference in fill volume is 0.2 fluid ounces should not exceed 0.1, what sample sizes must be used? Use $\alpha = 0.05$.

5-6. Two different formulations of an oxygenated motor fuel are being tested to study their road octane numbers. The variance of road octane number for formulation 1 is $\sigma_1^2 = 1.5$, and for formulation 2 it is $\sigma_2^2 = 1.2$. Two random samples of size $n_1 = 15$ and $n_2 = 20$ are tested, and the mean road octane numbers observed are $\bar{x}_1 = 88.85$ and $\bar{x}_2 = 92.54$. Assume normality.

(a) Construct a 95% two-sided confidence interval on the difference in mean road octane number.

(b) If formulation 2 produces a higher road octane number than formulation 1, the manufacturer would like to detect it. Formulate and test an appropriate hypothesis, using $\alpha = 0.05$.

(c) What is the P-value for the test you conducted in part (b)?

5-7. Consider the situation described in Exercise 5-3. What sample size would be required in each population if we wanted the error in estimating the difference in mean burning rates to be less than 4 cm/s with 99% confidence?

5-8. Consider the road octane test situation described in Exercise 5-6. What sample size would be required in each population if we wanted to be 95% confident that the error in estimating the difference in mean road octane number is less than 1?

5-9. A polymer is manufactured in a batch chemical process. Viscosity measurements are normally made on each batch, and long experience with the process has indicated that the variability in the process is fairly stable with $\sigma = 20$. Fifteen batch viscosity measurements are given as follows: 724, 718, 776, 760, 745, 759, 795, 756, 742, 740, 761, 749, 739, 747, 742. A process change is made that involves switching the type of catalyst used in the process. Following the process change, eight batch viscosity measurements are taken: 735, 775, 729, 755, 783, 760, 738, 780. Assume that process variability is unaffected by the catalyst change. Find a 90% confidence interval on the difference in mean batch viscosity resulting from the process change.

5-10. The concentration of active ingredient in a liquid laundry detergent is thought to be affected by the type of catalyst used in the process. The standard deviation of active concentration is known to be 3 g/l, regardless of the catalyst type. Ten observations on concentration are taken with each catalyst, and the data are shown here:

Catalyst 1: 57.9, 66.2, 65.4, 65.4, 65.2, 62.6, 67.6, 63.7, 67.2, 71.0

Catalyst 2: 66.4, 71.7, 70.3, 69.3, 64.8, 69.6, 68.6, 69.4, 65.3, 68.8

(a) Find a 95% confidence interval on the difference in mean active concentrations for the two catalysts.
(b) Is there any evidence to indicate that the mean active concentrations depend on the choice of catalyst? Base your answer on the results of part (a).

5-11. Consider the polymer batch viscosity data in Exercise 5-9. If the difference in mean batch viscosity is 10 or less, the manufacturer would like to detect it with a high probability.

(a) Formulate and test an appropriate hypothesis using $\alpha = 0.10$. What are your conclusions?
(b) Calculate the P-value for this test.
(c) Compare the results of parts (a) and (b) to the length of the 90% confidence interval obtained in Exercise 5-9 and discuss your findings.

5-12. For the laundry detergent problem in Exercise 5-10, test the hypothesis that the mean active concentrations are the same for both types of catalyst. Use $\alpha = 0.05$. What is the P-value for this test? Compare your answer to that found in part (b) of Exercise 5-10, and comment on why they are the same or different.

5-3 INFERENCE ON THE MEANS OF TWO POPULATIONS, VARIANCES UNKNOWN

We now extend the results of the previous section to the difference in means of the two distributions in Figure 5-1 when the variances of both distributions σ_1^2 and σ_2^2 are unknown. If the sample sizes n_1 and n_2 exceed 30, then the normal distribution procedures in Section 5-2 could be used. However, when small samples are taken, we will assume that the populations are normally distributed and base our hypotheses tests and confidence intervals on the t distribution. This nicely parallels the case of inference on the mean of a single sample with unknown variance.

5-3.1 Hypothesis Testing on the Difference in Means

We now consider tests of hypotheses on the difference in means $\mu_1 - \mu_2$ of two normal distributions where the variances σ_1^2 and σ_2^2 are unknown. A t-statistic will be used to test these hypotheses. As noted above and in Section 4-6, the normality assumption is required to develop the test procedure, but moderate departures from normality do not adversely affect the procedure. Two different situations must be treated. In the first case, we assume that the variances of the two normal distributions are unknown but equal; that is, $\sigma_1^2 = \sigma_2^2 = \sigma^2$. In the second, we assume that σ_1^2 and σ_2^2 are unknown and not necessarily equal.

Case 1: $\sigma_1^2 = \sigma_2^2 = \sigma^2$

Suppose we have two independent normal populations with unknown means μ_1 and μ_2, and unknown but equal variances, $\sigma_1^2 = \sigma_2^2 = \sigma^2$. We wish to test

$$H_0: \mu_1 - \mu_2 = \Delta_0$$
$$H_1: \mu_1 - \mu_2 \neq \Delta_0 \tag{5-6}$$

Let $X_{11}, X_{12}, \ldots, X_{1n_1}$ be a random sample of n_1 observations from the first population and $X_{21}, X_{22}, \ldots, X_{2n_2}$ be a random sample of n_2 observations from the second population. Let $\overline{X}_1, \overline{X}_2, S_1^2, S_2^2$ be the sample means and sample variances, respectively. Now the expected value of the difference in sample means $\overline{X}_1 - \overline{X}_2$ is $E(\overline{X}_1 - \overline{X}_2) = \mu_1 - \mu_2$, so $\overline{X}_1 - \overline{X}_2$ is an unbiased estimator of the difference in means. The variance of $\overline{X}_1 - \overline{X}_2$ is

$$V(\overline{X}_1 - \overline{X}_2) = \frac{\sigma^2}{n_1} + \frac{\sigma^2}{n_2} = \sigma^2\left(\frac{1}{n_1} + \frac{1}{n_2}\right)$$

It seems reasonable to combine the two sample variances S_1^2 and S_2^2 to form an estimator of σ^2. The **pooled estimator** of σ^2 is defined as follows.

> The **pooled estimator** of σ^2, denoted by S_p^2, is defined by
>
> $$S_p^2 = \frac{(n_1 - 1)S_1^2 + (n_2 - 1)S_2^2}{n_1 + n_2 - 2} \tag{5-7}$$

It is easy to see that the pooled estimator S_p^2 can be written as

$$S_p^2 = \frac{n_1 - 1}{n_1 + n_2 - 2} S_1^2 + \frac{n_2 - 1}{n_1 + n_2 - 2} S_2^2$$
$$= wS_1^2 + (1 - w)S_2^2$$

where $0 < w \leq 1$. Thus S_p^2 is a **weighted average** of the two sample variances S_1^2 and S_2^2, where the weights w and $1 - w$ depend on the two sample sizes n_1 and n_2. Obviously, if $n_1 = n_2 = n$, then $w = 0.5$ and S_p^2 is just the arithmetic average of S_1^2 and S_2^2. If $n_1 = 10$ and $n_2 = 20$ (say), then $w = 0.32$ and $1 - w = 0.68$. The first sample contributes $n_1 - 1$ degrees of freedom to S_p^2 and the second sample contributes $n_2 - 1$ degrees of freedom. Therefore, S_p^2 has $n_1 + n_2 - 2$ degrees of freedom.

Now we know that

$$Z = \frac{\overline{X}_1 - \overline{X}_2 - (\mu_1 - \mu_2)}{\sigma\sqrt{\dfrac{1}{n_1} + \dfrac{1}{n_2}}}$$

has an $N(0, 1)$ distribution. Replacing σ by S_p gives the following.

Given the assumptions of this section, the quantity

$$T = \frac{\overline{X}_1 - \overline{X}_2 - (\mu_1 - \mu_2)}{S_p\sqrt{\dfrac{1}{n_1} + \dfrac{1}{n_2}}} \tag{5-8}$$

has a t distribution with $n_1 + n_2 - 2$ degrees of freedom.

The use of this information to test the hypotheses in equation 5-6 is now straightforward: Simply replace $\mu_1 - \mu_2$ by Δ_0, and the resulting **test statistic** has a t distribution with $n_1 + n_2 - 2$ degrees of freedom under H_0: $\mu_1 - \mu_2 = \Delta_0$. The location of the critical region for both two-sided and one-sided alternatives parallels those in the one-sample case. This procedure is often called the **pooled t-test.**

Testing Hypotheses on the Difference in Means of Two Normal Distributions, Variances Unknown and Equal[1]

Null hypothesis: H_0: $\mu_1 - \mu_2 = \Delta_0$

Test statistic: $$T_0 = \frac{\overline{X}_1 - \overline{X}_2 - \Delta_0}{S_p\sqrt{\dfrac{1}{n_1} + \dfrac{1}{n_2}}} \tag{5-9}$$

Alternative Hypothesis	Rejection Criterion
H_1: $\mu_1 - \mu_2 \neq \Delta_0$	$t_0 > t_{\alpha/2, n_1 + n_2 - 2}$ or $t_0 < -t_{\alpha/2, n_1 + n_2 - 2}$
H_1: $\mu_1 - \mu_2 > \Delta_0$	$t_0 > t_{\alpha, n_1 + n_2 - 2}$
H_1: $\mu_1 - \mu_2 < \Delta_0$	$t_0 < -t_{\alpha, n_1 + n_2 - 2}$

---

[1] Although we have given the development of this procedure for the case in which the sample sizes could be different, there is an advantage to using equal sample sizes $n_1 = n_2 = n$. When the sample sizes are the same from both populations, the t-test is very robust or insensitive to the assumption of equal variances.

EXAMPLE 5-4

Two catalysts are being analyzed to determine how they affect the mean yield of a chemical process. Specifically, catalyst 1 is currently in use, but catalyst 2 is acceptable. Since catalyst 2 is cheaper, it should be adopted, providing it does not change the process yield. A test is run in the pilot plant and results in the data shown in Table 5-2. Is there any difference between the mean yields? Use $\alpha = 0.05$. Assume equal variances.

The solution using the eight-step hypothesis testing procedure is as follows:

1. The parameters of interest are μ_1 and μ_2, the mean process yield using catalysts 1 and 2, respectively, and we want to know whether $\mu_1 - \mu_2 = 0$.

2. H_0: $\mu_1 - \mu_2 = 0$, or H_0: $\mu_1 = \mu_2$

3. H_1: $\mu_1 \neq \mu_2$ CASE 2

4. $\alpha = 0.05$

5. The test statistic is

$$t_0 = \frac{\bar{x}_1 - \bar{x}_2 - 0}{s_p\sqrt{\dfrac{1}{n_1} + \dfrac{1}{n_2}}}$$

6. Reject H_0 if $t_0 > t_{0.025,14} = 2.145$ or if $t_0 < -t_{0.025,14} = -2.145$.

7. Computation: From Table 5-2 we have $\bar{x}_1 = 92.255$, $s_1 = 2.39$, $n_1 = 8$, $\bar{x}_2 = 92.733$, $s_2 = 2.98$, and $n_2 = 8$. Therefore,

$$s_p^2 = \frac{(n_1 - 1)s_1^2 + (n_2 - 1)s_2^2}{n_1 + n_2 - 2} = \frac{(7)(2.39)^2 + 7(2.98)^2}{8 + 8 - 2} = 7.30$$

$$s_p = \sqrt{7.30} = 2.70$$

Table 5-2   Catalyst Yield Data, Example 5-4

Observation Number	Catalyst 1	Catalyst 2
1	91.50	89.19
2	94.18	90.95
3	92.18	90.46
4	95.39	93.21
5	91.79	97.19
6	89.07	97.04
7	94.72	91.07
8	89.21	92.75
	$\bar{x}_1 = 92.255$	$\bar{x}_2 = 92.733$
	$s_1 = 2.39$	$s_2 = 2.98$

and

$$t_0 = \frac{\bar{x}_1 - \bar{x}_2}{2.70\sqrt{\dfrac{1}{n_1} + \dfrac{1}{n_2}}} = \frac{92.255 - 92.733}{2.70\sqrt{\dfrac{1}{8} + \dfrac{1}{8}}} = -0.35$$

8. Conclusion: Since $t_0 = -2.145 < -0.35 < 2.145$, the null hypothesis cannot be rejected. That is, at the 0.05 level of significance, we do not have strong evidence to conclude that catalyst 2 results in a mean yield that differs from the mean yield when catalyst 1 is used.

A *P*-value could also be used for decision making in this example. From Appendix A Table II we find that $t_{0.40,14} = 0.258$ and $t_{0.25,14} = 0.692$. Therefore, since $0.258 < 0.35 < 0.692$, we conclude that lower and upper bounds on the *P*-value are $0.50 < P < 0.80$. In fact, the actual value is $P = 0.7315$. (This value was obtained from computer software.) Therefore, since the *P*-value exceeds $\alpha = 0.05$, the null hypothesis cannot be rejected.

Case 2: $\sigma_1^2 \neq \sigma_2^2$

In some situations, we cannot reasonably assume that the unknown variances σ_1^2 and σ_2^2 are equal. There is not an exact *t*-statistic available for testing H_0: $\mu_1 - \mu_2 = \Delta_0$ in this case. However, if H_0: $\mu_1 - \mu_2 = \Delta_0$ is true, then the following statistic is used.

Test Statistic for the Difference in Means of Two Normal Distributions, Variances Unknown and Not Necessarily Equal

$$T_0^* = \frac{\bar{X}_1 - \bar{X}_2 - \Delta_0}{\sqrt{\dfrac{S_1^2}{n_1} + \dfrac{S_2^2}{n_2}}} \qquad (5\text{-}10)$$

is distributed approximately as *t* with degrees of freedom given by

$$\nu = \frac{\left(\dfrac{S_1^2}{n_1} + \dfrac{S_2^2}{n_2}\right)^2}{\dfrac{(S_1^2/n_1)^2}{n_1 + 1} + \dfrac{(S_2^2/n_2)^2}{n_2 + 1}} - 2 \qquad (5\text{-}11)$$

if the null hypothesis H_0: $\mu_1 - \mu_2 = \Delta_0$ is true.

Therefore, if $\sigma_1^2 \neq \sigma_2^2$, the hypotheses on differences in the means of two normal distributions are tested as in the equal variances case, except that T_0^* is used as the test statistic and $n_1 + n_2 - 2$ is replaced by v in determining the degrees of freedom for the test.

EXAMPLE 5-5

A manufacturer of video display units is testing two microcircuit designs to determine whether they produce equivalent mean current flow. Development engineering has obtained the following data:

Design 1:	$n_1 = 15$	$\bar{x}_1 = 24.2$	$s_1^2 = 10$
Design 2:	$n_2 = 10$	$\bar{x}_2 = 23.9$	$s_2^2 = 20$

Using $\alpha = 0.10$, we wish to determine whether there is any difference in mean current flow between the two designs, where both populations are assumed to be normal, but we are unwilling to assume that the unknown variances σ_1^2 and σ_2^2 are equal.

Applying the eight-step procedure gives the following:

1. The parameters of interest are the mean current flows for the two circuit designs— say, μ_1 and μ_2—and we are interested in determining whether $\mu_1 - \mu_2 = 0$.
2. $H_0: \mu_1 - \mu_2 = 0$, or $H_0: \mu_1 = \mu_2$
3. $H_1: \mu_1 \neq \mu_2$
4. $\alpha = 0.10$
5. The test statistic is

$$t_0^* = \frac{\bar{x}_1 - \bar{x}_2 - 0}{\sqrt{\dfrac{s_1^2}{n_1} + \dfrac{s_2^2}{n_2}}}$$

6. The degrees of freedom on t_0^* are found from equation 5-11 as

$$v = \frac{\left(\dfrac{s_1^2}{n_1} + \dfrac{s_2^2}{n_2}\right)^2}{\dfrac{(s_1^2/n_1)^2}{n_1 + 1} + \dfrac{(s_2^2/n_2)^2}{n_2 + 1}} - 2 = \frac{\left(\dfrac{10}{15} + \dfrac{20}{10}\right)^2}{\dfrac{(10/15)^2}{16} + \dfrac{(20/10)^2}{11}} - 2 = 16.17 \approx 16$$

Therefore, since $\alpha = 0.10$, we would reject $H_0: \mu_1 = \mu_2$ if $t_0^* > t_{0.05,16} = 1.746$ or if $t_0^* < -t_{0.05,16} = -1.746$.

7. Computation: Using the sample data, we have

$$t_0^* = \frac{\bar{x}_1 - \bar{x}_2}{\sqrt{\frac{s_1^2}{n_1} + \frac{s_2^2}{n_2}}} = \frac{24.2 - 23.9}{\sqrt{\frac{10}{15} + \frac{20}{10}}} = 0.18$$

8. Conclusion: Since $-1.746 < 0.18 < 1.746$, we are unable to reject H_0: $\mu_1 = \mu_2$ at the $\alpha = 0.10$ level of significance. That is, there is no strong evidence indicating that the mean current flow is different for the two designs. The P-value for $t_0^* = 0.18$ is approximately 0.859.

5-3.2   Type II Error and Choice of Sample Size

The operating characteristic curves in Appendix A Charts Va, Vb, Vc, and Vd are used to evaluate the type II error for the case in which $\sigma_1^2 = \sigma_2^2 = \sigma^2$. Unfortunately, when $\sigma_1^2 \neq \sigma_2^2$, the distribution of T_0^* is unknown if the null hypothesis is false, and no operating characteristic curves are available for this case.

For the two-sided alternative H_1: $\mu_1 - \mu_2 \neq \Delta_0$, when $\sigma_1^2 = \sigma_2^2 = \sigma^2$ and $n_1 = n_2 = n$, Charts Va and Vb are used with

$$d = \frac{|\Delta - \Delta_0|}{2\sigma} \tag{5-12}$$

where Δ is the true difference in means that is of interest. To use these curves, they must be entered with the sample size $n^* = 2n - 1$. For the one-sided alternative hypothesis, we use Charts Vc and Vd and define d and Δ as in equation 5-12. It is noted that the parameter d is a function of σ, which is unknown. As in the single-sample t-test, we may have to rely on a prior estimate of σ or use a subjective estimate. Alternatively, we could define the differences in the mean that we wish to detect relative to σ.

EXAMPLE 5-6

Consider the catalyst experiment in Example 5-4. Suppose that, if catalyst 2 produces a mean yield that differs from the mean yield of catalyst 1 by 4.0%, we would like to reject the null hypothesis with probability at least 0.85. What sample size is required?

Using $s_p = 2.70$ as a rough estimate of the common standard deviation σ, we have $d = |\Delta|/2\sigma = |4.0|/[(2)(2.70)] = 0.74$. From Appendix A Chart Va with $d = 0.74$ and $\beta = 0.15$, we find $n^* = 20$, approximately. Therefore, since $n^* = 2n - 1$,

$$n = \frac{n^* + 1}{2} = \frac{20 + 1}{2} = 10.5 \approx 11 \text{(say)}$$

and we would use sample sizes of $n_1 = n_2 = n = 11$.

5-3.3 Confidence Interval on the Difference in Means

Case 1: $\sigma_1^2 = \sigma_2^2 = \sigma^2$

To develop the confidence interval for the difference in means $\mu_1 - \mu_2$ when both variances are equal, note that the distribution of the statistic

$$T = \frac{\overline{X}_1 - \overline{X}_2 - (\mu_1 - \mu_2)}{S_p\sqrt{\dfrac{1}{n_1} + \dfrac{1}{n_2}}}$$

is the t distribution with $n_1 + n_2 - 2$ degrees of freedom. Therefore,

$$P(-t_{\alpha/2,n_1+n_2-2} \le T \le t_{\alpha/2,n_1+n_2-2}) = 1 - \alpha$$

or

$$P\left(-t_{\alpha/2,n_1+n_2-2} \le \frac{\overline{X}_1 - \overline{X}_2 - (\mu_1 - \mu_2)}{S_p\sqrt{\dfrac{1}{n_1} + \dfrac{1}{n_2}}} \le t_{\alpha/2,n_1-n_2-2}\right) = 1 - \alpha$$

Manipulation of the quantities inside the probability statement leads to the $100(1 - \alpha)\%$ confidence interval on $\mu_1 - \mu_2$.

Case 1: Confidence Interval on the Difference in Means of Two Normal Distributions, Variances Unknown and Equal

If $\overline{x}_1$, $\overline{x}_2$, s_1^2, and s_2^2 are the means and variances of two random samples of sizes n_1 and n_2, respectively, from two independent normal populations with unknown but equal variances, then a $100(1 - \alpha)\%$ confidence interval on the difference in means $\mu_1 - \mu_2$ is

$$\overline{x}_1 - \overline{x}_2 - t_{\alpha/2,n_1+n_2-2}\, s_p\sqrt{\frac{1}{n_1} + \frac{1}{n_2}}$$

$$\le \mu_1 - \mu_2 \le \overline{x}_1 - \overline{x}_2 + t_{\alpha/2,n_1+n_2-2}\, s_p\sqrt{\frac{1}{n_1} + \frac{1}{n_2}} \qquad (5\text{-}13)$$

where $s_p = \sqrt{[(n_1 - 1)s_1^2 + (n_2 - 1)s_2^2]/(n_1 + n_2 - 2)}$ is the pooled estimate of the common population standard deviation, and $t_{\alpha/2,n_1+n_2-2}$ is the upper $100\,\alpha/2$ percentage point of the t distribution with $n_1 + n_2 - 2$ degrees of freedom.

EXAMPLE 5-7

An article in the journal *Hazardous Waste and Hazardous Materials* (Vol. 6, 1989) reported the results of an analysis of the weight of calcium in standard cement and cement doped with lead. Reduced levels of calcium would indicate that the hydration mechanism in the cement is blocked and would allow water to attack various locations in the cement structure. Ten samples of standard cement had an average weight percent calcium of $\bar{x}_1 = 90.0$, with a sample standard deviation of $s_1 = 5.0$, and 15 samples of the lead-doped cement had an average weight percent calcium of $\bar{x}_2 = 87.0$, with a sample standard deviation of $s_2 = 4.0$.

We will assume that weight percent calcium is normally distributed and find a 95% confidence interval on the difference in means, $\mu_1 - \mu_2$, for the two types of cement. Furthermore, we will assume that both normal populations have the same standard deviation.

The pooled estimate of the common standard deviation is found using equation 5-7 as follows:

$$s_p^2 = \frac{(n_1 - 1)s_1^2 + (n_2 - 1)s_2^2}{n_1 + n_2 - 2}$$
$$= \frac{9(5.0)^2 + 14(4.0)^2}{10 + 15 - 2}$$
$$= 19.52$$

Therefore, the pooled standard deviation estimate is $s_p = \sqrt{19.52} = 4.4$. The 95% confidence interval is found using equation 5-13:

$$\bar{x}_1 - \bar{x}_2 - t_{0.025,23}s_p\sqrt{\frac{1}{n_1} + \frac{1}{n_2}} \leq \mu_1 - \mu_2 \leq \bar{x}_1 - \bar{x}_2 + t_{0.025,23}s_p\sqrt{\frac{1}{n_1} + \frac{1}{n_2}}$$

or, upon substituting the sample values and using $t_{0.025,23} = 2.069$,

$$90.0 - 87.0 - 2.069(4.4)\sqrt{\frac{1}{10} + \frac{1}{15}} \leq \mu_1 - \mu_2$$
$$\leq 90.0 - 87.0 + 2.069(4.4)\sqrt{\frac{1}{10} + \frac{1}{15}}$$

which reduces to

$$-0.72 \leq \mu_1 - \mu_2 \leq 6.72$$

Note that the 95% confidence interval includes zero; therefore, at this level of confidence we cannot conclude that there is a difference in the means. Put another way, there is no evidence that doping the cement with lead affected the mean weight percent of calcium; therefore, we cannot claim that the presence of lead affects this aspect of the hydration mechanism at the 95% level of confidence.

Case 2: $\sigma_1^2 \neq \sigma_2^2$

In many situations it is not reasonable to assume that $\sigma_1^2 = \sigma_2^2$. When this assumption is unwarranted, we may still find a $100(1 - \alpha)\%$ confidence interval on $\mu_1 - \mu_2$ using the fact that

$$T^* = \frac{\overline{X}_1 - \overline{X}_2 - (\mu_1 - \mu_2)}{\sqrt{S_1^2/n_1 + S_2^2/n_2}}$$

is distributed approximately as t with degrees of freedom v given by equation 5-11. Therefore,

$$P(-t_{\alpha/2,v} \leq T^* \leq t_{\alpha/2,v}) \cong 1 - \alpha$$

and, if we substitute for T^* in this expression and isolate $\mu_1 - \mu_2$ between the inequalities, we can obtain the confidence interval for $\mu_1 - \mu_2$.

**Case 2:   Confidence Interval on the Difference in Means
of Two Normal Distributions, Variances Unknown and Unequal**

If $\overline{x}_1$, $\overline{x}_2$, s_1^2, and s_2^2 are the means and variances of two random samples of sizes n_1 and n_2, respectively, from two independent normal populations with unknown and unequal variances, then an approximate $100(1 - \alpha)\%$ confidence interval on the difference in means $\mu_1 - \mu_2$ is

$$\overline{x}_1 - \overline{x}_2 - t_{\alpha/2,v}\sqrt{\frac{s_1^2}{n_1} + \frac{s_2^2}{n_2}} \leq \mu_1 - \mu_2 \leq \overline{x}_1 - \overline{x}_2 + t_{\alpha/2,v}\sqrt{\frac{s_1^2}{n_1} + \frac{s_2^2}{n_2}} \quad (5\text{-}14)$$

where v is given by equation 5-11 and $t_{\alpha/2,v}$ is the upper $100\ \alpha/2$ percentage point of the t distribution with v degrees of freedom.

5-3.4   Computer Solution

The two-sample t-test can be performed using most statistics software packages. Table 5-3 presents the output from the Minitab Two-Sample t-test routine for the catalyst yield data in Example 5-4. The output includes summary statistics for each sample, confidence intervals on the difference in means, and the hypothesis testing results. This analysis was performed assuming equal variances. Minitab has an option to perform the analysis

Table 5-3   Minitab Two-Sample *t*-Test Output for Example 5-4

Two Sample T-Test and Confidence Interval

Catalyst 1	Catalyst 2
91.50	89.19
94.18	90.95
92.18	90.46
95.39	93.21
91.79	97.19
89.07	97.04
94.72	91.07
89.21	92.75

Two sample T for Catalyst 1 vs Catalyst 2

	N	Mean	StDev	SE Mean
Catalyst 1	8	92.26	2.39	0.84
Catalyst 2	8	92.73	2.98	1.1

95% CI for mu Catalyst 1 − mu Catalyst 2: (−3.39, 2.4)
T-Test mu Catalyst 1 = mu Catalyst 2 (vs not =): T= −0.35 P=0.73 DF = 13

assuming unequal variances. The confidence levels and α-value may be specified by the user. The hypothesis testing procedure indicates that we cannot reject the hypothesis that the mean yields are equal, which agrees with the conclusions we reached originally in Example 5-4.

Figure 5-2 shows comparative box plots for the yield data for the two types of catalysts in Example 5-4. These comparative box plots indicate that there is no obvious

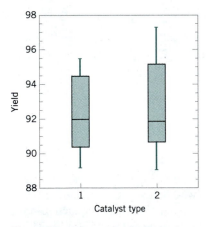

Figure 5-2   Comparative box plots for the catalyst yield data in Example 5-4.

difference in the median of the two samples, although the second sample has a slightly larger sample dispersion or variance. There are no exact rules for comparing two samples with box plots; their primary value is in the visual impression they provide as a tool for explaining the results of a hypothesis test, as well as in verification of assumptions.

Figure 5-3 presents a normal probability plot of the two samples in Example 5-4. Note that both samples plot approximately along straight lines, and the straight lines for each sample have similar slopes. Therefore, we conclude that the normality and equal variances assumptions are reasonable.

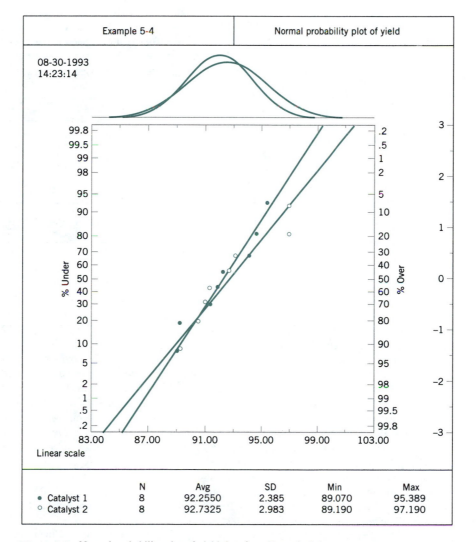

	N	Avg	SD	Min	Max
• Catalyst 1	8	92.2550	2.385	89.070	95.389
○ Catalyst 2	8	92.7325	2.983	89.190	97.190

Figure 5-3   Normal probability plot of yield data from Example 5-4.

EXERCISES FOR SECTION 5-3

5-13. The diameter of steel rods manufactured on two different extrusion machines is being investigated. Two random samples of sizes $n_1 = 15$ and $n_2 = 17$ are selected, and the sample means and sample variances are $\bar{x}_1 = 8.73$, $s_1^2 = 0.35$, $\bar{x}_2 = 8.68$, and $s_2^2 = 0.40$, respectively. Assume that $\sigma_1^2 = \sigma_2^2$ and that the data are drawn from a normal distribution.

(a) Is there evidence to support the claim that the two machines produce rods with different mean diameters? Use $\alpha = 0.05$ in arriving at this conclusion.

(b) Find the P-value for the t-statistic you calculated in part (a).

(c) Construct a 95% confidence interval for the difference in mean rod diameter. Interpret this interval.

5-14. An article in *Fire Technology* investigated two different foam expanding agents that can be used in the nozzles of fire-fighting spray equipment. A random sample of five observations with an aqueous film-forming foam (AFFF) had a sample mean of 4.340 and a standard deviation of 0.508. A random sample of five observations with alcohol-type concentrates (ATC) had a sample mean of 7.091 and a standard deviation of 0.430. Find a 95% confidence interval on the difference in mean foam expansion of these two agents. Can you draw any conclusions about which agent produces the greatest mean foam expansion? Assume that both populations are well represented by normal distributions with the same standard deviations.

5-15. Two catalysts may be used in a batch chemical process. Twelve batches were prepared using catalyst 1, resulting in an average yield of 86.20 and a sample standard deviation of 2.91. Fifteen batches were prepared using catalyst 2, and they resulted in an average yield of 89.38 with a standard deviation of 2.07.

Assume that yield measurements are approximately normally distributed with the same standard deviation.

(a) Is there evidence to support a claim that catalyst 2 produces a higher mean yield than catalyst 1? Use $\alpha = 0.01$.

(b) Find a 95% confidence interval on the difference in mean yields.

5-16. The deflection temperature under load for two different types of plastic pipe is being investigated. Two random samples of 15 pipe specimens are tested, and the deflection temperatures observed are reported here (in °F):

Type 1		
206	193	192
188	207	210
205	185	194
187	189	178
194	213	205

Type 2		
177	176	198
197	185	188
206	200	189
201	197	203
180	192	192

(a) Construct box plots and normal probability plots for the two samples. Do these plots provide support of the assumptions of normality and equal variances? Write a practical interpretation for these plots.

(b) Do the data support the claim that the deflection temperature under load for type 1 pipe exceeds that of type 2? In reaching your conclusions, use $\alpha = 0.05$.

(c) Calculate a P-value for the test in part (b).

(d) Suppose that if the mean deflection temperature for type 2 pipe exceeds that of type 1 by as much as 5°F it is important to detect this difference with probability at least 0.90. Is the choice of $n_1 = n_2 = 15$ in part (a) of this problem adequate?

5-17. In semiconductor manufacturing, wet chemical etching is often used to remove silicon from the backs of wafers prior to metalization. The etch rate is an important characteristic in this process and known to follow a normal distribution. Two different etching solutions have been compared, using two random samples of 10 wafers for each solution. The observed etch rates are as follows (in mils/min):

Solution 1		Solution 2	
9.9	10.6	10.2	10.0
9.4	10.3	10.6	10.2
9.3	10.0	10.7	10.7
9.6	10.3	10.4	10.4
10.2	10.1	10.5	10.3

(a) Do the data support the claim that the mean etch rate is the same for both solutions? In reaching your conclusions, use $\alpha = 0.05$ and assume that both population variances are equal.

(b) Calculate a P-value for the test in part (a).

(c) Find a 95% confidence interval on the difference in mean etch rates.

(d) Construct normal probability plots for the two samples. Do these plots provide support for the assumptions of normality and equal variances? Write a practical interpretation for these plots.

5-18. Two suppliers manufacture a plastic gear used in a laser printer. The impact strength of these gears measured in foot-pounds is an important characteristic. A random sample of 10 gears from supplier 1 results in $\bar{x}_1 = 289.30$ and $s_1 = 22.5$, and another random sample of 16 gears from the second supplier results in $\bar{x}_2 = 321.50$ and $s_2 = 21$.

(a) Is there evidence to support the claim that supplier 2 provides gears with higher mean impact strength? Use $\alpha = 0.05$, and assume that both populations are normally distributed but the variances are not equal.

(b) What is the P-value for this test?

(c) Do the data support the claim that the mean impact strength of gears from supplier 2 is at least 25 foot-pounds higher than that of supplier 1? Make the same assumptions as in part (a).

5-19. A photoconductor film is manufactured at a nominal thickness of 25 mils. The product engineer wishes to decrease the energy absorption of the film, and he believes this can be achieved by reducing the thickness of the film to 20 mils. Eight samples of each film thickness are manufactured in a pilot production process, and the film absorption (in $\mu J/in^2$) is measured. For the 25-mil film the sample data result is $\bar{x}_1 = 1.179$ and $s_1 = 0.088$, and for the 20-mil film, the data yield $\bar{x}_2 = 1.036$ and $s_2 = 0.093$.

(a) Do the data support the claim that reducing the film thickness decreases the energy absorption of the film? Use $\alpha = 0.10$ and assume that the two population variances are equal and the underlying population is normally distributed.

(b) What is the P-value for this test?

(c) Find a 95% confidence interval on the difference in the two means.

5-20. The melting points of two alloys used in formulating solder were investigated by melting 21 samples of each material. The sample mean and standard deviation for alloy 1 was $\bar{x}_1 = 420.48°F$ and $s_1 = 2.34°F$, and for alloy 2 they were $\bar{x}_2 = 425°F$ and $s_2 = 2.5°F$. Do the sample data support the claim that both alloys have the same melting point? Use $\alpha = 0.05$ and assume that both populations are normally distributed and have the same standard deviation. Find the P-value for the test.

5-21. Referring to the melting point experiment in Exercise 5-20, suppose that the true mean difference in melting points is 3°F. How large a sample would be required to detect this difference using an $\alpha = 0.05$ level test with probability at least 0.9? Use $\sigma_1 = \sigma_2 = 4$ as an initial estimate of the common standard deviation.

5-22. Two companies manufacture a rubber material intended for use in an automotive application. The part will be subjected to abrasive wear in the field application, so we decide to compare the material produced by each company in a test. Twenty-five samples of material from each company are tested in an abrasion test, and the amount of wear after 1000 cycles is observed. For company 1, the sample mean and standard deviation of wear are $\bar{x}_1 = 20.12$ mg/1000 cycles and $s_1 = 1.9$ mg/1000 cycles, and for company 2 we obtain $\bar{x}_2 = 11.64$ mg/1000 cycles and $s_2 = 7.9$ mg/1000 cycles.

(a) Do the data support the claim that the two companies produce material with different mean wear? Use $\alpha = 0.05$, and assume each population is normally distributed but that their variances are not equal.

(b) What is the P-value for this test?

(c) Do the data support a claim that the material from company 1 has higher mean wear than the material from company 2? Use the same assumptions as in part (a).

5-23. The thickness of a plastic film (in mils) on a substrate material is thought to be influenced by the temperature at which the coating is applied. A completely randomized experiment is carried out. Eleven substrates are coated at 125°F, resulting in a sample mean coating thickness of $\bar{x}_1 = 101.28$ and a sample standard deviation of $s_1 = 5.08$. Another 13 substrates are coated at 150°F, for which $\bar{x}_2 = 101.70$ and $s_2 = 20.15$ are observed. It was originally suspected that raising the process temperature would reduce mean coating thickness. Do the data support this claim? Use $\alpha = 0.01$ and assume that the two population standard deviations are not equal. Calculate the P-value for this test.

5-24. Reconsider the coating thickness experiment in Exercise 5-23. How could you have answered the question posed regarding the effect of temperature on coating thickness by using a confidence interval? Explain your answer.

5-25. Reconsider the abrasive wear test in Exercise 5-22. Construct a confidence interval that will address the questions in parts (a) and (c) in that exercise.

5-4   THE PAIRED t-TEST

A special case of the two-sample t-tests of Section 5-3 occurs when the observations on the two populations of interest are collected in **pairs.** Each pair of observations—say, (X_{1j}, X_{2j})—are taken under homogeneous conditions, but these conditions may change from one pair to another. For example, suppose that we are interested in comparing two different types of tips for a hardness-testing machine. This machine presses the tip into a metal specimen with a known force. By measuring the depth of the depression caused by the tip, the hardness of the specimen can be determined. If several specimens were selected at random, half tested with tip 1, half tested with tip 2, and the pooled or independent *t*-test in Section 5-3 was applied, the results of the test could be erroneous. The metal specimens could have been cut from bar stock that was produced in different heats, or they might not be homogeneous in some other way that might affect hardness. Then the observed difference between mean hardness readings for the two tip types also includes hardness differences between specimens.

A more powerful experimental procedure is to collect the data in pairs—that is, to make two hardness readings on each specimen, one with each tip. The test procedure would then consist of analyzing the *differences* between hardness readings on each specimen. If there is no difference between tips, then the mean of the differences should be zero. This test procedure is called the **paired *t*-test.**

Let $(X_{11}, X_{21}), (X_{12}, X_{22}), \ldots, (X_{1n}, X_{2n})$ be a set of n paired observations where we assume that the mean and variance of the population represented by X_1 are μ_1 and σ_1^2, and the mean and variance of the population represented by X_2 are μ_2 and σ_2^2. Define the differences between each pair of observations as $D_j = X_{1j} - X_{2j}, j = 1, 2, \ldots, n$. The D_j's are assumed to be normally distributed with mean

$$\mu_D = E(X_1 - X_2) = E(X_1) - E(X_2) = \mu_1 - \mu_2$$

and variance σ_D^2, so testing hypotheses about the difference between μ_1 and μ_2 can be accomplished by performing a one-sample *t*-test on μ_D. Specifically, testing $H_0: \mu_1 - \mu_2 = \Delta_0$ against $H_1: \mu_1 - \mu_2 \neq \Delta_0$ is equivalent to testing

$$H_0: \mu_D = \Delta_0$$

$$H_1: \mu_D \neq \Delta_0 \tag{5-15}$$

The test statistic is given next.

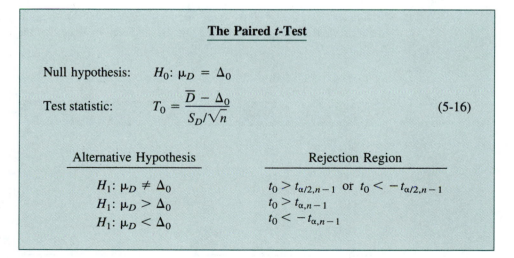

The Paired *t*-Test

Null hypothesis: $H_0: \mu_D = \Delta_0$

Test statistic: $T_0 = \dfrac{\overline{D} - \Delta_0}{S_D/\sqrt{n}}$ (5-16)

Alternative Hypothesis	Rejection Region
$H_1: \mu_D \neq \Delta_0$	$t_0 > t_{\alpha/2, n-1}$ or $t_0 < -t_{\alpha/2, n-1}$
$H_1: \mu_D > \Delta_0$	$t_0 > t_{\alpha, n-1}$
$H_1: \mu_D < \Delta_0$	$t_0 < -t_{\alpha, n-1}$

In equation 5-16, $\overline{D}$ is the sample average of the n differences $D_1, D_2, \ldots, D_n$, and S_D is the sample standard deviation of these differences.

EXAMPLE 5-8

An article in the *Journal of Strain Analysis* (1983, Vol. 18, No. 2) compares several methods for predicting the shear strength for steel plate girders. Data for two of these methods, the Karlsruhe and Lehigh procedures, when applied to nine specific girders, are shown in Table 5-4. We wish to determine whether there is any difference (on the average) between the two methods.

The eight-step procedure is applied as follows:

1. The parameter of interest is the difference in mean shear strength between the two methods—say, $\mu_D = \mu_1 - \mu_2 = 0$.

2. $H_0: \mu_D = 0$

3. $H_1: \mu_D \neq 0$

4. $\alpha = 0.05$

5. The test statistic is

$$t_0 = \frac{\overline{d}}{s_d/\sqrt{n}}$$

6. Reject H_0 if $t_0 > t_{0.025,8} = 2.306$ or if $t_0 < -t_{0.025,8} = -2.306$.

7. Computation: The sample average and standard deviation of the differences d_j are $\overline{d} = 0.2736$ and $s_d = 0.1356$, so the test statistic is

$$t_0 = \frac{\overline{d}}{s_d/\sqrt{n}} = \frac{0.2739}{0.1351/\sqrt{9}} = 6.08$$

8. Conclusion: Since $t_0 = 6.05 > 2.306$, we conclude that the strength prediction methods yield different results. Specifically, the data indicate that the Karlsruhe

Table 5-4   Strength Predictions for Nine Steel Plate Girders (Predicted Load/Observed Load)

Girder	Karlsruhe Method	Lehigh Method	Difference d_j
S1/1	1.186	1.061	0.125
S2/1	1.151	0.992	0.159
S3/1	1.322	1.063	0.259
S4/1	1.339	1.062	0.277
S5/1	1.200	1.065	0.135
S2/1	1.402	1.178	0.224
S2/2	1.365	1.037	0.328
S2/3	1.537	1.086	0.451
S2/4	1.559	1.052	0.507

method produces, on the average, higher strength predictions than does the Lehigh method. The P-value for $t_0 = 6.05$ is $P = 0.0002$, so the test statistic is well into the critical region.

Paired Versus Unpaired Comparisons

In performing a comparative experiment, the investigator can sometimes choose between the paired experiment and the two-sample (or unpaired) experiment. If n measurements are to be made on each population, the two-sample t-statistic is

$$T_0 = \frac{\overline{X}_1 - \overline{X}_2 - \Delta_0}{S_p\sqrt{\dfrac{1}{n} + \dfrac{1}{n}}}$$

which would be compared to t_{2n-2}, and of course, the paired t-statistic is

$$T_0 = \frac{\overline{D} - \Delta_0}{S_D/\sqrt{n}}$$

which is compared to t_{n-1}. Note that since

$$\overline{D} = \sum_{j=1}^{n} \frac{D_j}{n} = \sum_{j=1}^{n} \frac{(X_{1j} - X_{2j})}{n} = \sum_{j=1}^{n} \frac{X_{1j}}{n} - \sum_{j=1}^{n} \frac{X_{2j}}{n} = \overline{X}_1 - \overline{X}_2$$

the numerators of both statistics are identical. However, the denominator of the two-sample t-test is based on the assumption that X_1 and X_2 are *independent*. In many paired experiments, a strong positive correlation ρ exists between X_1 and X_2. Then it can be shown that

$$\begin{aligned}
V(\overline{D}) &= V(\overline{X}_1 - \overline{X}_2 - \Delta_0) \\
&= V(\overline{X}_1) + V(\overline{X}_2) - 2\,\mathrm{cov}(\overline{X}_1, \overline{X}_2) \\
&= \frac{2\sigma^2(1 - \rho)}{n}
\end{aligned}$$

assuming that both populations X_1 and X_2 have identical variances σ^2. Furthermore, S_D^2/n estimates the variance of $\overline{D}$. Whenever there is positive correlation within the pairs, the denominator for the paired t-test will be smaller than the denominator of the two-sample t-test. This can cause the two-sample t-test to considerably understate the significance of the data if it is incorrectly applied to paired samples.

Although pairing will often lead to a smaller value of the variance of $\overline{X}_1 - \overline{X}_2$, it does have a disadvantage—namely, the paired t-test leads to a loss of $n - 1$ degrees of freedom in comparison to the two-sample t-test. Generally, we know that increasing the degrees of freedom of a test increases the power against any fixed alternative values of the parameter.

So how do we decide to conduct the experiment? Should we pair the observations or not? Although there is no general answer to this question, we can give some guidelines based on the previous discussion.

1. If the experimental units are relatively homogeneous (small σ) and the correlation within pairs is small, the gain in precision attributable to pairing will be offset by the loss of degrees of freedom, so an independent-sample experiment should be used.

2. If the experimental units are relatively heterogeneous (large σ) and there is large positive correlation within pairs, the paired experiment should be used. Typically, this case occurs when the experimental units are the *same* for both treatments; as in Example 5-8, the same girders were used to test the two methods.

Implementing the rules still requires judgment, because σ and ρ are never known precisely. Furthermore, if the number of degrees of freedom is large (say, 40 or 50), then the loss of $n - 1$ of them for pairing may not be serious. However, if the number of degrees of freedom is small (say, 10 or 20), then losing half of them is potentially serious if not compensated for by increased precision from pairing.

Confidence Interval for μ_D

To construct the confidence interval for μ_D, note that

$$T = \frac{\overline{D} - \mu_D}{S_D/\sqrt{n}}$$

follows a t distribution with $n - 1$ degrees of freedom. Then, since

$$P(-t_{\alpha/2,n-1} \leq T \leq t_{\alpha/2,n-1}) = 1 - \alpha$$

we can substitute for T in the above expression and perform the necessary steps to isolate $\mu_D = \mu_1 - \mu_2$ between the inequalities. This leads to the following $100(1 - \alpha)\%$ confidence interval on $\mu_D = \mu_1 - \mu_2$.

Confidence Interval on μ_D for Paired Observations

If $\overline{d}$ and s_d are the sample mean and standard deviation, respectively, of the difference of n random pairs of normally distributed measurements, then a $100(1 - \alpha)\%$ confidence interval on the difference in means $\mu_D = \mu_1 - \mu_2$ is

$$\overline{d} - t_{\alpha/2,n-1}s_d/\sqrt{n} \leq \mu_D \leq \overline{d} + t_{\alpha/2,n-1}s_d/\sqrt{n} \qquad (5\text{-}17)$$

where $t_{\alpha/2,n-1}$ is the upper $100\ \alpha/2$ percentage point of the t-distribution with $n - 1$ degrees of freedom.

Table 5-5   Time in Seconds to Parallel Park Two Automobiles

Subject	Automobile 1 (x_{1j})	Automobile 2 (x_{2j})	Difference (d_j)
1	37.0	17.8	19.2
2	25.8	20.2	5.6
3	16.2	16.8	−0.6
4	24.2	41.4	−17.2
5	22.0	21.4	0.6
6	33.4	38.4	−5.0
7	23.8	16.8	7.0
8	58.2	32.2	26.0
9	33.6	27.8	5.8
10	24.4	23.2	1.2
11	23.4	29.6	−6.2
12	21.2	20.6	0.6
13	36.2	32.2	4.0
14	29.8	53.8	−24.0

This confidence interval is also valid for the case where $\sigma_1^2 \neq \sigma_2^2$, because s_D^2 estimates $\sigma_D^2 = V(X_1 - X_2)$. Also, for large samples (say, $n \geq 30$ pairs), the explicit assumption of normality is unnecessary because of the central limit theorem.

EXAMPLE 5-9

The journal *Human Factors* (1962, pp. 375–380) reports a study in which $n = 14$ subjects were asked to parallel park two cars having very different wheel bases and turning radii. The time in seconds for each subject was recorded and is given in Table 5-5. From the column of observed differences we calculate $\bar{d} = 1.21$ and $s_d = 12.68$. The 90% confidence interval for $\mu_D = \mu_1 - \mu_2$ is found from equation 5-17 as follows:

$$\bar{d} - t_{0.05,13}s_d/\sqrt{n} \leq \mu_D \leq \bar{d} + t_{0.05,13}s_d/\sqrt{n}$$
$$1.21 - 1.771(12.68)/\sqrt{14} \leq \mu_D \leq 1.21 + 1.771(12.68)/\sqrt{14}$$
$$-4.79 \leq \mu_D \leq 7.21$$

Note that the confidence interval on μ_D includes zero. This implies that, at the 90% level of confidence, the data do not support the claim that the two cars have different mean parking times μ_1 and μ_2. That is, the value $\mu_D = \mu_1 - \mu_2 = 0$ is not inconsistent with the observed data.

EXERCISES FOR SECTION 5-4

5-26.   Consider the shear strength experiment described in Example 5-8. Construct a 95% confidence interval on the difference in mean shear strength for the two methods. Is the result you obtained consistent with the findings in Example 5-8? Explain why.

5-27. Reconsider the shear strength experiment described in Example 5-8. Does each of the individual shear strengths have to be normally distributed for the paired *t*-test to be appropriate, or is it only the difference in shear strengths that must be normal? Use a normal probability plot to investigate the normality assumption.

5-28. Consider the parking data in Example 5-9. Use the paired *t*-test to investigate the claim that the two types of cars have different levels of difficulty to parallel park. Use $\alpha = 0.10$. Compare your results with the confidence interval constructed in Example 5-9 and comment on why they are the same or different.

5-29. Reconsider the parking data in Example 5-9. Investigate the assumption that the differences in parking times are normally distributed.

5-30. The manager of a fleet of automobiles is testing two brands of radial tires. He assigns one tire of each brand at random to the two rear wheels of eight cars and runs the cars until the tires wear out. The data are shown here (in kilometers). Find a 99% confidence interval on the difference in mean life. Which brand would you prefer, based on this calculation?

Car	Brand 1	Brand 2
1	36,925	34,318
2	45,300	42,280
3	36,240	35,500
4	32,100	31,950
5	37,210	38,015
6	48,360	47,800
7	38,200	37,810
8	33,500	33,215

5-31. A computer scientist is investigating the usefulness of two different design languages in improving programming tasks. Twelve expert programmers, familiar with both languages, are asked to code a standard function in both languages, and the time in minutes is recorded. The data are shown here:

	Time	
Programmer	Design Language 1	Design Language 2
1	17	18
2	16	14
3	21	19
4	14	11
5	18	23
6	24	21
7	16	10
8	14	13
9	21	19
10	23	24
11	13	15
12	18	20

(a) Find a 95% confidence interval on the difference in mean coding times. Is there any indication that one design language is preferable?

(b) Is it reasonable to assume that the difference in coding time is normally distributed? Show evidence to support your answer.

5-32. Fifteen adult males between the ages of 35 and 50 participated in a study to evaluate the effect of diet and exercise on blood cholesterol levels. The total cholesterol was measured in each subject initially, and then 3 months after participating in an aerobic exercise program and switching to a low-fat diet, as shown in the accompanying table. Do the data support the claim that low-fat diet and aerobic exercise are of value in producing a mean reduction in blood cholesterol levels? Use $\alpha = 0.05$.

Blood Cholesterol Level		
Subject	Before	After
1	265	229
2	240	231
3	258	227
4	295	240
5	251	238
6	245	241
7	287	234
8	314	256
9	260	247
10	279	239
11	283	246
12	240	218
13	238	219
14	225	226
15	247	233

5-33. An article in the *Journal of Aircraft* (Vol. 23, 1986, pp. 859–864) describes a new equivalent plate analysis method formulation that is capable of modeling aircraft structures such as cranked wing boxes and that produces results similar to the more computationally intensive finite element analysis method. Natural vibration frequencies for the cranked wing box structure are calculated using both methods, and results for the first seven natural frequencies are shown here.

Car	Finite Element, Cycle/s	Equivalent Plate, Cycle/s
1	14.58	14.76
2	48.52	49.10
3	97.22	99.99
4	113.99	117.53
5	174.73	181.22
6	212.72	220.14
7	277.38	294.80

(a) Do the data suggest that the two methods prove the same mean value for natural vibration frequency? Use $\alpha = 0.05$.

(b) Find a 95% confidence interval on the mean difference between the two methods.

5-34. Ten individuals have participated in a diet-modification program to stimulate weight loss. Their weight both before and after participation in the program is shown in the following list. Is there evidence to support the claim that this particular diet-modification program is effective in producing a mean weight reduction? Use $\alpha = 0.05$.

Subject	Before	After
1	195	187
2	213	195
3	247	221
4	201	190
5	187	175
6	210	197
7	215	199
8	246	221
9	294	278
10	310	285

5-35. Two different analytical tests can be used to determine the impurity level in steel alloys. Eight specimens are tested using both procedures, and the results are shown in the following tabulation. Is there sufficient evidence to conclude that both tests give the same mean impurity level, using $\alpha = 0.01$?

Specimen	Test 1	Test 2
1	1.2	1.4
2	1.3	1.7
3	1.5	1.5
4	1.4	1.3
5	1.7	2.0
6	1.8	2.1
7	1.4	1.7
8	1.3	1.6

5-36. Consider the weight-loss data in Exercise 5-34. Is there evidence to support the claim that this particular diet-modification program will result in a mean weight loss of at least 10 pounds? Use $\alpha = 0.05$.

5-37. Consider the weight-loss experiment in Exercise 5-34. Suppose that, if the diet-modification program results in mean weight loss of at least 10 pounds, it is important to detect this with probability of at least 0.90. Was the use of 10 subjects an adequate sample size? If not, how many subjects should have been used?

5-5 INFERENCE ON THE RATIO OF VARIANCES OF TWO NORMAL POPULATIONS

We now introduce tests and confidence intervals for the two population variances shown in Figure 5-1. We will assume that both populations are normal. Both the hypothesis testing and confidence interval procedures are relatively sensitive to the normality assumption.

5-5.1 Hypothesis Testing on the Ratio of Two Variances

Suppose that two independent normal populations are of interest, where the population means and variances—say, μ_1, σ_1^2, μ_2, and σ_2^2—are unknown. We wish to test hypotheses about the equality of the two variances—say, $H_0: \sigma_1^2 = \sigma_2^2$. Assume that two random samples of size n_1 from population 1 and of size n_2 from population 2 are available, and let S_1^2 and S_2^2 be the sample variances. We wish to test the hypotheses

$$H_0: \sigma_1^2 = \sigma_2^2$$
$$H_1: \sigma_1^2 \neq \sigma_2^2$$

The development of a test procedure for these hypotheses requires a new probability distribution.

The F Distribution

One of the most useful distributions in statistics is the F distribution. The random variable F is defined to be the ratio of two independent chi-square random variables, each divided by its number of degrees of freedom. That is,

$$F = \frac{W/u}{Y/v}$$

where W and Y are independent chi-square random variables with u and v degrees of freedom, respectively. We now formally state the sampling distribution of F.

The *F* Distribution

Let W and Y be independent chi-square random variables with u and v degrees of freedom, respectively. Then the ratio

$$F = \frac{W/u}{Y/v} \tag{5-18}$$

has the probability density function

$$f(x) = \frac{\Gamma\left(\dfrac{u+v}{2}\right)\left(\dfrac{u}{v}\right)^{u/2} x^{(u/2)-1}}{\Gamma\left(\dfrac{u}{2}\right)\Gamma\left(\dfrac{v}{2}\right)\left[\left(\dfrac{u}{v}\right)x+1\right]^{(u+v)/2}}, \quad 0 < x < \infty \tag{5-19}$$

and is said to follow the *F* distribution with u degrees of freedom in the numerator and v degrees of freedom in the denominator. It is usually abbreviated as $F_{u,v}$.

The mean and variance of the *F* distribution are $\mu = v/(v-2)$ for $v > 2$, and

$$\sigma^2 = \frac{2v^2(u+v-2)}{u(v-2)^2(v-4)}, \quad v > 4$$

Two *F* distributions are shown in Fig. 5-4. The *F* random variable is nonnegative, and the distribution is skewed to the right. The *F* distribution looks very similar to the chi-

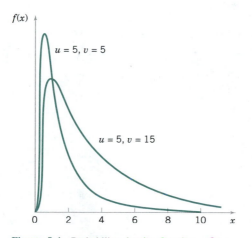

Figure 5-4 Probability density functions of two *F* distributions.

square distribution in Fig. 4-17; however, the two parameters u and v provide extra flexibility regarding shape.

The percentage points of the F distribution are given in Table IV of Appendix A. Let $f_{\alpha,u,v}$ be the percentage point of the F distribution, with numerator degrees of freedom u and denominator degrees of freedom v such that the probability that the random variable F exceeds this value is

$$P(F > f_{\alpha,u,v}) = \int_{f_{\alpha,u,v}}^{\infty} f(x)\, dx = \alpha$$

This is illustrated in Fig. 5-5. For example, if $u = 5$ and $v = 10$, we find from Table IV of Appendix A that

$$P(F > f_{0.05,5,10}) = P(F_{5,10} > 3.33) = 0.05$$

That is, the upper 5 percentage point of $F_{5,10}$ is $f_{0.05,5,10} = 3.33$.

Table IV contains only upper-tail percentage points (for selected values of $f_{\alpha,u,v}$ for $\alpha \le 0.25$) of the F distribution. The lower-tail percentage points $f_{1-\alpha,u,v}$ can be found as follows:

$$f_{1-\alpha,u,v} = \frac{1}{f_{\alpha,v,u}} \tag{5-20}$$

For example, to find the lower-tail percentage point $f_{0.95,5,10}$, note that

$$f_{0.95,5,10} = \frac{1}{f_{0.05,10,5}} = \frac{1}{4.74} = 0.211$$

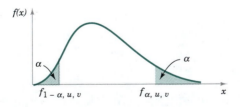

Figure 5-5 Upper and lower percentage points of the F distribution.

The Test Procedure

A hypothesis testing procedure for the equality of two variances is based on the following result.

Let $X_{11}, X_{12}, \ldots, X_{1n_1}$ be a random sample from a normal population with mean μ_1 and variance σ_1^2, and let $X_{21}, X_{22}, \ldots, X_{2n_2}$ be a random sample from a second normal population with mean μ_2 and variance σ_2^2. Assume that both normal populations are independent. Let S_1^2 and S_2^2 be the sample variances. Then the ratio

$$F = \frac{S_1^2/\sigma_1^2}{S_2^2/\sigma_2^2}$$

has an F distribution with $n_1 - 1$ numerator degrees of freedom and $n_2 - 1$ denominator degrees of freedom.

This result is based on the fact that $(n_1 - 1)S_1^2/\sigma_1^2$ is a chi-square random variable with $n_1 - 1$ degrees of freedom, that $(n_2 - 1)S_2^2/\sigma_2^2$ is a chi-square random variable with $n_2 - 1$ degrees of freedom, and that the two normal populations are independent. Clearly under the null hypothesis $H_0: \sigma_1^2 = \sigma_2^2$ the ratio $F_0 = S_1^2/S_2^2$ has an F_{n_1-1,n_2-1} distribution. This is the basis of the following test procedure.

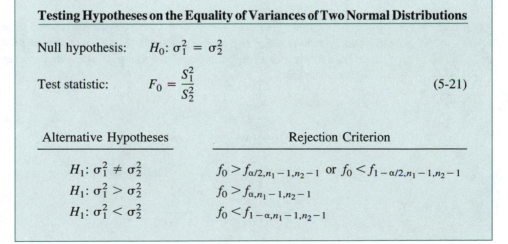

Testing Hypotheses on the Equality of Variances of Two Normal Distributions

Null hypothesis: $H_0: \sigma_1^2 = \sigma_2^2$

Test statistic: $F_0 = \dfrac{S_1^2}{S_2^2}$ (5-21)

Alternative Hypotheses	Rejection Criterion
$H_1: \sigma_1^2 \neq \sigma_2^2$	$f_0 > f_{\alpha/2,n_1-1,n_2-1}$ or $f_0 < f_{1-\alpha/2,n_1-1,n_2-1}$
$H_1: \sigma_1^2 > \sigma_2^2$	$f_0 > f_{\alpha,n_1-1,n_2-1}$
$H_1: \sigma_1^2 < \sigma_2^2$	$f_0 < f_{1-\alpha,n_1-1,n_2-1}$

EXAMPLE 5-10

Oxide layers on semiconductor wafers are etched in a mixture of gases to achieve the proper thickness. The variability in the thickness of these oxide layers is a critical characteristic of the wafer, and low variability is desirable for subsequent processing steps. Two different mixtures of gases are being studied to determine whether one is superior in reducing the variability of the oxide thickness. Sixteen wafers are etched in each gas. The sample standard deviations of oxide thickness are $s_1 = 1.96$ angstroms and $s_2 = 2.13$ angstroms, respectively. Is there any evidence to indicate that either gas is preferable? Use $\alpha = 0.05$.

The eight-step hypothesis testing procedure may be applied to this problem as follows:

1. The parameters of interest are the variances of oxide thickness σ_1^2 and σ_2^2. We will assume that oxide thickness is a normal random variable for both gas mixtures.

2. $H_0: \sigma_1^2 = \sigma_2^2$

3. $H_1: \sigma_1^2 \neq \sigma_2^2$

4. $\alpha = 0.05$

5. The test statistic is given by equation 5-21:

$$f_0 = \frac{s_1^2}{s_2^2}$$

6. Since $n_1 = n_2 = 16$, we will reject $H_0: \sigma_1^2 = \sigma_2^2$ if $f_0 > f_{0.025,15,15} = 2.86$ or if $f_0 < f_{0.975,15,15} = 1/f_{0.025,15,15} = 1/2.86 = 0.35$.

7. Computation: Since $s_1^2 = (1.96)^2 = 3.84$ and $s_2^2 = (2.13)^2 = 4.54$, the test statistic is

$$f_0 = \frac{s_1^2}{s_2^2} = \frac{3.84}{4.54} = 0.85$$

8. Conclusion: Since $f_{0.975,15,15} = 0.35 < 0.85 < f_{0.025,15,15} = 2.86$, we cannot reject the null hypothesis $H_0: \sigma_1^2 = \sigma_2^2$ at the 0.05 level of significance. Therefore, there is no strong evidence to indicate that either gas results in a smaller variance of oxide thickness.

5-5.2   Confidence Interval on the Ratio of Two Variances

To find the confidence interval, recall that the sampling distribution of

$$F = \frac{S_2^2/\sigma_2^2}{S_1^2/\sigma_1^2}$$

is an F with $n_2 - 1$ and $n_1 - 1$ degrees of freedom. *Note:* We start with S_2^2 in the numerator and S_1^2 in the denominator to simplify the algebra used to obtain an interval for σ_1^2/σ_2^2. Therefore,

$$P(f_{1-\alpha/2,n_2-1,n_1-1} \le F \le f_{\alpha/2,n_2-1,n_1-1}) = 1 - \alpha$$

Substitution for F and manipulation of the inequalities will lead to the $100(1 - \alpha)\%$ confidence interval for σ_1^2/σ_2^2.

Confidence Interval on the Ratio of Variances of Two Normal Distributions

If s_1^2 and s_2^2 are the sample variances of random samples of sizes n_1 and n_2, respectively, from two independent normal populations with unknown variances σ_1^2 and σ_2^2, then a $100(1 - \alpha)\%$ confidence interval on the ratio σ_1^2/σ_2^2 is

$$\frac{s_1^2}{s_2^2} f_{1-\alpha/2,n_2-1,n_1-1} \le \frac{\sigma_1^2}{\sigma_2^2} \le \frac{s_1^2}{s_2^2} f_{\alpha/2,n_2-1,n_1-1} \tag{5-22}$$

where $f_{\alpha/2,n_2-1,n_1-1}$ and $f_{1-\alpha/2,n_2-1,n_1-1}$ are the upper and lower $100\,\alpha/2$ percentage points of the F distribution with $n_2 - 1$ numerator and $n_1 - 1$ denominator degrees of freedom, respectively.

EXAMPLE 5-11

A company manufactures impellers for use in jet-turbine engines. One of the operations involves grinding a particular surface finish on a titanium alloy component. Two different grinding processes can be used, and both processes can produce parts at identical mean surface roughness. The manufacturing engineer would like to select the process having the least variability in surface roughness. A random sample of $n_1 = 11$ parts from the first process results in a sample standard deviation $s_1 = 5.1$ microinches, and a random sample of $n_2 = 16$ parts from the second process results in a sample standard deviation of $s_2 = 4.7$ microinches. We will find a 90% confidence interval on the ratio of the two variances σ_1^2/σ_2^2.

Assuming that the two processes are independent and that surface roughness is normally distributed, we can use equation 5-22 as follows:

$$\frac{s_1^2}{s_2^2} f_{0.95,15,10} \le \frac{\sigma_1^2}{\sigma_2^2} \le \frac{s_1^2}{s_2^2} f_{0.05,15,10}$$

$$\frac{(5.1)^2}{(4.7)^2} 0.39 \le \frac{\sigma_1^2}{\sigma_2^2} \le \frac{(5.1)^2}{(4.7)^2} 2.85$$

or

$$0.46 \leq \frac{\sigma_1^2}{\sigma_2^2} \leq 3.36$$

Note that we have used equation 5-20 to find $f_{0.95,15,10} = 1/f_{0.05,10,15} = 1/2.54 = 0.39$. Since this confidence interval includes unity, we cannot claim that the standard deviations of surface roughness for the two processes are different at the 90% level of confidence.

EXERCISES FOR SECTION 5-5

5-38. For an F distribution, find the following:

(a) $f_{0.25,5,10}$ (d) $f_{0.75,5,10}$

(b) $f_{0.10,24,9}$ (e) $f_{0.90,24,9}$

(c) $f_{0.05,8,15}$ (f) $f_{0.95,8,15}$

5-39. For an F distribution, find the following:

(a) $f_{0.25,7,15}$ (d) $f_{0.75,7,15}$

(b) $f_{0.10,10,12}$ (e) $f_{0.90,10,12}$

(c) $f_{0.01,20,10}$ (f) $f_{0.99,20,10}$

5-40. Eleven resilient modulus observations of a ceramic mixture of type A are measured and found to have a sample average of 18.42 psi and sample standard deviation of 2.77 psi. Ten resilient modulus observations of a ceramic mixture of type B are measured and found to have a sample average of 19.28 psi and sample standard deviation of 2.41 psi. Is there sufficient evidence to support the investigator's claim that type A ceramic has larger variability than type B? Use $\alpha = 0.05$.

5-41. Consider the etch rate data in Exercise 5-17. Test the hypothesis $H_0: \sigma_1^2 = \sigma_2^2$ against $H_1: \sigma_1^2 \neq \sigma_2^2$ using $\alpha = 0.05$, and draw conclusions.

5-42. Consider the diameter data in Exercise 5-13. Construct the following:

(a) A 90% two-sided confidence interval on σ_1/σ_2.

(b) A 95% two-sided confidence interval on σ_1/σ_2. Comment on the comparison of the width of this interval with the width of the interval in part (a).

(c) A 90% lower-confidence interval on σ_1/σ_2.

5-43. Consider the foam data in Exercise 5-14. Construct the following:

(a) A 90% two-sided confidence interval on σ_1^2/σ_2^2.

(b) A 95% two-sided confidence interval on σ_1^2/σ_2^2. Comment on the comparison of the width of this interval with the width of the interval in part (a).

(c) A 90% lower-confidence interval on σ_1/σ_2.

5-44. Consider the film data in Exercise 5-19. Test $H_0: \sigma_1^2 = \sigma_2^2$ versus $H_1: \sigma_1^2 \neq \sigma_2^2$ using $\alpha = 0.02$.

5-45. Consider the gear impact strength data in Exercise 5-18. Is there sufficient evidence to conclude that the variance of impact strength is different for the two suppliers? Use $\alpha = 0.05$.

5-46. Consider the melting point data in Exercise 5-20. Do the sample data support a claim that both alloys have the same variance of melting point? Use $\alpha = 0.05$ in reaching your conclusion.

5-47. Exercise 5-23 presented measurements of plastic coating thickness at two different application temperatures. Test $H_0: \sigma_1^2 = \sigma_2^2$ against $H_1: \sigma_1^2 \neq \sigma_2^2$ using $\alpha = 0.10$.

5-48. A study was performed to determine whether men and women differ in their repeatability in assembling components on printed circuit boards. Two samples of 25 men and 21 women were selected, and each subject assembled the units. The two sample standard deviations of assembly time were $s_{men} = 0.914$ min and $s_{women} = 1.093$ min. Is there evidence to support the claim that men and women differ in repeatability for this assembly task? Use $\alpha = 0.02$ and state any necessary assumptions about the underlying distribution of the data.

5-49. Reconsider the assembly repeatability experiment described in Exercise 5-48. Find a 98% confidence interval on the ratio of the two variances. Provide an interpretation of the interval.

5-6 INFERENCE ON TWO POPULATION PROPORTIONS

We now consider the case in which there are two binomial parameters of interest—say, p_1 and p_2—and we wish to draw inferences about these proportions. We will present large-sample hypothesis testing and confidence interval procedures based on the normal approximation to the binomial.

5-6.1 Hypothesis Testing on the Equality of Two Binomial Proportions

Suppose that the two independent random samples of sizes n_1 and n_2 are taken from two populations, and let X_1 and X_2 represent the number of observations that belong to the class of interest in samples 1 and 2, respectively. Furthermore, suppose that the normal approximation to the binomial is applied to each population, so that the estimators of the population proportions $\hat{P}_1 = X_1/n_1$ and $\hat{P}_2 = X_2/n_2$ have approximate normal distributions. We are interested in testing the hypotheses

$$H_0: p_1 = p_2$$
$$H_1: p_1 \neq p_2$$

The quantity

$$Z = \frac{\hat{P}_1 - \hat{P}_2 - (p_1 - p_2)}{\sqrt{\dfrac{p_1(1 - p_1)}{n_1} + \dfrac{p_2(1 - p_2)}{n_2}}} \tag{5-23}$$

has approximately a standard normal distribution, $N(0, 1)$.

This result is the basis of a test for H_0: $p_1 = p_2$. Specifically, if the null hypothesis H_0: $p_1 = p_2$ is true, then using the fact that $p_1 = p_2 = p$, the random variable

$$Z = \frac{\hat{P}_1 - \hat{P}_2}{\sqrt{p(1-p)\left(\dfrac{1}{n_1} + \dfrac{1}{n_2}\right)}}$$

is distributed approximately $N(0, 1)$. An estimator of the common parameter p is

$$\hat{P} = \frac{X_1 + X_2}{n_1 + n_2}$$

The test statistic for H_0: $p_1 = p_2$ is then

$$Z_0 = \frac{\hat{P}_1 - \hat{P}_2}{\sqrt{\hat{P}(1-\hat{P})\left(\dfrac{1}{n_1} + \dfrac{1}{n_2}\right)}}$$

This leads to the test procedures described next.

Testing Hypotheses on the Equality of Two Binomial Proportions

Null hypothesis: H_0: $p_1 = p_2$

Test statistic: $$Z_0 = \frac{\hat{P}_1 - \hat{P}_2}{\sqrt{\hat{P}(1-\hat{P})\left(\dfrac{1}{n_1} + \dfrac{1}{n_2}\right)}} \qquad (5\text{-}24)$$

Alternative Hypotheses	Rejection Criterion
H_1: $p_1 \neq p_2$	$z_0 > z_{\alpha/2}$ or $z_0 < -z_{\alpha/2}$
H_1: $p_1 > p_2$	$z_0 > z_\alpha$
H_1: $p_1 < p_2$	$z_0 < -z_\alpha$

EXAMPLE 5-12

Two different types of polishing solution are being evaluated for possible use in a tumble-polish operation for manufacturing interocular lenses used in the human eye following cataract surgery. Three hundred lenses were tumble-polished using the first polishing solution, and of this number 253 had no polishing-induced defects. Another 300 lenses were tumble-polished using the second polishing solution, and 196 lenses were satisfactory on completion. Is there any reason to believe that the two polishing solutions differ? Use $\alpha = 0.01$.

The eight-step hypothesis procedure leads to the following results:

1. The parameters of interest are p_1 and p_2, the proportion of lenses that are satisfactory following tumble-polishing with polishing fluids 1 or 2.

2. $H_0: p_1 = p_2$

3. $H_1: p_1 \neq p_2$

4. $\alpha = 0.01$

5. The test statistic is

$$z_0 = \frac{\hat{p}_1 - \hat{p}_2}{\sqrt{\hat{p}(1 - \hat{p})\left(\dfrac{1}{n_1} + \dfrac{1}{n_2}\right)}}$$

where $\hat{p}_1 = 253/300 = 0.8433$, $\hat{p}_2 = 196/300 = 0.6533$, $n_1 = n_2 = 300$, and

$$\hat{p} = \frac{x_1 + x_2}{n_1 + n_2} = \frac{253 + 196}{300 + 300} = 0.7483$$

6. Reject $H_0: p_1 = p_2$ if $z_0 > z_{0.005} = 2.58$ or if $z_0 < -z_{0.005} = -2.58$.

7. Computation: The value of the test statistic is

$$z_0 = \frac{0.8433 - 0.6533}{\sqrt{0.7483(0.2517)\left(\dfrac{1}{300} + \dfrac{1}{300}\right)}} = 5.36$$

8. Conclusion: Since $z_0 = 5.36 > z_{0.005} = 2.58$, we reject the null hypothesis. Note that the P-value is $P \simeq 8.32 \times 10^{-8}$. There is strong evidence to support the claim that the two polishing fluids are different. Fluid 1 produces a higher fraction of nondefective lenses.

5-6.2 Type II Error and Choice of Sample Size

The computation of the β-error for the foregoing test is somewhat more involved than in the single-sample case. The problem is that the denominator of Z_0 is an estimate of the standard deviation of $\hat{P}_1 - \hat{P}_2$ under the assumption that $p_1 = p_2 = p$. When $H_0: p_1 = p_2$ is false, the standard deviation of $\hat{P}_1 - \hat{P}_2$ is

$$\sigma_{\hat{P}_1 - \hat{P}_2} = \sqrt{\frac{p_1(1 - p_1)}{n_1} + \frac{p_2(1 - p_2)}{n_2}} \tag{5-25}$$

If the alternative hypothesis is two sided, the β-error is

$$\beta = \Phi\left(\frac{z_{\alpha/2}\sqrt{\bar{p}\bar{q}(1/n_1 + 1/n_2)} - (p_1 - p_2)}{\sigma_{\hat{P}_1 - \hat{P}_2}}\right)$$

$$- \Phi\left(\frac{-z_{\alpha/2}\sqrt{\bar{p}\bar{q}(1/n_1 + 1/n_2)} - (p_1 - p_2)}{\sigma_{\hat{P}_1 - \hat{P}_2}}\right) \tag{5-26}$$

where

$$\bar{p} = \frac{n_1 p_1 + n_2 p_2}{n_1 + n_2}$$

$$\bar{q} = \frac{n_1(1 - p_1) + n_2(1 - p_2)}{n_1 + n_2} = 1 - \bar{p}$$

and $\sigma_{\hat{P}_1 - \hat{P}_2}$ is given by equation 5-25.

If the alternative hypothesis is $H_1: p_1 > p_2$, then

$$\beta = \Phi\left(\frac{z_\alpha\sqrt{\bar{p}\bar{q}(1/n_1 + 1/n_2)} - (p_1 - p_2)}{\sigma_{\hat{P}_1 - \hat{P}_2}}\right) \tag{5-27}$$

and if the alternative hypothesis is $H_1: p_1 < p_2$, then

$$\beta = 1 - \Phi\left(\frac{-z_\alpha\sqrt{\bar{p}\bar{q}(1/n_1 + 1/n_2)} - (p_1 - p_2)}{\sigma_{\hat{P}_1 - \hat{P}_2}}\right) \tag{5-28}$$

For a specified pair of values p_1 and p_2, we can find the sample sizes $n_1 = n_2 = n$ required to give the test of size α that has specified type II error β.

**Sample Size for a Two-Sided Hypothesis Test on the
Difference in Two Binomial Proportions**

For the two-sided alternative, the common sample size is

$$n = \frac{\left(z_{\alpha/2}\sqrt{(p_1 + p_2)(q_1 + q_2)/2} + z_\beta\sqrt{p_1 q_1 + p_2 q_2}\right)^2}{(p_1 - p_2)^2} \qquad (5\text{-}29)$$

where $q_1 = 1 - p_1$ and $q_2 = 1 - p_2$.

For a one-sided alternative, replace $z_{\alpha/2}$ in equation 5-29 by z_α.

5-6.3   Confidence Interval on the Difference in Binomial Proportions

The confidence interval for $p_1 - p_2$ can be found directly, since we know that

$$z = \frac{\hat{P}_1 - \hat{P}_2 - (p_1 - p_2)}{\sqrt{\dfrac{p_1(1 - p_1)}{n_1} + \dfrac{p_2(1 - p_2)}{n_2}}}$$

is a standard normal random variable. Thus

$$P(-z_{\alpha/2} \leq Z \leq z_{\alpha/2}) \simeq 1 - \alpha$$

so we can substitute for Z in this last expression and use an approach similar to the one employed previously to find an approximate $100(1 - \alpha)\%$ confidence interval for $p_1 - p_2$.

Confidence Interval on the Difference in Binomial Proportions

If $\hat{p}_1$ and $\hat{p}_2$ are the sample proportions of observation in two independent random samples of sizes n_1 and n_2 that belong to a class of interest, then an approximate $100(1 - \alpha)\%$ confidence interval on the difference in the true proportions $p_1 - p_2$ is

$$\hat{p}_1 - \hat{p}_2 - z_{\alpha/2}\sqrt{\frac{\hat{p}_1(1 - \hat{p}_1)}{n_1} + \frac{\hat{p}_2(1 - \hat{p}_2)}{n_2}}$$

$$\leq p_1 - p_2 \leq \hat{p}_1 - \hat{p}_2 + z_{\alpha/2}\sqrt{\frac{\hat{p}_1(1 - \hat{p}_1)}{n_1} + \frac{\hat{p}_2(1 - \hat{p}_2)}{n_2}} \qquad (5\text{-}30)$$

where $z_{\alpha/2}$ is the upper $\alpha/2$ percentage point of the standard normal distribution.

EXAMPLE 5-13

Consider the process manufacturing crankshaft bearings described in Example 4-14. Suppose that a modification is made in the surface finishing process and that, subsequently, a second random sample of 85 axle shafts is obtained. The number of defective shafts in this second sample is 8. Therefore, since $n_1 = 85$, $\hat{p}_1 = 0.12$, $n_2 = 85$, and $\hat{p}_2 = 8/85 = 0.09$, we can obtain an approximate 95% confidence interval on the difference in the proportion of defective bearings produced under the two processes from equation 5-30 as follows:

$$\hat{p}_1 - \hat{p}_2 - z_{0.025}\sqrt{\frac{\hat{p}_1(1 - \hat{p}_1)}{n_1} + \frac{\hat{p}_2(1 - \hat{p}_2)}{n_2}}$$

$$\leq p_1 - p_2 \leq \hat{p}_1 - \hat{p}_2 + z_{0.025}\sqrt{\frac{\hat{p}_1(1 - \hat{p}_1)}{n_1} + \frac{\hat{p}_2(1 - \hat{p}_2)}{n_2}}$$

or

$$0.12 - 0.09 - 1.96\sqrt{\frac{0.12(0.88)}{85} + \frac{0.09(0.91)}{85}}$$

$$\leq p_1 - p_2 \leq 0.12 - 0.09 + 1.96\sqrt{\frac{0.12(0.88)}{85} + \frac{0.09(0.91)}{85}}$$

This simplifies to

$$-0.06 \leq p_1 - p_2 \leq 0.12$$

This confidence interval includes zero, so, based on the sample data, it seems unlikely that the changes made in the surface finish process have reduced the proportion of defective crankshaft bearings being produced.

EXERCISES FOR SECTION 5-6

5-50. Two different types of injection-molding machines are used to form plastic parts. A part is considered defective if it has excessive shrinkage or is discolored. Two random samples, each of size 300, are selected, and 15 defective parts are found in the sample from machine 1 whereas 8 defective parts are found in the sample from machine 2. Is it reasonable to conclude that both machines produce the same fraction of defective parts, using $\alpha = 0.05$? Find the P-value for this test.

5-51. Consider the situation described in Exercise 5-50. Suppose that $p_1 = 0.05$ and $p_2 = 0.01$.
(a) With the sample sizes given here, what is the power of the test for this two-sided alternative?
(b) Determine the sample size needed to detect this difference with a probability of at least 0.9. Use $\alpha = 0.05$.

5-52. Consider the situation described in Exercise 5-50. Suppose that $p_1 = 0.05$ and $p_2 = 0.02$.
(a) With the sample sizes given here,

what is the power of the test for this two-sided alternative?

(b) Determine the sample size needed to detect this difference with a probability of at least 0.9. Use $\alpha = 0.05$.

5-53. In a survey of 500 teens in years 1992 and 1997, the number of teens reported to have used drugs changed from 35 to 41. Is there a statistically significant

difference in percentages of reported drug use? Use $\alpha = 0.1$.

5-54. Construct a 95% confidence interval on the difference in the two fractions defective for Exercise 5-50.

5-55. Construct a 95% confidence interval on the difference in the two proportions for Exercise 5-53. Provide a practical interpretation of this interval.

5-7 SUMMARY TABLES FOR INFERENCE PROCEDURES FOR TWO SAMPLES

The tables on the inside back cover of the book summarize all of the two-sample inference procedures given in this chapter. The tables contain the null hypothesis statements, the test statistics, the criteria for rejection of the various alternative hypotheses, and the formulas for constructing the $100(1 - \alpha)$% confidence intervals.

5-8 WHAT IF WE HAVE MORE THAN TWO SAMPLES?

As this chapter and Chapter 4 have illustrated, testing and experimentation are a natural part of the engineering analysis and decision-making process. Suppose, for example, that a civil engineer is investigating the effect of different curing methods on the mean compressive strength of concrete. The experiment would consist of making up several test specimens of concrete using each of the proposed curing methods and then testing the compressive strength of each specimen. The data from this experiment could be used to determine which curing method should be used to provide maximum mean compressive strength.

If there are only two curing methods of interest, this experiment could be designed and analyzed using the two-sample t-test presented in this chapter. That is, the experimenter has a **single factor** of interest—curing methods—and there are only two **levels** of the factor.

Many single-factor experiments require that more than two levels of the factor be considered. For example, the civil engineer may want to investigate five different curing methods. In this chapter we show how the **analysis of variance** can be used for comparing means when there are more than two levels of a single factor. We will also discuss **randomization** of the experimental runs and the important role this concept plays in the overall experimentation strategy. In Chapter 7, we will show how to design and analyze experiments with several factors.

5-8.1 Completely Randomized Experiment and Analysis of Variance

A manufacturer of paper used for making grocery bags is interested in improving the tensile strength of the product. Product engineering thinks that tensile strength is a function of the hardwood concentration in the pulp and that the range of hardwood concentrations of practical interest is between 5% and 20%. A team of engineers responsible for the study decides to investigate four levels of hardwood concentration: 5%, 10%, 15%, and 20%. They decide to make up six test specimens at each concentration level, using a pilot plant. All 24 specimens are tested on a laboratory tensile tester, in random order. The data from this experiment are shown in Table 5-6.

This is an example of a completely randomized single-factor experiment with four levels of the factor. The levels of the factor are sometimes called **treatments,** and each treatment has six observations or **replicates.** The role of **randomization** in this experiment is extremely important. By randomizing the order of the 24 runs, the effect of any nuisance variable that may influence the observed tensile strength is approximately balanced out. For example, suppose that there is a warm-up effect on the tensile testing machine; that is, the longer the machine is on, the greater the observed tensile strength. If all 24 runs are made in order of increasing hardwood concentration (that is, all six 5% concentration specimens are tested first, followed by all six 10% concentration specimens, etc.), then any observed differences in tensile strength could also be due to the warm-up effect.

It is important to graphically analyze the data from a designed experiment. Figure 5-6a presents box plots of tensile strength at the four hardwood concentration levels. This figure indicates that changing the hardwood concentration has an effect on tensile strength; specifically, higher hardwood concentrations produce higher observed tensile strength. Furthermore, the distribution of tensile strength at a particular hardwood level is reasonably symmetric, and the variability in tensile strength does not change dramatically as the hardwood concentration changes.

Graphical interpretation of the data is always a good idea. Box plots show the variability of the observations *within* a treatment (factor level) and the variability *between* treatments. We now discuss how the data from a single-factor randomized experiment can be analyzed statistically.

Analysis of Variance

Suppose we have a different levels of a single factor that we wish to compare. Sometimes each factor level is called a treatment, a very general term that can be traced to the early

Table 5-6   Tensile Strength of Paper (psi)

Hardwood Concentration (%)	Observations						Totals	Averages
	1	2	3	4	5	6		
5	7	8	15	11	9	10	60	10.00
10	12	17	13	18	19	15	94	15.67
15	14	18	19	17	16	18	102	17.00
20	19	25	22	23	18	20	127	21.17
							383	15.96

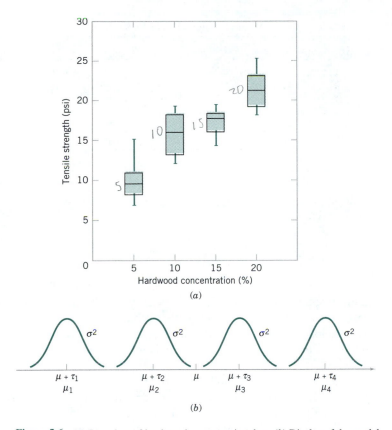

Figure 5-6   (*a*) Box plots of hardwood concentration data. (*b*) Display of the model in equation 5-31 for the completely randomized single-factor experiment.

applications of experimental design methodology in the agricultural sciences. The response for each of the *a* treatments is a random variable. The observed data would appear as shown in Table 5-7. An entry in Table 5-7—say, y_{ij}—represents the *j*th observation taken under treatment *i*. We initially consider the case in which there are an equal number of observations, *n*, on each treatment.

We may describe the observations in Table 5-7 by the **linear statistical model**

$$Y_{ij} = \mu + \tau_i + \epsilon_{ij} \begin{cases} i = 1, 2, \ldots, a \\ j = 1, 2, \ldots, n \end{cases} \tag{5-31}$$

where Y_{ij} is a random variable denoting the (*ij*)th observation, μ is a parameter common to all treatments called the **overall mean**, τ_i is a parameter associated with the *i*th treatment

Table 5-7   Typical Data for a Single-Factor Experiment

Treatment	Observations				Totals	Averages
1	y_{11}	y_{12}	$\cdots$	y_{1n}	$y_{1\cdot}$	$\bar{y}_{1\cdot}$
2	y_{21}	y_{22}	$\cdots$	y_{2n}	$y_{2\cdot}$	$\bar{y}_{2\cdot}$
.	.	.	.	.	.	.
.	.	.	.	.	.	.
.	.	.	.	.	.	.
a	y_{a1}	y_{a2}	$\cdots$	y_{an}	$y_{a\cdot}$	$\bar{y}_{a\cdot}$
					$y_{\cdot\cdot}$	$\bar{y}_{\cdot\cdot}$

called the ith **treatment effect,** and ϵ_{ij} is a random error component. Notice that the model could have been written as

$$Y_{ij} = \mu_i + \epsilon_{ij} \begin{cases} i = 1, 2, \ldots, a \\ j = 1, 2, \ldots, n \end{cases}$$

where $\mu_i = \mu + \tau_i$ is the mean of the ith treatment. In this form of the model, we see that each treatment defines a population that has mean μ_i, consisting of the overall mean μ plus an effect τ_i that is due to that particular treatment. We will assume that the errors ϵ_{ij} are normally and independently distributed with mean zero and variance σ^2. Therefore, each treatment can be thought of as a normal population with mean μ_i and variance σ^2. See Fig. 5-6b.

Equation 5-31 is the underlying model for a single-factor experiment. Furthermore, since we require that the observations are taken in random order and that the environment (often called the experimental units) in which the treatments are used is as uniform as possible, this design is called a **completely randomized experiment.**

We now present the analysis of variance for testing the equality of a population means. However, the analysis of variance is a far more useful and general technique; it will be used extensively in the next two chapters. In this section we show how it can be used to test for equality of treatment effects. In our application the treatment effects τ_i are usually defined as deviations from the overall mean μ, so that

$$\sum_{i=1}^{a} \tau_i = 0 \tag{5-32}$$

Let $y_{i\cdot}$ represent the total of the observations under the ith treatment and $\bar{y}_{i\cdot}$ represent the average of the observations under the ith treatment. Similarly, let $y_{\cdot\cdot}$ represent the grand total of all observations and $\bar{y}_{\cdot\cdot}$ represent the grand mean of all observations. Expressed mathematically,

$$y_{i\cdot} = \sum_{j=1}^{n} y_{ij} \qquad \bar{y}_{i\cdot} = y_{i\cdot}/n \qquad i = 1, 2, \ldots, a$$

$$y_{\cdot\cdot} = \sum_{i=1}^{a} \sum_{j=1}^{n} y_{ij} \qquad \bar{y}_{\cdot\cdot} = y_{\cdot\cdot}/N \tag{5-33}$$

where $N = an$ is the total number of observations. Thus the "dot" subscript notation implies summation over the subscript that it replaces.

We are interested in testing the equality of the a treatment means $\mu_1, \mu_2, \ldots, \mu_a$. Using equation 5-32, we find that this is equivalent to testing the hypotheses

$$H_0: \tau_1 = \tau_2 = \cdots = \tau_a = 0$$

$$H_1: \tau_i \neq 0 \quad \text{for at least one } i \tag{5-34}$$

Thus, if the null hypothesis is true, each observation consists of the overall mean μ plus a realization of the random error component ϵ_{ij}. This is equivalent to saying that all N observations are taken from a normal distribution with mean μ and variance σ^2. Therefore, if the null hypothesis is true, changing the levels of the factor has no effect on the mean response.

The analysis of variance partitions the total variability in the sample data into two component parts. Then, the test of the hypothesis in equation 5-34 is based on a comparison of two independent estimates of the population variance. The total variability in the data is described by the **total sum of squares**

$$SS_T = \sum_{i=1}^{a} \sum_{j=1}^{n} (y_{ij} - \bar{y}..)^2$$

The partition of the total sum of squares is given in the following definition.

$$s^2 = \frac{\left(y_i - \bar{y}\right)^2}{n-1}$$

The **sum of squares identity** is

$$\sum_{i=1}^{a} \sum_{j=1}^{n} (y_{ij} - \bar{y}..)^2 = n \sum_{i=1}^{a} (\bar{y}_{i\cdot} - \bar{y}..)^2 + \sum_{i=1}^{a} \sum_{j=1}^{n} (y_{ij} - \bar{y}_{i\cdot})^2 \tag{5-35}$$

The proof of this identity is straightforward, and is provided in Montgomery and Runger (1999).

The identity in equation 5-35 shows that the total variability in the data, measured by the total sum of squares, can be partitioned into a sum of squares of differences between treatment means and the grand mean and a sum of squares of differences of observations within a treatment from the treatment mean. Differences between observed treatment means and the grand mean measure the differences between treatments, whereas differences of observations within a treatment from the treatment mean can be due only to random error. Therefore, we write equation 5-35 symbolically as

$$SS_T = SS_{\text{Treatments}} + SS_E \qquad (5\text{-}36)$$

where

$$SS_T = \sum_{i=1}^{a} \sum_{j=1}^{n} (y_{ij} - \bar{y}..)^2 = \text{total sum of squares}$$

$$SS_{\text{Treatments}} = n \sum_{i=1}^{a} (\bar{y}_{i\cdot} - \bar{y}..)^2 = \text{treatment sum of squares}$$

and

$$SS_E = \sum_{i=1}^{a} \sum_{j=1}^{n} (y_{ij} - \bar{y}_{i\cdot})^2 = \text{error sum of squares}$$

We can gain considerable insight into how the analysis of variance works by examining the expected values of $SS_{\text{Treatments}}$ and SS_E. This will lead us to an appropriate statistic for testing the hypothesis of no differences among treatment means (or $\tau_i = 0$).

It can be shown that

$$E\left(\frac{SS_{\text{Treatments}}}{a - 1}\right) = \sigma^2 + \frac{n \sum\limits_{i=1}^{a} \tau_i^2}{a - 1} \qquad (5\text{-}37)$$

The ratio

$$MS_{\text{Treatments}} = SS_{\text{Treatments}}/(a - 1)$$

is called the **mean square for treatments.** Thus, if H_0 is true, $MS_{\text{Treatments}}$ is an unbiased estimator of σ^2 because under H_0 each $\tau_i = 0$. If H_1 is true, $MS_{\text{Treatments}}$ estimates σ^2 plus a positive term that incorporates variation due to the systematic difference in treatment means.

We can also show that the expected value of the error sum of squares is $E(SS_E) = a(n - 1)\sigma^2$. Therefore, the **error mean square**

$$MS_E = SS_E/[a(n - 1)]$$

is an unbiased estimator of σ^2 regardless of whether or not H_0 is true.

There is also a partition of the number of degrees of freedom that corresponds to the sum of squares identity in equation 5-35. That is, there are $an = N$ observations; thus,

SS_T has $an - 1$ degrees of freedom. There are a levels of the factor, so $SS_{\text{Treatments}}$ has $a - 1$ degrees of freedom. Finally, within any treatment there are n replicates providing $n - 1$ degrees of freedom with which to estimate the experimental error. Since there are a treatments, we have $a(n - 1)$ degrees of freedom for error. Therefore, the degrees of freedom partition is

$$an - 1 = a - 1 + a(n - 1)$$

Now assume that each of the a populations can be modeled as a normal distribution. Using this assumption we can show that if the null hypothesis H_0 is true, the ratio

$$F_0 = \frac{SS_{\text{Treatments}}/(a - 1)}{SS_E/[a(n - 1)]} = \frac{MS_{\text{Treatments}}}{MS_E} \tag{5-38}$$

has an F distribution with $a - 1$ and $a(n - 1)$ degrees of freedom. Furthermore, from the expected mean squares, we know that MS_E is an unbiased estimator of σ^2. Also, under the null hypothesis, $MS_{\text{Treatments}}$ is an unbiased estimator of σ^2. However, if the null hypothesis is false, then the expected value of $MS_{\text{Treatments}}$ is greater than σ^2. Therefore, under the alternative hypothesis, the expected value of the numerator of the test statistic (equation 5-38) is greater than the expected value of the denominator. Consequently, we should reject H_0 if the statistic is large. This implies an upper-tail, one-tail critical region. Therefore, we would reject H_0 if $f_0 > f_{\alpha, a-1, a(n-1)}$ where f_0 is computed from equation 5-38. These results are summarized as follows.

Testing Hypotheses on More Than Two Means

$$MS_{\text{Treatments}} = \frac{SS_{\text{Treatments}}}{a - 1} \qquad\qquad E(MS_{\text{Treatments}}) = \sigma^2 + \frac{n \sum_{i=1}^{a} \tau_i^2}{a - 1}$$

$$MS_E = \frac{SS_E}{a(n - 1)} \qquad\qquad E(MS_E) = \sigma^2$$

Null hypothesis: $\qquad H_0: \tau_1 = \tau_2 = \cdots = \tau_a = 0$

Alternative hypothesis: $\quad H_1: \tau_i \neq 0 \qquad$ for at least one i

Test statistic: $\qquad\qquad F_0 = \dfrac{MS_{\text{Treatments}}}{MS_E}$

Rejection criterion: $\qquad f_0 > f_{\alpha, a-1, a(n-1)}$

Efficient computational formulas for the sums of squares may be obtained by expanding and simplifying the definitions of $SS_{\text{Treatments}}$ and SS_T. This yields the following results.

Completely Randomized Experiment with Equal Sample Sizes

The computing formulas for the sums of squares in the analysis of variance for a completely randomized experiment with equal sample sizes in each treatment are

$$\text{✶} \quad SS_T = \sum_{i=1}^{a} \sum_{j=1}^{n} y_{ij}^2 - \frac{y_{..}^2}{N}$$

and

$$SS_{\text{Treatments}} = \sum_{i=1}^{a} \frac{y_{i.}^2}{n} - \frac{y_{..}^2}{N}$$

The error sum of squares is usually obtained by subtraction as

$$SS_E = SS_T - SS_{\text{Treatments}}$$

The computations for this test procedure are usually summarized in tabular form as shown in Table 5-8. This is called an **analysis of variance table.**

EXAMPLE 5-14

Consider the paper tensile strength experiment described in Section 5-8.1. We can use the analysis of variance to test the hypothesis that different hardwood concentrations do not affect the mean tensile strength of the paper.

Table 5-8   The Analysis of Variance for a Single-Factor Experiment

Source of Variation	Sum of Squares	Degrees of Freedom	Mean Square	F_0
Treatments	$SS_{\text{Treatments}}$	$a - 1$	$MS_{\text{Treatments}}$	$\dfrac{MS_{\text{Treatments}}}{MS_E}$
Error	SS_E	$a(n - 1)$	MS_E	
Total	SS_T	$an - 1$		

NOTE: $\sqrt{MS_{\text{error}}} = \hat{\sigma}$ for Y

The hypotheses are

$$H_0: \tau_1 = \tau_2 = \tau_3 = \tau_4 = 0$$

$$H_1: \tau_i \neq 0 \text{ for at least one } i.$$

We will use $\alpha = 0.01$. The sums of squares for the analysis of variance are computed from equations 5-39, 5-40, and 5-41 as follows:

$$SS_T = \sum_{i=1}^{4} \sum_{j=1}^{6} y_{ij}^2 - \frac{y_{..}^2}{N}$$

$$= (7)^2 + (8)^2 + \cdots + (20)^2 - \frac{(383)^2}{24} = 512.96$$

$$SS_{\text{Treatments}} = \sum_{i=1}^{4} \frac{y_{i.}^2}{n} - \frac{y_{..}^2}{N}$$

$$= \frac{(60)^2 + (94)^2 + (102)^2 + (127)^2}{6} - \frac{(383)^2}{24} = 382.79$$

$$SS_E = SS_T - SS_{\text{Treatments}}$$

$$= 512.96 - 382.79 = 130.17$$

We usually do not perform these calculations by hand. The analysis of variance computed by Minitab is presented in Table 5-9. Since $f_{0.01,3,20} = 4.94$, we reject H_0 and conclude

Table 5-9   Minitab Analysis of Variance Output for the Paper Tensile Strength Experiment

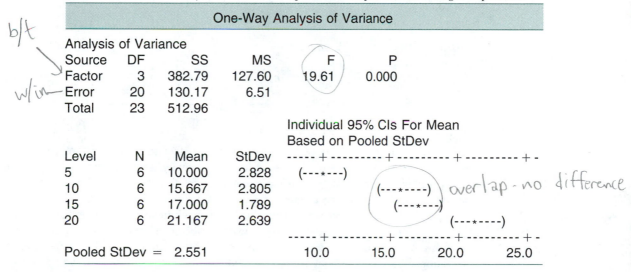

b/t

w/in

One-Way Analysis of Variance

Analysis of Variance

Source	DF	SS	MS	F	P
Factor	3	382.79	127.60	19.61	0.000
Error	20	130.17	6.51		
Total	23	512.96			

Individual 95% CIs For Mean
Based on Pooled StDev

Level	N	Mean	StDev
5	6	10.000	2.828
10	6	15.667	2.805
15	6	17.000	1.789
20	6	21.167	2.639

overlap - no difference

Pooled StDev = 2.551

```
----- + --------- + --------- + --------- + -
     (---*---)
                 (---*----)
                 (---*---)
                        (---*----)
----- + --------- + --------- + --------- + -
    10.0        15.0        20.0        25.0
```

that hardwood concentration in the pulp significantly affects the strength of the paper. Note that the computer output reports a P-value of 0 for the test statistic $F = 19.61$ in Table 5-9. This is a truncated value; the actual P-value is $P = 3.59 \times 10^{-6}$. However, since the P-value is considerably smaller than $\alpha = 0.01$, we have strong evidence to conclude that H_0 is not true. Note that Minitab also provides some summary information about each level of hardwood concentration, including the confidence interval on each mean.

In some single-factor experiments, the number of observations taken under each treatment may be different. We then say that the design is **unbalanced.** The analysis of variance described earlier is still valid, but slight modifications must be made in the sums of squares formulas. Let n_i observations be taken under treatment $i(i = 1, 2, \ldots, a)$, and let the total number of observations $N = \sum_{i=1}^{a} n_i$. The computational formulas for SS_T and $SS_{\text{Treatments}}$ are as shown in the following definition.

Completely Randomized Experiment with Unequal Sample Sizes

The computing formulas for the sums of squares in the analysis of variance for a completely randomized experiment with unequal sample sizes n_i in each treatment are

$$SS_T = \sum_{i=1}^{a} \sum_{j=1}^{n_i} y_{ij}^2 - \frac{y_{..}^2}{N}$$

$$SS_{\text{Treatments}} = \sum_{i=1}^{a} \frac{y_{i.}^2}{n_i} - \frac{y_{..}^2}{N}$$

and

$$SS_E = SS_T - SS_{\text{Treatments}}$$

Which Means Differ?

Finally, note that the analysis of variance tells us whether there is a difference among means. It does not tell us which means differ. If the analysis of variance indicates that there is a statistically significant difference among means, there is a simple graphical procedure that can be used to isolate the specific differences. Suppose that $\bar{y}_{1.}, \bar{y}_{2.}, \ldots, \bar{y}_{a.}$ are the observed averages for these factor levels. Each treatment average has standard deviation $\sigma/\sqrt{n}$, where σ is the standard deviation of an individual observation. If all treatment means are equal, the observed means $\bar{y}_{i.}$ would behave as if they were a set of observations drawn at random from a normal distribution with mean μ and standard deviation $\sigma/\sqrt{n}$.

Visualize this normal distribution capable of being slid along an axis below which the treatment means $\bar{y}_{1.}, \bar{y}_{2.}, \ldots, \bar{y}_{a.}$ are plotted. If all treatment means are equal, there should be some position for this distribution that makes it obvious that the $\bar{y}_{i.}$ values were drawn from the same distribution. If this is not the case, then the $\bar{y}_{i.}$ values that do not appear to have been drawn from this distribution are associated with treatments that produce different mean responses.

The only flaw in this logic is that σ is unknown. However, we can use $\sqrt{MS_E}$ from the analysis of variance to estimate σ. This implies that a t distribution should be used instead of the normal in making the plot, but since the t looks so much like the normal, sketching a normal curve that is approximately $6\sqrt{MS_E/n}$ units wide will usually work very well.

Figure 5-7 shows this arrangement for the hardwood concentration experiment. The standard deviation of this normal distribution is

$$\sqrt{MS_E/n} = \sqrt{6.51/6} = 1.04$$

If we visualize sliding this distribution along the horizontal axis, we note that there is no location for the distribution that would suggest that all four observations (the plotted means) are typical, randomly selected values from that distribution. This, of course, should be expected, because the analysis of variance has indicated that the means differ, and the display in Fig. 5-7 is only a graphical representation of the analysis of variance results. The figure does indicate that treatment 4 (20% hardwood) produces paper with higher mean tensile strength than do the other treatments and that treatment 1 (5% hardwood) results in lower mean tensile strength than do the other treatments. The means of treatments 2 and 3 (10 and 15% hardwood, respectively) do not differ.

This simple procedure is a rough but very useful and effective technique for comparing means following an analysis of variance. There are more quantitative techniques, called **multiple comparison procedures,** for testing for differences between specific means following an analysis of variance. Since these procedures typically involve a series of tests, the type I error compounds to produce an **experiment-wise** or **family error rate.** For more details on these procedures, see Montgomery (1997).

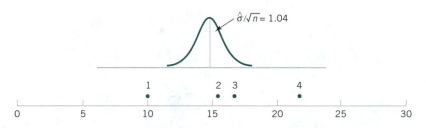

Figure 5-7 Tensile strength averages from the hardwood concentration experiment in relation to a normal distribution with standard deviation $\sqrt{MS_E/n} = \sqrt{6.51/6} = 1.04$.

Residual Analysis and Model Checking

The one-way model analysis of variance assumes that the observations are normally and independently distributed with the same variance for each treatment or factor level. These assumptions should be checked by examining the residuals. A residual is the difference between an observation y_{ij} and its estimated (or fitted) value from the statistical model being studied, denoted as $\hat{y}_{ij}$. For the completely randomized design $\hat{y}_{ij} = \bar{y}_{i\cdot}$ and each residual is $e_{ij} = y_{ij} - \bar{y}_{i\cdot}$—that is, the difference between an observation and the corresponding observed treatment mean. The residuals for the hardware percentage experiment are shown in Table 5-10. Using $\bar{y}_{i\cdot}$ to calculate each residual essentially removes the effect of hardware concentration from those data; consequently, the residuals contain information about unexplained variability.

The normality assumption can be checked by constructing a normal probability plot of the residuals. To check the assumption of equal variances at each factor level, plot the residuals against the factor levels and compare the spread in the residuals. It is also useful to plot the residuals against $\bar{y}_{i\cdot}$ (sometimes called the fitted value); the variability in the residuals should not depend in any way on the value of $\bar{y}_{i\cdot}$. Most statistics software packages will construct these plots on request. When a pattern appears in these plots, it usually suggests the need for a transformation—that is, analyzing the data in a different metric. For example, if the variability in the residuals increases with $\bar{y}_{i\cdot}$, then a transformation such as $\log y$ or $\sqrt{y}$ should be considered. In some problems, the dependency of residual scatter on the observed mean $\bar{y}_{i\cdot}$ is very important information. It may be desirable to select the factor level that results in maximum response; however, this level may also cause more variation in response from run to run.

The independence assumption can be checked by plotting the residuals against the time or run order in which the experiment was performed. A pattern in this plot, such as sequences of positive and negative residuals, may indicate that the observations are not independent. This suggests that time or run order is important or that variables that change over time are important and have not been included in the experimental design.

A normal probability plot of the residuals from the paper tensile strength experiment is shown in Fig. 5-8. Figures 5-9 and 5-10 present the residuals plotted against the factor levels and the fitted value $\bar{y}_{i\cdot}$ respectively. These plots do not reveal any model inadequacy or unusual problem with the assumptions.

Table 5-10   Residuals for the Tensile Strength Experiment

Hardwood Concentration	Residuals					
5%	−3.00	−2.00	5.00	1.00	−1.00	0.00
10%	−3.67	1.33	−2.67	2.33	3.33	−0.67
15%	−3.00	1.00	2.00	0.00	−1.00	1.00
20%	−2.17	3.83	0.83	1.83	−3.17	−1.17

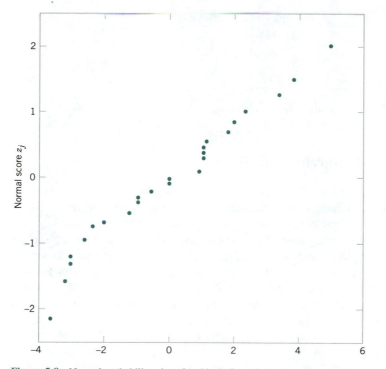

Figure 5-8   Normal probability plot of residuals from the hardwood concentration experiment.

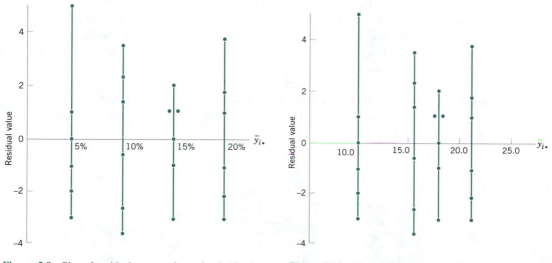
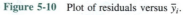

Figure 5-9   Plot of residuals versus factor levels (hardwood concentration).

Figure 5-10   Plot of residuals versus $\bar{y}_i$.

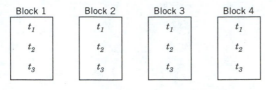

Figure 5-11   A randomized complete block design.

5-8.2   Randomized Complete Block Experiment

In many experimental design problems, it is necessary to design the experiment so that the variability arising from a nuisance factor can be controlled. For example, consider the situation of Example 5-8, where two different methods were used to predict the shear strength of steel plate girders. Because each girder has different strength (potentially) and this variability in strength was not of direct interest, we designed the experiment by using the two test methods on each girder and then comparing the average difference in strength readings on each girder to zero using the paired t-test. The paired t-test is a procedure for comparing two treatment means when all experimental runs cannot be made under homogeneous conditions. Alternatively, we can view the paired t-test as a method for reducing the background noise in the experiment by blocking out a **nuisance factor** effect. The block is the nuisance factor, and in this case, the nuisance factor is the actual **experimental unit**—the steel girder specimens used in the experiment.

The randomized block design is an extension of the paired t-test to situations where the factor of interest has more than two levels; that is, more than two treatments must be compared. For example, suppose that three methods could be used to evaluate the strength readings on steel plate girders. We may think of these as three treatments—say, t_1, t_2, and t_3. If we use four girders as the experimental units, then a **randomized complete block design** would appear as shown in Fig. 5-11. The design is called a randomized complete block design because each block is large enough to hold all the treatments and because the actual assignment of each of the three treatments within each block is done randomly. Once the experiment has been conducted, the data are recorded in a table, such as is shown in Table 5-11. The observations in this table—say, y_{ij}—represent the response obtained when method i is used on girder j.

Table 5-11   A Randomized Complete Block Design

Treatment (Method)	Block (Girder)			
	1	2	3	4
1	y_{11}	y_{12}	y_{13}	y_{14}
2	y_{21}	y_{22}	y_{23}	y_{24}
3	y_{31}	y_{32}	y_{33}	y_{34}

The general procedure for a randomized complete block design consists of selecting b blocks and running a complete replicate of the experiment in each block. The data that result from running a randomized complete block design for investigating a single factor with a levels and b blocks are shown in Table 5-12. There will be a observations (one per factor level) in each block, and the order in which these observations are run is randomly assigned within the block.

We will now describe the statistical analysis for a randomized complete block design. Suppose that a single factor with a levels is of interest and that the experiment is run in b blocks. The observations may be represented by the linear statistical model.

$$Y_{ij} = \mu + \tau_i + \beta_j + \epsilon_{ij} \begin{cases} i = 1, 2, \ldots, a \\ j = 1, 2, \ldots, b \end{cases} \qquad (5\text{-}39)$$

where μ is an overall mean, τ_i is the effect of the ith treatment, β_j is the effect of the jth block, and ϵ_{ij} is the random error term, which is assumed to be normally and independently distributed with mean zero and variance σ^2. Treatments and blocks will initially be considered as fixed factors. Furthermore, the treatment and block effects are defined as deviations from the overall mean, so that $\sum_{i=1}^{a} \tau_i = 0$ and $\sum_{j=1}^{b} \beta_j = 0$. We also assume that treatments and blocks do not interact; that is, the effect of treatment i is the same regardless of which block (or blocks) it is tested in. We are interested in testing the equality of the treatment effects; that is,

$$H_0: \tau_1 = \tau_2 = \cdots = \tau_a = 0 \qquad (5\text{-}40)$$

$$H_1: \tau_i \neq 0 \text{ at least one } i$$

As in the completely randomized experiment, testing the hypothesis that all the treatment effects τ_i are equal to zero is equivalent to testing the hypothesis that the treatment means are equal.

Table 5-12 A Randomized Complete Block Design with a Treatments and b Blocks

Treatments	Blocks				Totals	Averages
	1	2	. . .	b		
1	y_{11}	y_{12}	. . .	y_{1b}	$y_{1\cdot}$	$\bar{y}_{1\cdot}$
2	y_{21}	y_{22}	. . .	y_{2b}	$y_{2\cdot}$	$\bar{y}_{2\cdot}$
$\vdots$	$\vdots$	$\vdots$		$\vdots$	$\vdots$	$\vdots$
a	y_{a1}	y_{a2}	. . .	y_{ab}	$y_{a\cdot}$	$\bar{y}_{a\cdot}$
Totals	$y_{\cdot 1}$	$y_{\cdot 2}$	. . .	$y_{\cdot b}$	$y_{\cdot\cdot}$	
Averages	$\bar{y}_{\cdot 1}$	$\bar{y}_{\cdot 2}$	. . .	$\bar{y}_{\cdot b}$		$\bar{y}_{\cdot\cdot}$

The analysis of variance can be extended to the randomized complete block design. The procedure uses a sum of squares identity that partitions the total sum of squares into three components.

The **sum of squares identity for the randomized complete block design** is

$$\sum_{i=1}^{a} \sum_{j=1}^{b} (y_{ij} - \bar{y}..)^2 = b \sum_{i=1}^{a} (\bar{y}_{i.} - \bar{y}..)^2 + a \sum_{j=1}^{b} (\bar{y}_{.j} - \bar{y}..)^2$$

$$+ \sum_{i=1}^{a} \sum_{j=1}^{b} (y_{ij} - \bar{y}_{.j} - \bar{y}_{i.} + \bar{y}..)^2 \qquad (5\text{-}41)$$

The sum of squares identity may be represented symbolically as

$$SS_T = SS_{\text{Treatments}} + SS_{\text{Blocks}} + SS_E$$

where

$$SS_T = \sum_{i=1}^{a} \sum_{j=1}^{b} (y_{ij} - \bar{y}..)^2 = \text{total sum of squares}$$

$$SS_{\text{Treatments}} = b \sum_{i=1}^{a} (\bar{y}_{i.} - \bar{y}..)^2 = \text{treatment sum of squares}$$

$$SS_{\text{Blocks}} = a \sum_{j=1}^{b} (\bar{y}_{.j} - \bar{y}..)^2 = \text{block sum of squares}$$

$$SS_E = \sum_{i=1}^{a} \sum_{j=1}^{b} (\bar{y}_{ij} - \bar{y}_{.j} - \bar{y}_{i.} + \bar{y}..)^2 = \text{error sum of squares}$$

Furthermore, the degree-of-freedom breakdown corresponding to these sums of squares is

$$ab - 1 = (a - 1) + (b - 1) + (a - 1)(b - 1)$$

For the randomized block design, the relevant mean squares are

$$MS_{\text{Treatments}} = \frac{SS_{\text{Treatments}}}{a - 1} \qquad MS_{\text{Blocks}} = \frac{SS_{\text{Blocks}}}{b - 1} \qquad MS_E = \frac{SS_E}{(a - 1)(b - 1)} \qquad (5\text{-}42)$$

The expected values of these mean squares can be shown to be as follows:

$$E(MS_{\text{Treatments}}) = \sigma^2 + \frac{b \sum\limits_{i=1}^{a} \tau_i^2}{a - 1}$$

$$E(MS_{\text{Blocks}}) = \sigma^2 + \frac{a \sum\limits_{j=1}^{b} \beta_j^2}{b - 1}$$

$$E(MS_E) = \sigma^2$$

Therefore, if the null hypothesis H_0 is true so that all treatment effects $\tau_i = 0$, then $MS_{\text{Treatments}}$ is an unbiased estimator of σ^2, whereas if H_0 is false, $MS_{\text{Treatments}}$ overestimates σ^2. The mean square for error is always an unbiased estimate of σ^2. To test the null hypothesis that the treatment effects are all zero, we compute the ratio

$$F_0 = \frac{MS_{\text{Treatments}}}{MS_E} \tag{5-43}$$

which has an F distribution with $a - 1$ and $(a - 1)(b - 1)$ degrees of freedom if the null hypothesis is true. We would reject the null hypothesis at the α level of significance if the computed value of the test statistic in equation 5-43

$$f_0 > f_{\alpha, a-1, (a-1)(b-1)}$$

In practice, we compute SS_T, $SS_{\text{Treatments}}$, and SS_{Blocks}, and then obtain the error sum of squares SS_E by subtraction. The appropriate computing formulas are as follows:

Randomized Complete Block Experiment

The computing formulas for the sums of squares in the analysis of variance for a randomized complete block design are

$$SS_T = \sum_{i=1}^{a} \sum_{j=1}^{b} y_{ij}^2 - \frac{y_{..}^2}{ab}$$

$$SS_{\text{Treatments}} = \frac{1}{b} \sum_{i=1}^{a} y_{i.}^2 - \frac{y_{..}^2}{ab}$$

$$SS_{\text{Blocks}} = \frac{1}{a} \sum_{j=1}^{b} y_{.j}^2 - \frac{y_{..}^2}{ab}$$

and

$$SS_E = SS_T - SS_{\text{Treatments}} - SS_{\text{Blocks}}$$

The computations are usually arranged in an analysis of variance table, such as is shown in Table 5-13. Generally, a computer software package will be used to perform the analysis of variance for the randomized complete block design.

EXAMPLE 5-15

An experiment was performed to determine the effect of four different chemicals on the strength of a fabric. These chemicals are used as part of the permanent press finishing process. Five fabric samples were selected, and a randomized complete block design was run by testing each chemical type once in random order on each fabric sample. The data are shown in Table 5-14. We will test for differences in means using the analysis of variance with $\alpha = 0.01$.

The sums of squares for the analysis of variance are computed as follows:

$$SS_T = \sum_{i=1}^{4} \sum_{j=1}^{5} y_{ij}^2 - \frac{y_{..}^2}{ab}$$

$$= (1.3)^2 + (1.6)^2 + \cdots + (3.4)^2 - \frac{(39.2)^2}{20} = 25.69$$

$$SS_{\text{Treatments}} = \sum_{i=1}^{4} \frac{y_{i.}^2}{b} - \frac{y_{..}^2}{ab}$$

$$= \frac{(5.7)^2 + (8.8)^2 + (6.9)^2 + (17.8)^2}{5} - \frac{(39.2)^2}{20} = 18.04$$

$$SS_{\text{Blocks}} = \sum_{j=1}^{5} \frac{y_{.j}^2}{a} - \frac{y_{..}^2}{ab}$$

$$= \frac{(9.2)^2 + (10.1)^2 + (3.5)^2 + (8.8)^2 + (7.6)^2}{4} - \frac{(39.2)^2}{20} = 6.69$$

$$SS_E = SS_T - SS_{\text{Blocks}} - SS_{\text{Treatments}}$$

$$= 25.69 - 6.69 - 18.04 = 0.96$$

Table 5-13   Analysis of Variance for a Randomized Complete Block Design

Source of Variation	Sum of Squares	Degrees of Freedom	Mean Square	F_0
Treatments	$SS_{\text{Treatments}}$	$a - 1$	$\dfrac{SS_{\text{Treatments}}}{a - 1}$	$\dfrac{MS_{\text{Treatments}}}{MS_E}$
Blocks	SS_{Blocks}	$b - 1$	$\dfrac{SS_{\text{Blocks}}}{b - 1}$	
Error	SS_E (by subtraction)	$(a - 1)(b - 1)$	$\dfrac{SS_E}{(a - 1)(b - 1)}$	
Total	SS_T	$ab - 1$		

Table 5-14   Fabric Strength Data—Randomized Complete Block Design

Chemical Type	Fabric Sample					Treatment Totals $y_{i \cdot}$	Treatment Averages $\bar{y}_{i \cdot}$
	1	2	3	4	5		
1	1.3	1.6	0.5	1.2	1.1	5.7	1.14
2	2.2	2.4	0.4	2.0	1.8	8.8	1.76
3	1.8	1.7	0.6	1.5	1.3	6.9	1.38
4	3.9	4.4	2.0	4.1	3.4	17.8	3.56
Block totals $y_{\cdot j}$	9.2	10.1	3.5	8.8	7.6	39.2($y_{\cdot \cdot}$)	
Block averages $\bar{y}_{\cdot j}$	2.30	2.53	0.88	2.20	1.90		1.96($\bar{y}_{\cdot \cdot}$)

The analysis of variance is summarized in Table 5-15. Since $f_0 = 75.13 > f_{0.01,3,12} = 5.95$ (the P-value is 4.79×10^{-8}) we conclude that there is a significant difference in the chemical types so far as their effect on mean fabric strength is concerned.

When Is Blocking Necessary?

Suppose an experiment was conducted as a randomized block design, and blocking was not really necessary. There are ab observations and $(a - 1)(b - 1)$ degrees of freedom for error. If the experiment had been run as a completely randomized single-factor design with b replicates, we would have had $a(b - 1)$ degrees of freedom for error. Therefore, blocking has cost $a(b - 1) - (a - 1)(b - 1) = b - 1$ degrees of freedom for error. Thus, since the loss in error degrees of freedom is usually small, if there is a reasonable chance that block effects may be important, the experimenter should use the randomized block design.

Computer Solution

Table 5-16 presents the computer output from Minitab for the randomized complete block design example. We used the analysis of variance menu for balanced designs to solve this problem. The results agree closely with the hand calculations from

Table 5-15   Analysis of Variance for the Randomized Complete Block Experiment

Source of Variation	Sum of Squares	Degrees of Freedom	Mean Square	f_0	P-value
Chemical types (treatments)	18.04	3	6.01	75.13	4.79 E-8
Fabric samples (blocks)	6.69	4	1.67		
Error	0.96	12	0.08		
Total	25.69	19			

Table 5-16   Minitab Analysis of Variance for the Randomized Complete Block Design in Example 5-15

Analysis of Variance (Balanced Designs)						
Factor	Type	Levels	Values			
Chemical	fixed	4	1	2	3	4
Fabric S	fixed	5	1	2	3	4	5

Analysis of Variance for strength

Source	DF	SS	MS	F	P
Chemical	3	18.0440	6.0147	75.89	0.000
Fabric S	4	6.6930	1.6733	21.11	0.000
Error	12	0.9510	0.0792		
Total	19	25.6880			

F-test with denominator: Error
Denominator MS = 0.079250 with 12 degrees of freedom

Numerator	DF	MS	F	P
Chemical	3	6.015	75.89	0.000
Fabric S	4	1.673	21.11	0.000

Table 5-15. Note that Minitab computes an F-statistic for the blocks (the fabric samples). The validity of this ratio as a test statistic for the null hypothesis of no block effects is doubtful, as the blocks represent a **restriction on randomization;** that is, we have only randomized within the blocks. If the blocks are not chosen at random, or if they are not run in random order, then the F-ratio for blocks may not provide reliable information about block effects. For more discussion see Montgomery (1997, Chapter 5).

Which Means Differ?

When the analysis of variance indicates that a difference exists between the treatment means, we may need to perform some followup tests to isolate the specific differences. The graphical method previously described can be used for this purpose. The four chemical type averages are:

$$\bar{y}_{1.} = 1.14 \qquad \bar{y}_{2.} = 1.76 \qquad \bar{y}_{3.} = 1.38 \qquad \bar{y}_{4.} = 3.56$$

Each treatment average uses $b = 5$ observations (one from each block). Therefore, the standard deviation of a treatment average is $\sigma/\sqrt{b}$. The estimate of σ is $\sqrt{MS_E}$. Thus, the standard deviation used for the normal distribution is

$$\sqrt{MS_E/b} = \sqrt{0.0792/5} = 0.126$$

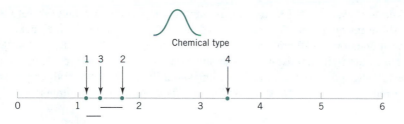

Figure 5-12   Strength averages from the fabric experiment in relation to a normal distribution with standard deviation $\sqrt{MS_E/b} = \sqrt{0.0792/5} = 0.126$.

A sketch of a normal distribution that is $6\sqrt{MS_E/b} = 0.755$ units wide is shown in Figure 5-12. If we visualize sliding this distribution along the horizontal axis, we note that there is no location for the distribution that would suggest that all four means are typical, randomly selected values from that distribution. This should be expected, because the analysis of variance has indicated that the means differ. The underlined pairs of means are not different. Chemical type 4 results in significantly different strengths than the other three types. Chemical types 2 and 3 do not differ, and types 1 and 3 do not differ. There may be a small difference in strength between types 1 and 2.

Residual Analysis and Model Checking

In any designed experiment, it is always important to examine the residuals and to check for violation of basic assumptions that could invalidate the results. As usual, the residuals for the randomized complete block design are only the difference between the observed and estimated (or fitted) values from the statistical model—say,

$$e_{ij} = y_{ij} - \hat{y}_{ij}$$

and the fitted values are

$$\hat{y}_{ij} = \bar{y}_{i\cdot} + \bar{y}_{\cdot j} - \bar{y}_{\cdot\cdot} \tag{5-44}$$

The fitted value represents the estimate of the mean response when the ith treatment is run in the jth block. The residuals from the chemical type experiment are shown in Table 5-17.

Table 5-17   Residuals from the Randomized Complete Block Design

Chemical Type	Fabric Sample				
	1	2	3	4	5
1	−0.18	−0.10	0.44	−0.18	0.02
2	0.10	0.08	−0.28	0.00	0.10
3	0.08	−0.24	0.30	−0.12	−0.02
4	0.00	0.28	−0.48	0.30	−0.10

Figures 5-13, 5-14, 5-15, and 5-16 present the important residual plots for the experiment. These residual plots are usually constructed by computer software packages. There is some indication that fabric sample (block) 3 has greater variability in strength when treated with the four chemicals than the other samples. Chemical type 4, which provides the greatest strength, also has somewhat more variability in strength. Followup experiments may be necessary to confirm these findings, if they are potentially important.

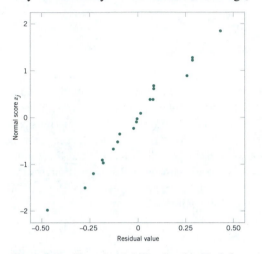

Figure 5-13   Normal probability plot of residuals from the randomized complete block design.

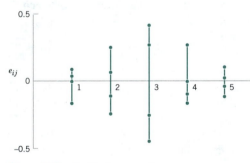

Figure 5-14   Residuals versus $\hat{y}_{ij}$.

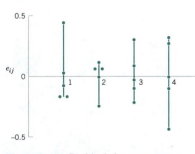

Figure 5-15   Residuals by treatment.

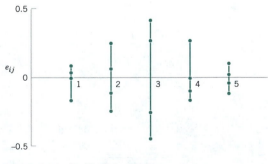

Figure 5-16   Residuals by block.

EXERCISES FOR SECTION 5-8

5-56. In "Orthogonal Design for Process Optimization and Its Application to Plasma Etching" (*Solid State Technology,* May 1987), G. Z. Yin and D. W. Jillie describe an experiment to determine the effect of C_2F_6 flow rate on the uniformity of the etch on a silicon wafer used in integrated circuit manufacturing. Three flow rates are used in the experiment, and the resulting uniformity (in percent) for six replicates is shown here.

C_2F_6 Flow (SCCM)	Observations					
	1	2	3	4	5	6
125	2.7	4.6	2.6	3.0	3.2	3.8
160	4.9	4.6	5.0	4.2	3.6	4.2
200	4.6	3.4	2.9	3.5	4.1	5.1

(a) Does C_2F_6 flow rate affect etch uniformity? Construct box plots to compare the factor levels and perform the analysis of variance. Use $\alpha = 0.1$.

(b) Which gas flow rates produce different mean etch uniformities?

5-57. In *Design and Analysis of Experiments,* 4th edition (John Wiley & Sons, 1997) D. C. Montgomery describes an experiment in which the tensile strength of a synthetic fiber is of interest to the manufacturer. It is suspected that strength is related to the percentage of cotton in the fiber. Five levels of cotton percentage are used, and five replicates are run in random order, resulting in the data that follow.

Cotton Percentage	Observations				
	1	2	3	4	5
15	7	7	15	11	9
20	12	17	12	18	18
25	14	18	18	19	19
30	19	25	22	19	23
35	7	10	11	15	11

(a) Does cotton percentage affect breaking strength? Draw comparative box plots and perform an analysis of variance. Use $\alpha = 0.05$.

(b) Plot average tensile strength against cotton percentage and interpret the results.

(c) Which specific means are different?

5-58. An experiment was run to determine whether four specific firing temperatures affect the density of a certain type of brick. The experiment led to the following data.

	Temperature (°F)		
100	125	150	175
	Density		
21.8	21.7	21.9	21.9
21.9	21.4	21.8	21.7
21.7	21.5	21.8	21.8
21.6	21.5	21.6	21.7
21.7	—	21.5	21.6
21.5	—	—	21.8
21.8	—	—	—

(a) Does the firing temperature affect the density of the bricks? Use $\alpha = 0.05$.

(b) Find the *P*-value for the *F*-statistic computed in part (a).

5-59. The compressive strength of concrete is being studied, and four different mixing techniques are being investigated. The following data have been collected.

Mixing Technique	Compressive Strength (psi)			
1	3129	3000	2865	2890
2	3200	3300	2975	3150
3	2800	2900	2985	3050
4	2600	2700	2600	2765

(a) Test the hypothesis that mixing techniques affect the strength of the concrete. Use $\alpha = 0.05$.

(b) Find the *P*-value for the *F*-statistic computed in part (a).

5-60. An electronics engineer is interested in the effect on tube conductivity of five different types of coating for cathode ray tubes in a telecommunications system display device. The following conductivity data are obtained. If $\alpha = 0.05$, can you isolate any differences in mean conductivity due to the coating type?

Coating Type	Conductivity			
1	143	141	150	146
2	152	149	137	143
3	134	133	132	127
4	129	127	132	129
5	147	148	144	142

5-61. In "The Effect of Nozzle Design on the Stability and Performance of Turbulent Water Jets" (*Fire Safety Journal*, Vol. 4, August 1981), C. Theobald describes an experiment in which a shape measurement was determined for several different nozzle types at different levels of jet efflux velocity. Interest in this experiment focuses primarily on nozzle type, and velocity is a nuisance factor. The data are as follows.

Nozzle Type	Jet Efflux Velocity (m/s)					
	11.73	14.37	16.59	20.43	23.46	28.74
1	0.78	0.80	0.81	0.75	0.77	0.78
2	0.85	0.85	0.92	0.86	0.81	0.83
3	0.93	0.92	0.95	0.89	0.89	0.83
4	1.14	0.97	0.98	0.88	0.86	0.83
5	0.97	0.86	0.78	0.76	0.76	0.75

(a) Does nozzle type affect shape measurement? Compare the nozzles with box plots and the analysis of variance.

(b) Use the graphical method from Section 5-8.1 to determine specific differences between the nozzles. Does a graph of the average (or standard deviation) of the shape measurements versus nozzle type assist with the conclusions?

(c) Analyze the residuals from this experiment.

5-62. In *Design and Analysis of Experiments,* 4th edition (John Wiley & Sons, 1997), D. C. Montgomery describes an experiment that determined the effect of four different types of tips in a hardness tester on the observed hardness of a metal alloy. Four specimens of the alloy were obtained, and each tip was tested once on each specimen, producing the following data:

Type of Tip	Specimen			
	1	2	3	4
1	9.3	9.4	9.6	10.0
2	9.4	9.3	9.8	9.9
3	9.2	9.4	9.5	9.7
4	9.7	9.6	10.0	10.2

(a) Is there any difference in hardness measurements between the tips?

(b) Use the graphical method from Section 5-8.1 to investigate specific differences between the tips.

(c) Analyze the residuals from this experiment.

5-63. An article in the *American Industrial Hygiene Association Journal* (Vol. 37, 1976, pp. 418–422) describes a field test for detecting the presence of arsenic in urine samples. The test has been proposed for use among forestry workers because of the increasing use of organic arsenics in that industry. The experiment compared the test as performed by both a trainee and an experienced trainer to an analysis at a remote laboratory. Four subjects were selected for testing and are considered as blocks. The response variable is arsenic content (in ppm) in the subject's urine. The data are as follows.

Test	Subject			
	1	2	3	4
Trainee	0.05	0.05	0.04	0.15
Trainer	0.05	0.05	0.04	0.17
Lab	0.04	0.04	0.03	0.10

(a) Is there any difference in the arsenic test procedure?
(b) Analyze the residuals from this experiment.

5-64. An article in the *Food Technology Journal* (Vol. 10, 1956, pp. 39–42) describes a study on the protopectin content of tomatoes during storage. Four storage times were selected, and samples from nine lots of tomatoes were analyzed. The protopectin content (expressed as hydrochloric acid soluble fraction mg/kg) is in the table below.

(a) The researchers in this study hypothesized that mean protopectin content would be different at different storage times. Can you confirm this hypothesis with a statistical test using $\alpha = 0.05$?
(b) Find the *P*-value for the test in part (a).
(c) Which specific storage times are different? Would you agree with the statement that protopectin content decreases as storage time increases?

(d) Analyze the residuals from this experiment.

5-65. An experiment was conducted to investigate leaking current in a near-micron SOS MOSFETS device. The purpose of the experiment was to investigate how leakage current varies as the channel length changes. Four channel lengths were selected. For each channel length, five different widths were also used, and width is to be considered a nuisance factor. The data are as follows.

Channel Length	Width				
	1	2	3	4	5
1	0.7	0.8	0.8	0.9	1.0
2	0.8	0.8	0.9	0.9	1.0
3	0.9	1.0	1.7	2.0	4.0
4	1.0	1.5	2.0	3.0	20.0

(a) Test the hypothesis that mean leakage voltage does not depend on the channel length, using $\alpha = 0.05$.
(b) Analyze the residuals from this experiment. Comment on the residual plots.

5-66. Consider the leakage voltage experiment described in Exercise 5-65. The observed leakage voltage for channel length 4 and width 5 was erroneously recorded. The correct observation is 4.0. Analyze the corrected data from this experiment. Is there evidence to conclude that mean leakage voltage increases with channel length?

Storage Time (days)	Lot								
	1	2	3	4	5	6	7	8	9
0	1694.0	989.0	917.3	346.1	1260.0	965.6	1123.0	1106.0	1116.0
7	1802.0	1074.0	278.8	1375.0	544.0	672.2	818.0	406.8	461.6
14	1568.0	646.2	1820.0	1150.0	983.7	395.3	422.3	420.0	409.5
21	415.5	845.4	377.6	279.4	447.8	272.1	394.1	356.4	351.2

SUPPLEMENTAL EXERCISES

5-67. A procurement specialist has purchased 25 resistors from vendor 1 and 35 resistors from vendor 2. Each resistor's resistance is measured with the following results:

Vendor 1

96.8 100.0 100.3 98.5 98.3 98.2 99.6
99.4 99.9 101.1 103.7 97.7 99.7 101.1
97.7 98.6 101.9 101.0 99.4 99.8 99.1
99.6 101.2 98.2 98.6

Vendor 2

106.8 106.8 104.7 104.7 108.0 102.2
103.2 103.7 106.8 105.1 104.0 106.2
102.6 100.3 104.0 107.0 104.3 105.8
104.0 106.3 102.2 102.8 104.2 103.4
104.6 103.5 106.3 109.2 107.2 105.4
106.4 106.8 104.1 107.1 107.7

(a) What distributional assumption is needed to test the claim that the variance of resistance of product from vendor 1 is not significantly different from the variance of resistance of product from vendor 2? Perform a graphical procedure to check this assumption.

(b) Perform an appropriate statistical hypothesis testing procedure to determine whether the procurement specialist can claim that the variance of resistance of product from vendor 1 is significantly different from the variance of resistance of product from vendor 2.

5-68. An article in the *Journal of Materials Engineering* (1989, Vol 11, No. 4, pp. 275–282) reported the results of an experiment to determine failure mechanisms for plasma-sprayed thermal barrier coatings. The failure stress for one particular coating (NiCrAlZr) under two different test conditions is as follows:

Failure stress ($\times\ 10^6$ Pa) after nine 1-hr cycles: 19.8, 18.5, 17.6, 16.7, 16.7, 14.8, 15.4, 14.1, 13.6

Failure stress ($\times\ 10^6$ Pa) after six 1-hr cycles: 14.9, 12.7, 11.9, 11.4, 10.1, 7.9

(a) What assumptions are needed to construct confidence intervals for the difference in mean failure stress under the two different test conditions? Use normal probability plots of the data to check these assumptions.

(b) Find a 99% confidence interval on the difference in mean failure stress under the two different test conditions.

(c) Using the confidence interval constructed in part (b), does the evidence support the claim that the first test conditions yield higher results, on the average, than the second? Explain your answer.

5-69. Consider Supplemental Exercise 5-68.

(a) Construct a 95% confidence interval on the ratio of the variances, σ_1^2/σ_2^2, of failure stress under the two different test conditions.

(b) Use your answer in part (a) to determine whether there is a significant difference in variances of the two different test conditions. Explain your answer.

5-70. A liquid dietary product implies in its advertising that use of the product for 1 month results in an average weight loss of at least 3 pounds. Eight subjects

Subject	Initial Weight (lb)	Final Weight (lb)
1	165	161
2	201	195
3	195	192
4	198	193
5	155	150
6	143	141
7	150	146
8	187	183

use the product for 1 month, and the resulting weight loss data are reported here. Use hypothesis testing procedures to answer the following questions.

(a) Do the data support the claim of the producer of the dietary product with the probability of a type I error set to 0.05?

(b) Do the data support the claim of the producer of the dietary product with the probability of a type I error set to 0.01?

(c) In an effort to improve sales, the producer is considering changing its claim from "at least 3 pounds" to "at least 5 pounds." Repeat parts (a) and (b) to test this new claim.

5-71. The breaking strength of yarn supplied by two manufacturers is being investigated. We know from experience with the manufacturers' processes that $\sigma_1 = 5$ psi and $\sigma_2 = 4$ psi. A random sample of 20 test specimens from each manufacturer results in $\bar{x}_1 = 88$ psi and $\bar{x}_2 = 91$ psi, respectively.

(a) Using a 90% confidence interval on the difference in mean breaking strength, comment on whether or not there is evidence to support the claim that manufacturer 2 produces yarn with higher mean breaking strength.

(b) Using a 98% confidence interval on the difference in mean breaking strength, comment on whether or not there is evidence to support the claim that manufacturer 2 produces yarn with higher mean breaking strength.

(c) Comment on why the results from parts (a) and (b) are different or the same. Which would you choose to make your decision and why?

5-72. Consider the previous supplemental exercise. Suppose that prior to collecting the data, you decide that you want the error in estimating $\mu_1 - \mu_2$ by $\bar{x}_1 - \bar{x}_2$ to be less than 1.5 psi. Specify the sample size for the following percentage confidence:

(a) 90%

(b) 98%

(c) Comment on the effect of increasing the percentage confidence on the sample size needed.

(d) Repeat parts (a)–(c) with an error of less than 0.75 psi instead of 1.5 psi.

(e) Comment on the effect of decreasing the error on the sample size needed.

5-73. The Salk polio vaccine experiment in 1954 focused on the effectiveness of the vaccine in combating paralytic polio. Because it was felt that without a control group of children there would be no sound basis for evaluating the efficacy of the Salk vaccine, the vaccine was administered to one group, and a placebo (visually identical to the vaccine but known to have no effect) was administered to a second group. For ethical reasons, and because it was suspected that knowledge of vaccine administration would affect subsequent diagnosis, the experiment was conducted in a double-blind fashion. That is, neither the subjects nor the administrators knew who received the vaccine and who received the placebo. The actual data for this experiment are as follows:

Placebo group: $n = 201,299$: 110 cases of polio observed

Vaccine group: $n = 200,745$: 33 cases of polio observed

(a) Use a hypothesis testing procedure to determine whether the proportion of children in the two groups who contracted paralytic polio is statistically different. Use a probability of a type I error equal to 0.05.

(b) Repeat part (a) using a probability of a type I error equal to 0.01.

(c) Compare your conclusions from parts (a) and (b) and explain why they are the same or different.

5-74. A study was carried out to determine the accuracy of Medicaid claims. In a sample of 1095 physician-filed claims, 942 claims exactly matched information in medical records. In a sample of 1042 hospital-filed claims, the corresponding number was 850.

(a) Is there a difference in the accuracy between these two sources? Use $\alpha = 0.05$. What is the P-value of the test?

(b) Suppose a second study was conducted. Of the 550 physician-filed claims examined, 473 were accurate, whereas of the 550 hospital-filed claims examined, 451 were accurate. Is there a statistically significant difference in the accuracy in the second study's data? Again, use $\alpha = 0.05$.

(c) Note that the estimated accuracy percentages are nearly identical for the first and second studies; however, the results of the hypothesis tests in parts (a) and (b) are different. Explain why this occurs.

(d) Construct a 95% confidence interval on the difference of the two proportions for part (a). Then construct a 95% confidence interval on the difference of the two proportions for part (b). Explain why the estimated accuracy percentages for the two studies are nearly identical but the lengths of the confidence intervals are different.

5-75. In a random sample of 200 Phoenix residents who drive a domestic car, 165 reported wearing their seat belt regularly, whereas another sample of 250 Phoenix residents who drive a foreign car revealed 198 who regularly wore their seat belt.

(a) Perform a hypothesis testing procedure to determine whether there is a statistically significant difference in seat belt usage between domestic and foreign car drivers. Set your probability of a type I error to 0.05.

(b) Perform a hypothesis testing procedure to determine whether there is a statistically significant difference in seat belt usage between domestic and foreign car drivers. Set your probability of a type I error to 0.1.

(c) Compare your answers for parts (a) and (b) and explain why they are the same or different.

(d) Suppose that all the numbers in the problem description were doubled. That is, in a random sample of 400 Phoenix residents who drive a domestic car, 330 reported wearing their seat belt regularly, whereas another sample of 500 Phoenix residents who drive a foreign car revealed 396 who regularly wore their seat belt. Repeat parts (a) and (b) and comment on the effect of increasing the sample size without changing the proportions on your results.

5-76. Consider the previous exercise, which summarized data collected from drivers about their seat belt usage.

(a) Do you think there is a reason not to believe these data? Explain your answer.

(b) Is it reasonable to use the hypothesis testing results from the previous problem to draw an inference about the difference in proportion of seat belt usage

(i) Of the spouses of these drivers of domestic and foreign cars? Explain your answer.

(ii) Of the children of these drivers of domestic and foreign cars? Explain your answer.

(iii) Of all drivers of domestic and foreign cars? Explain your answer.

(iv) Of all drivers of domestic and foreign trucks? Explain your answer.

5-77. Consider Example 5-12 in the text.

(a) Redefine the parameters of interest to be the proportion of lenses that are unsatisfactory following tumble polishing with polishing fluids 1 or 2. Test the hypothesis that the two polishing solutions give different results using $\alpha = 0.01$.

(b) Compare your answer in part (a) with that in the example. Explain why they are the same or different.

5-78. A manufacturer of a new pain relief tablet would like to demonstrate that his product works twice as fast as the competitor's product. Specifically, he would like to test

$$H_0: \mu_1 = 2\mu_2$$
$$H_1: \mu_1 > 2\mu_2$$

where μ_1 is the mean absorption time of the competitive product and μ_2 is the mean absorption time of the new product. Assuming that the variances σ_1^2 and σ_2^2 are known, develop a procedure for testing this hypothesis.

5-79. Suppose that we are testing $H_0: \mu_1 = \mu_2$ versus $H_1: \mu_1 \neq \mu_2$, and we plan to use equal sample sizes from the two populations. Both populations are assumed to be normal with unknown but equal variances. If we use $\alpha = 0.05$ and if the true mean $\mu_1 = \mu_2 + \sigma$, what sample size must be used for the power of this test to be at least 0.90?

5-80. A fuel-economy study was conducted for two German automobiles, Mercedes and Volkswagen. One vehicle of each brand was selected, and the mileage performance was observed for 10 tanks of fuel in each car. The data are as follows (in mpg):

Mercedes		Volkswagen	
24.7	24.9	41.7	42.8
24.8	24.6	42.3	42.4
24.9	23.9	41.6	39.9
24.7	24.9	39.5	40.8
24.5	24.8	41.9	29.6

(a) Construct a normal probability plot of each of the data sets. Based on

these plots, is it reasonable to assume that they are each drawn from a normal population?

(b) Suppose it was determined that the lowest observation of the Mercedes data was erroneously recorded and should be 24.6. Furthermore, the lowest observation of the Volkswagen data was also mistaken and should be 39.6. Again construct normal probability plots of each of the data sets with the corrected values. Based on these new plots, is it reasonable to assume that they are each drawn from a normal population?

(c) Compare your answers from parts (a) and (b) and comment on the effect of these mistaken observations on the normality assumption.

(d) Using the corrected data from part (b) and a 95% confidence interval, is there evidence to support the claim that the variability in mileage performance is greater for a Volkswagen than for a Mercedes?

5-81. Consider the fire-fighting foam expanding agents investigated in Exercise 5-14, in which five observations of each agent were recorded. Suppose that, if agent 1 produces a mean expansion that differs from the mean expansion of agent 2 by 1.5, we would like to reject the null hypothesis with probability at least 0.95.

(a) What sample size is required?

(b) Do you think that the original sample size in Exercise 5-14 was appropriate to detect this difference? Explain your answer.

5-82. An experiment was conducted to compare the filling capability of packaging equipment at two different wineries. Ten bottles of Pinot Noir from Ridgecrest Vineyards were randomly selected and measured, along with 10 bottles of Pinot Noir from Valley View Vineyards. The data are as follows (fill volume is in ml):

Ridgecrest	Valley View
755 751 752 753	756 754 757 756
753 753 753 754	755 756 756 755
752 751	755 756

(a) What assumptions are necessary to perform a hypothesis testing procedure for equality of means of these data? Check these assumptions.

(b) Perform the appropriate hypothesis testing procedure to determine whether the data support the claim that both wineries will fill bottles to the same mean volume.

5-83. A materials engineer performs an experiment to investigate whether there is a difference among five types of foam pads used under carpeting. A mechanical device is constructed to simulate "walkers" on the pad and four samples of each pad are randomly tested on the simulator. After a certain amount of time, the foam pad is removed from the simulator, examined, and scored for wear quality. The data are compiled into the following partially complete analysis of variance table.

Source of Variation	Sum of Squares	Degrees of Freedom	Mean Square	F_0
Treatments	95.129			
Error	86.752			
Total		19		

(a) Complete the analysis of variance table.

(b) Use the analysis of variance table to test the hypothesis that wear quality differs among the types of foam pads. Use $\alpha = 0.01$.

5-84. A Rockwell hardness-testing machine presses a tip into a test coupon and uses the depth of the resulting depression to indicate hardness. Two different tips are being compared to determine whether they provide the same Rockwell C-scale hardness readings. Nine coupons are

tested, with both tips being tested on each coupon. The data are shown next.

Coupon	Tip 1	Tip 2
1	47	46
2	42	40
3	43	45
4	40	41
5	42	43
6	41	41
7	45	46
8	45	46
9	49	48

(a) State any assumptions necessary to test the claim that both tips produce the same Rockwell C-scale hardness readings. Check those assumptions for which you have the data.

(b) Apply an appropriate statistical method to determine whether the data support the claim that the difference in Rockwell C-scale hardness readings of the two tips is significantly different from zero.

(c) Suppose that if the two tips differ in mean hardness readings by as much as 1.0, we want the power of the test to be at least 0.9. For an $\alpha = 0.01$, how many coupons should have been used in the test?

5-85. Two different gauges can be used to measure the depth of bath material in a Hall cell used in smelting aluminum. Each gauge is used once in 15 cells by the same operator. Depth measurements from both gauges for 15 cells are shown on the next page.

(a) State any assumptions necessary to test the claim that both gauges produce the same mean bath depth readings. Check those assumptions for which you have the data.

(b) Apply an appropriate statistical method to determine whether the data support the claim that the two gauges produce different bath depth readings.

(c) Suppose that if the two gauges differ in mean bath depth readings by as much as 1.65 inch, we want the power of the test to be at least 0.8. For $\alpha = 0.01$, how many cells should have been used?

Cell	Gauge 1	Gauge 2
1	46 in.	47 in.
2	50	53
3	47	45
4	53	50
5	49	51
6	48	48
7	53	54
8	56	53
9	52	51
10	47	45
11	49	51
12	45	45
13	47	49
14	46	43
15	50	51

5-86. An article in the *Materials Research Bulletin* (Vol. 26, No. 11, 1991) investigated four different methods of preparing the superconducting compound $PbMo_6S_8$. The authors contend that the presence of oxygen during the preparation process affects the material's superconducting transition temperature T_c. Preparation methods 1 and 2 use techniques that are designed to eliminate the presence of oxygen, whereas methods 3 and 4 allow oxygen to be present. Five observations on T_c (in kelvins, K) were made for each method, and the results are as follows.

Preparation Method	Transition Temperature T_c (K)				
1	14.8	14.8	14.7	14.8	14.9
2	14.6	15.0	14.9	14.8	14.7
3	12.7	11.6	12.4	12.7	12.1
4	14.2	14.4	14.4	12.2	11.7

(a) Is there evidence to support the claim that the presence of oxygen during preparation affects the mean transition temperature? Use $\alpha = 0.05$.

(b) What is the *P*-value for the *F*-test in part (a)?

5-87. A paper in the *Journal of the Association of Asphalt Paving Technologists* (Vol. 59, 1990) describes an experiment to determine the effect of air voids on percentage retained strength of asphalt. For purposes of the experiment, air voids are controlled at three levels: low (2–4%), medium (4–6%), and high (6–8%). The data are shown in the following table.

Air Voids	Retained Strength (%)							
Low	106	90	103	90	79	88	92	95
Medium	80	69	94	91	70	83	87	83
High	78	80	62	69	76	85	69	85

(a) Do the different levels of air voids significantly affect mean retained strength? Use $\alpha = 0.01$.

(b) Find the *P*-value for the *F*-statistic in part (a).

5-88. An article in *Environment International* (Vol. 18, No. 4, 1992) describes an experiment in which the amount of radon released in showers was investigated. Radon-enriched water was used in the experiment, and six different orifice diameters were tested in shower heads. The data from the experiment are shown in the following table.

Orifice Diameter	Radon Released (%)			
0.37	80	83	83	85
0.51	75	75	79	79
0.71	74	73	76	77
1.02	67	72	74	74
1.40	62	62	67	69
1.99	60	61	64	66

(a) Does the size of the orifice affect the mean percentage of radon released? Use $\alpha = 0.05$.

(b) Find the P-value for the F-statistic in part (a).

Team Exercises

5-89.   Construct a data set for which the paired t-test statistic is very large, indicating that when this analysis is used the two population means are different, but t_0 for the two-sample t-test is very small so the incorrect analysis would indicate that there is no significant difference between the means.

5-90.   Identify an example in which a comparative standard or claim is made about two independent populations. For example, "Car Type A gets more average miles per gallon in urban driving than Car Type B." The standard or claim may be expressed as a mean (average), variance, standard deviation or proportion. Collect two appropriate random samples of data and perform a hypothesis test. Report on your results. Be sure to include in your report the comparison expressed as a hypothesis test, a description of the data collected, the analysis performed, and the conclusion reached.

Building Empirical Models

CHAPTER OUTLINE

6-1 INTRODUCTION TO EMPIRICAL MODELS

Engineers frequently use **models** in problem formulation and solution. Sometimes these models are based on our physical, chemical, or engineering science knowledge of the phenomenon, and in such cases we call these models **mechanistic models.** Examples of mechanistic models include Ohm's law, the gas laws, and Kirchhoff's laws. However, there are many situations in which two or more variables of interest are related, and the mechanistic model relating these variables is unknown. In these cases it is necessary to build a model relating the variables based on observed data. This type of model is called an **empirical model.** An empirical model can be manipulated and analyzed just as a mechanistic model can.

As an illustration, consider the data in Table 6-1. In this table y is the purity of oxygen produced in a chemical distillation process, and x is the percentage hydrocarbons that are present in the main condenser of the distillation unit. Figure 6-1 presents a **scatter diagram** of the data in Table 6-1. This is simply a graph on which each (x_i, y_i) pair is represented as a point plotted in a two-dimensional coordinate system. Note that there is no obvious mechanistic model that relates purity to the hydrocarbon level. However, inspection of the scatter diagram indicates that, although no simple curve will pass exactly through all the points, there is a strong indication that the points lie scattered randomly around a straight line. Therefore, it is probably reasonable to assume that the mean of the random variable Y is related to x by the following straight-line relationship:

$$E(Y|x) = \mu_{Y|x} = \beta_0 + \beta_1 x$$

where the slope and intercept of the line are unknown parameters. The notation $E(Y|x)$ represents the expected value of the response variable at a particular value of the regressor x. Although the mean Y is a linear function of x, the actual observed value y does not

Table 6-1   Oxygen and Hydrocarbon Levels

Observation Number	Hydrocarbon Level x (%)	Purity y (%)
1	0.99	90.01
2	1.02	89.05
3	1.15	91.43
4	1.29	93.74
5	1.46	96.73
6	1.36	94.45
7	0.87	87.59
8	1.23	91.77
9	1.55	99.42
10	1.40	93.65
11	1.19	93.54
12	1.15	92.52
13	0.98	90.56
14	1.01	89.54
15	1.11	89.85
16	1.20	90.39
17	1.26	93.25
18	1.32	93.41
19	1.43	94.98
20	0.95	87.33

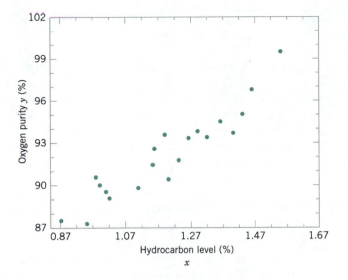

Figure 6-1 Scatter diagram of oxygen purity versus hydrocarbon level from Table 6-1.

fall exactly on a straight line. The appropriate way to generalize this to a **probabilistic linear model** is to assume that the expected value of Y is a linear function of x, but that for a fixed value of x the actual value of Y is determined by the mean value function (the linear model) plus a random error term ϵ.

Simple Linear Regression Model

The dependent variable or response is related to one independent or regressor variable, as

$$Y = \beta_0 + \beta_1 x + \epsilon \qquad (6\text{-}1)$$

where ϵ is the random error term. The parameters β_0 and β_1 are called **regression coefficients**.

To gain more insight into this model, suppose that we can fix the value of x and observe the value of the random variable Y. Now if x is fixed, the random component ϵ on the right-hand side of the model in equation 6-1 determines the properties of Y. Suppose that the mean and variance of ϵ are 0 and σ^2, respectively. Then

$$E(Y|x) = E(\beta_0 + \beta_1 x + \epsilon) = \beta_0 + \beta_1 x + E(\epsilon) = \beta_0 + \beta_1 x$$

Note that this is the same relationship we initially wrote down empirically from inspection of the scatter diagram in Fig. 6-1. The variance of Y given x is

$$V(Y|x) = V(\beta_0 + \beta_1 x + \epsilon) = V(\beta_0 + \beta_1 x) + V(\epsilon) = 0 + \sigma^2 = \sigma^2$$

Thus the true regression model $\mu_{Y|x} = \beta_0 + \beta_1 x$ is a line of mean values; that is, the height of the regression line at any value of x is simply the expected value of Y for that x. The slope, β_1, can be interpreted as the change in the mean of Y for a unit change in x. Furthermore, the variability of Y at a particular value of x is determined by the error variance σ^2. This implies that there is a distribution of Y-values at each x and that the variance of this distribution is the same at each x.

For example, suppose that the true regression model relating oxygen purity to hydrocarbon level is $\mu_{Y|x} = 75 + 15x$, and suppose that the variance is $\sigma^2 = 2$. Figure 6-2 illustrates this situation. Note that we have used a normal distribution to describe the random variation in ϵ. Since Y is the sum of a constant $\beta_0 + \beta_1 x$ (the mean) and a normally distributed random variable, Y is a normally distributed random variable. The variance σ^2 determines the variability in the observations Y on oxygen purity. Thus, when σ^2 is small, the observed values of Y will fall close to the line, and when σ^2 is large, the observed values of Y may deviate considerably from the line. Because σ^2 is constant, the variability in Y at any value of x is the same.

The regression model describes the relationship between oxygen purity Y and hydrocarbon level x. Thus, for any value of hydrocarbon level, oxygen purity has a normal distribution with mean $75 + 15x$ and variance 2. For example, if $x = 1.25$, then Y has mean value $\mu_{Y|x} = 75 + 15(1.25) = 93.75$ and variance 2.

There are many empirical model building situations in which there is more than one regressor variable. Once again, a regression model can be used to describe the relationship. A regression model that contains more than one regressor variable is called a **multiple regression model.**

As an example, suppose that the effective life of a cutting tool depends on the cutting speed and the tool angle. A multiple regression model that might describe this relationship is

$$Y = \beta_0 + \beta_1 x_1 + \beta_2 x_2 + \epsilon \tag{6-2}$$

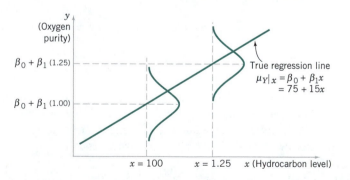

Figure 6-2   The distribution of Y for a given value of x for the oxygen purity–hydrocarbon data.

where Y represents the tool life, x_1 represents the cutting speed, x_2 represents the tool angle, and ϵ is a random error term. This is a **multiple linear regression model** with two regressors. The term *linear* is used because equation 6-2 is a linear function of the unknown parameters β_0, β_1, and β_2.

The regression model in equation 6-2 describes a plane in the three-dimensional space of Y, x_1, and x_2. Figure 6-3a shows this plane for the regression model

$$E(Y) = 50 + 10x_1 + 7x_2$$

where we have assumed that the expected value of the error term is zero; that is, $E(\epsilon) = 0$. The parameter β_0 is the **intercept** of the plane. We sometimes call β_1 and β_2 **partial regression coefficients,** because β_1 measures the expected change in Y per unit change in x_1 when x_2 is held constant, and β_2 measures the expected change in Y per unit change in x_2 when x_1 is held constant. Figure 6-3b shows a **contour plot** of the regression model—that is, lines of constant $E(Y)$ as a function of x_1 and x_2. Note that the contour lines in this plot are straight lines.

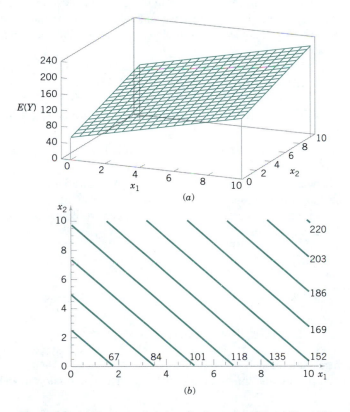

Figure 6-3 (a) The regression plane for the model $E(Y) = 50 + 10x_1 + 7x_2$. (b) The contour plot.

> ### Multiple Linear Regression Model
>
> In general, the **dependent variable** or **response** is related to k **independent** or **regressor variables.** The model
>
> $$Y = \beta_0 + \beta_1 x_1 + \beta_2 x_2 + \cdots + \beta_k x_k + \epsilon \qquad (6\text{-}3)$$
>
> is called a multiple linear regression model with k regressor variables.

The parameters β_j, $j = 0, 1, \ldots, k$, are called the regression coefficients. This model describes a hyperplane in the k-dimensional space of the regressor variables $\{x_j\}$. The parameter β_j represents the expected change in response Y per unit change in x_j when all the remaining regressors x_i ($i \neq j$) are held constant.

Multiple linear regression models are often used as empirical models. That is, the mechanistic model that relates Y and $x_1, x_2, \ldots, x_k$ is unknown, but over certain ranges of the independent variables the linear regression model is an adequate approximation.

Models that are more complex in structure than equation 6-3 often may still be analyzed by multiple linear regression techniques. For example, consider the cubic polynomial model in one regressor variable.

$$Y = \beta_0 + \beta_1 x + \beta_2 x^2 + \beta_3 x^3 + \epsilon \qquad (6\text{-}4)$$

If we let $x_1 = x$, $x_2 = x^2$, $x_3 = x^3$, then equation 6-4 can be written as

$$Y = \beta_0 + \beta_1 x_1 + \beta_2 x_2 + \beta_3 x_3 + \epsilon \qquad (6\text{-}5)$$

which is a multiple linear regression model with three regressor variables.

Models that include **interaction** effects may also be analyzed by multiple linear regression methods. An interaction between two variables can be represented by a cross-product term in the model, such as

$$Y = \beta_0 + \beta_1 x_1 + \beta_2 x_2 + \beta_{12} x_1 x_2 + \epsilon \qquad (6\text{-}6)$$

If we let $x_3 = x_1 x_2$ and $\beta_3 = \beta_{12}$, then equation 6-6 can be written as

$$Y = \beta_0 + \beta_1 x_1 + \beta_2 x_2 + \beta_3 x_3 + \epsilon$$

which is a linear regression model.

Figures 6-4a and b show the three-dimensional plot of the regression model

$$Y = 50 + 10x_1 + 7x_2 + 5x_1 x_2$$

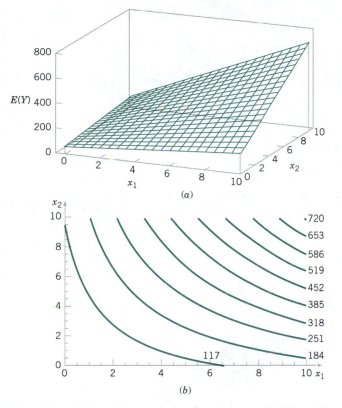

Figure 6-4   (a) Three-dimensional plot of regression model $E(Y) = 50 + 10x_1 + 7x_2 + 5x_1x_2$. (b) The contour plot.

and the corresponding two-dimensional contour plot. Note that, although this model is a linear regression model, the shape of the surface that is generated by the model is not linear. In general, **any regression model that is linear in parameters** (the β's) **is a linear regression model, regardless of the shape of the surface that it generates.**

Figure 6-4 provides a nice graphical interpretation of an interaction. Generally, interaction implies that the effect produced by changing one variable (x_1, say) depends on the level of the other variable (x_2). For example, Fig. 6-4 shows that changing x_1 from 2 to 8 produces a much smaller change in $E(Y)$ when $x_2 = 2$ than when $x_2 = 10$. Interaction effects occur frequently in the study and analysis of real-world systems, and regression methods are one of the techniques that we can use to describe them.

As a final example, consider the second-order model with interaction

$$Y = \beta_0 + \beta_1 x_1 + \beta_2 x_2 + \beta_{11} x_1^2 + \beta_{22} x_2^2 + \beta_{12} x_1 x_2 + \epsilon \tag{6-7}$$

If we let $x_3 = x_1^2$, $x_4 = x_2^2$, $x_5 = x_1 x_2$, $\beta_3 = \beta_{11}$, $\beta_4 = \beta_{22}$, and $\beta_5 = \beta_{12}$, then equation 6-7 can be written as a multiple linear regression model as follows:

$$Y = \beta_0 + \beta_1 x_1 + \beta_2 x_2 + \beta_3 x_3 + \beta_4 x_4 + \beta_5 x_5 + \epsilon$$

Figures 6-5*a* and *b* show the three-dimensional plot and the corresponding contour plot for

$$E(Y) = 800 + 10x_1 + 7x_2 - 8.5x_1^2 - 5x_2^2 + 4x_1x_2$$

These plots indicate that the expected change in Y when x_1 is changed by one unit (say) is a function of *both* x_1 and x_2. The quadratic and interaction terms in this model produce a mound-shaped function. Depending on the values of the regression coefficients, the second-order model with interaction is capable of assuming a wide variety of shapes; thus it is a very flexible regression model.

In most real-world problems, the values of the parameters (the regression coefficients β_i) and the error variance σ^2 will not be known, and they must be estimated from sample data. **Regression analysis** is a collection of statistical tools for finding estimates of the parameters in the regression model. Then this fitted regression equation or model is typically used in prediction of future observations of Y or for estimating the mean response at a particular level of x. To illustrate with the simple linear regression model example, a chemical engineer might be interested in estimating the mean purity of oxygen produced

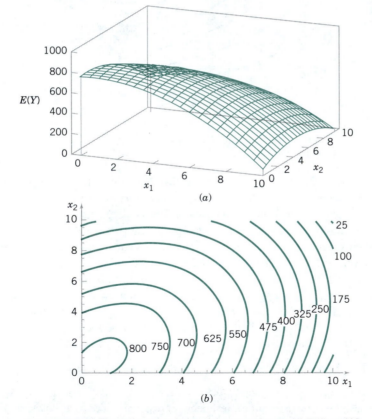

Figure 6-5   (a) Three-dimensional plot of the regression model $E(Y) = 800 + 10x_1 + 7x_2 - 8.5x_1^2 - 5x_2^2 + 4x_1x_2$. (b) The contour plot.

when the hydrocarbon level is $x = 1.25\%$. This chapter discusses such procedures and applications for linear regression models.

Correlation

A regression equation is an empirical model relating a response variable to one or more regressors. Sometimes it is useful to give a descriptive measure of linear association known as **correlation.** Consider the purity (y) and hydrocarbon level (x) data given in Table 6-1 and in the scatter plot of Fig. 6-1. The sample correlation coefficient between x and y is defined as

$$r_{xy} = \frac{\sum_{i=1}^{n} (y_i - \bar{y})(x_i - \bar{x})}{\sqrt{\sum_{i=1}^{n} (y_i - \bar{y})^2 \sum_{i=1}^{n} (x_i - \bar{x})^2}}$$

Strictly speaking, correlation is only defined between two **random variables,** so we assume that both purity and hydrocarbon level are random variables. It can be shown that $-1 \le r_{xy} \le +1$, and if $r_{xy} = +1$ (or -1) the observations lie exactly on a straight line with positive (or negative) slope. In our example, $r_{xy} = 0.936$. This is a dimensionless quantity that expresses the linear relationship between a pair of variables that were originally measured in possibly different units. The sample correlation coefficient is an estimator of the correlation coefficient ρ_{xy}. It is possible to develop procedures for testing hypotheses and constructing confidence intervals for ρ_{xy}. See Montgomery and Runger (1999) for details.

6-2 LEAST SQUARES ESTIMATION OF THE PARAMETERS

6-2.1 Simple Linear Regression

The case of **simple linear regression** considers a *single regressor* or *predictor* x and a dependent or *response* variable Y. Suppose that the true relationship between Y and x is a straight line and that the observation Y at each level of x is a random variable. As noted previously, the expected value of Y for each value of x is

$$E(Y|x) = \beta_0 + \beta_1 x$$

where the intercept β_0 and the slope β_1 are unknown regression coefficients. We assume that each observation, Y, can be described by the model

$$Y = \beta_0 + \beta_1 x + \epsilon \tag{6-8}$$

where ϵ is a random error with mean zero and variance σ^2. The random errors corresponding to different observations are also assumed to be uncorrelated random variables.

Suppose that we have n pairs of observations (x_1, y_1), (x_2, y_2), ... (x_n, y_n). Figure 6-6 shows a typical scatter plot of observed data and a candidate for the estimated regression line. The estimates of β_0 and β_1 should result in a line that is (in some sense) a "best fit" to the data. The German scientist Karl Gauss (1777–1855) proposed estimating the parameters β_0 and β_1 in equation 6-8 in order to minimize the sum of the squares of the vertical deviations in Fig. 6-6.

We call this criterion for estimating the regression coefficients the **method of least squares.** Using equation 6-8, we may express the n observations in the sample as

$$y_i = \beta_0 + \beta_1 x_i + \epsilon_i, \qquad i = 1, 2, \dots, n \tag{6-9}$$

and the sum of the squares of the deviations of the observations from the true regression line is

$$L = \sum_{i=1}^{n} \epsilon_i^2 = \sum_{i=1}^{n} (y_i - \beta_0 - \beta_1 x_i)^2 \tag{6-10}$$

The least squares estimators of β_0 and β_1, say, $\hat{\beta}_0$ and $\hat{\beta}_1$, must satisfy

$$\left. \frac{\partial L}{\partial \beta_0} \right|_{\hat{\beta}_0, \hat{\beta}_1} = -2 \sum_{i=1}^{n} (y_i - \hat{\beta}_0 - \hat{\beta}_1 x_i) = 0 \tag{6-11}$$

$$\left. \frac{\partial L}{\partial \beta_1} \right|_{\hat{\beta}_0, \hat{\beta}_1} = -2 \sum_{i=1}^{n} (y_i - \hat{\beta}_0 - \hat{\beta}_1 x_i) x_i = 0$$

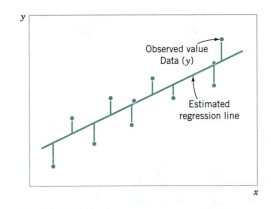

Figure 6-6   Deviations of the data from the estimated regression model.

Simplifying these two equations yields

$$n\hat{\beta}_0 + \hat{\beta}_1 \sum_{i=1}^{n} x_i = \sum_{i=1}^{n} y_i \tag{6-12}$$

$$\hat{\beta}_0 \sum_{i=1}^{n} x_i + \hat{\beta}_1 \sum_{i=1}^{n} x_i^2 = \sum_{i=1}^{n} y_i x_i$$

Equations 6-12 are called the **least squares normal equations.** The solution to the normal equations results in the least squares estimates $\hat{\beta}_0$ and $\hat{\beta}_1$.

Computing Formulas for Simple Linear Regression

The **least squares estimates** of the intercept and slope in the simple linear regression model are

$$\hat{\beta}_0 = \bar{y} - \hat{\beta}_1 \bar{x} \tag{6-13}$$

$$\hat{\beta}_1 = \frac{\displaystyle\sum_{i=1}^{n} y_i x_i - \frac{\left(\displaystyle\sum_{i=1}^{n} y_i\right)\left(\displaystyle\sum_{i=1}^{n} x_i\right)}{n}}{\displaystyle\sum_{i=1}^{n} x_i^2 - \frac{\left(\displaystyle\sum_{i=1}^{n} x_i\right)^2}{n}} \tag{6-14}$$

where $\bar{y} = (1/n)\sum_{i=1}^{n} y_i$ and $\bar{x} = (1/n)\sum_{i=1}^{n} x_i$.

The **fitted** or **estimated regression line** is therefore

$$\hat{y} = \hat{\beta}_0 + \hat{\beta}_1 x \tag{6-15}$$

Note that each pair of observations satisfies the relationship

$$y_i = \hat{\beta}_0 + \hat{\beta}_1 x_i + e_i, \qquad i = 1, 2, \ldots, n$$

where $e_i = y_i - \hat{y}_i$ is called the **residual.** The residual describes the error in the fit of the model to the ith observation y_i. Later in this chapter we will use the residuals to provide information about the adequacy of the fitted model.

Notationally, it is occasionally convenient to give special symbols to the numerator and denominator of equation 6-14. Given data $(x_1, y_1), (x_2, y_2), \ldots, (x_n, y_n)$, let

$$S_{xx} = \sum_{i=1}^{n} (x_i - \bar{x})^2 = \sum_{i=1}^{n} x_i^2 - \frac{\left(\sum_{i=1}^{n} x_i\right)^2}{n} \tag{6-16}$$

and

$$S_{xy} = \sum_{i=1}^{n} (x_i - \bar{x})(y_i - \bar{y}) = \sum_{i=1}^{n} x_i y_i - \frac{\left(\sum_{i=1}^{n} x_i\right)\left(\sum_{i=1}^{n} y_i\right)}{n} \tag{6-17}$$

EXAMPLE 6-1

We will fit a simple linear regression model to the oxygen purity data in Table 6-1. The following quantities may be computed:

$$n = 20 \quad \sum_{i=1}^{20} x_i = 23.92 \quad \sum_{i=1}^{20} y_i = 1{,}843.21 \quad \bar{x} = 1.1960 \quad \bar{y} = 92.1605$$

$$\sum_{i=1}^{20} y_i^2 = 170{,}044.5321 \quad \sum_{i=1}^{20} x_i^2 = 29.2892 \quad \sum_{i=1}^{20} x_i y_i = 2{,}214.6566$$

$$S_{xx} = \sum_{i=1}^{20} x_i^2 - \frac{\left(\sum_{i=1}^{20} x_i\right)^2}{20} = 29.2892 - \frac{(23.92)^2}{20} = 0.68088$$

and

$$S_{xy} = \sum_{i=1}^{20} x_i y_i - \frac{\left(\sum_{i=1}^{20} x_i\right)\left(\sum_{i=1}^{20} y_i\right)}{20} = 2{,}214.6566 - \frac{(23.92)(1{,}843.21)}{20} = 10.17744$$

Therefore, the least squares estimates of the slope and intercept are

$$\hat{\beta}_1 = \frac{S_{xy}}{S_{xx}} = \frac{10.17744}{0.68088} = 14.94748$$

and

$$\hat{\beta}_0 = \bar{y} - \hat{\beta}_1 \bar{x} = 92.1605 - (14.94748)1.196 = 74.28331$$

The fitted simple linear regression model is

$$\hat{y} = 74.3 + 14.9x$$

This model is plotted in Fig. 6-7, along with the sample data. We have carried five decimal places in the calculation of these regression coefficients and then rounded the coefficients for the final equation. It is important to use sufficient decimal places in intermediate calculations to minimize the effect of round-off errors.

Computer software programs are widely used in regression modeling. Table 6-2 shows a portion of the Minitab output for this problem. In subsequent sections we will provide explanations for the other information provided in this computer output.

Using the regression model of Example 6-1 we would predict oxygen purity of $\hat{y} = 89.2\%$ when the hydrocarbon level is $x = 1.00\%$. The purity 89.2% may be interpreted as an estimate of the true population mean purity when $x = 1.00\%$, or as an estimate of a new observation when $x = 1.00\%$. These estimates are, of course, subject to error; that is, it is unlikely that a future observation on purity would be exactly 89.2% when the hydrocarbon level is 1.00%. In subsequent sections we will see how to use confidence intervals and prediction intervals to describe the error in estimation from a regression model.

6-2.2 Multiple Linear Regression

The method of least squares may be used to estimate the regression coefficients in the multiple regression model, equation 6-3. Suppose that $n > k$ observations are available, and let x_{ij} denote the ith observation or level of variable x_j. The observations are

$$(x_{i1}, x_{i2}, \ldots, x_{ik}, y_i) \qquad i = 1, 2, \ldots, n \text{ and } n > k$$

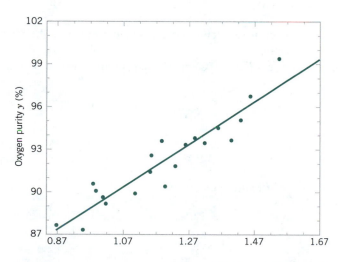

Figure 6-7 Scatter plot of oxygen purity y versus hydrocarbon level x and regression model $\hat{y} = 74.3 + 14.9x$.

Table 6-2   Minitab Output for the Oxygen Purity Data in Example 6-1

Regression Analysis				

The regression equation is
y = 74.3 + 14.9 x

Predictor	Coef	StDev	T	P
Constant	74.283	1.593	46.62	0.000
x	14.947	1.317	11.35	0.000

S = 1.087   R−Sq = 87.7%   R−Sq(adj) = 87.1%

Analysis of Variance

Source	DF	SS	MS	F	P
Regression	1	152.13	152.13	128.86	0.000
Error	18	21.25	1.18		
Total	19	173.38			

It is customary to present the data for multiple regression in a table such as Table 6-3. Each observation $(x_{i1}, x_{i2}, \ldots, x_{ik}, y_i)$ satisfies the model in equation 6-3, or

$$y_i = \beta_0 + \beta_1 x_{i1} + \beta_2 x_{i2} + \cdots + \beta_k x_{ik} + \epsilon_i \qquad i = 1, 2, \ldots, n \qquad (6\text{-}18)$$

In fitting this regression model, it is much more convenient to express the mathematical operations using matrix notation.

This model in equation 6-18 is a system of n equations that can be expressed in matrix notation as

$$\mathbf{y} = \mathbf{X}\boldsymbol{\beta} + \boldsymbol{\epsilon} \qquad (6\text{-}19)$$

Table 6-3   Data for Multiple Linear Regression

y	x_1	x_2	$\cdots$	x_k
y_1	x_{11}	x_{12}	$\cdots$	x_{1k}
y_2	x_{21}	x_{22}	$\cdots$	x_{2k}
$\vdots$	$\vdots$	$\vdots$		$\vdots$
y_n	x_{n1}	x_{n2}	$\cdots$	x_{nk}

where

$$\mathbf{y} = \begin{bmatrix} y_1 \\ y_2 \\ \vdots \\ y_n \end{bmatrix} \qquad \mathbf{X} = \begin{bmatrix} 1 & x_{11} & x_{12} & \cdots & x_{1k} \\ 1 & x_{21} & x_{22} & \cdots & x_{2k} \\ \vdots & \vdots & \vdots & & \vdots \\ 1 & x_{n1} & x_{n2} & \cdots & x_{nk} \end{bmatrix}$$

$$\boldsymbol{\beta} = \begin{bmatrix} \beta_0 \\ \beta_1 \\ \vdots \\ \beta_k \end{bmatrix} \quad \text{and} \quad \boldsymbol{\epsilon} = \begin{bmatrix} \epsilon_1 \\ \epsilon_2 \\ \vdots \\ \epsilon_n \end{bmatrix}$$

In general, $\mathbf{y}$ is an $(n \times 1)$ vector of the observations, $\mathbf{X}$ is an $(n \times p)$ matrix of the levels of the independent variables, $\boldsymbol{\beta}$ is a $(p \times 1)$ vector of the regression coefficients, and $\boldsymbol{\epsilon}$ is an $(n \times 1)$ vector of random errors.

We wish to find the vector of least squares estimators, $\hat{\boldsymbol{\beta}}$, that minimizes

$$L = \sum_{i=1}^{n} \epsilon_i^2 = \boldsymbol{\epsilon}'\boldsymbol{\epsilon} = (\mathbf{y} - \mathbf{X}\boldsymbol{\beta})'(\mathbf{y} - \mathbf{X}\boldsymbol{\beta})$$

where the prime $(')$ denotes transpose. The least squares estimator $\hat{\boldsymbol{\beta}}$ is the solution for $\boldsymbol{\beta}$ in the equations

$$\frac{\partial L}{\partial \boldsymbol{\beta}} = \mathbf{0}$$

We will not give the details of taking the derivatives above; however, the resulting equations that must be solved are

$$\mathbf{X}'\mathbf{X}\hat{\boldsymbol{\beta}} = \mathbf{X}'\mathbf{y} \tag{6-20}$$

Equations 6-20 are the least squares normal equations in matrix form. To solve the normal equations, multiply both sides of equations 6-20 by the inverse of $\mathbf{X}'\mathbf{X}$, and obtain the least squares estimate of $\boldsymbol{\beta}$.

The least squares estimate of $\boldsymbol{\beta}$ is

$$\hat{\boldsymbol{\beta}} = (\mathbf{X}'\mathbf{X})^{-1}\mathbf{X}'\mathbf{y} \tag{6-21}$$

Note that there are $p = k + 1$ normal equations in $p = k + 1$ unknowns (the values of $\hat{\beta}_0, \hat{\beta}_1, \ldots, \hat{\beta}_k$). Furthermore, the matrix $\mathbf{X}'\mathbf{X}$ is generally nonsingular, as was assumed above, so that the methods described in textbooks on determinants and methods for inverting these matrices can be used to find $(\mathbf{X}'\mathbf{X})^{-1}$. In practice, multiple regression calculations are almost always performed using a computer.

Writing out equation 6-20 in detail, we obtain

$$
\begin{bmatrix}
n & \sum\limits_{i=1}^{n} x_{i1} & \sum\limits_{i=1}^{n} x_{i2} & \cdots & \sum\limits_{i=1}^{n} x_{ik} \\
\sum\limits_{i=1}^{n} x_{i1} & \sum\limits_{i=1}^{n} x_{i1}^2 & \sum\limits_{i=1}^{n} x_{i1}x_{i2} & \cdots & \sum\limits_{i=1}^{n} x_{i1}x_{ik} \\
\vdots & \vdots & \vdots & & \vdots \\
\sum\limits_{i=1}^{n} x_{ik} & \sum\limits_{i=1}^{n} x_{ik}x_{i1} & \sum\limits_{i=1}^{n} x_{ik}x_{i2} & \cdots & \sum\limits_{i=1}^{n} x_{ik}^2
\end{bmatrix}
\begin{bmatrix}
\hat{\beta}_0 \\ \hat{\beta}_1 \\ \vdots \\ \hat{\beta}_k
\end{bmatrix}
=
\begin{bmatrix}
\sum\limits_{i=1}^{n} y_i \\ \sum\limits_{i=1}^{n} x_{i1}y_i \\ \vdots \\ \sum\limits_{i=1}^{n} x_{ik}y_i
\end{bmatrix}
$$

It is easy to see that $\mathbf{X}'\mathbf{X}$ is a $(p \times p)$ symmetric matrix and $\mathbf{X}'\mathbf{y}$ is a $(p \times 1)$ column vector. Note the special structure of the $\mathbf{X}'\mathbf{X}$ matrix. The diagonal elements of $\mathbf{X}'\mathbf{X}$ are the sums of squares of the elements in the columns of $\mathbf{X}$, and the off-diagonal elements are the sums of cross-products of the elements in the columns of $\mathbf{X}$. Furthermore, note that the elements of $\mathbf{X}'\mathbf{y}$ are the sums of cross-products of the columns of $\mathbf{X}$ and the observations $\{y_i\}$.

The fitted regression model is

$$\hat{y}_i = \hat{\beta}_0 + \sum_{j=1}^{k} \hat{\beta}_j x_{ij} \qquad i = 1, 2, \ldots, n \tag{6-22}$$

In matrix notation, the fitted model is

$$\hat{\mathbf{y}} = \mathbf{X}\hat{\boldsymbol{\beta}}$$

The difference between the observation y_i and the fitted value $\hat{y}_i$ is a **residual—** say, $e_i = y_i - \hat{y}_i$. The $(n \times 1)$ vector of residuals is denoted by

$$\mathbf{e} = \mathbf{y} - \hat{\mathbf{y}} \tag{6-23}$$

EXAMPLE 6-2

The pull strength of a wire bond in a semiconductor product is an important characteristic. We want to investigate the suitability of using a multiple regression model to predict pull strength (y) as a function of wire length (x_1) and die height (x_2). We will fit the multiple regression model

$$y = \beta_0 + \beta_1 x_1 + \beta_2 x_2 + \epsilon$$

where y is the observed wire bond pull strength, x_1 is the wire length, and x_2 is the die height. The 25 observations are in Table 6-4. We will use the matrix approach to fit the regression model above to these data. The $\mathbf{X}$ matrix and $\mathbf{y}$ vector for this model are

$$
\mathbf{X} = \begin{bmatrix}
1 & 2 & 50 \\
1 & 8 & 110 \\
1 & 11 & 120 \\
1 & 10 & 550 \\
1 & 8 & 295 \\
1 & 4 & 200 \\
1 & 2 & 375 \\
1 & 2 & 52 \\
1 & 9 & 100 \\
1 & 8 & 300 \\
1 & 4 & 412 \\
1 & 11 & 400 \\
1 & 12 & 500 \\
1 & 2 & 360 \\
1 & 4 & 205 \\
1 & 4 & 400 \\
1 & 20 & 600 \\
1 & 1 & 585 \\
1 & 10 & 540 \\
1 & 15 & 250 \\
1 & 15 & 290 \\
1 & 16 & 510 \\
1 & 17 & 590 \\
1 & 6 & 100 \\
1 & 5 & 400
\end{bmatrix}
\qquad
\mathbf{y} = \begin{bmatrix}
9.95 \\
24.45 \\
31.75 \\
35.00 \\
25.02 \\
16.86 \\
14.38 \\
9.60 \\
24.35 \\
27.50 \\
17.08 \\
37.00 \\
41.95 \\
11.66 \\
21.65 \\
17.89 \\
69.00 \\
10.30 \\
34.93 \\
46.59 \\
44.88 \\
54.12 \\
56.63 \\
22.13 \\
21.15
\end{bmatrix}
$$

The $\mathbf{X}'\mathbf{X}$ matrix is

$$
\mathbf{X}'\mathbf{X} = \begin{bmatrix}
1 & 1 & \cdots & 1 \\
2 & 8 & \cdots & 5 \\
50 & 110 & \cdots & 400
\end{bmatrix}
\begin{bmatrix}
1 & 2 & 50 \\
1 & 8 & 110 \\
\vdots & \vdots & \vdots \\
1 & 5 & 400
\end{bmatrix}
\begin{bmatrix}
25 & 206 & 8{,}294 \\
206 & 2{,}396 & 77{,}177 \\
8{,}294 & 77{,}177 & 3{,}531{,}848
\end{bmatrix}
$$

and the $\mathbf{X}'\mathbf{y}$ vector is

$$
\mathbf{X}'\mathbf{y} = \begin{bmatrix}
1 & 1 & \cdots & 1 \\
2 & 8 & \cdots & 5 \\
50 & 110 & \cdots & 400
\end{bmatrix}
\begin{bmatrix}
9.95 \\
24.45 \\
\vdots \\
21.15
\end{bmatrix}
= \begin{bmatrix}
725.82 \\
8{,}008.37 \\
274{,}811.31
\end{bmatrix}
$$

Table 6-4   Wire Bond Pull Strength Data for Example 6-2

Observation Number	Pull Strength y	Wire Length x_1	Die Height x_2
1	9.95	2	50
2	24.45	8	110
3	31.75	11	120
4	35.00	10	550
5	25.02	8	295
6	16.86	4	200
7	14.38	2	375
8	9.60	2	52
9	24.35	9	100
10	27.50	8	300
11	17.08	4	412
12	37.00	11	400
13	41.95	12	500
14	11.66	2	360
15	21.65	4	205
16	17.89	4	400
17	69.00	20	600
18	10.30	1	585
19	34.93	10	540
20	46.59	15	250
21	44.88	15	290
22	54.12	16	510
23	56.63	17	590
24	22.13	6	100
25	21.15	5	400

The least squares estimates are found from equation 6-21 as

$$\hat{\boldsymbol{\beta}} = (\mathbf{X}'\mathbf{X})^{-1}\mathbf{X}'\mathbf{y}$$

or

$$\begin{bmatrix} \hat{\beta}_0 \\ \hat{\beta}_1 \\ \hat{\beta}_2 \end{bmatrix} = \begin{bmatrix} 25 & 206 & 8{,}294 \\ 206 & 2{,}396 & 77{,}177 \\ 8{,}294 & 77{,}177 & 3{,}531{,}848 \end{bmatrix}^{-1} \begin{bmatrix} 725.82 \\ 8{,}008.37 \\ 274{,}811.31 \end{bmatrix}$$

$$= \begin{bmatrix} 0.21653 & -0.007491 & -0.000340 \\ -0.007491 & 0.001671 & -0.000019 \\ -0.000340 & -0.000019 & 0.0000015 \end{bmatrix} \begin{bmatrix} 725.82 \\ 8{,}008.47 \\ 274{,}811.31 \end{bmatrix}$$

$$= \begin{bmatrix} 2.26379143 \\ 2.74426964 \\ 0.01252781 \end{bmatrix}$$

Therefore, the fitted regression model with the regression coefficients rounded to five decimal places is

$$\hat{y} = 2.26379 + 2.74427x_1 + 0.01253x_2$$

This regression model can be used to predict values of pull strength for various values of wire length (x_1) and die height (x_2). We can also obtain the **fitted values** $\hat{y}_i$ by substituting each observation (x_{i1}, x_{i2}), $i = 1, 2, \ldots, n$, into the equation. For example, the first observation has $x_{11} = 2$ and $x_{12} = 50$, and the fitted value is

$$\begin{aligned}
\hat{y}_1 &= 2.26379 + 2.74427x_{11} + 0.01253x_{12} \\
&= 2.26379 + 2.74427(2) + 0.01253(50) \\
&= 8.38
\end{aligned}$$

The corresponding observed value is $y_1 = 9.95$. The *residual* corresponding to the first observation is

$$\begin{aligned}
e_1 &= y_1 - \hat{y}_1 \\
&= 9.95 - 8.38 \\
&= 1.57
\end{aligned}$$

Table 6-5 displays all 25 fitted values $\hat{y}_i$ and the corresponding residuals. The fitted values and residuals are calculated to the same accuracy as the original data.

Table 6-5 Observations, Fitted Values, and Residuals for Example 6-2

Observation Number	y_i	$\hat{y}_i$	$e_i = y_i - \hat{y}_i$
1	9.95	8.38	1.57
2	24.45	25.60	-1.15
3	31.75	33.95	-2.20
4	35.00	36.60	-1.60
5	25.02	27.91	-2.89
6	16.86	15.75	1.11
7	14.38	12.45	1.93
8	9.60	8.40	1.20
9	24.35	28.21	-3.86
10	27.50	27.98	-0.48
11	17.08	18.40	-1.32
12	37.00	37.46	-0.46
13	41.95	41.46	0.49
14	11.66	12.26	-0.60
15	21.65	15.81	5.84
16	17.89	18.25	-0.36
17	69.00	64.67	4.33
18	10.30	12.34	-2.04
19	34.93	36.47	-1.54
20	46.59	46.56	-0.03
21	44.88	47.06	-2.18
22	54.12	52.56	1.56
23	56.63	56.31	0.32
24	22.13	19.98	2.15
25	21.15	21.00	0.15

Table 6-6 Minitab Output for the Wire Bond Pull Strength Data in Example 6-2

Regression Analysis				

The regression equation is
y = 2.26 + 2.74 x1 + 0.0125 x2

Predictor	Coef	StDev	T	P
Constant	2.264	1.060	2.14	0.044
x1	2.74427	0.09352	29.34	0.000
x2	0.012528	0.002798	4.48	0.000

S = 2.288   R$-$Sq = 98.1%   R$-$Sq(adj) = 97.9%

Analysis of Variance

Source	DF	SS	MS	F	P
Regression	2	5990.8	2995.4	572.17	0.000
Error	22	115.2	5.2		
Total	24	6105.9			

Source	DF	Seq SS
x1	1	5885.9
x2	1	104.9

Computers are almost always used in fitting multiple regression models. Table 6-6 presents some of the Minitab output for the wire bond pull strength data in Example 6-2. Note that the upper part of the table contains the numerical estimates of the regression coefficients. The computer also calculates many other quantities that reflect important information about the regression model. In subsequent sections, we will define and explain the quantities in this output.

EXERCISES FOR SECTION 6-2

6-1. An article in *Concrete Research* ("Near Surface Characteristics of Concrete: Intrinsic Permeability," Vol. 41, 1989), presented data on compressive strength x and intrinsic permeability y of various concrete mixes and cures. Summary quantities are $n = 14$, $\Sigma y_i = 572$, $\Sigma y_i^2 = 23,530$, $\Sigma x_i = 43$, $\Sigma x_i^2 = 157.42$, and $\Sigma x_i y_i = 1,697.80$. Assume that the two variables are related according to the simple linear regression model.

(a) Calculate the least squares estimates of the slope and intercept.

(b) Use the equation of the fitted line to predict what mean permeability would be observed when the compressive strength is $x = 4.3$.

(c) Give a point estimate of the mean permeability when compressive strength is $x = 3.7$.

(d) Suppose that the observed value of permeability at $x = 3.7$ is $y = 46.1$. Calculate the value of the corresponding residual.

6-2. Regression methods were used to analyze the data from a study investigating the relationship between roadway surface temperature (x) and pavement deflection (y). Summary quantities were $n = 20$, $\Sigma y_i = 12.75$, $\Sigma y_i^2 = 8.86$, $\Sigma x_i = 1478$, $\Sigma x_i^2 = 143,215.8$, and $\Sigma x_i y_i = 1083.67$.

(a) Calculate the least squares estimates of the slope and intercept of a simple linear regression model. Graph the regression line.

(b) Use the equation of the fitted line to predict what mean pavement deflection would be observed when the surface temperature is 85°F.

(c) What is the mean pavement deflection when the surface temperature is 90°F?

(d) What change in mean pavement deflection would be expected for a 1°F change in surface temperature?

6-3. Consider the regression model developed in Exercise 6-2.

(a) Suppose that temperature is measured in °C rather than °F. Write the new regression model that results.

(b) What change in expected pavement deflection is associated with a 1°C change in surface temperature?

6-4. Montgomery and Peck (1992) present data concerning the performance of the 28 National Football League teams in 1976. It is suspected that the number of games won (y) is related to the number of yards gained rushing by an opponent (x). The data are shown in the following table.

(a) Calculate the least squares estimates of the slope and intercept of a simple linear regression model. Graph the regression model.

(b) Find an estimate of the mean number of games won if the opponent's rushing yard total is 1800.

(c) What change in the expected number of games won is associated with a decrease of 100 yards rushing by an opponent?

(d) To increase by one the mean number of games won, how much decrease in rushing yards must be generated by the defense?

(e) Given that $x = 1917$ yards (Cincinnati), find the fitted value of y and the corresponding residual.

Teams	Games Won (y)	Yards Rushing by Opponent (x)
Washington	10	2205
Minnesota	11	2096
New England	11	1847
Oakland	13	1903
Pittsburgh	10	1457
Baltimore	11	1848
Los Angeles	10	1564
Dallas	11	1821
Atlanta	4	2577
Buffalo	2	2476
Chicago	7	1984
Cincinnati	10	1917
Cleveland	9	1761
Denver	9	1709
Detroit	6	1901
Green Bay	5	2288
Houston	5	2072
Kansas City	5	2861
Miami	6	2411
New Orleans	4	2289
New York Giants	3	2203
New York Jets	3	2592
Philadelphia	4	2053
St. Louis	10	1979
San Diego	6	2048
San Francisco	8	1786
Seattle	2	2876
Tampa Bay	0	2560

6-5. Establishing the properties of materials is an important problem in identifying a suitable substitute for biodegradable materials in the fast-food packaging industry. Consider the following data on product density (g/cm^3) and thermal conductivity K-factor (W/mK) published in *Materials Research and Innovation* (1999, pp. 2–8).

Thermal Conductivity	Product Density
0.0480	0.1750
0.0525	0.2200
0.0540	0.2250
0.0535	0.2260
0.0570	0.2500
0.0610	0.2765

(a) Obtain the least squares fit relating product density (regressor variable) to thermal conductivity (response variable).

(b) Find the mean thermal conductivity given that the product density is 0.2350.

(c) Calculate the fitted value of y corresponding to $x = 0.2260$.

(d) Calculate the fitted value y for each of x used to fit the model. Then construct a graph of $\hat{y}$ versus the corresponding observed value y and comment on what this plot would look like if the relationship between y and x was a deterministic (no random error) straight line. Does the plot actually obtained indicate that product density is an effective regressor variable in predicting thermal conductivity?

6-6. The number of pounds of steam used per month by a chemical plant is thought to be related to the average ambient temperature (in °F) for that month. The past year's usage and temperature are shown in the following table.

Month	Temp.	Usage/1000
Jan.	21	185.79
Feb.	24	214.47
Mar.	32	288.03
Apr.	47	424.84
May	50	454.58
June	59	539.03
July	68	621.55
Aug.	74	675.06
Sept.	62	562.03
Oct.	50	452.93
Nov.	41	369.95
Dec.	30	273.98

(a) Assuming that a simple linear regression model is appropriate, fit the regression model relating steam usage (y) to the average temperature (x).

(b) What is the estimate of expected steam usage when the average temperature is 55°F?

(c) What change in mean steam usage is expected when the monthly average temperature changes by 1°F?

(d) Suppose the monthly average temperature is 47°F. Calculate the fitted value of y and the corresponding residual.

6-7. A regression model is to be developed for predicting the ability of soil to absorb chemical contaminants. Ten observations have been taken on a soil absorption index (y) and two regressors: $x_1 = $ amount of extractable iron ore and $x_2 = $ amount of bauxite. We wish to fit the model $Y = \beta_0 + \beta_1 x_1 + \beta_2 x_2 + \epsilon$. Some necessary quantities are:

$$(\mathbf{X}'\mathbf{X})^{-1}$$
$$= \begin{bmatrix} 1.17991 & -7.30982 \text{ E-3} & 7.3006 \text{ E-4} \\ -7.30982 \text{ E-3} & 7.9799 \text{ E-5} & -1.23713 \text{ E-4} \\ 7.3006 \text{ E-4} & -1.23713 \text{ E-4} & 4.6576 \text{ E-4} \end{bmatrix}$$

$$\mathbf{X}'\mathbf{y} = \begin{bmatrix} 220 \\ 36{,}768 \\ 9{,}965 \end{bmatrix}$$

(a) Estimate the regression coefficients in the specified model.

(b) What is the predicted value of the absorption index y when $x_1 = 200$ and $x_2 = 50$?

6-8. A study was performed on wear of a bearing y and its relationship to $x_1 = $ oil viscosity and $x_2 = $ load. The following data were obtained.

y	x_1	x_2
193	1.6	851
230	15.5	816
172	22.0	1058
91	43.0	1201
113	33.0	1357
125	40.0	1115

(a) Fit a multiple linear regression model without an interaction term to these data.

(b) Use the model to predict wear, when $x_1 = 25$ and $x_2 = 1000$.

(c) Fit a multiple linear regression model with an interaction term to these data.

(d) Use the model in (c) to predict wear when $x_1 = 25$ and $x_2 = 1000$. Compare this prediction with the predicted value from part (b).

6-9. A chemical engineer is investigating how the amount of conversion of a product from a raw material (y) depends on reaction temperature (x_1) and the reaction time (x_2). He has developed the following regression models:

1. $\hat{y} = 100 + 2x_1 + 4x_2$
2. $\hat{y} = 95 + 1.5x_1 + 3x_2 + 2x_1x_2$

Both models have been built over the range $0.5 \le x_2 \le 10$.

(a) Using both models, what is the predicted value of conversion when $x_2 = 2$ in terms of x_1? Repeat this calculation for $x_2 = 8$. Draw a graph of the predicted values for both conversion models. Comment on the effect of the interaction term in model 2.

(b) Find the expected change in the mean conversion for a unit change in temperature x_1 for model 1 when $x_2 = 5$. Does this quantity depend on the specific value of reaction time selected? Why?

(c) Find the expected change in the mean conversion for a unit change in temperature x_1 for model 2 when $x_2 = 5$. Repeat this calculation for $x_2 = 2$ and $x_2 = 8$. Does the result depend on the value selected for x_2? Why?

6-10. An engineer at a semiconductor company wants to model the relationship between the device HFE (y) and three parameters: Emitter-RS (x_1), Base-RS (x_2), and Emitter-to-Base RS (x_3). The data are shown in the following table.

x_1 Emitter-RS	x_2 Base-RS	x_3 E-B-RS	y HFE
14.620	226.00	7.000	128.40
15.630	220.00	3.375	52.62
14.620	217.40	6.375	113.90
15.000	220.00	6.000	98.01
14.500	226.50	7.625	139.90
15.250	224.10	6.000	102.60
16.120	220.50	3.375	48.14
15.130	223.50	6.125	109.60
15.500	217.60	5.000	82.68
15.130	228.50	6.625	112.60
15.500	230.20	5.750	97.52
16.120	226.50	3.750	59.06
15.130	226.60	6.125	111.80
15.630	225.60	5.375	89.09
15.380	229.70	5.875	101.00
14.380	234.00	8.875	171.90
15.500	230.00	4.000	66.80
14.250	224.30	8.000	157.10
14.500	240.50	10.870	208.40
14.620	223.70	7.375	133.40

(a) Fit a multiple linear regression model without interaction terms to the data.

(b) Predict HFE when $x_1 = 14.5$, $x_2 = 220$, and $x_3 = 5.0$.

6-11. The electric power consumed each month by a chemical plant is thought to be related to the average ambient temperature (x_1), the number of days in the month (x_2), the average product purity (x_3), and the tons of product produced (x_4). The past year's historical data are available and are presented in the following table:

y	x_1	x_2	x_3	x_4
240	25	24	91	100
236	31	21	90	95
290	45	24	88	110
274	60	25	87	88
301	65	25	91	94
316	72	26	94	99
300	80	25	87	97
296	84	25	86	96
267	75	24	88	110
276	60	25	91	105
288	50	25	90	100
261	38	23	89	98

(a) Fit a multiple linear regression model without interaction terms to these data.

(b) Predict power consumption for a month in which $x_1 = 75°F$, $x_2 = 24$ days, $x_3 = 90\%$, and $x_4 = 98$ tons.

6-3 PROPERTIES OF THE LEAST SQUARES ESTIMATORS AND ESTIMATION OF σ^2

The statistical properties of the least squares estimators $\hat{\beta}_0, \hat{\beta}_1, \ldots, \hat{\beta}_k$ may be easily found, under certain assumptions on the error terms $\epsilon_1, \epsilon_2, \ldots, \epsilon_n$, in the regression model. We assume that the errors ϵ_i are statistically independent with mean zero and variance σ^2. Under these assumptions, the least squares estimators $\hat{\beta}_0, \hat{\beta}_1, \ldots, \hat{\beta}_k$ are **unbiased estimators** of the regression coefficients $\beta_0, \beta_1, \ldots, \beta_k$. This property may be shown as follows:

$$\begin{aligned} E(\hat{\boldsymbol{\beta}}) &= E[(\mathbf{X}'\mathbf{X})^{-1}\mathbf{X}'\mathbf{Y}] \\ &= E[(\mathbf{X}'\mathbf{X})^{-1}\mathbf{X}'(\mathbf{X}\boldsymbol{\beta} + \boldsymbol{\epsilon})] \\ &= E[(\mathbf{X}'\mathbf{X})^{-1}\mathbf{X}'\mathbf{X}\boldsymbol{\beta} + (\mathbf{X}'\mathbf{X})^{-1}\mathbf{X}'\boldsymbol{\epsilon}] \\ &= \boldsymbol{\beta} \end{aligned}$$

since $E(\boldsymbol{\epsilon}) = \mathbf{0}$ and $(\mathbf{X}'\mathbf{X})^{-1}\mathbf{X}'\mathbf{X} = \mathbf{I}$, the identity matrix. Thus $\hat{\boldsymbol{\beta}}$ is an unbiased estimator of $\boldsymbol{\beta}$.

The variances of the $\hat{\beta}$'s are expressed in terms of the elements of the inverse of the $\mathbf{X}'\mathbf{X}$ matrix. The inverse of $\mathbf{X}'\mathbf{X}$ times the constant σ^2 represents a $(p \times p)$ symmetric **covariance matrix** of the regression coefficients $\hat{\boldsymbol{\beta}}$. The diagonal elements of $\sigma^2 (\mathbf{X}'\mathbf{X})^{-1}$ are the variances of $\hat{\beta}_0, \hat{\beta}_1, \ldots, \hat{\beta}_k$, and the off-diagonal elements of this matrix are the covariances. For example, if we have $k = 2$ regressors, such as in the wire bond pull strength problem of Example 6-2, then

$$\begin{aligned} \mathbf{C} &= (\mathbf{X}'\mathbf{X})^{-1} \\ &= \begin{bmatrix} C_{00} & C_{01} & C_{02} \\ C_{10} & C_{11} & C_{12} \\ C_{20} & C_{21} & C_{22} \end{bmatrix} \end{aligned}$$

which is symmetric ($C_{10} = C_{01}$, $C_{20} = C_{02}$, and $C_{21} = C_{12}$) because $(\mathbf{X}'\mathbf{X})^{-1}$ is symmetric, and we have

$$\begin{aligned} V(\hat{\beta}_j) &= \sigma^2 C_{jj}, \; j = 0, 1, 2 \\ \text{cov}(\hat{\beta}_i, \hat{\beta}_j) &= \sigma^2 C_{ij}, \; i \neq j \end{aligned}$$

In general, the covariance matrix of $\hat{\boldsymbol{\beta}}$ is a $(p \times p)$ symmetric matrix whose jjth element is the variance of $\hat{\beta}_j$ and whose ijth element is the covariance between $\hat{\beta}_i$ and $\hat{\beta}_j$; that is,

$$\text{cov}(\hat{\boldsymbol{\beta}}) = \sigma^2(\mathbf{X}'\mathbf{X})^{-1} = \sigma^2 \mathbf{C}$$

The estimates of the variances of these regression coefficients are obtained by replacing σ^2 with an appropriate estimate. When σ^2 is replaced by an estimate $\hat{\sigma}^2$, the square root of the estimated variance of the jth regression coefficient is called the **estimated standard error** of $\hat{\beta}_j$, or

$$se(\hat{\beta}_j) = \sqrt{\hat{\sigma}^2 C_{jj}}$$

The estimate of σ^2 is obtained from the residuals. Define the residual sum of squares as

$$SS_E = \sum_{i=1}^{n} (y_i - \hat{y}_i)^2 = \sum_{i=1}^{n} e_i^2$$

Now it can be shown that the expected value of the quantity SS_E is $\sigma^2(n - p)$. Therefore an unbiased estimator of σ^2 can be obtained.

An unbiased estimator of σ^2 is given by the **residual mean square** (or **error mean square**)

$$\hat{\sigma}^2 = MS_E = \frac{SS_E}{n - p} \qquad (6\text{-}24)$$

For simple linear regression, $k = 1$, a convenient way to calculate SS_E is

$$SS_E = SS_T - \hat{\beta}_1 S_{xy} \qquad (6\text{-}25)$$

where $\hat{\beta}_1$ was defined in equation 6-14, S_{xy} was defined in equation 6-17,

$$SS_T = \sum_{i=1}^{n} y_i^2 - \frac{\left(\sum y_i\right)^2}{n} \qquad (6\text{-}26)$$

and

$$\hat{\sigma}^2 = MS_E = \frac{SS_E}{n - 2} \qquad (6\text{-}27)$$

EXAMPLE 6-3

We will find the estimate of the variance σ^2 using the data from Example 6-1. Now $S_{xy} = 10.18$, $\hat{\beta}_1 = 14.97$, and

$$SS_T = \sum_{i=1}^{20} y_i^2 - \frac{\left(\sum_{i=1}^{20} y_i\right)^2}{20} = 170,044.53 - \frac{(1,843.21)^2}{20}$$

$$= 173.37$$

Therefore,

$$\hat{\sigma}^2 = \frac{SS_E}{n-2}$$

$$= \frac{SS_T - \hat{\beta}_1 S_{xy}}{n-2}$$

$$= \frac{173.37 - (14.97)(10.18)}{20 - 2}$$

$$= 1.17$$

Regression computer programs calculate the residual mean square. For example, consider the lower panel of the computer output for the oxygen purity data in Table 6-2. The mean square error is given as 1.18, slightly different from our calculated value because of rounding. Similarly, consider the computer output for the wire bond pull strength data in Table 6-6. The lower panel of the table reports the mean square error as 5.2. Therefore, we would use $\hat{\sigma}^2 = 5.2$ as the estimate of σ^2 in this problem.

We typically use computer software to provide the estimated standard error of $\hat{\beta}_j$. However, for the simple linear regression, it is convenient to have the following formula. They are the diagonal elements of $\hat{\sigma}^2(\mathbf{X}'\mathbf{X})^{-1}$ for $k = 1$ ($p = 2$).

Estimated Standard Errors for Simple Linear Regression

In simple linear regression the **estimated standard error of the slope** is

$$se(\hat{\beta}_1) = \sqrt{\frac{\hat{\sigma}^2}{S_{xx}}} \tag{6-28}$$

and the **estimated standard error of the intercept** is

$$se(\hat{\beta}_0) = \sqrt{\hat{\sigma}^2\left[\frac{1}{n} + \frac{\bar{x}^2}{S_{xx}}\right]} \tag{6-29}$$

where $\hat{\sigma}^2$ is computed from equation 6-27.

EXAMPLE 6-4

We will find the estimated standard error of the intercept and the slope for the simple linear regression model of the oxygen purity data in Examples 6-1 and 6-3. Now, $\bar{x} = 1.20$, so

$$se(\hat{\beta}_1) = \sqrt{\frac{\hat{\sigma}^2}{S_{xx}}} = \sqrt{\frac{1.17}{0.68}} = 1.31$$

and

$$se(\hat{\beta}_0) = \sqrt{\hat{\sigma}^2\left[\frac{1}{n} + \frac{\bar{x}^2}{S_{xx}}\right]} = \sqrt{1.17[0.05 + 2.12]}$$
$$= 1.59$$

The standard error of the parameter estimates can also be found in computer printouts. The upper panel of Table 6-2 gives the values 1.317 and 1.593 for $se(\hat{\beta}_1)$ and $se(\hat{\beta}_0)$, respectively. Again, the difference in computed and calculated values occurs due to rounding. For the multiple linear regression model for the wire bond pull strength problem, the upper panel of Table 6-6 gives the standard errors of the coefficients. We can also compute these values using $\sqrt{\hat{\sigma}^2 C_{jj}}$ and $(\mathbf{X}'\mathbf{X})^{-1}$ from Example 6-2, specifically,

$$se(\hat{\beta}_0) = \sqrt{(5.2)(0.21653)} = 1.06111$$
$$se(\hat{\beta}_1) = \sqrt{(5.2)(0.00167)} = 0.09322$$
$$se(\hat{\beta}_2) = \sqrt{(5.2)(0.0000015)} = 0.00280$$

Again, these values differ from those given in the upper panel of the Minitab output in Table 6-6 due to rounding.

6-4 HYPOTHESIS TESTING IN LINEAR REGRESSION

In multiple regression problems, certain tests of hypotheses about the model parameters are useful in measuring model adequacy. In this section, we describe several important hypothesis testing procedures. Hypothesis testing requires that the error terms ϵ_i in the regression model are normally and independently distributed with mean zero and variance σ^2.

6-4.1 Test for Significance of Regression

The test for significance of regression is a test to determine whether a linear relationship exists between the response variable y and a subset of the regressor variables $x_1, x_2, \ldots, x_k$. The appropriate hypotheses are

$$H_0: \beta_1 = \beta_2 = \cdots = \beta_k = 0$$
$$H_1: \beta_j \neq 0 \quad \text{for at least one } j \tag{6-30}$$

Rejection of H_0: $\beta_1 = \beta_2 = \cdots = \beta_k = 0$ implies that at least one of the regressor variables $x_1, x_2, \ldots, x_k$ contributes significantly to the model.

In the case of simple linear regression, the hypotheses in equation 6-30 become

$$H_0: \beta_1 = 0$$

$$H_1: \beta_1 \neq 0 \qquad (6\text{-}31)$$

Failure to reject H_0: $\beta_1 = 0$ is equivalent to concluding that there is no linear relationship between x and Y. This situation is illustrated in Fig. 6-8. Note that this may imply either that x is of little value in explaining the variation in Y and that the best estimator of Y for any x is $\hat{Y} = \overline{Y}$ (Fig. 6-8a) or that the true relationship between x and Y is not linear (Fig. 6-8b). Alternatively, if H_0: $\beta_1 = 0$ is rejected, this implies that x is of value in explaining the variability in Y (see Fig. 6-9). Rejecting H_0: $\beta_1 = 0$ could mean either that the straight-line model is adequate (Fig. 6-9a) or that, although there is a linear effect of x, better results could be obtained with the addition of higher order polynomial terms in x (Fig. 6-9b).

The **analysis of variance** introduced in Chapter 5 can be used to test for significance of regression. The procedure partitions the total variability in the response variable into meaningful components as the basis for the test.

The **analysis of variance identity** is

$$\sum_{i=1}^{n} (y_i - \overline{y})^2 = \sum_{i=1}^{n} (\hat{y}_i - \overline{y})^2 + \sum_{i=1}^{n} (y_i - \hat{y}_i)^2 \qquad (6\text{-}32)$$

The two components of the right-hand side of equation 6-32 measure, respectively, the amount of variability in y_i accounted for by the regression line and the residual variation left unexplained by the regression line. We recognize $SS_E = \sum_{i=1}^{n} (y_i - \hat{y}_i)^2$, the **error**

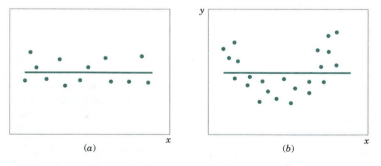

(a)   (b)

Figure 6-8   The hypothesis H_0: $\beta_1 = 0$ is not rejected.

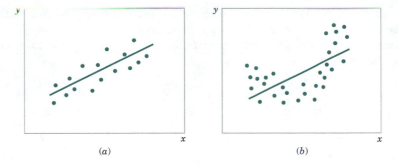

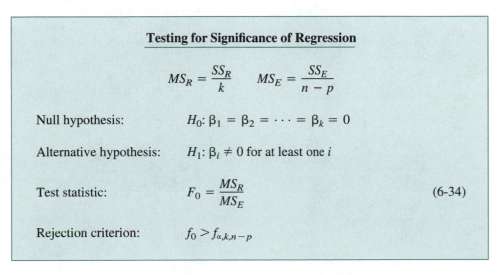

Figure 6-9   The hypothesis H_0: $\beta_1 = 0$ is rejected.

or **residual sum of squares** introduced in Section 6-3, and we will call $SS_R = \sum_{i=1}^{n} (\hat{y}_i - \bar{y})^2$ the **regression sum of squares.** Symbolically, equation 6-32 may be written as

$$SS_T = SS_R + SS_E \qquad (6\text{-}33)$$

where $SS_T = \sum_{i=1}^{n} (y_i - \bar{y})^2$ is the **total corrected sum of squares** of y. The total corrected sum of squares has $n - 1$ degrees of freedom. Furthermore, if the null hypothesis H_0: $\beta_1 = \beta_2 = \cdots = \beta_k = 0$ is true, then SS_R/σ^2 is a chi-squared random variable with k degrees of freedom. Note that the number of degrees of freedom for this chi-squared random variable is equal to the number of regressor variables in the model. We can also show that SS_E/σ^2 is a chi-squared random variable with $n - p$ degrees of freedom, and that SS_E and SS_R are independent.

Testing for Significance of Regression

$$MS_R = \frac{SS_R}{k} \qquad MS_E = \frac{SS_E}{n - p}$$

Null hypothesis: H_0: $\beta_1 = \beta_2 = \cdots = \beta_k = 0$

Alternative hypothesis: H_1: $\beta_i \neq 0$ for at least one i

Test statistic: $F_0 = \dfrac{MS_R}{MS_E}$ $\qquad (6\text{-}34)$

Rejection criterion: $f_0 > f_{\alpha,k,n-p}$

The quantity MS_R is called the **regression** (or **model**) **mean square.** We should reject H_0 if the computed value of the test statistic in equation 6-34, f_0, is greater than $f_{\alpha,k,n-p}$. The procedure is usually summarized in an analysis of variance table such as Table 6-7.

For simple linear regression

$$SS_R = \hat{\beta}_1 S_{xy} \tag{6-35}$$

For multiple linear regression

$$SS_R = \hat{\beta}'\mathbf{X}'\mathbf{y} - \frac{\left(\sum\limits_{i=1}^{n} y_i\right)^2}{n} \tag{6-36}$$

The SS_T is given in equation 6-26 and SS_E is found by subtraction.

Most multiple regression computer programs provide the test for significance of regression in their output display. Recalling the computer program output for the oxygen purity data in Table 6-2, note that the lower panel contains the analysis of variance table to test the hypotheses

$$H_0: \beta_1 = 0$$

$$H_1: \beta_1 \neq 0$$

The output indicates that $f_0 = 128.86$, which is compared to $f_{0.05,1,18} = 4.41$. Since $f_0 > 4.41$, we reject H_0 and conclude that oxygen purity is linearly related to hydrocarbon level.

Similarly, the bottom portion of Table 6-6 is the Minitab analysis of variance for the wire bond pull strength data in Example 6-2. In this problem, the hypotheses in equation 6-30 become

$$H_0: \beta_1 = \beta_2 = 0$$

$$H_1: \text{above statement not true}$$

To test these hypotheses, we use the F-ratio in equation 6-34, which we find in Table 6-6 to be

$$f_0 = \frac{MS_R}{MS_E} = \frac{2995.4}{5.2} = 572.17$$

Table 6-7   Analysis of Variance for Testing Significance of Regression

Source of Variation	Sum of Squares	Degrees of Freedom	Mean Square	F_0
Regression	SS_R	k	MS_R	MS_R/MS_E
Error or residual	SS_E	$n - p$	MS_E	
Total	SS_T	$n - 1$		

Since $f_0 > f_{0.05,2,22} = 3.44$ (or since the P-value is considerably smaller than $\alpha = 0.05$), we reject the null hypothesis and conclude that pull strength is linearly related to either wire length or die height, or both. However, we note that this does not necessarily imply that the relationship found is an appropriate model for predicting pull strength as a function of wire length or die height. Further tests of model adequacy are required before we can be comfortable using this model in practice.

6-4.2   Tests on Individual Regression Coefficients

In multiple regression, we are usually interested in testing hypotheses on the individual regression coefficients. Such tests would be useful in determining the potential value of each of the regressor variables in the regression model. For example, the model might be more effective with the inclusion of additional variables, or perhaps with the deletion of one or more of the regressors presently in the model.

Adding a variable to a regression model always causes the sum of squares for regression to increase and the error sum of squares to decrease. We must decide whether the increase in the regression sum of squares is large enough to justify using the additional variable in the model. Furthermore, adding an unimportant variable to the model can actually increase the error mean square, indicating that adding such a variable has actually made the model a poorer fit to the data.

Testing the Significance of Any Individual Regression Coefficient β_j

Null hypothesis:	$H_0: \beta_j = 0$	
Alternative hypothesis:	$H_1: \beta_j \neq 0$	(6-37)
Test statistic:	$T_0 = \dfrac{\hat{\beta}_j}{\sqrt{\hat{\sigma}^2 C_{jj}}}$	(6-38)
	where C_{jj} is the diagonal element of $(\mathbf{X'X})^{-1}$ corresponding to $\hat{\beta}_j$.	
Rejection criterion:	$t_0 > t_{\alpha/2,n-p}$ or $t_0 < -t_{\alpha/2,n-p}$	

Note that the denominator of equation 6-38 is the estimated standard error of the regression coefficient $\hat{\beta}_j$.

If $H_0: \beta_j = 0$ is not rejected, then this indicates that the regressor x_j can be deleted from the model. This test is called a **partial** or **marginal test,** because the regression coefficient $\hat{\beta}_j$ depends on all the other regressor variables $x_i (i \neq j)$ that are in the model. More will be said about this in Example 6-6.

EXAMPLE 6-5

Consider the oxygen purity data and suppose we want to test the hypothesis that the regression coefficient for x is zero. The hypotheses are

$$H_0: \beta_1 = 0$$
$$H_1: \beta_1 \neq 0$$

From Examples 6-1 and 6-4 we found $\hat{\beta}_1 = 14.97$ and $se(\hat{\beta}_1) = 1.31$, so the t-statistic in equation 6-38 is

$$t_0 = \frac{14.97}{1.31} = 11.43$$

Since $t_{0.025,18} = 2.101$, we reject $H_0: \beta_1 = 0$ and conclude that there is a significant linear relationship between oxygen purity and hydrocarbon levels. Note that the computed value of t_0 found in the upper panel of Table 6-2 is 11.35; again the difference is due to rounding. It is also worth noting that t_0^2 is equal (except due to rounding) to the f_0 ratio 128.86. It is true, in general, that the square of a t-random variable with v degrees of freedom is an F-random variable with one and v degrees of freedom in the numerator and denominator, respectively. Thus, in simple linear regression, the test using T_0 is equivalent to the test based on F_0.

EXAMPLE 6-6

Consider again the wire bond pull strength data, and suppose that we want to test the hypothesis that the regression coefficient for x_2 (die height) is zero. The hypotheses are

$$H_0: \beta_2 = 0$$
$$H_1: \beta_2 \neq 0$$

The main diagonal element of the $(\mathbf{X'X})^{-1}$ matrix corresponding to $\hat{\beta}_2$ is $C_{22} = 0.0000015$, so the t-statistic in equation 6-38 is

$$t_0 = \frac{\hat{\beta}_2}{\sqrt{\hat{\sigma}^2 C_{22}}} = \frac{0.01253}{\sqrt{(5.2352)(0.0000015)}} = 4.4767$$

Since $t_{0.025,22} = 2.074$, we reject $H_0: \beta_2 = 0$ and conclude that the variable x_2 (die height) contributes significantly to the model. We could also have used a P-value to draw conclusions. The P-value for $t_0 = 4.4767$ is $P = 0.0002$, so with $\alpha = 0.05$ we would reject the null hypothesis. Note that this test measures the marginal or partial contribution of x_2 given that x_1 is in the model. That is, the t-test measures the contribution of adding the variable x_2 = die height to a model that already contains x_1 = wire length. Table 6-6 shows the value of the t-test computed by Minitab. Note that the computer produces a t-test for each regression coefficient in the model. These t-tests indicate that both regressors contribute to the model.

EXERCISES FOR SECTIONS 6-3 AND 6-4

6-12. Consider the data from Exercise 6-1 on concrete mixes and cures.

(a) Test for significance of regression using α = 0.05. Find the P-value for this test. Can you conclude that the model specifies a useful linear relationship between these two variables?

(b) Estimate σ^2.

6-13. Consider the data from Exercise 6-2 on x = roadway surface temperature and y = pavement deflection.

(a) Test for significance of regression using α = 0.05. Find the P-value for this test. What conclusions can you draw?

(b) Estimate σ^2.

6-14. Consider the National Football League data in Exercise 6-4.

(a) Estimate σ^2.

(b) Test for significance of regression using α = 0.01. Or, equivalently, test (using α = 0.01) H_0: β_1 = 0.0 versus H_1: β_1 ≠ 0.0. Would you agree with the statement that this is a test of the claim that if you can decrease the opponent's rushing yardage by 100 yards the team will win one more game?

6-15. Consider the data from Exercise 6-5 on y = thermal conductivity and x = product density.

(a) Test H_0: β_1 = 0 using the t-test; use α = 0.05.

(b) Test H_0: β_1 = 0 using the analysis of variance with α = 0.05. Do you see any relationship of this test to

the test in part (a)? Explain your answer.

(c) Estimate σ^2.

6-16. Consider the data from Exercise 6-6 on y = steam usage and x = average temperature.

(a) Test for significance of regression using α = 0.01. What is the P-value for this test? State the conclusions that result from this test.

(b) Estimate σ^2.

6-17. Consider the absorption index data in Exercise 6-7. The total sum of squares for y is SS_T = 742.00.

(a) Test for significance of regression using α = 0.01. What is the P-value for this test?

(b) Estimate σ^2.

(c) Test the hypothesis H_0: β_1 = 0 versus H_1: β_1 ≠ 0 using α = 0.01. What is the P-value for this test? What conclusion can you draw about the usefulness of x_1 as a regressor in this model?

6-18. A regression model $Y = \beta_0 + \beta_1 x_1 + \beta_2 x_2 + \beta_3 x_3 + \epsilon$ has been fit to a sample of n = 25 observations. The calculated t-ratios $\hat{\beta}_j / se(\hat{\beta}_j)$, j = 1, 2, 3, are as follows: for β_1, t_0 = 4.82; for β_2, t_0 = 8.21; and for β_3, t_0 = 0.98.

(a) Find P-values for each of the t-statistics.

(b) Using α = 0.05, what conclusions can you draw about the regressor x_3? Does it seem likely that this regressor contributes significantly to the model?

6-19. Consider the electric power consumption data in Exercise 6-11.

(a) Test for significance of regression using $\alpha = 0.01$. What is the P-value for this test?

(b) Estimate σ^2.

(c) Use the t-test to assess the contribution of each regressor to the model. Using $\alpha = 0.01$, what conclusions can you draw?

6-20. Consider the bearing wear data in Exercise 6-8 with no interaction.

(a) Test for significance of regression using $\alpha = 0.05$. What is the P-value for this test? What are your conclusions?

(b) Compute the t-statistics for each regression coefficient. Using $\alpha = 0.05$, what conclusions can you draw?

6-21. Reconsider the bearing wear data from Exercises 6-8 and 6-20.

(a) Refit the model with an interaction term. Test for significance of regression using $\alpha = 0.05$.

(b) Use the t-test to determine whether the interaction term contributes significantly to the model. Use $\alpha = 0.05$.

(c) Estimate σ^2 for the interaction model. Compare this to the estimate of σ^2 from the model in Exercise 6-20.

6-22. Reconsider the semiconductor data in Exercise 6-10.

(a) Test for significance of regression using $\alpha = 0.05$. What conclusions can you draw?

(b) Estimate σ^2 for this model.

(c) Calculate the t-test statistic for each regression coefficient. Using $\alpha = 0.05$, what conclusions can you draw?

6-5 CONFIDENCE INTERVALS IN LINEAR REGRESSION

6-5.1 Confidence Intervals on Individual Regression Coefficients

In linear regression models, it is often useful to construct confidence interval estimates for the regression coefficients $\{\beta_j\}$. The development of a procedure for obtaining these confidence intervals requires that the errors $\{\epsilon_i\}$ are normally and independently distributed with mean zero and variance σ^2. This is the same assumption required in hypothesis testing. Therefore, the observations $\{Y_i\}$ are normally and independently distributed with mean $\beta_0 + \sum_{j=1}^{k} \beta_j x_{ij}$ and variance σ^2. Since the least squares estimator $\hat{\boldsymbol{\beta}}$ is a linear combination of the observations, it follows that $\hat{\boldsymbol{\beta}}$ is normally distributed with mean vector $\boldsymbol{\beta}$ and covariance matrix $\sigma^2(\mathbf{X'X})^{-1}$. Then each of the statistics

$$\frac{\hat{\beta}_j - \beta_j}{\sqrt{\hat{\sigma}^2 C_{jj}}} \qquad j = 0, 1, \ldots, k \qquad (6\text{-}39)$$

has a t distribution with $n - p$ degrees of freedom, where C_{jj} is the jjth element of the $(\mathbf{X'X})^{-1}$ matrix, and $\hat{\sigma}^2$ is the estimate of the error variance, obtained from equation 6-24. This leads to the following $100(1 - \alpha)\%$ confidence interval for the regression coefficient $\beta_j, j = 0, 1, \ldots, k$.

Multiple Linear Regression

A $100(1 - \alpha)\%$ **confidence interval on the regression coefficient** β_j, $j = 0, 1, \ldots, k$, in the multiple linear regression model is given by

$$\hat{\beta}_j - t_{\alpha/2,n-p}\sqrt{\hat{\sigma}^2 C_{jj}} \leq \beta_j \leq \hat{\beta}_j + t_{\alpha/2,n-p}\sqrt{\hat{\sigma}^2 C_{jj}} \qquad (6\text{-}40)$$

For the case of simple linear regression, the standard error of the estimates of β_0 and β_1 are given in equations (6-28) and (6-29) and the following equations result from substitution.

Simple Linear Regression

Under the assumption that the observations are normally and independently distributed, a $100(1 - \alpha)\%$ **confidence interval on the slope** β_1 in simple linear regression is

$$\hat{\beta}_1 - t_{\alpha/2,n-2}\sqrt{\frac{\hat{\sigma}^2}{S_{xx}}} \leq \beta_1 \leq \hat{\beta}_1 + t_{\alpha/2,n-2}\sqrt{\frac{\hat{\sigma}^2}{S_{xx}}} \qquad (6\text{-}41)$$

Similarly, a $100(1 - \alpha)\%$ **confidence interval on the intercept** β_0 is

$$\hat{\beta}_0 - t_{\alpha/2,n-2}\sqrt{\hat{\sigma}^2\left[\frac{1}{n} + \frac{\bar{x}^2}{S_{xx}}\right]} \leq \beta_0 \leq \hat{\beta}_0 + t_{\alpha/2,n-2}\sqrt{\hat{\sigma}^2\left[\frac{1}{n} + \frac{\bar{x}^2}{S_{xx}}\right]} \qquad (6\text{-}42)$$

Using the results from Example 6-4, we leave it to the reader to confirm that the lower and upper bounds of the 95% confidence interval on the slope of the regression line β_1 of the oxygen purity data are 12.21 and 17.73, respectively. We now illustrate the confidence interval construction for the multiple regression case.

EXAMPLE 6-7

We will construct a 95% confidence interval on the parameter β_1 in the wire bond pull strength problem. Note that the point estimate of β_1 is $\hat{\beta}_1 = 2.74427$ and that the diagonal element of $(\mathbf{X'X})^{-1}$ corresponding to β_1 is $C_{11} = 0.001671$. The estimate of σ^2 is 5.2352, and $t_{0.025,22} = 2.074$. Therefore, the 95% confidence interval on β_1 is computed from equation 6-40 as

$$2.74427 - (2.074)\sqrt{(5.2352)(.001671)} \leq \beta_1 \leq 2.74427 + (2.074)\sqrt{(5.2352)(.001671)}$$

which reduces to

$$2.55029 \leq \beta_1 \leq 2.93825$$

6-5.2   Confidence Interval on the Mean Response

We may also obtain a confidence interval on the mean response at a particular point—say, $x_{01}, x_{02}, \ldots, x_{0k}$. To estimate the mean response at this point, define the vector

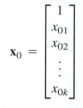

$$\mathbf{x}_0 = \begin{bmatrix} 1 \\ x_{01} \\ x_{02} \\ \vdots \\ x_{0k} \end{bmatrix}$$

The mean response at this point is $E(Y|\mathbf{x}_0) = \mu_{Y|\mathbf{x}_0} = \mathbf{x}_0'\boldsymbol{\beta}$, which is estimated by

$$\hat{\mu}_{Y|\mathbf{x}_0} = \mathbf{x}_0'\hat{\boldsymbol{\beta}} \tag{6-43}$$

This estimator is unbiased, since $E(\mathbf{x}_0'\hat{\boldsymbol{\beta}}) = \mathbf{x}_0'\boldsymbol{\beta} = E(Y|\mathbf{x}_0) = \mu_{Y|\mathbf{x}_0}$ and the variance of $\hat{\mu}_{Y|\mathbf{x}_0}$ can be shown to be

$$V(\hat{\mu}_{Y|\mathbf{x}_0}) = \sigma^2\mathbf{x}_0'(\mathbf{X}'\mathbf{X})^{-1}\mathbf{x}_0 \tag{6-44}$$

A $100(1 - \alpha)\%$ confidence interval on $\mu_{Y|\mathbf{x}_0}$ can be constructed from the statistic

$$\frac{\hat{\mu}_{Y|\mathbf{x}_0} - \mu_{Y|\mathbf{x}_0}}{\sqrt{\hat{\sigma}^2\mathbf{x}_0'(\mathbf{X}'\mathbf{X})^{-1}\mathbf{x}_0}} \tag{6-45}$$

Multiple Linear Regression

For the multiple linear regression model, a $100(1 - \alpha)\%$ **confidence interval on the mean response** at the point $x_{01}, x_{02}, \ldots, x_{0k}$ is

$$\hat{\mu}_{Y|\mathbf{x}_0} - t_{\alpha/2,n-p}\sqrt{\hat{\sigma}^2\mathbf{x}_0'(\mathbf{X}'\mathbf{X})^{-1}\mathbf{x}_0}$$
$$\leq \mu_{Y|\mathbf{x}_0} \leq \hat{\mu}_{Y|\mathbf{x}_0} + t_{\alpha/2,n-p}\sqrt{\hat{\sigma}^2\mathbf{x}_0'(\mathbf{X}'\mathbf{X})^{-1}\mathbf{x}_0} \tag{6-46}$$

In the case of simple linear regression, equation 6-46 produces a confidence interval about the mean response at a particular point on the regression line. Let x_0 be the value of the regressor x that our interest is focused on. The point estimate of $\mu_{Y|x_0}$ is

$$\hat{\mu}_{Y|x_0} = \hat{\beta}_0 + \hat{\beta}_1 x_0$$

We can show that equation 6-46 reduces to the following expression.

Simple Linear Regression

For the simple linear regression model, a $100(1 - \alpha)\%$ **confidence interval about the mean response** at the value of $x = x_0$—say, $\mu_{Y|x_0}$—is given by

$$\hat{\mu}_{Y|x_0} - t_{\alpha/2, n-2}\sqrt{\hat{\sigma}^2\left[\frac{1}{n} + \frac{(x_0 - \bar{x})^2}{S_{xx}}\right]}$$

$$\leq \mu_{Y|x_0} \leq \hat{\mu}_{Y|x_0} + t_{\alpha/2, n-2}\sqrt{\hat{\sigma}^2\left[\frac{1}{n} + \frac{(x_0 - \bar{x})^2}{S_{xx}}\right]} \quad (6\text{-}47)$$

where $\hat{\mu}_{Y|x_0} = \hat{\beta}_0 + \hat{\beta}_1 x_0$ is computed from the fitted regression model.

Note that the width of the confidence interval for $\mu_{Y|x_0}$ is a function of the value specified for x_0. The interval width is a minimum for $x_0 = \bar{x}$ and widens as $|x_0 - \bar{x}|$ increases.

EXAMPLE 6-8

We will construct a 95% confidence interval about the mean response for the oxygen purity data in Example 6-1. The fitted model for x_0 is $\hat{\mu}_{Y|x_0} = 74.3 + 14.9x_0$, the estimate of σ^2 is $\hat{\sigma}^2 = 1.18$ (from Table 6-2), and the 95% confidence interval on $\mu_{Y|x_0}$ is found from equation 6-47 as

$$\hat{\mu}_{Y|x_0} \pm 2.101\sqrt{1.18\left[\frac{1}{20} + \frac{(x_0 - 1.20)^2}{0.68}\right]}$$

Suppose that we are interested in predicting mean oxygen purity when $x_0 = 1.00\%$. Then

$$\hat{\mu}_{Y|x_0} = 74.3 + 14.9(1.00) = 89.2$$

and the 95% confidence interval is

$$\left[89.17 \pm 2.101\sqrt{1.18\left[\frac{1}{20} + \frac{(1.00 - 1.20)^2}{0.68}\right]}\right]$$

or

$$89.2 \pm 0.75$$

Therefore, the 95% confidence interval on $\mu_{Y|1.00}$ is

$$88.45 \leq \mu_{Y|1.00} \leq 89.95$$

By repeating these calculations for several different values for x_0 we can obtain confidence limits for each corresponding value of $\mu_{Y|x_0}$. Figure 6-10 displays the scatter diagram with the fitted model and the corresponding 95% confidence limits plotted as the upper and lower lines. The 95% confidence level applies only to the interval obtained at one value of x, and not to the entire set of x-levels. Note that the width of the confidence interval on $\mu_{Y|x_0}$ increases as $|x_0 - \bar{x}|$ increases.

We will now illustrate the construction of this confidence interval in the multiple regression case.

EXAMPLE 6-9

The engineer in Example 6-2 would like to construct a 95% confidence interval on the mean wire bond pull strength when $x_1 = 8$ and $x_2 = 275$. Therefore,

$$\mathbf{x}_0 = \begin{bmatrix} 1 \\ 8 \\ 275 \end{bmatrix}$$

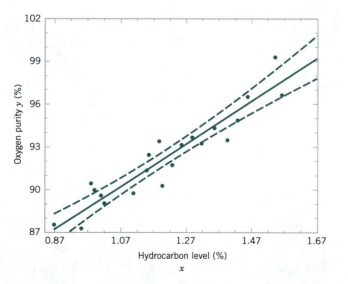

Figure 6-10   Scatter diagram of oxygen purity data from Example 6-8 with fitted regression line and 95% confidence limits on $\mu_{Y|x_0}$.

The estimated mean response at this point is found from equation 6-43 as

$$\hat{\mu}_{Y|x_0} = x_0'\hat{\beta} = \begin{bmatrix} 1 & 8 & 275 \end{bmatrix} \begin{bmatrix} 2.26379 \\ 2.74427 \\ 0.01253 \end{bmatrix} = 27.66 \text{ psi}$$

The variance of $\hat{\mu}_{Y|x_0}$ is estimated by

$$\hat{\sigma}^2 x_0'(X'X)^{-1}x_0 = 5.2352 \begin{bmatrix} 1 & 8 & 275 \end{bmatrix}$$

$$\times \begin{bmatrix} 0.214653 & -0.007491 & -0.000340 \\ -0.007491 & 0.001671 & -0.000019 \\ -0.000340 & -0.000019 & 0.0000015 \end{bmatrix} \begin{bmatrix} 1 \\ 8 \\ 275 \end{bmatrix}$$

$$= 5.2352 (0.04444) = 0.23266$$

Therefore, a 95% confidence interval on the mean pull strength at this point is found from equation 6-46 as

$$27.66 - 2.074\sqrt{0.23266} \le \mu_{Y|x_0} \le 27.66 + 2.074\sqrt{0.23266}$$

which reduces to

$$26.66 \le \mu_{Y|x_0} \le 28.66$$

6-6   PREDICTION OF NEW OBSERVATIONS

A regression model can be used to predict future observations on the response variable Y corresponding to particular values of the independent variables—say, $x_{01}, x_{02}, \ldots, x_{0k}$. If $x_0' = [1, x_{01}, x_{02}, \ldots, x_{0k}]$, then a point estimate of the future observation Y_0 at the point $x_{01}, x_{02}, \ldots, x_{0k}$ is

$$\hat{y}_0 = x_0'\hat{\beta} \tag{6-48}$$

We now present a $100(1 - \alpha)\%$ prediction interval for this future observation.

Multiple Linear Regression

A $100(1 - \alpha)\%$ **prediction interval on the future observation** of Y at the point $x = x_0$ in multiple linear regression is

$$\hat{y}_0 - t_{\alpha/2, n-p}\sqrt{\hat{\sigma}^2(1 + x_0'(X'X)^{-1}x_0)}$$

$$\le y_0 \le \hat{y}_0 + t_{\alpha/2, n-p}\sqrt{\hat{\sigma}^2(1 + x_0'(X'X)^{-1}x_0)} \tag{6-49}$$

If you compare the prediction interval equation 6-49 with the expression for the confidence interval on the mean, equation 6-46, you will observe that the prediction interval is always wider than the confidence interval. The confidence interval expresses the error in estimating the mean of a distribution, whereas the prediction interval expresses the error in predicting a future observation from the distribution at the point x_0. The prediction interval must include the error in estimating the mean at that point, as well as the inherent variability in the random variable Y at the same value $x = x_0$.

In predicting new observations and in estimating the mean response at a given point $x_{01}, x_{02}, \ldots, x_{0k}$, we must be careful about extrapolating beyond the region containing the original observations. It is very possible that a model that fits well in the region of the original data will no longer fit well outside of that region. In multiple regression it is often easy to inadvertently extrapolate, since the levels of the variables $(x_{i1}, x_{i2}, \ldots, x_{ik})$, $i = 1, 2, \ldots, n$, jointly define the region containing the data. As an example, consider Fig. 6-11, which illustrates the region containing the observations for a two-variable regression model. Note that the point (x_{01}, x_{02}) lies within the ranges of both regressor variables x_1 and x_2, but it is outside the region that is actually spanned by the original observations. Thus, either predicting the value of a new observation or estimating the mean response at this point is an extrapolation of the original regression model.

In the case of simple linear regression, the prediction interval in equation 6-49 reduces to the following.

Simple Linear Regression

A $100(1 - \alpha)\%$ **prediction interval on a future observation** y_0 at the value x_0 in simple linear regression is given by

$$\hat{y}_0 - t_{\alpha/2, n-2}\sqrt{\hat{\sigma}^2\left[1 + \frac{1}{n} + \frac{(x_0 - \bar{x})^2}{S_{xx}}\right]}$$

$$\leq y_0 \leq \hat{y}_0 + t_{\alpha/2, n-2}\sqrt{\hat{\sigma}^2\left[1 + \frac{1}{n} + \frac{(x_0 - \bar{x})^2}{S_{xx}}\right]} \quad (6\text{-}50)$$

The value $\hat{y}_0$ is computed from the regression model $\hat{y}_0 = \hat{\beta}_0 + \hat{\beta}_1 x_0$.

Note that the prediction interval is of minimum width at $x_0 = \bar{x}$ and widens as $|x_0 - \bar{x}|$ increases. By comparing equation 6-50 with equation 6-47, we observe that the prediction interval at the point x_0 is always wider than the confidence interval at x_0. This results because the prediction interval depends on both the error from the fitted model and the error associated with future observations.

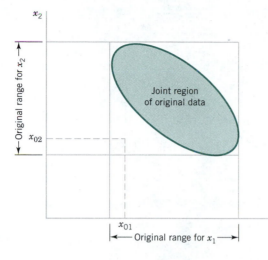

Figure 6-11 An example of extrapolation in multiple regression.

EXAMPLE 6-10

To illustrate the construction of a prediction interval, suppose we use the data in Example 6-1 and find a 95% prediction interval on the next observation of oxygen purity at $x_0 = 1.00\%$. Using equation 6-50 and recalling from Example 6-8 that $\hat{y}_0 = 89.2$, we find that the prediction interval is

$$89.2 - 2.101 \sqrt{1.18 \left[1 + \frac{1}{20} + \frac{(1.00 - 1.20)^2}{0.68} \right]}$$

$$\le y_0 \le 89.2 + 2.101 \sqrt{1.18 \left[1 + \frac{1}{20} + \frac{(1.00 - 1.20)^2}{0.68} \right]}$$

which simplifies to

$$86.80 \le y_0 \le 91.60$$

By repeating the foregoing calculations at different levels of x_0, we may obtain the 95% prediction intervals shown graphically as the lower and upper lines about the fitted regression model in Fig. 6-12. Note that this graph also shows the 95% confidence limits on $\mu_{Y|x_0}$ calculated in Example 6-8, and it illustrates that the prediction limits are always wider than the confidence limits.

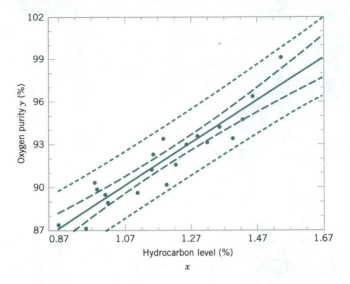

Figure 6-12   Scatter diagram of oxygen purity data from Example 6-1 with fitted regression line, 95% prediction limits (outer lines) and 95% confidence limits on $\mu_{Y|x_0}$.

EXAMPLE 6-11

Suppose that the engineer in Example 6-1 wishes to construct a 95% prediction interval on wire bond pull strength where $x_1 = 8$ and $x_2 = 275$. Note that $\mathbf{x}_0' = [1 \quad 8 \quad 275]$, and the point estimate of the pull strength is $\hat{y}_0 = \mathbf{x}_0'\hat{\boldsymbol{\beta}} = 27.66$. Also, in Example 6-9 we calculated $\mathbf{x}_0'(\mathbf{X}'\mathbf{X})^{-1}\mathbf{x}_0 = 0.04444$. Therefore, from equation 6-49 we have

$$27.66 - 2.074\sqrt{5.2352(1 + 0.04444)} \leq y_0 \leq 27.66 + 2.074\sqrt{5.2352(1 + 0.04444)}$$

and the 95% prediction interval is

$$22.81 \leq y_0 \leq 32.51$$

Note that the prediction interval is wider than the confidence interval on the mean response at the same point, calculated in Example 6-9.

EXERCISES FOR SECTIONS 6-5 AND 6-6

6-23.   Refer to the data in Exercise 6-1 on concrete mixtures and cures.
(a) Find a 95% confidence interval on the slope.
(b) Find a 95% confidence interval on the intercept.
(c) Find a 95% confidence interval for the mean value of y when $x = 2.5$.

(d) Find a 95% prediction interval on y when $x = 2.5$. Explain why this interval is wider than the interval in part (c).

6-24.   Exercise 6-2 presented data on roadway surface temperature x and pavement deflection y.

(a) Find a 99% confidence interval on the slope.

(b) Find a 99% confidence interval on the intercept.

(c) Find a 99% confidence interval on mean deflection when temperature $x = 85°F$.

(d) Find a 99% prediction interval on pavement deflection when the temperature is 90°F.

6-25. Exercise 6-4 presented data on the number of games won by NFL teams in 1976.

(a) Find a 95% confidence interval for the slope and the intercept. (Use the standard errors from your computer analysis.)

(b) Find a 95% confidence interval on the mean number of games won when the opponent's rushing yardage is limited to $x = 1800$.

(c) Find a 95% prediction interval on the number of games won when the opponent's rushing yardage is 1800.

6-26. Refer to the data on $y =$ thermal conductivity and $x =$ product density in Exercise 6-5.

(a) Find a 95% confidence interval for β_1 and β_0. (Use the standard errors from your computer analysis.)

(b) Find a 95% confidence interval for the mean thermal conductivity when the product density is $x = 0.2350$.

(c) Compute the 95% prediction interval for thermal conductivity when the product density is $x = 0.2350$.

6-27. Exercise 6-6 presented data on $y =$ steam usage and $x =$ monthly average temperature.

(a) Find a 99% confidence interval for β_1.

(b) Find a 99% confidence interval for β_0.

(c) Find a 95% confidence interval on mean steam usage when the average temperature is 55°F.

(d) Find a 95% prediction interval on steam usage when temperature is 55°F. Explain why this interval is wider than the interval in part (c).

6-28. Consider the soil absorption data in Exercise 6-7.

(a) Find a 95% confidence interval on the regression coefficient β_1.

(b) Find a 95% confidence interval on mean soil absorption index when $x_1 = 200$ and $x_2 = 50$.

(c) Find a 95% prediction interval on the soil absorption index when $x_1 = 200$ and $x_2 = 50$.

6-29. Consider the electric power consumption data in Exercise 6-11.

(a) Find 95% confidence intervals on β_1, β_2, β_3, and β_4.

(b) Find a 95% confidence interval on the mean of Y when $x_1 = 75$, $x_2 = 24$, $x_3 = 90$, and $x_4 = 98$.

(c) Find a 95% prediction interval on the power consumption when $x_1 = 75$, $x_2 = 24$, $x_3 = 90$, and $x_4 = 98$.

6-30. Consider the bearing wear data in Exercise 6-8.

(a) Find 99% confidence intervals on β_1 and β_2.

(b) Recompute the confidence intervals in part (a) after the interaction term x_1x_2 is added to the model. Compare the lengths of these confidence intervals with those computed in part (a). Do the lengths of these intervals provide any information about the contribution of the interaction term in the model?

6-31. Consider the semiconductor data in Exercise 6-10.

(a) Find 99% confidence intervals on the regression coefficients.

(b) Find a 99% prediction interval on HFE when $x_1 = 14.5$, $x_2 = 220$, and $x_3 = 5.0$.

(c) Find a 99% confidence interval on mean HFE when $x_1 = 14.5$, $x_2 = 220$, and $x_3 = 5.0$.

6-7   ASSESSING THE ADEQUACY OF THE REGRESSION MODEL

Fitting a regression model requires several assumptions. Estimation of the model parameters requires the assumption that the errors are uncorrelated random variables with mean zero and constant variance. Tests of hypotheses and interval estimation require that the errors be normally distributed. In addition, we assume that the order of the model is correct; that is, if we fit a simple linear regression model, then we are assuming that the phenomenon actually behaves in a linear or first-order manner.

The analyst should always consider the validity of these assumptions to be doubtful and conduct analyses to examine the adequacy of the model that has been tentatively entertained. In this section we discuss methods useful in this respect.

6-7.1   Residual Analysis

The residuals from a regression model are $e_i = y_i - \hat{y}_i$, $i = 1, 2, \ldots, n$, where y_i is an actual observation and $\hat{y}_i$ is the corresponding fitted value from the regression model. Analysis of the residuals is frequently helpful in checking the assumption that the errors are approximately normally distributed with constant variance, as well as in determining whether additional terms in the model would be useful.

As an approximate check of normality, the experimenter can construct a frequency histogram of the residuals or a **normal probability plot of residuals.** Many computer programs will produce a normal probability plot of residuals, and since the sample sizes in regression are often too small for a histogram to be meaningful, the normal probability plotting method is preferred. It requires judgment to assess the abnormality of such plots. (Refer to the discussion of the "fat pencil" method in Chapter 3.)

We may also **standardize** the residuals by computing $d_i = e_i / \sqrt{\hat{\sigma}^2}$, $i = 1, 2, \ldots, n$. If the errors are normally distributed, then approximately 95% of the standardized residuals should fall in the interval $(-2, +2)$. Residuals that are far outside this interval may indicate the presence of an **outlier;** that is, an observation that is not typical of the rest of the data. Various rules have been proposed for discarding outliers. However, outliers sometimes provide important information about unusual circumstances of interest to experimenters and should not be discarded. For further discussion of outliers, see Montgomery and Peck (1992).

It is frequently helpful to plot the residuals (1) in time sequence (if known), (2) against the $\hat{y}_i$, and (3) against the independent variable x. These graphs will usually look like one of the four general patterns shown in Fig. 6-13. Pattern (a) in Fig. 6-13 represents the ideal situation, while patterns (b), (c), and (d) represent anomalies. If the residuals appear as in (b), the variance of the observations may be increasing with time or with the magnitude of y_i or x_i. Data transformation on the response y is often used to eliminate this problem. Widely used variance-stabilizing transformations include the use of $\sqrt{y}$, $ln\ y$, or $1/y$ as the response. See Montgomery and Peck (1992) for more details regarding methods for selecting an appropriate transformation. If a plot of the residuals against time has the appearance of (b), then the variance of the observations is increasing with time.

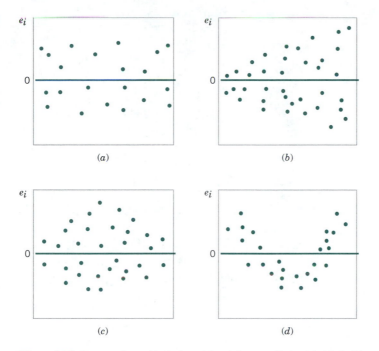

Figure 6-13 Patterns for residual plots: (a) satisfactory, (b) funnel, (c) double bow, (d) nonlinear. Horizontal axis may be time, y_i, $\hat{y}_i$, or x_i. [Adapted from Montgomery and Peck (1992).]

Plots of residuals against $\hat{y}_i$ and x_i that look like (c) also indicate inequality of variance. Residual plots that look like (d) indicate model inadequacy; that is, higher order terms should be added to the model, a transformation on the x-variable or the y-variable (or both) should be considered, or other regressors should be considered.

EXAMPLE 6-12

The residuals for the model from Example 6-2 are shown in Table 6-5. A normal probability plot of these residuals is shown in Fig. 6-14. No severe deviations from normality are obviously apparent, although the two largest residuals ($e_{15} = 5.88$ and $e_{17} = 4.33$) do not fall extremely close to a straight line drawn through the remaining residuals.

As noted above, the standardized residuals $d_i = e_i/\sqrt{MS_E}$ are often more useful than the ordinary residuals when assessing residual magnitude. The standardized residuals corresponding to e_{15} and e_{17} are $d_{15} = 5.84/\sqrt{5.2} = 2.58$ and $d_{17} = 4.33/\sqrt{5.2} = 2.56$, and they do not seem unusually large. Inspection of the data does not reveal any error in collecting observations 15 and 17, or any other reason to discard or modify these two points.

The residuals are plotted against $\hat{y}$ in Fig. 6-15, and against x_1 and x_2 in Figs. 6-16 and 6-17, respectively. The two largest residuals, e_{15} and e_{17}, are apparent. Figure 6-15 gives some indication that the model overpredicts the intermediate pull strength values

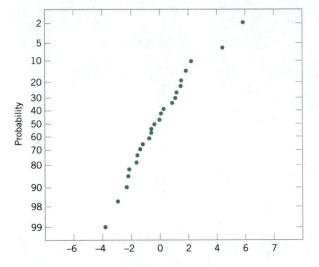

Figure 6-14   Normal probability plot of residuals.

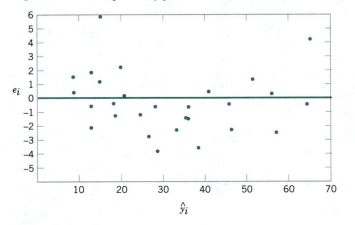

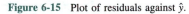

Figure 6-15   Plot of residuals against $\hat{y}$.

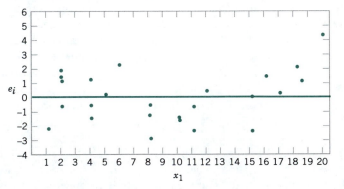

Figure 6-16   Plot of residuals against x_1.

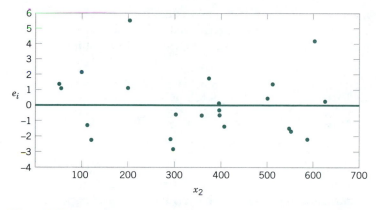

Figure 6-17 Plot of residuals against x_2.

and the lower and higher values of strength. Figure 6-16 suggests a nonlinear effect of wire length. The relationship between pull strength and wire length may not be linear (requiring that a term involving x_1^2, say, be added to the model), or other regressor variables not presently in the model affected the response.

In Example 6-12 we used the standardized residuals $d_i = e_i/\sqrt{\hat{\sigma}^2}$ as a measure of residual magnitude. Some analysts prefer to plot standardized residuals instead of ordinary residuals, because the standardized residuals are scaled so that their standard deviation is approximately unity. Consequently, large residuals (that may indicate possible outliers or unusual observations) will be more obvious from inspection of the residual plots.

Many regression computer programs compute other types of scaled residuals. One of the most popular is the **studentized residual,** defined as follows.

The studentized residual is

$$r_i = \frac{e_i}{\sqrt{\hat{\sigma}^2(1 - h_{ii})}} \qquad i = 1, 2, \ldots, n \qquad (6\text{-}51)$$

where h_{ii} is the ith diagonal element of the matrix

$$\mathbf{H} = \mathbf{X}(\mathbf{X}'\mathbf{X})^{-1}\mathbf{X}'$$

The **H** matrix is sometimes called the "hat" matrix, since

$$\begin{aligned}\hat{\mathbf{y}} &= \mathbf{X}\hat{\boldsymbol{\beta}} \\ &= \mathbf{X}(\mathbf{X}'\mathbf{X})^{-1}\mathbf{X}'\mathbf{y} \\ &= \mathbf{H}\mathbf{y}\end{aligned}$$

Thus $\mathbf{H}$ transforms the observed values of $\mathbf{y}$ into a vector of fitted values $\hat{\mathbf{y}}$.

Since each row of the matrix $\mathbf{X}$ corresponds to a vector, say $\mathbf{x}_i' = [1, x_{i1}, x_{i2}, \ldots, x_{ik}]$, then another way to write the diagonal elements of the hat matrix is

$$h_{ii} = \mathbf{x}_i'(\mathbf{X'X})^{-1}\mathbf{x}_i \tag{6-52}$$

Note that apart from σ^2, h_{ii} is the variance of the fitted value $\hat{y}_i$. The quantities h_{ii} were used in the computation of the confidence interval on the mean response in Section 6-5.2.

Under the usual assumptions that the model errors are independently distributed with mean zero and variance σ^2, we can show that the variance of the ith residual e_i is

$$V(e_i) = \sigma^2(1 - h_{ii}) \qquad i = 1, 2, \ldots, n$$

Furthermore, the h_{ii} elements must fall in the interval

$$0 < h_{ii} \le 1$$

This implies that the standardized residuals understate the true residual magnitude; thus, the studentized residuals would be a better statistic to examine in evaluating potential outliers.

6-7.2   Coefficient of Multiple Determination

The coefficient of multiple determination R^2 is defined as follows.

<u>R^2</u>

The coefficient of multiple determination is

$$R^2 = \frac{SS_R}{SS_T} = 1 - \frac{SS_E}{SS_T} \tag{6-53}$$

R^2 is a measure of the fraction of variability in the observations y obtained by the regression equation using the variables $x_1, x_2, \ldots, x_k$. Because of the analysis of variance identity, we must have $0 \leq R^2 \leq 1$. However, a large value of R^2 does not necessarily imply that the regression model is a good one. Adding a variable to the model will always increase R^2, regardless of whether or not the additional variable is statistically significant. Thus, models that have large values of R^2 may yield poor predictions of new observations or estimates of the mean response.

The positive square root of R^2 is called the **multiple correlation coefficient** between y and the set of regressor variables $x_1, x_2, \ldots, x_k$. That is, R is a measure of the linear association between y and $x_1, x_2, \ldots, x_k$. When $k = 1$, this becomes r_{xy}, the simple correlation between y and x mentioned in Section 6-1.

EXAMPLE 6-13

The coefficient of multiple determination for the regression model fit to the wire bond pull strength data in Example 6-2 is

$$R^2 = \frac{SS_R}{SS_T} = \frac{5990.8}{6105.9} = 0.981$$

That is, about 98.1% of the variability in pull strength y has been explained when the two regressors variables, wire length (x_1) and die height (x_2), are used. To illustrate how R^2 increases as variables are added to a regression model, suppose we consider a regression equation involving y (strength) and only x_1 (wire length). We can show that the value of R^2 for this model is $R^2 = 0.964$. Therefore, adding the variable x_2 to the model has increased R^2 from 0.964 to 0.981.

Because R^2 always increases when a regressor variable is added to a regression model, it is not always a good indicator of model adequacy. Many computer programs report an **adjusted R^2** statistic, defined as follows.

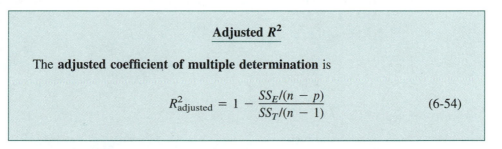

Adjusted R^2

The **adjusted coefficient of multiple determination** is

$$R_{adjusted}^2 = 1 - \frac{SS_E/(n - p)}{SS_T/(n - 1)} \qquad (6\text{-}54)$$

The adjusted R^2 is a better reflection of the proportion of variability explained by a regression model because it takes the number of regressor variables into account. In general, the adjusted R^2 statistic will not always increase when a variable is added to a model. $R_{adjusted}^2$ will increase only if adding the variable produces a large enough reduction in the residual sum of squares to compensate for the loss of one residual degree of freedom.

To illustrate these points, consider a regression model for the wire bond pull strength data with only one regressor, x_1 (wire length). The value of the residual sum of squares for this model is $SS_E = 220.09$. From equation 6-54 the adjusted R^2 is

$$R^2_{\text{adjusted}} = 1 - \frac{SS_E/(n - p)}{SS_T/(n - 1)}$$

$$= 1 - \frac{220.09/(25 - 2)}{6105.94/(25 - 1)}$$

$$= 0.9624$$

The adjusted R^2 for the wire bond pull strength model with both regressors x_1 (wire length) and x_2 (die height) is given in Table 6-6 as $R^2_{\text{adjusted}} = 97.9\%$. Since the adjusted R^2 has increased with the addition of the regressor x_2, we would conclude that adding this variable to the model is a good idea. The t-test on the variable x_2 in Table 6-6 confirms that x_2 is a useful regressor, since the P-value is small enough for us to conclude that $\beta_2 \neq 0$. Furthermore, for the simple linear regression model with x_1 the residual mean square is 9.57, whereas from Table 6-6 we see that for the model with both x_1 and x_2, the residual mean square is 5.2. When comparing regression models, the residual mean square is a useful measure, because the model with the smallest residual mean square explains more of the variability in y.

6-7.3 Influential Observations

When using multiple regression, we occasionally find that some subset of the observations is unusually influential. Sometimes these influential observations are relatively far away from the vicinity where the rest of the data were collected. A hypothetical situation for two variables is depicted in Fig. 6-18, where one observation in x-space is remote from the rest of the data. The disposition of points in the x-space is important in determining the properties of the model. For example, point (x_{i1}, x_{i2}) in Fig. 6-18 may be very influential

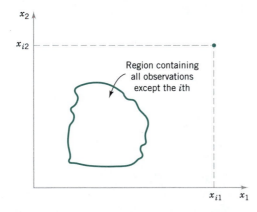

Figure 6-18   A point that is remote in x-space.

in determining R^2, the estimates of the regression coefficients, and the magnitude of the error mean square.

We would like to examine the influential points to determine whether they control many model properties. If these influential points are "bad" points, or erroneous in any way, then they should be eliminated. On the other hand, there may be nothing wrong with these points, but at least we would like to determine whether or not they produce results consistent with the rest of the data. In any event, even if an influential point is a valid one, if it controls important model properties, we would like to know this, since it could have an impact on the use of the model.

One useful method is to inspect the h_{ii}'s, the diagonal elements of the hat matrix. The value of h_{ii} can be interpreted as a measure of the distance of the point $\mathbf{x}_i = (x_{i1}, x_{i2}, \ldots x_{ik})$ from the average of all of the $\mathbf{x}$ points in the data set. The value of h_{ii} is not the usual distance measure but it has similar properties. Consequently, a large value for h_{ii} implies that $\mathbf{x}_i$ is remote from the center of the data (as in Figure 6-18). A rule of thumb is that h_{ii}'s greater than $2p/n$ should be investigated. An $\mathbf{x}_i$ that exceeds this value is considered a **leverage point.** Because it is remote, it has substantial leverage, or potential to change the regression analysis. With some matrix algebra, it can be shown that the average value of h_{ii} in any data set is p/n. Therefore, the rule flags values greater than two times the average. Notice that the h_{ii}'s are calculated exclusively from the predicator variables (the $\mathbf{x}_i$'s). The y's do not enter the calculation.

Montgomery and Peck (1992) and Myers (1990) describe several other methods for detecting influential observations. An excellent diagnostic is the distance measure developed by Dennis R. Cook. This is a measure of the squared distance between the usual least squares estimate $\hat{\boldsymbol{\beta}}$ based on all n observations and the estimate obtained when the ith point is removed—say, $\hat{\boldsymbol{\beta}}_{(i)}$.

Cook's distance measure is

$$D_i = \frac{(\hat{\boldsymbol{\beta}}_{(i)} - \hat{\boldsymbol{\beta}})\mathbf{X'X}(\hat{\boldsymbol{\beta}}_{(i)} - \hat{\boldsymbol{\beta}})}{p\hat{\sigma}^2} \qquad i = 1, 2, \ldots, n \qquad (6\text{-}55)$$

$$= \frac{r_i^2}{p} \frac{h_{ii}}{(1 - h_{ii})} \qquad i = 1, 2, \ldots, n$$

Clearly, if the ith point is influential, its removal will result in $\hat{\boldsymbol{\beta}}_{(i)}$ changing considerably from the value $\hat{\boldsymbol{\beta}}$. Thus, a large value of D_i implies that the ith point is influential. The statistic D_i is actually computed using equation 6-55. We see that D_i consists of the squared studentized residual, which reflects how well the model fits the ith observation y_i [recall that $r_i = e_i/\sqrt{\hat{\sigma}^2(1 - h_{ii})}$ in equation 6-51] and a component that measures how far that point is from the rest of the data [$h_{ii}/(1 - h_{ii})$ is a measure of the distance of the ith point from the centroid of the remaining $n - 1$ points]. A value of $D_i > 1$ would indicate that the point is influential. Either component of D_i (or both) may contribute to a large value.

EXAMPLE 6-14

Table 6-8 lists the values of the hat matrix diagonals h_{ii} and Cook's distance measure D_i for the wire bond pull strength data in Example 6-2. To illustrate the calculations, consider the first observation:

$$
\begin{aligned}
D_1 &= \frac{r_1^2}{p} \cdot \frac{h_{11}}{(1 - h_{11})} \\
&= \frac{\left[e_1/\sqrt{MS_E(1 - h_{11})}\right]^2}{p} \cdot \frac{h_{11}}{(1 - h_{11})} \\
&= \frac{\left[1.57/\sqrt{5.2352(1 - 0.1573)}\right]^2}{3} \cdot \frac{0.1573}{(1 - 0.1573)} \\
&= 0.035
\end{aligned}
$$

The Cook distance measure D_i does not identify any potentially influential observations in the data, for no value of D_i exceeds unity.

Table 6-8   Influence Diagnostics for the Pull Strength Data in Example 6-2

Observations i	h_{ii}	Cook's Distance Measure D_i
1	0.1573	0.035
2	0.1116	0.012
3	0.1419	0.060
4	0.1019	0.021
5	0.0418	0.024
6	0.0749	0.007
7	0.1181	0.036
8	0.1561	0.020
9	0.1280	0.160
10	0.0413	0.001
11	0.0925	0.013
12	0.0526	0.001
13	0.0820	0.001
14	0.1129	0.003
15	0.0737	0.187
16	0.0879	0.001
17	0.2593	0.565
18	0.2929	0.155
19	0.0962	0.018
20	0.1473	0.000
21	0.1296	0.052
22	0.1358	0.028
23	0.1824	0.002
24	0.1091	0.040
25	0.0729	0.000

EXERCISES FOR SECTION 6-7

6-32. Refer to the NFL team performance data in Exercise 6-4.

(a) Calculate R^2 and adjusted R^2 for this model and provide a practical interpretation of these quantities.

(b) Prepare a normal probability plot of the residuals from the least squares model. Does the normality assumption seem to be satisfied?

(c) Plot the residuals versus $\hat{y}$ and versus x. Interpret these graphs.

6-33. Refer to the data in Exercise 6-5 on thermal conductivity y and product density x.

(a) Find the residuals for the least squares model.

(b) Prepare a normal probability plot of the residuals and interpret this display.

(c) Plot the residuals versus $\hat{y}$ and versus x. Does the assumption of constant variance seem to be satisfied?

(d) What proportion of total variability is explained by the regression model?

6-34. Exercise 6-6 presents data on $y =$ steam usage and $x =$ average monthly temperature.

(a) What proportion of total variability is accounted for by the simple linear regression model?

(b) Prepare a normal probability plot of the residuals and interpret this graph.

(c) Plot residuals versus $\hat{y}$ and versus x. Do the regression assumptions appear to be satisfied?

6-35. Consider the wear data in Exercise 6-8.

(a) Find the values of R^2 and adjusted R^2 when the model uses the regressors x_1 and x_2.

(b) What happens to the values of R^2 and adjusted R^2 when an interaction term $x_1 x_2$ is added to the model? Does this necessarily imply that adding the interaction term is a good idea?

6-36. For the regression model for the power consumption data in Exercise 6-11:

(a) Plot the residuals versus $\hat{y}$ and versus the regressors used in the model. What information do these plots provide?

(b) Construct a normal probability plot of the residuals. Are there reasons to doubt the normality assumption for this model?

(c) Use the standardized residuals to determine whether there are any indications of influential observations in the data.

6-37. Consider the semiconductor HFE data in Exercise 6-10.

(a) Plot the residuals from this model versus $\hat{y}$. Comment on the information in this plot.

(b) What are the values of R^2 and adjusted R^2 for this model?

(c) Refit the model using log HFE as the response variable.

(d) Plot the residuals versus predicted log HFE for the model in part (c). Does this give any information about which model is preferable?

(e) Plot the residuals from the model in part (d) versus the regressor x_3. Comment on this plot.

(f) Refit the model to log HFE using x_1, x_2, and $1/x_3$, as the regressors. Comment on the effect of this change in the model.

SUPPLEMENTAL EXERCISES

6-38. An industrial engineer at a furniture manufacturing plant wishes to investigate how the plant's electricity usage depends on the amount of the plant's production. He suspects that there is a simple linear relationship between production measured as the value in million dollar units of furniture produced in that month (x) and the electrical usage in units of kWh (kilowatt-hours, y). The following data were collected:

Dollars	kWh
4.70	2.59
4.00	2.61
4.59	2.66
4.70	2.58
4.65	2.32
4.19	2.31
4.69	2.52
3.95	2.32
4.01	2.65
4.31	2.64
4.51	2.38
4.46	2.41
4.55	2.35
4.14	2.55
4.25	2.34

(a) Draw a scatter diagram of these data. Does a straight-line relationship seem plausible?
(b) Fit a simple linear regression model to these data.
(c) Test for significance of regression using $\alpha = 0.05$. What is the P-value for this test?
(d) Find a 95% confidence interval estimate on the slope.
(e) Test the hypothesis $H_0: \beta_1 = 0$ versus $H_1: \beta_1 \neq 0$ using $\alpha = 0.05$. What conclusion can you draw about the slope coefficient?
(f) Test the hypothesis $H_0: \beta_0 = 0$ versus $H_1: \beta_0 \neq 0$ using $\alpha = 0.05$. What conclusions can you draw about the best model?

6-39. Show that an equivalent way to define the test for significance of regression in simple linear regression is to base the test on R^2 as follows: To test $H_0: \beta_1 = 0$ versus $H_1: \beta_1 \neq 0$, calculate

$$F_0 = \frac{R^2(n-2)}{1 - R^2}$$

and to reject $H_0: \beta_1 = 0$ if the computed value $f_0 > f_{\alpha,1,n-2}$.

6-40. Suppose that the simple linear regression model has been fit to $n = 25$ observations and $R^2 = 0.90$.
(a) Test for significance of regression at $\alpha = 0.05$. Use the results of Exercise 6-39.
(b) What is the smallest value of R^2 that would lead to the conclusion of a significant regression if $\alpha = 0.05$?

6-41. **Studentized Residuals.** Show that in a simple linear regression model the variance of the ith residual is

$$V(e_i) = \sigma^2 \left[1 - \left(\frac{1}{n} + \frac{(x_i - \bar{x})^2}{S_{xx}} \right) \right]$$

Hint:

$$\text{cov}(Y_i, \hat{Y}_i) = \sigma^2 \left[\frac{1}{n} + \frac{(x_i - \bar{x})^2}{S_{xx}} \right]$$

The ith studentized residual for this model is defined as

$$r_i = \frac{e_i}{\sqrt{\hat{\sigma}^2 \left[1 - \left(\frac{1}{n} + \frac{(x_i - \bar{x})^2}{S_{xx}} \right) \right]}}$$

(a) Explain why r_i has unit standard deviation.
(b) Do the standardized residuals have unit standard deviation?
(c) Discuss the behavior of the standardized residual when the sample value x_i is very close to the middle of the range of x.
(d) Discuss the behavior of the studentized residual when the sample value x_i is very near one end of the range of x.

6-42. The following data are DC output from a windmill (y) and wind velocity (x).
(a) Draw a scatter diagram of these data. What type of relationship seems appropriate in relating y to x?
(b) Fit a simple linear regression model to these data.
(c) Test for significance of regression using $\alpha = 0.05$. What conclusions can you draw?

(d) Plot the residuals from the simple linear regression model versus $\hat{y}_i$ and versus wind velocity x. What do you conclude about model adequacy?

(e) Based on the analysis, propose another model relating y to x. Justify why this model seems reasonable.

(f) Fit the regression model you have proposed in part (e). Test for significance of regression (use $\alpha = 0.05$), and graphically analyze the residuals from this model. What can you conclude about model adequacy?

Observation Number	Wind Velocity (MPH), x_i	DC Output y_i
1	5.00	1.582
2	6.00	1.822
3	3.40	1.057
4	2.70	0.500
5	10.00	2.236
6	9.70	2.386
7	9.55	2.294
8	3.05	0.558
9	8.15	2.166
10	6.20	1.866
11	2.90	0.653
12	6.35	1.930
13	4.60	1.562
14	5.80	1.737
15	7.40	2.088
16	3.60	1.137
17	7.85	2.179
18	8.80	2.112
19	7.00	1.800
20	5.45	1.501
21	9.10	2.303
22	10.20	2.310
23	4.10	1.194
24	3.95	1.144
25	2.45	0.123

6-43. The diagonal elements of the hat matrix are often used to denote **leverage**—that is, a point that is unusual in its location in the x-space and that may be influential. Generally, the ith point is called a **leverage point** if its hat matrix diagonal h_{ii} exceeds $2p/n$, which is twice the aver-

age size of all the hat diagonals. Recall that $p = k + 1$.

(a) Table 6-8 contains the hat matrix diagonal for the wire bond pull strength data used in Example 6-2. Find the average size of these elements.

(b) Based on the criterion given, are there any observations that are leverage points in the data set?

6-44. The data shown in Table 6-9 represent the thrust of a jet-turbine engine (y) and six candidate regressors: $x_1 =$ primary speed of rotation, $x_2 =$ secondary speed of rotation, $x_3 =$ fuel flow rate, $x_4 =$ pressure, $x_5 =$ exhaust temperature, and $x_6 =$ ambient temperature at time of test.

(a) Fit a multiple linear regression model using $x_3 =$ fuel flow rate, $x_4 =$ pressure, and $x_5 =$ exhaust temperature as the regressors.

(b) Test for significance of regression using $\alpha = 0.01$. Find the P-value for this test. What are your conclusions?

(c) Find the t-test statistic for each regressor. Using $\alpha = 0.01$, explain carefully the conclusion you can draw from these statistics.

(d) Find R^2 and the adjusted statistic for this model. Comment on the meaning of each value and its usefulness in assessing the model.

(e) Construct a normal probability plot of the residuals and interpret this graph.

(f) Plot the residuals versus $\hat{y}$. Are there any indications of inequality of variance or nonlinearity?

(g) Plot the residuals versus x_3. Is there any indication of nonlinearity?

(h) Predict the thrust for an engine for which $x_3 = 1670$, $x_4 = 170$, and $x_5 = 1589$.

6-45. Consider the engine thrust data in Exercise 6-44. Refit the model using $y^* = \ln y$ as the response variable and $x_3^* = \ln x_3$ as the regressor (along with x_4 and x_5).

Table 6-9   Data for Exercise 6-44

Observation Number	y	x_1	x_2	x_3	x_4	x_5	x_6
1	4540	2140	20640	30250	205	1732	99
2	4315	2016	20280	30010	195	1697	100
3	4095	1905	19860	29780	184	1662	97
4	3650	1675	18980	29330	164	1598	97
5	3200	1474	18100	28960	144	1541	97
6	4833	2239	20740	30083	216	1709	87
7	4617	2120	20305	29831	206	1669	87
8	4340	1990	19961	29604	196	1640	87
9	3820	1702	18916	29088	171	1572	85
10	3368	1487	18012	28675	149	1522	85
11	4445	2107	20520	30120	195	1740	101
12	4188	1973	20130	29920	190	1711	100
13	3981	1864	19780	29720	180	1682	100
14	3622	1674	19020	29370	161	1630	100
15	3125	1440	18030	28940	139	1572	101
16	4560	2165	20680	30160	208	1704	98
17	4340	2048	20340	29960	199	1679	96
18	4115	1916	19860	29710	187	1642	94
19	3630	1658	18950	29250	164	1576	94
20	3210	1489	18700	28890	145	1528	94
21	4330	2062	20500	30190	193	1748	101
22	4119	1929	20050	29960	183	1713	100
23	3891	1815	19680	29770	173	1684	100
24	3467	1595	18890	29360	153	1624	99
25	3045	1400	17870	28960	134	1569	100
26	4411	2047	20540	30160	193	1746	99
27	4203	1935	20160	29940	184	1714	99
28	3968	1807	19750	29760	173	1679	99
29	3531	1591	18890	29350	153	1621	99
30	3074	1388	17870	28910	133	1561	99
31	4350	2071	20460	30180	198	1729	102
32	4128	1944	20010	29940	186	1692	101
33	3940	1831	19640	29750	178	1667	101
34	3480	1612	18710	29360	156	1609	101
35	3064	1410	17780	28900	136	1552	101
36	4402	2066	20520	30170	197	1758	100
37	4180	1954	20150	29950	188	1729	99
38	3973	1835	19750	29740	178	1690	99
39	3530	1616	18850	29320	156	1616	99
40	3080	1407	17910	28910	137	1569	100

(a) Test for significance of regression using $\alpha = 0.01$. Find the P-value for this test and state your conclusions.

(b) Use the t-statistic to test $H_0: \beta_j = 0$ versus $H_1: \beta_j \neq 0$ for each variable in the model. If $\alpha = 0.01$, what conclusions can you draw?

(c) Plot the residuals versus $\hat{y}^*$ and versus x_3^*. Comment on these plots. How do they compare with their counterparts obtained in Exercise 6-44 parts (f) and (g)?

6-46. Following are data on y = green liquor (g/l) and x = paper machine speed (ft/min) from a Kraft paper machine. (The data were read from a graph in an article in the *Tappi Journal,* March 1986.)

y	16.0	15.8	15.6	15.5	14.8
x	1700	1720	1730	1740	1750

y	14.0	13.5	13.0	12.0	11.0
x	1760	1770	1780	1790	1795

(a) Fit the model $Y = \beta_0 + \beta_1 x + \beta_2 x^2 + \epsilon$ using least squares.

(b) Test for significance of regression using $\alpha = 0.05$. What are your conclusions?

(c) Test the contribution of the quadratic term to the model, over the contribution of the linear term, using a *t*-test. If $\alpha = 0.05$, what conclusion can you draw?

(d) Plot the residuals from the model in part (a) versus $\hat{y}$. Does the plot reveal any inadequacies?

(e) Construct a normal probability plot of the residuals. Comment on the normality assumption.

6-47. An article in the *Journal of Environmental Engineering* (Vol. 115, No. 3, 1989, pp. 608–619) reported the results of a study on the occurrence of sodium and chloride in surface streams in central Rhode Island. The data shown are chloride concentration y (in mg/l) and roadway area in the watershed x (in %).

y	4.4	6.6	9.7	10.6	10.8
x	0.19	0.15	0.57	0.70	0.67

y	10.9	11.8	12.1	14.3	14.7
x	0.63	0.47	0.70	0.60	0.78

y	15.0	17.3	19.2	23.1	27.4
x	0.81	0.78	0.69	1.30	1.05

y	27.7	31.8	39.5
x	1.06	1.74	1.62

(a) Draw a scatter diagram of the data. Does a simple linear regression model seem appropriate here?

(b) Fit the simple linear regression model using the method of least squares.

(c) Estimate the mean chloride concentration for a watershed that has 1% roadway area.

(d) Find the fitted value corresponding to $x = 0.47$ and the associated residual.

(e) Suppose we wish to fit a regression model for which the true regression line passes through the point (0, 0). The appropriate model is $Y = \beta x + \epsilon$. Assume that we have n pairs of data $(x_1, y_1), (x_2, y_2), \ldots, (x_n, y_n)$. Show that the least squares estimate of β is $\Sigma y_i x_i / \Sigma x_i^2$.

(f) Use the results of part (e) to fit the model $Y = \beta x + \epsilon$ to the chloride concentration–roadway area data in this exercise. Plot the fitted model on a scatter diagram of the data and comment on the appropriateness of the model.

6-48. Consider the no-intercept model $Y = \beta x + \epsilon$ with the ϵ's NID(0, σ^2). The estimate of σ^2 is $s^2 = \Sigma_{i=1}^{n}(y_i - \hat{\beta}x_i)^2 / (n - 1)$ and $V(\hat{\beta}) = \sigma^2 / \Sigma_{i=1}^{n} x_i^2$.

(a) Devise a test statistic for H_0: $\beta = 0$ versus H_1: $\beta \neq 0$.

(b) Apply the test in (a) to the model from Exercise 6-47 part (f).

6-49. A rocket motor is manufactured by bonding together two types of propellants, an igniter and a sustainer. The shear strength of the bond y is thought to be a linear function of the age of the propellant x when the motor is cast. Twenty observations are shown in the following table.

(a) Draw a scatter diagram of the data. Does the straight-line regression model seem to be plausible?

(b) Find the least squares estimates of the slope and intercept in the simple linear regression model.

(c) Estimate the mean shear strength of a motor made from propellant that is 20 weeks old.

(d) Obtain the fitted values $\hat{y}_i$ that correspond to each observed value y_i. Plot $\hat{y}_i$ versus y_i, and comment on what this plot would look like if the linear relationship between shear strength and age were perfectly deterministic (no error). Does this plot indicate that age is a reasonable choice of regressor variable in this model?

Observation Number	Strength y (psi)	Age x (weeks)
1	2158.70	15.50
2	1678.15	23.75
3	2316.00	8.00
4	2061.30	17.00
5	2207.50	5.00
6	1708.30	19.00
7	1784.70	24.00
8	2575.00	2.50
9	2357.90	7.50
10	2277.70	11.00
11	2165.20	13.00
12	2399.55	3.75
13	1779.80	25.00
14	2336.75	9.75
15	1765.30	22.00
16	2053.50	18.00
17	2414.40	6.00
18	2200.50	12.50
19	2654.20	2.00
20	1753.70	21.50

6-50. Consider the simple linear regression model $Y = \beta_0 + \beta_1 x + \epsilon$. Suppose that the analyst wants to use $z = x - \bar{x}$ as the regressor variable.

(a) Using the data in Exercise 6-49, construct one scatter plot of the (x_i, y_i) points and then another of the $(z_i = x_i - \bar{x}, y_i)$ points. Use the two plots to explain intuitively how the two models, $Y = \beta_0 + \beta_1 x + \epsilon$ and $Y = \beta_0^* + \beta_1^* z + \epsilon$, are related.

(b) Find the least squares estimates of β_0^* and β_1^* in the model $Y = \beta_0^* + \beta_1^* z + \epsilon$. How do they relate to the least squares estimates $\hat{\beta}_0$ and $\hat{\beta}_1$?

6-51. Suppose that each value of x_i is multiplied by a positive constant a, and each value of y_i is multiplied by another positive constant b. Show that the t-sta-

tistic for testing $H_0: \beta_1 = 0$ versus $H_1: \beta_1 \neq 0$ is unchanged in value.

6-52. Recall that the hypothesis test $H_0: \beta_1 = 0.0$ versus $H_1: \beta_1 \neq 0.0$ was performed on the regression coefficient β_1 of the simple regression model of the National Football League data in Exercise 6-4.

(a) Propose a new test statistic to test $H_0: \beta_1 = -0.01$ versus $H_1: \beta_1 \neq -0.01$. Carry out the hypothesis test using $\alpha = 0.01$. Also, find the P-value for this test.

(b) Would you agree with the statement that this is a test of the claim that if you can decrease the opponent's rushing yardage by 100 yards the team will win one more game?

6-53. Similarly to Exercise 6-52 part (a), test the hypothesis $H_0: \beta_1 = 10$ versus $H_1: \beta_1 \neq 10$ (using $\alpha = 0.01$) for the steam usage data in Exercise 6-6 using a new test statistic. Also, find the P-value for this test.

Team Exercise

6-54. Identify a situation in which two or more variables of interest may be related but the mechanistic model relating the variables is unknown. Collect a random sample of data for these variables and perform the following analyses.

(a) Build a simple or multiple linear regression model.

(b) Test for significance of the regression model.

(c) Test for significance of the individual regression coefficients.

(d) Construct confidence intervals on the individual regression coefficients.

(e) Construct a confidence interval on the mean response.

(f) Select two values for the regressor variables and use the model to make predictions.

(g) Perform a residual analysis and compute the coefficient of multiple determination.

7

Design
of Engineering
Experiments

CHAPTER OUTLINE

7-1 THE STRATEGY OF EXPERIMENTATION

Recall from Chapter 1 that engineers conduct tests or **experiments** as a natural part of their work. Statistically based experimental design techniques are particularly useful in the engineering world for improving the performance of a manufacturing process. They also have extensive application in the development of new processes. Most processes can be described in terms of several **controllable variables,** such as temperature, pressure, and feed rate. By using designed experiments, engineers can determine which subset of the process variables have the most influence on process performance. The results of such an experiment can lead to

1. Improved process yield
2. Reduced variability in the process and closer conformance to nominal or target requirements
3. Reduced design and development time
4. Reduced cost of operation

Experimental design methods are also useful in **engineering design** activities, where new products are developed and existing ones improved. Some typical applications of statistically designed experiments in engineering design include

1. Evaluation and comparison of basic design configurations
2. Evaluation of different materials
3. Selection of design parameters so that the product will work well under a wide variety of field conditions (or so that the design will be robust)
4. Determination of key product design parameters that affect product performance

The use of experimental design in the engineering design process can result in products that are easier to manufacture, products that have better field performance and reliability than their competitors, and products that can be designed, developed, and produced in less time.

Designed experiments are usually employed **sequentially.** That is, the first experiment with a complex system (perhaps a manufacturing process) that has many controllable variables is often a **screening experiment** designed to determine which variables are most important. Subsequent experiments are then used to refine this information and determine which adjustments to these critical variables are required to improve the process. Finally, the objective of the experimenter is optimization—that is, to determine which levels of the critical variables result in the best process performance. This is the KISS principle: "keep it small and sequential." When small steps are completed, the knowledge gained can improve the subsequent experiments.

Every experiment involves a sequence of activities:

1. **Conjecture**—the original hypothesis that motivates the experiment.
2. **Experiment**—the test performed to investigate the conjecture.
3. **Analysis**—the statistical analysis of the data from the experiment.
4. **Conclusion**—what has been learned about the original conjecture from the experiment will lead to a revised conjecture, a new experiment, and so forth.

Statistical methods are essential to good experimentation. All experiments are designed experiments; some of them are poorly designed, and as a result, valuable resources are used ineffectively. Statistically designed experiments permit efficiency and economy in the experimental process, and the use of statistical methods in examining the data results in **scientific objectivity** when drawing conclusions.

In this chapter we focus on experiments that include two or more factors that the experimenter thinks may be important. The **factorial experimental design** will be introduced as a powerful technique for this type of problem. Generally, in a factorial experimental design, experimental trials (or runs) are performed at all combinations of factor levels. For example, if a chemical engineer is interested in investigating the effects of reaction time and reaction temperature on the yield of a process, and if two levels of time (1 hr and 1.5 hrs) and two levels of temperature (125°F and 150°F) are considered important, then a factorial experiment would consist of making the experimental runs at each of the four possible combinations of these levels of reaction time and reaction temperature.

Most of the statistical concepts introduced previously can be extended to the factorial experiments of this chapter. We will also introduce several graphical methods that are useful in analyzing the data from designed experiments.

7-2 SOME APPLICATIONS OF EXPERIMENTAL DESIGN TECHNIQUES

Experimental design is an extremely important tool for engineers and scientists who are interested in improving the performance of a manufacturing process. It also has extensive application in the development of new processes and in new product design. We now give some examples.

EXAMPLE 7-1

A Process Characterization Experiment

A team of development engineers is working on a new process for soldering electronic components to printed circuit boards. Specifically, the team is working with a new type of flow solder machine that should reduce the number of defective solder joints. (A flow solder machine preheats printed circuit boards and then moves them into contact with a wave of liquid solder. This machine makes all the electrical and most of the mechanical connections of the components to the printed circuit board. Solder defects require touchup or rework, which adds cost and often damages the boards.) The process will have several (perhaps many) variables, and not all of them are equally important. The initial list of candidate variables to be included in the experiment is constructed by combining the knowledge and information about the process from all team members. In this example, the engineers conducted a *brainstorming* session and invited manufacturing personnel with experience using various types of flow soldering equipment to participate. The team determined that the flow solder machine has several variables that can be controlled. They are

1. Solder temperature
2. Preheat temperature

   3. Conveyor speed

   4. Flux type

   5. Flux-specific gravity

   6. Solder wave depth

   7. Conveyor angle

In addition to these controllable factors, there are several other factors that cannot be easily controlled, once the machine enters routine manufacturing, including

   1. Thickness of the printed circuit board

   2. Types of components used on the board

   3. Layout of the components on the board

   4. Operator

   5. Environmental factors

   6. Production rate

Sometimes we call the uncontrollable factors *noise* factors. A schematic representation of the process is shown in Fig. 7-1.

In this situation the engineer is interested in **characterizing** the flow solder machine; that is, he is interested in determining which factors (both controllable and uncontrollable) affect the occurrence of defects on the printed circuit boards. To determine these factors, he can design an experiment that will enable him to estimate the magnitude and direction of the factor effects. Sometimes we call such an experiment a **screening experiment.** The information from this characterization study or screening experiment can help determine the critical process variables, as well as the direction of adjustment for these factors in order to reduce the number of defects, and assist in determining which process variables should be carefully controlled during manufacturing to prevent high defect levels and erratic process performance.

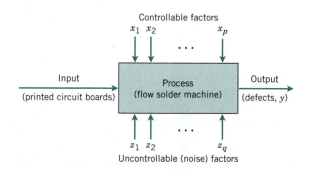

Figure 7-1   The flow solder experiment.

EXAMPLE 7-2

An Optimization Experiment

In a characterization experiment, we are interested in determining which factors affect the response. A logical next step is to determine the region in the important factors that leads to an **optimum response.** For example, if the response is cost, we will look for a region of minimum cost.

As an illustration, suppose that the yield of a chemical process is influenced by the operating temperature and the reaction time. We are currently operating the process at 155°F and 1.7 hours of reaction time, and the current process yield is around 75%. Figure 7-2 shows a view of the time–temperature space from above. In this graph we have connected points of constant yield with lines. These curves are yield **contours,** and we have shown the contours at 60, 70, 80, 90, and 95% yield. To locate the optimum, we might begin with a factorial experiment such as we described in Example 7-1, with the two factors time and temperature run at two levels each at 10°F and 0.5 hours above and below the current operating conditions. This two-factor factorial design is shown in Fig. 7-2. The average responses observed at the four points in the experiment (145°F, 1.2 hrs; 145°F, 2.2 hrs; 165°F, 1.2 hrs; and 165°F, 2.2 hrs) indicate that we should move in

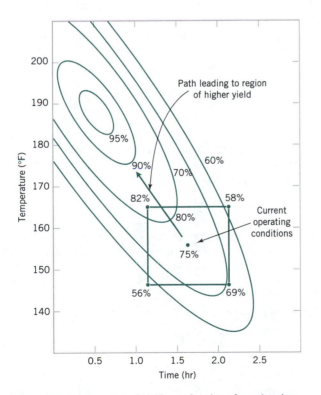

Figure 7-2 Contour plot of yield as a function of reaction time and reaction temperature, illustrating an optimization experiment.

the general direction of increased temperature and lower reaction time to increase yield. A few additional runs could be performed in this direction to locate the region of maximum yield.

EXAMPLE 7-3

A Product Design Example

We can also use experimental design in the development of new products. For example, suppose that a group of engineers are designing a door hinge for an automobile. The product characteristic is the check effort, or the holding ability of the latch that prevents the door from swinging closed when the vehicle is parked on the hill. The check mechanism consists of a leaf spring and a roller. When the door is opened, the roller travels through an arc causing the leaf spring to be compressed. To close the door, the spring must be forced aside, and this creates the check effort. The engineering team thinks that check effort is a function of the following factors:

1. Roller travel distance
2. Spring height from pivot to base
3. Horizontal distance from pivot to spring
4. Free height of the reinforcement spring
5. Free height of the main spring

The engineers can build a prototype hinge mechanism in which all these factors can be varied over certain ranges. Once appropriate levels for these five factors have been identified, an experiment can be designed consisting of various combinations of the factor levels, and the prototype can be tested at these combinations. This will produce information concerning which factors are most influential on the latch check effort, and through analysis of this information, the latch design can be improved.

These examples illustrate only three applications of experimental design methods. In the engineering environment, experimental design applications are numerous. Some potential areas of use are

1. Process troubleshooting
2. Process development and optimization
3. Evaluation of material and alternatives
4. Reliability and life testing
5. Performance testing
6. Product design configuration
7. Component tolerance determination

Experimental design methods allow these problems to be solved efficiently during the early stages of the product cycle. This has the potential to dramatically lower overall product cost and reduce development lead time.

7-3 FACTORIAL EXPERIMENTS

When several factors are of interest in an experiment, a factorial experiment should be used. As noted previously, in these experiments factors are varied together.

> By a **factorial experiment** we mean that in each complete replicate of the experiment all possible combinations of the levels of the factors are investigated.

Thus, if there are two factors A and B with a levels of factor A and b levels of factor B, then each replicate contains all ab treatment combinations.

The effect of a factor is defined as the change in response produced by a change in the level of the factor. It is called a **main effect** because it refers to the primary factors in the study. For example, consider the data in Table 7-1. This is a factorial experiment with two factors, A and B, each at two levels (A_{low}, A_{high}, and B_{low}, B_{high}). The main effect of factor A is the difference between the average response at the high level of A and the average response at the low level of A, or

$$A = \frac{30 + 40}{2} - \frac{10 + 20}{2} = 20$$

That is, changing factor A from the low level to the high level causes an average response increase of 20 units. Similarly, the main effect of B is

$$B = \frac{20 + 40}{2} - \frac{10 + 30}{2} = 10$$

Table 7-1 A Factorial Experiment without Interaction

	Factor B	
Factor A	B_{low}	B_{high}
A_{low}	10	20
A_{high}	30	40

In some experiments, the difference in response between the levels of one factor is not the same at all levels of the other factors. When this occurs, there is an **interaction** between the factors. For example, consider the data in Table 7-2. At the low level of factor B, the A effect is

$$A = 30 - 10 = 20$$

and at the high level of factor B, the A effect is

$$A = 0 - 20 = -20$$

Since the effect of A depends on the level chosen for factor B, there is interaction between A and B.

When an interaction is large, the corresponding main effects have very little practical meaning. For example, by using the data in Table 7-2, we find the main effect of A as

$$A = \frac{30 + 0}{2} - \frac{10 + 20}{2} = 0$$

and we would be tempted to conclude that there is no factor A effect. However, when we examined the effects of A at *different levels of factor B*, we saw that this was not the case. The effect of factor A depends on the levels of factor B. Thus, knowledge of the AB interaction is more useful than knowledge of the main effect. A significant interaction can mask the significance of main effects. Consequently, when interaction is present, the main effects of the factors involved in the interaction may not have much meaning.

It is easy to estimate the interaction effect in factorial experiments such as those illustrated in Tables 7-1 and 7-2. In this type of experiment, when both factors have two levels, the AB interaction effect is the difference in the diagonal averages. This represents one-half the difference between the A effects at the two levels of B. For example, in Table 7-1, we find the AB interaction effect to be

$$AB = \frac{20 + 30}{2} - \frac{10 + 40}{2} = 0$$

Thus, there is no interaction between A and B. In Table 7-2, the AB interaction effect is

$$AB = \frac{20 + 30}{2} - \frac{10 + 0}{2} = 20$$

As we noted before, the interaction effect in these data is very large.

Table 7-2 A Factorial Experiment with Interaction

Factor A	Factor B	
	B_{low}	B_{high}
A_{low}	10	20
A_{high}	30	0

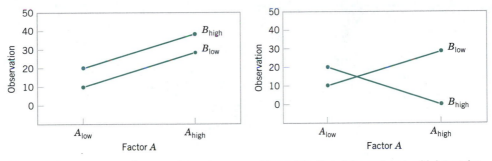

Figure 7-3 Factorial experiment, no interaction.   **Figure 7-4** Factorial experiment, with interaction.

The concept of interaction can be illustrated graphically in several ways. Figure 7-3 is a plot of the data in Table 7-1 against the levels of A for both levels of B. Note that the B_{low} and B_{high} lines are approximately parallel, indicating that factors A and B do not interact significantly. Figure 7-4 presents a similar plot for the data in Table 7-2. In this graph, the B_{low} and B_{high} lines are not parallel, indicating the interaction between factors A and B. Such graphical displays are called **two-factor interaction plots.** They are often useful in presenting the results of experiments, and many computer software programs used for analyzing data from designed experiments will construct these graphs automatically.

Figures 7-5 and 7-6 present another graphical illustration of the data from Tables 7-1 and 7-2. In Fig. 7-5 we have shown a **three-dimensional surface plot** of the data from Table 7-1, where the low and high levels are set at -1 and 1, respectively, for both A and B. The equations for these surfaces are discussed later in the chapter. These data contain no interaction, and the surface plot is a plane lying above the A–B space. The slope of the plane in the A and B directions is proportional to the main effects of factors

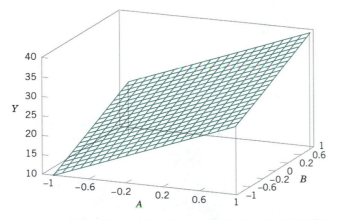

Figure 7-5 Three-dimensional surface plot for the data from Table 7-1, showing main effects of the two factors A and B.

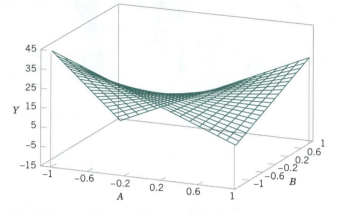

Figure 7-6   Three-dimensional surface plot for the data from Table 7-2, showing the effect of the *A* and *B* interaction.

A and *B,* respectively. Figure 7-6 is a surface plot for the data from Table 7-2. Note that the effect of the interaction in these data is to "twist" the plane, so that there is curvature in the response function. **Factorial experiments are the only way to discover interactions between variables.**

An alternative to the factorial design that is (unfortunately) used in practice is to change the factors *one at a time* rather than to vary them simultaneously. To illustrate this one-factor-at-a-time procedure, consider the optimization experiment described in Example 7-2. The engineer is interested in finding the values of temperature and pressure that maximize yield. Suppose that we fix temperature at 155°F (the current operating level) and perform five runs at different levels of time—say, 0.5 hour, 1.0 hour, 1.5 hours, 2.0 hours, and 2.5 hours. The results of this series of runs are shown in Fig. 7-7. This figure indicates that maximum yield is achieved at about 1.7 hours of reaction time. To optimize temperature, the engineer then fixes time at 1.7 hours (the apparent optimum)

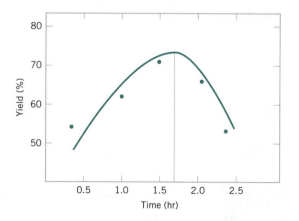

Figure 7-7   Yield versus reaction time with temperature constant at 155°F.

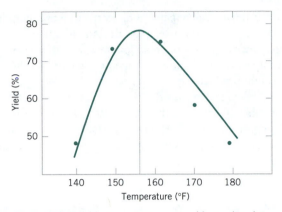

Figure 7-8 Yield versus temperature with reaction time constant at 1.7 hours.

and performs five runs at different temperatures—say, 140°F, 150°F, 160°F, 170°F, and 180°F. The results of this set of runs are plotted in Fig. 7-8. Maximum yield occurs at about 155°F. Therefore, we would conclude that running the process at 155°F and 1.7 hours is the best set of operating conditions, resulting in yields of around 75%.

Figure 7-9 displays the contour plot of yield as a function of temperature and time with the one-factor-at-a-time experiments superimposed on the contours. Clearly, this

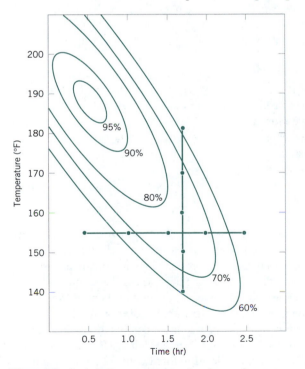

Figure 7-9 Optimization experiment using the one-factor-at-a-time method.

one-factor-at-a-time approach has failed dramatically here, as the true optimum is at least 20 yield points higher and occurs at much lower reaction times and higher temperatures. The failure to discover the importance of the shorter reaction times is particularly important as this could have significant impact on production volume or capacity, production planning, manufacturing cost, and total productivity.

The one-factor-at-a-time approach has failed here because it cannot detect the interaction between temperature and time. Factorial experiments are the only way to detect interactions. Furthermore, the one-factor-at-a-time method is inefficient. It will require more experimentation than a factorial, and as we have just seen, there is no assurance that it will produce the correct results.

7-4 2^k FACTORIAL DESIGN

Factorial designs are frequently used in experiments involving several factors where it is necessary to study the joint effect of the factors on a response. However, several special cases of the general factorial design are important because they are widely employed in research work and because they form the basis of other designs of considerable practical value.

The most important of these special cases is that of k factors, each at only two levels. These levels may be quantitative, such as two values of temperature, pressure, or time; or they may be qualitative, such as two machines, two operators, the "high" and "low" levels of a factor, or perhaps the presence and absence of a factor. A complete replicate of such a design requires $2 \times 2 \times \cdots \times 2 = 2^k$ observations and is called a **2^k factorial design.**

The 2^k design is particularly useful in the early stages of experimental work, when many factors are likely to be investigated. It provides the smallest number of runs for which k factors can be studied in a complete factorial design. Because there are only two levels for each factor, we must assume that the response is approximately linear over the range of the factor levels chosen. The 2^k design is a basic building block that is used to begin the study of a system.

7-4.1 2^2 Example

The simplest type of 2^k design is the 2^2—that is, two factors A and B, each at two levels. We usually think of these levels as the low and high levels of the factor. The 2^2 design is shown in Fig. 7-10. Note that the design can be represented geometrically as a square with the 2^2 = 4 runs, or treatment combinations, forming the corners of the square. In the 2^2 design it is customary to denote the low and high levels of the factors A and B by the signs − and +, respectively. This is sometimes called the **geometric notation** for the design.

A special notation is used to label the treatment combinations. In general, a treatment combination is represented by a series of lowercase letters. If a letter is present, then the corresponding factor is run at the high level in that treatment combination; if it is absent, the factor is run at its low level. For example, treatment combination a indicates that

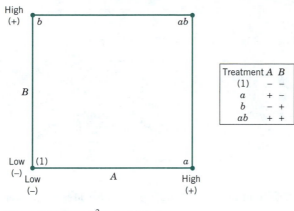

Figure 7-10 The 2^2 factorial design.

factor A is at the high level and factor B is at the low level. The treatment combination with both factors at the low level is represented by (1). This notation is used throughout the 2^k design series. For example, the treatment combination in a 2^4 with A and C at the high level and B and D at the low level is denoted by ac.

The effects of interest in the 2^2 design are the main effects A and B and the two-factor interaction AB. Let the letters (1), a, b, and ab also represent the totals of all n observations taken at each of these design points. It is easy to estimate the effects of these factors. To estimate the main effect of A, we would average the observations on the right side of the square in Fig. 7-10 where A is at the high level, and subtract from this the average of the observations on the left side of the square, where A is at the low level, or

---

Main Effect of A

$$A = \bar{y}_{A+} - \bar{y}_{A-} = \frac{a + ab}{2n} - \frac{b + (1)}{2n} = \frac{1}{2n}[a + ab - b - (1)] \quad (7\text{-}1)$$

Similarly, the main effect of B is found by averaging the observations on the top of the square, where B is at the high level, and subtracting the average of the observations on the bottom of the square, where B is at the low level:

Main Effect of B

$$B = \bar{y}_{B+} - \bar{y}_{B-} = \frac{b + ab}{2n} - \frac{a + (1)}{2n} = \frac{1}{2n}[b + ab - a - (1)] \quad (7\text{-}2)$$

Finally, the *AB* interaction is estimated by taking the difference in the diagonal averages in Fig. 7-10, or

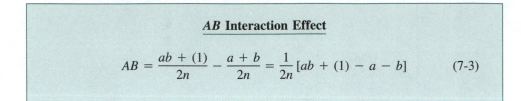

AB Interaction Effect

$$AB = \frac{ab + (1)}{2n} - \frac{a + b}{2n} = \frac{1}{2n}[ab + (1) - a - b] \qquad (7\text{-}3)$$

The quantities in brackets in equations 7-1, 7-2, and 7-3 are called **contrasts.** For example, the *A* contrast is

$$\text{Contrast}_A = a + ab - b - (1)$$

In these equations, the contrast coefficients are always either $+1$ or -1. A table of plus and minus signs, such as Table 7-3, can be used to determine the sign on each treatment combination for a particular contrast. The column headings for Table 7-3 are the main effects *A* and *B*, the *AB* interaction, and *I,* which represents the total. The row headings are the treatment combinations. Note that the signs in the *AB* column are the product of signs from columns *A* and *B*. To generate a contrast from this table, multiply the signs in the appropriate column of Table 7-3 by the treatment combinations listed in the rows and add. For example, contrast$_{AB}$ = $[(1)] + [-a] + [-b] + [ab] = ab + (1) - a - b$.

EXAMPLE 7-4

An article in the *AT&T Technical Journal* (Vol. 65, March/April 1986, pp. 39–50) describes the application of two-level factorial designs to integrated circuit manufacturing. A basic processing step in this industry is to grow an epitaxial layer on polished silicon wafers. The wafers are mounted on a susceptor and positioned inside a bell jar. Chemical vapors are introduced through nozzles near the top of the jar. The susceptor is rotated, and heat is applied. These conditions are maintained until the epitaxial layer is thick enough.

Table 7-3 Signs for Effects in the 2^2 Design

Treatment Combination	Factorial Effect			
	I	*A*	*B*	*AB*
(1)	+	−	−	+
a	+	+	−	−
b	+	−	+	−
ab	+	+	+	+

Table 7-4   The 2^2 Design for the Epitaxial Process Experiment

Treatment Combination	Factorial Effect			Thickness (μm)				Thickness (μm)		
	A	B	AB					Total	Average	Variance
(1)	−	−	+	14.037	14.165	13.972	13.907	56.081	14.020	0.0121
a	+	−	−	14.821	14.757	14.843	14.878	59.299	14.825	0.0026
b	−	+	−	13.880	13.860	14.032	13.914	55.686	13.922	0.0059
ab	+	+	+	14.888	14.921	14.415	14.932	59.156	14.789	0.0625

Table 7-4 presents the results of a 2^2 factorial design with $n = 4$ replicates using the factors A = deposition time and B = arsenic flow rate. The two levels of deposition time are $-$ = short and $+$ = long, and the two levels of arsenic flow rate are $-$ = 55% and $+$ = 59%. The response variable is epitaxial layer thickness (μm). We may find the estimates of the effects using equations 7-1, 7-2, and 7-3 as follows:

$$A = \frac{1}{2n}[a + ab - b - (1)]$$

$$= \frac{1}{2(4)}[59.299 + 59.156 - 55.686 - 56.081] = 0.836$$

$$B = \frac{1}{2n}[b + ab - a - (1)]$$

$$= \frac{1}{2(4)}[55.686 + 59.156 - 59.299 - 56.081] = -0.067$$

$$AB = \frac{1}{2n}[ab + (1) - a - b]$$

$$= \frac{1}{2(4)}[59.156 + 56.081 - 59.299 - 55.686] = 0.032$$

The numerical estimates of the effects indicate that the effect of deposition time is large and has a positive direction (increasing deposition time increases thickness), since changing deposition time from low to high changes the mean epitaxial layer thickness by 0.836 μm. The effects of arsenic flow rate (B) and the AB interaction appear small.

7-4.2   Statistical Analysis

We present two related methods for determining which effects are significantly different from zero. In the first method, the magnitude of an effect is compared to its estimated standard error. In the second method, a regression model is used in which each effect is associated with a regression coefficient. Then the regression results developed in Chapter

6 can be used to conduct the analysis. The two methods produce identical results for two-level designs. One might choose the method that is easiest to interpret or the one that is used by the available computer software. A third method that uses normal probability plots is discussed later in this chapter.

Standard Errors of the Effects

The magnitude of the effects in Example 7-4 can be judged by comparing each effect to its estimated standard error. In a 2^k design with n replicates, there are a total of $N = n2^k$ measurements. An effect estimate is the difference between two means and each mean is calculated from half the measurements. Consequently, the variance of an effect estimate is

$$V(\text{Effect}) = \frac{\sigma^2}{N/2} + \frac{\sigma^2}{N/2} = \frac{2\sigma^2}{N/2} = \frac{\sigma^2}{n2^{k-2}} \tag{7-4}$$

To obtain the estimated standard error of an effect, replace σ^2 by an estimate $\hat{\sigma}^2$ and take the square root of equation 7-4.

If there are n replicates at each of the 2^k runs in the design, and if $y_{i1}, y_{i2}, \ldots, y_{in}$ are the observations at the ith run, then

$$\hat{\sigma}_i^2 = \frac{\sum_{j=1}^{n}(y_{ij} - \bar{y}_{i.})^2}{(n-1)} \qquad i = 1, 2, \ldots, 2^k$$

is an estimate of the variance at the ith run. The 2^k variance estimates can be pooled (averaged) to give an overall variance estimate

$$\hat{\sigma}^2 = \sum_{i=1}^{2^k} \frac{\hat{\sigma}_i^2}{2^k} \tag{7-5}$$

Each $\hat{\sigma}_i^2$ is associated with $n-1$ degrees of freedom and so $\hat{\sigma}^2$ is associated with $2^k(n-1)$ degrees of freedom.

To illustrate this approach for the epitaxial process experiment, we find that

$$\hat{\sigma}^2 = \frac{0.0121 + 0.0026 + 0.0059 + 0.0625}{4} = 0.0208$$

and the estimated standard error of each effect is

$$se(\text{Effect}) = \sqrt{[\hat{\sigma}^2/(n2^{k-2})]} = \sqrt{[0.0208/(4 \cdot 2^{2-2})]} = 0.072$$

In Table 7-5, each effect is divided by its estimated standard error and the resulting t ratio is compared to a t distribution with $2^2 \cdot 3 = 12$ degrees of freedom. Recall, the t ratio is used to judge whether the effect is significantly different from zero. The significant

Table 7-5   *t*-Tests of the Effects for Example 7-4

Effect	Effect Estimate	Estimated Standard Error	*t* Ratio	*P*-Value	Effect ± Two Estimated Standard Errors
A	0.836	0.072	11.61	0.00	0.836 ± 0.144
B	−0.067	0.072	−0.93	0.38	−0.067 ± 0.144
AB	0.032	0.072	0.44	0.67	0.032 ± 0.144

Degrees of freedom = 12.

effects are the important ones in the experiment. Two standard error limits on the effect estimates are also shown in Table 7-5. These intervals are approximate 95% confidence intervals.

The magnitude and direction of the effects were examined previously, and the analysis in Table 7-5 confirms those earlier, tentative conclusions. Deposition time is the only factor that significantly affects epitaxial layer thickness, and from the direction of the effect estimates we know that longer deposition times lead to thicker epitaxial layers.

Regression Analysis

In any designed experiment, it is important to examine a model for predicting responses. Furthermore, there is a close relationship between the analysis of a designed experiment and a regression analysis that can be used easily to obtain predictions from a 2^k experiment. For the epitaxial process experiment, an initial regression model is

$$Y = \beta_0 + \beta_1 x_1 + \beta_2 x_2 + \beta_{12} x_1 x_2 + \epsilon$$

The deposition time and arsenic flow are represented by coded variables x_1 and x_2, respectively. The low and high levels of deposition time are assigned values $x_1 = -1$ and $x_1 = +1$, respectively, and the low and high levels of arsenic flow are assigned values $x_2 = -1$ and $x_2 = +1$, respectively. The cross-product term $x_1 x_2$ represents the effect of the interaction between these variables.

The least squares fitted model is

$$\hat{y} = 14.389 + \left(\frac{0.836}{2}\right) x_1 + \left(\frac{-0.067}{2}\right) x_2 + \left(\frac{0.032}{2}\right) x_1 x_2$$

where the intercept $\hat{\beta}_0$ is the grand average of all 16 observations. The estimated coefficient of x_1 is one-half the effect estimate for deposition time. The estimated coefficient is one-half the effect estimate because regression coefficients measure the effect of a unit change in x_1 on the mean of Y, and the effect estimate is based on a two-unit change from -1 to $+1$. Similarly, the estimated coefficient of x_2 is one-half of the effect of arsenic flow and the estimated coefficient of the cross-product term is one-half of the interaction effect.

Table 7-6   Regression Analysis for Example 7-4. The regression equation is
Thickness $= 14.4 + 0.418A - 0.0336B + 0.0158A*B$

Analysis of Variance					
Source	Sum of Squares	Degrees of Freedom	Mean Square	f_0	P-Value
Model	2.81764	3	0.93921	45.18	0.000
Error	0.24948	12	0.02079		
Total	3.06712	15			

Independent Variable	Coefficient Estimate	Standard Error of Coefficient	t for H_0 Coefficient $= 0$	P-Value
Intercept	14.3889	0.0360	399.17	0.000
A	0.41800	0.03605	11.60	0.000
B	-0.03363	0.03605	-0.93	0.369
AB	0.01575	0.03605	0.44	0.670

The regression analysis is shown in Table 7-6. Note that mean square error equals the estimate of σ^2 calculated previously. Because the P-value associated with the F-test for the model is small (less than 0.05), we conclude that one or more of the effects are important. The t-test for the hypothesis $H_0: \beta_i = 0$ versus $H_1: \beta_i \neq 0$ (for each coefficient β_1, β_2, and β_3 in the regression analysis) is identical to the one computed from the standard error of the effects in Table 7-5. Consequently, the results in Table 7-5 can be interpreted as t-tests of regression coefficients. Because each estimated regression coefficient is one-half of the effect estimate, the standard errors in Table 7-6 are one-half of those in Table 7-5. The t-test from a regression analysis is identical to the t-test obtained from the standard error of an effect in a 2^k design whenever the estimate $\hat{\sigma}^2$ is the same in both analyses.

Similar to a regression analysis, a simpler model that uses only the important effects is the preferred choice to predict the response. Because the t-tests for the main effect of B and the AB interaction effect are not significant, these terms are removed from the model. The model then becomes

$$\hat{y} = 14.389 + \left(\frac{0.836}{2}\right)x_1$$

That is, the estimated regression coefficient for any effect is the same, regardless of the model considered. Although this is not true in general for a regression analysis, an estimated regression coefficient does not depend on the model in a factorial experiment. Consequently, it is easy to assess model changes when data are collected in one of these experiments. One may also revise the estimate of σ^2 by using the mean square error obtained from the ANOVA table for the simpler model.

These analysis methods are summarized as follows.

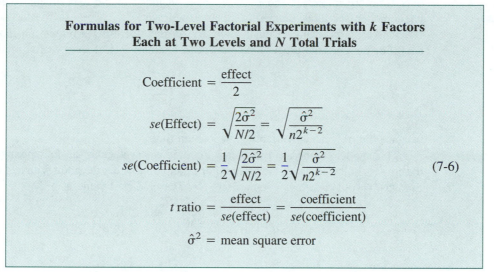

Formulas for Two-Level Factorial Experiments with k Factors Each at Two Levels and N Total Trials

$$\text{Coefficient} = \frac{\text{effect}}{2}$$

$$se(\text{Effect}) = \sqrt{\frac{2\hat{\sigma}^2}{N/2}} = \sqrt{\frac{\hat{\sigma}^2}{n2^{k-2}}}$$

$$se(\text{Coefficient}) = \frac{1}{2}\sqrt{\frac{2\hat{\sigma}^2}{N/2}} = \frac{1}{2}\sqrt{\frac{\hat{\sigma}^2}{n2^{k-2}}} \qquad (7\text{-}6)$$

$$t \text{ ratio} = \frac{\text{effect}}{se(\text{effect})} = \frac{\text{coefficient}}{se(\text{coefficient})}$$

$$\hat{\sigma}^2 = \text{mean square error}$$

7-4.3   Residual Analysis and Model Checking

The analysis of a 2^k design assumes that the observations are normally and independently distributed with the same variance in each treatment or factor level. These assumptions should be checked by examining the residuals. Residuals are calculated the same as in regression analysis. A **residual** is the difference between an observation y and its estimated (or fitted) value from the statistical model being studied, denoted as $\hat{y}$. Each residual is

$$e = y - \hat{y}$$

The normality assumption can be checked by constructing a normal probability plot of the residuals. To check the assumption of equal variances at each factor level, plot the residuals against the factor levels and compare the spread in the residuals. It is also useful to plot the residuals against $\hat{y}$; the variability in the residuals should not depend in any way on the value of $\hat{y}$. When a pattern appears in these plots, it usually suggests the need for transformation—that is, analyzing the data in a different metric. For example, if the variability in the residuals increases with $\hat{y}$, then a transformation such as log y or $\sqrt{y}$ should be considered. In some problems, the dependence of residual scatter on the fitted value $\hat{y}$ is very important information. It may be desirable to select the factor level that results in maximum response; however, this level may also cause more variation in response from run to run.

The independence assumption can be checked by plotting the residuals against the time or run order in which the experiment was performed. A pattern in this plot, such as sequences of positive and negative residuals, may indicate that the observations are not independent. This suggests that time or run order is important, or that variables that change over time are important and have not been included in the experimental design. This phenomenon should be studied in a new experiment.

It is easy to obtain residuals from a 2^k design by fitting a regression model to the data. For the epitaxial process experiment in Example 7-4, the regression model is

$$\hat{y} = 14.389 + \left(\frac{0.836}{2}\right)x_1$$

because the only active variable is deposition time.

This model can be used to obtain the predicted values at the four points that form the corners of the square in the design. For example, consider the point with low deposition time ($x_1 = -1$) and low arsenic flow rate. The predicted value is

$$\hat{y} = 14.389 + \left(\frac{0.836}{2}\right)(-1) = 13.971 \ \mu m$$

and the residuals are

$$e_1 = 14.037 - 13.971 = 0.066 \qquad e_3 = 13.972 - 13.971 = 0.001$$
$$e_2 = 14.165 - 13.971 = 0.194 \qquad e_4 = 13.907 - 13.971 = -0.064$$

It is easy to verify that the remaining predicted values and residuals are, for low deposition time ($x_1 = -1$) and high arsenic flow rate, $\hat{y} = 14.389 + (0.836/2)(-1) = 13.971 \ \mu m$

$$e_5 = 13.880 - 13.971 = -0.091 \qquad e_7 = 14.032 - 13.971 = 0.061$$
$$e_6 = 13.860 - 13.971 = -0.111 \qquad e_8 = 13.914 - 13.971 = -0.057$$

for high deposition time ($x_1 = +1$) and low arsenic flow rate, $\hat{y} = 14.389 + (0.836/2)(+1) = 14.807 \ \mu m$

$$e_9 = 14.821 - 14.807 = 0.014 \qquad e_{11} = 14.843 - 14.807 = 0.036$$
$$e_{10} = 14.757 - 14.807 = -0.050 \qquad e_{12} = 14.878 - 14.807 = 0.071$$

and for high deposition time ($x_1 = +1$) and high arsenic flow rate, $\hat{y} = 14.389 + (0.836/2)(+1) = 14.807 \ \mu m$

$$e_{13} = 14.888 - 14.807 = 0.081 \qquad e_{15} = 14.415 - 14.807 = -0.392$$
$$e_{14} = 14.921 - 14.807 = 0.114 \qquad e_{16} = 14.932 - 14.807 = 0.125$$

A normal probability plot of these residuals is shown in Fig. 7-11. This plot indicates that one residual $e_{15} = -0.392$ is an outlier. Examining the four runs with high deposition

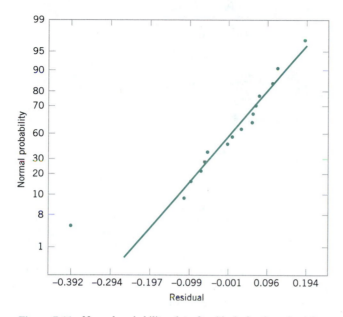

Figure 7-11   Normal probability plot of residuals for the epitaxial process experiment.

time and high arsenic flow rate reveals that observation $y_{15} = 14.415$ is considerably smaller than the other three observations at that treatment combination. This adds some additional evidence to the tentative conclusion that observation 15 is an outlier. Another possibility is that some process variables affect the *variability* in epitaxial layer thickness. If we could discover which variables produce this effect, then we could perhaps adjust these variables to levels that would minimize the variability in epitaxial layer thickness. This could have important implications in subsequent manufacturing stages. Figures 7-12 and 7-13 are plots of residuals versus deposition time and arsenic flow rate, respectively. Apart from that unusually large residual associated with y_{15}, there is no strong

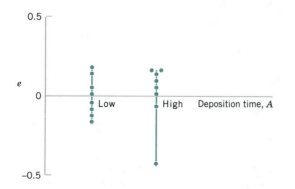

Figure 7-12   Plot of residuals versus deposition time.

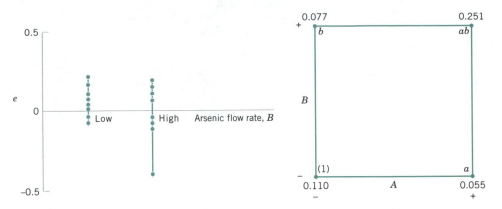

Figure 7-13 Plot of residuals versus arsenic flow rate.

Figure 7-14 The standard deviation of epitaxial layer thickness at the four runs in the 2^2 design.

evidence that either deposition time or arsenic flow rate influences the variability in epitaxial layer thickness.

Figure 7-14 shows the standard deviation of epitaxial layer thickness at all four runs in the 2^2 design. These standard deviations were calculated using the data in Table 7-4. Note that the standard deviation of the four observations with A and B at the high level is considerably larger than the standard deviations at any of the other three design points. Most of this difference is attributable to the unusually low thickness measurement associated with y_{15}. The standard deviation of the four observations with A and B at the low level is also somewhat larger than the standard deviations at the remaining two runs. This could indicate that other process variables not included in this experiment may affect the variability in epitaxial layer thickness. Another experiment to study this possibility, involving other process variables, could be designed and conducted. (The original paper in the *AT&T Technical Journal* shows that two additional factors, not considered in this example, affect process variability.)

EXERCISES FOR SECTION 7-4

For each of the following designs in Exercises 7-1 through 7-8, answer the following questions.

(a) Compute the estimates of the effects and their standard errors for this design.

(b) Construct two-factor interaction plots and comment on the interaction of the factors.

(c) Use the t ratio to determine the significance of each effect with $\alpha = 0.05$.

(d) Compute an approximate 95% confidence interval for each effect.

(e) Perform an analysis of variance of the appropriate regression model for this design. Include in your analysis hypothesis tests for each coefficient, as well as residual analysis. State your final conclusions about the adequacy of the model. Compare your results to part (c) and comment.

7-1. An experiment involves a storage battery used in the launching mechanism of a shoulder-fired ground-to-air missile. Two material types can be used to make the battery plates. The objective is to design a battery that is relatively unaffected by the ambient temperature. The output response from the battery is effective life in hours. Two temperature levels are selected, and a factorial experiment with four replicates is run. The data are as follows.

Material	Temperature (°F)			
	Low		High	
1	130	155	20	70
	74	180	82	58
2	138	110	96	104
	168	160	82	60

7-2. An engineer suspects that the surface finish of metal parts is influenced by the type of paint used and the drying time. He selects two drying times—20 and 30 minutes—and uses two types of paint. Three parts are tested with each combination of paint type and drying time. The data are as follows.

Paint	Drying Time (min)	
	20	30
1	74	78
	64	85
	50	92
2	92	66
	86	45
	68	85

7-3. An experiment was designed to identify a better ultrafiltration membrane for separating proteins and peptide drugs from fermentation broth. Two levels of an additive PVP (% wt) and time duration (hours) were investigated to determine the better membrane. The separation values (measured in %) resulting from these experimental runs are as follows.

PVP (% wt)	Time (hours)	
	1	3
2	69.6	80.0
	71.5	81.6
	70.0	83.0
	69.0	84.3
5	91.0	92.3
	93.2	93.4
	93.0	88.5
	87.2	95.6

7-4. An experiment was conducted to determine whether either firing temperature or furnace position affects the baked density of a carbon anode. The data are as follows.

Position	Temperature (°C)	
	800	825
1	570	1063
	565	1080
	583	1043
2	528	988
	547	1026
	521	1004

7-5. Johnson and Leone (*Statistics and Experimental Design in Engineering and the Physical Sciences,* John Wiley, 1977) describe an experiment conducted to investigate warping of copper plates. The two factors studied were temperature and the copper content of the plates. The response variable is the amount of warping. Some of the data are as follows.

Temperature (°C)	Copper Content (%)	
	40	80
50	17, 20	24, 22
100	16, 12	25, 23

7-6. An article in the *Journal of Testing and Evaluation* (Vol. 16, no. 6, 1988, pp. 508–515) investigated the effects of cyclic loading frequency and environment

		Environment	
		H_2O	Salt H_2O
Frequency	10	2.06	1.90
		2.05	1.93
		2.23	1.75
		2.03	2.06
	1	3.20	3.10
		3.18	3.24
		3.96	3.98
		3.64	3.24

conditions on fatigue crack growth at a constant 22 MPa stress for a particular material. Some of the data from the experiment are shown here. The response variable is fatigue crack growth rate.

7-7. An article in the *IEEE Transactions on Electron Devices* (November 1986, p. 1754) describes a study on the effects of two variables—polysilicon doping and anneal conditions (time and temperature)—on the base current of a bipolar transistor. Some of the data from this experiment are as follows.

		Anneal (temperature/ time)	
		900/180	1000/15
Polysilicon Doping	1×10^{20}	8.30 8.90	10.29 10.30
	2×10^{20}	7.81 7.75	10.19 10.10

7-8. An article in the *IEEE Transactions on Semiconductor Manufacturing* (Vol. 5, no. 3, 1992, pp. 214–222) describes an experiment to investigate the surface charge on a silicon wafer. The factors thought to influence induced surface charge are cleaning method (spin rinse dry or SRD and spin dry or SD) and the position on the wafer where the charge was measured. The surface charge ($\times 10^{11}$ q/cm^3) response data are as shown.

		Test Position	
		L	R
Cleaning Method	SD	1.66 1.90 1.92	1.84 1.84 1.62
	SRD	−4.21 −1.35 −2.08	−7.58 −2.20 −5.36

7-5 2^k DESIGN FOR $k \geq 3$ FACTORS

The methods presented in the previous section for factorial designs with $k = 2$ factors each at two levels can be easily extended to more than two factors. For example, consider $k = 3$ factors, each at two levels. This design is a 2^3 factorial design, and it has eight runs or treatment combinations. Geometrically, the design is a cube as shown in Fig. 7-15, with the eight runs forming the corners of the cube. This design allows three main effects to be estimated (A, B, and C) along with three two-factor interactions (AB, AC, and BC) and a three-factor interaction (ABC).

The main effects can easily be estimated. Remember that the lowercase letters (1), a, b, ab, c, ac, bc, and abc represent the total of all n replicates at each of the eight runs in the design. As seen in Fig. 7-16a, note that the main effect of A can be estimated by averaging the four treatment combinations on the right-hand side of the cube, where A is at the high level, and by subtracting from this quantity the average of the four treatment combinations on the left-hand side of the cube where A is at the low level. This gives

$$A = \bar{y}_{A+} - \bar{y}_{A-}$$
$$= \frac{a + ab + ac + abc}{4n} - \frac{(1) + b + c + bc}{4n}$$

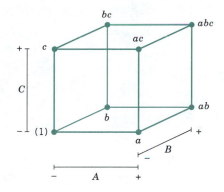

Figure 7-15 The 2^3 design.

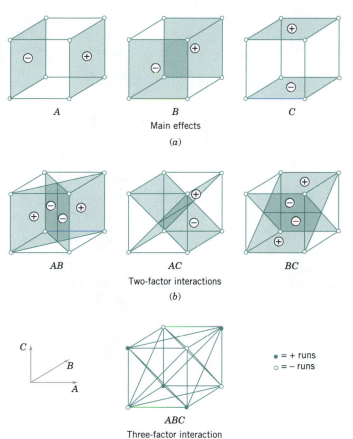

Figure 7-16 Geometric presentation of contrasts corresponding to the main effects and interaction in the 2^3 design. (a) Main effects. (b) Two-factor interactions. (c) Three-factor interaction.

This equation can be rearranged as

$$A = \bar{y}_{A+} - \bar{y}_{A-}$$
$$= \frac{1}{4n} [a + ab + ac + abc - (1) - b - c - bc] \qquad (7\text{-}7)$$

In a similar manner, the effect of B is the difference in averages between the four treatment combinations in the back face of the cube (Fig. 7-16a), and the four in the front. This yields

$$B = \bar{y}_{B+} - \bar{y}_{B-}$$
$$= \frac{1}{4n} [b + ab + bc + abc - (1) - a - c - ac] \qquad (7\text{-}8)$$

The effect of C is the difference in average response between the four treatment combinations in the top face of the cube (Fig. 7-16a) and the four in the bottom; that is,

$$C = \bar{y}_{C+} - \bar{y}_{C-}$$
$$= \frac{1}{4n} [c + ac + bc + abc - (1) - a - b - ab] \qquad (7\text{-}9)$$

The two-factor interaction effects may be computed easily. A measure of the AB interaction is the difference between the average A effects at the two levels of B. By convention, one-half of this difference is called the AB interaction. Symbolically,

B	Average A Effect
High $(+)$	$\dfrac{[(abc - bc) + (ab - b)]}{2n}$
Low $(-)$	$\dfrac{\{(ac - c) + [a - (1)]\}}{2n}$
Difference	$\dfrac{[abc - bc + ab - b - ac + c - a + (1)]}{2n}$

Since the AB interaction is one-half of this difference, then

$$AB = \frac{1}{4n}[abc - bc + ab - b - ac + c - a + (1)] \qquad (7\text{-}10)$$

We could write equation 7-10 as follows:

$$AB = \frac{abc + ab + c + (1)}{4n} - \frac{bc + b + ac + a}{4n}$$

In this form, the AB interaction is easily seen to be the difference in averages between runs on two diagonal planes in the cube in Fig. 7-16b. Using similar logic and referring to Fig. 7-16b, we find that the AC and BC interactions are

$$AC = \frac{1}{4n}[(1) - a + b - ab - c + ac - bc + abc] \qquad (7\text{-}11)$$

$$BC = \frac{1}{4n}[(1) + a - b - ab - c - ac + bc + abc] \qquad (7\text{-}12)$$

The ABC interaction is defined as the average difference between the AB interactions for the two different levels of C. Thus,

$$ABC = \frac{1}{4n}\{[abc - bc] - [ac - c] - [ab - b] + [a - (1)]\}$$

or

$$ABC = \frac{1}{4n}[abc - bc - ac + c - ab + b + a - (1)] \qquad (7\text{-}13)$$

As before, we can think of the ABC interaction as the difference in two averages. If the runs in the two averages are isolated, they define the vertices of the two tetrahedra that comprise the cube in Fig. 7-16c.

In equations 7-7 through 7-13, the quantities in brackets are contrasts in the treatment combinations. A table of plus and minus signs can be developed from the contrasts and is shown in Table 7-7. Signs for the main effects are determined by associating a plus with the high level and a minus with the low level. Once the signs for the main effects have been established, the signs for the remaining columns can be obtained by multiplying the appropriate preceding columns, row by row. For example, the signs in the AB column are the products of the A and B column signs in each row. The contrast for any effect can easily be obtained from this table.

Table 7-7 has several interesting properties:

1. Except for the identity column I, each column has an equal number of plus and minus signs.

2. The sum of products of signs in any two columns is zero; that is, the columns in the table are **orthogonal.**

3. Multiplying any column by column I leaves the column unchanged; that is, I is an **identity element.**

4. The product of any two columns yields a column in the table, for example $A \times B = AB$, and $AB \times ABC = A^2B^2C = C$, since any column multiplied by itself is the identity column.

The estimate of any main effect or interaction in a 2^k design is determined by multiplying the treatment combinations in the first column of the table by the signs in the corresponding main effect or interaction column, adding the result to produce a contrast, and then dividing the contrast by one-half the total number of runs in the experiment.

EXAMPLE 7-5

A mechanical engineer is studying the surface roughness of a part produced in a metal cutting operation. A 2^3 factorial design in the factors feed rate (A), depth of cut (B), and tool angle (C), with $n = 2$ replicates, is run. The levels for the three factors are: low $A = 20$ in./min, high $A = 30$ in./min; low $B = 0.025$ in., high $B = 0.040$ in.; low $C = 15°$, high $C = 25°$. Table 7-8 presents the observed surface roughness data.

Table 7-7   Algebraic Signs for Calculating Effects in the 2^3 Design

Treatment Combination	Factorial Effect							
	I	A	B	AB	C	AC	BC	ABC
(1)	+	−	−	+	−	+	+	−
a	+	+	−	−	−	−	+	+
b	+	−	+	−	−	+	−	+
ab	+	+	+	+	−	−	−	−
c	+	−	−	+	+	−	−	+
ac	+	+	−	−	+	+	−	−
bc	+	−	+	−	+	−	+	−
abc	+	+	+	+	+	+	+	+

The main effects may be estimated using equations 7-7 through 7-13. The effect of A, for example, is

$$A = \frac{1}{4n}[a + ab + ac + abc - (1) - b - c - bc]$$

$$= \frac{1}{4(2)}[22 + 27 + 23 + 30 - 16 - 20 - 21 - 18]$$

$$= \frac{1}{8}[27] = 3.375$$

It is easy to verify that the other effects are

$$B = 1.625 \qquad C = 0.875$$
$$AB = 1.375 \qquad AC = 0.125$$
$$BC = -0.625 \qquad ABC = 1.125$$

Examining the magnitude of the effects clearly shows that feed rate (factor A) is dominant, followed by depth of cut (B) and the AB interaction, although the interaction effect is relatively small.

For the surface roughness experiment, we find from pooling the variances at each of the eight treatments as in equation 7-5 that $\hat{\sigma}^2 = 2.4375$ and the estimated standard error of each effect is

$$se(\text{effect}) = \sqrt{\frac{\hat{\sigma}^2}{n2^{k-2}}} = \sqrt{\frac{2.4375}{2 \cdot 2^{3-2}}} = 0.78$$

Table 7-8 Surface Roughness Data for Example 7-5

Treatment Combinations	Design Factors			Surface Roughness	Total	Average	Variance
	A	B	C				
(1)	-1	-1	-1	9, 7	16	8	2.0
a	1	-1	-1	10, 12	22	11	2.0
b	-1	1	-1	9, 11	20	10	2.0
ab	1	1	-1	12, 15	27	13.5	4.5
c	-1	-1	1	11, 10	21	11.5	0.5
ac	1	-1	1	10, 13	23	12.5	4.5
bc	-1	1	1	10, 8	18	9	2.0
abc	1	1	1	16, 14	30	15	2.0
Average						11.065	2.4375

Therefore, two standard error limits on the effect estimates are

A:	3.375 ± 1.56	B:	1.625 ± 1.56
C:	0.875 ± 1.56	AB:	1.375 ± 1.56
AC:	0.125 ± 1.56	BC:	-0.625 ± 1.56
ABC:	1.125 ± 1.56		

These intervals are approximate 95% confidence intervals. They indicate that the two main effects A and B are important, but that the other effects are not, since the intervals for all effects except A and B include zero.

This confidence interval approach is a nice method of analysis. With relatively simple modifications, it can be used in situations where only a few of the design points have been replicated. Normal probability plots can also be used to judge the significance of effects. We will illustrate that method in the next section.

Regression Model and Residual Analysis

We may obtain the residuals from a 2^k design by using the method demonstrated earlier for the 2^2 design. As an example, consider the surface roughness experiment. The three largest effects are A, B, and the AB interaction. The regression model used to obtain the predicted values is

$$Y = \beta_0 + \beta_1 x_1 + \beta_2 x_2 + \beta_{12} x_1 x_2 + \epsilon$$

where x_1 represents factor A, x_2 represents factor B, and $x_1 x_2$ represents the AB interaction. The regression coefficients β_1, β_2, and β_{12} are estimated by one-half the corresponding effect estimates, and β_0 is the grand average. Thus

$$\hat{y} = 11.0625 + \left(\frac{3.375}{2}\right)x_1 + \left(\frac{1.625}{2}\right)x_2 + \left(\frac{1.375}{2}\right)x_1 x_2$$

and the predicted values would be obtained by substituting the low and high levels of A and B into this equation. To illustrate this, at the treatment combination where A, B, and C are all at the low level, the predicted value is

$$\hat{y} = 11.065 + \left(\frac{3.375}{2}\right)(-1) + \left(\frac{1.625}{2}\right)(-1) + \left(\frac{1.375}{2}\right)(-1)(-1)$$

$$= 9.25$$

Since the observed values at this run are 9 and 7, the residuals are $9 - 9.25 = -0.25$ and $7 - 9.25 = -2.25$. Residuals for the other 14 runs are obtained similarly.

A normal probability plot of the residuals is shown in Fig. 7-17. Since the residuals lie approximately along a straight line, we do not suspect any problem with normality in

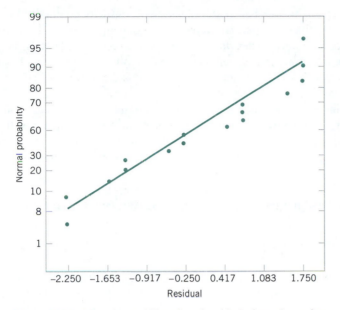

Figure 7-17 Normal probability plot of residuals from the surface roughness experiment.

the data. There are no indications of severe outliers. It would also be helpful to plot the residuals versus the predicted values and against each of the factors *A, B,* and *C.*

Projection of a 2^k Design

Any 2^k design will collapse or project into another 2^k design in fewer variables if one or more of the original factors are dropped. Sometimes this can provide additional insight into the remaining factors. For example, consider the surface roughness experiment. Since factor *C* and all its interactions are negligible, we could eliminate factor *C* from the design. The result is to collapse the cube in Fig. 7-15 into a square in the *A–B* plane; therefore, each of the four runs in the new design has four replicates. In general, if we delete *h* factors so that $r = k - h$ factors remain, the original 2^k design with *n* replicates will project into a 2^r design with $n2^h$ replicates.

7-6 SINGLE REPLICATE OF A 2^k DESIGN

As the number of factors in a factorial experiment grows, the number of effects that can be estimated also grows. For example, a 2^4 experiment has 4 main effects, 6 two-factor interactions, 4 three-factor interactions, and 1 four-factor interaction, and a 2^6 experiment has 6 main effects, 15 two-factor interactions, 20 three-factor interactions, 15 four-factor interactions, 6 five-factor interactions, and 1 six-factor interaction. In most situations the

sparsity of effects principle applies; that is, the system is usually dominated by the main effects and low-order interactions. The three-factor and higher order interactions are usually negligible. Therefore, when the number of factors is moderately large—say, $k \geq 4$ or 5—a common practice is to run only a single replicate of the 2^k design and then pool or combine the higher order interactions as an estimate of error. Sometimes a single replicate of a 2^k design is called an **unreplicated** 2^k factorial design.

When analyzing data from unreplicated factorial designs, occasionally real high-order interactions occur. The use of an error mean square obtained by pooling high-order interactions is inappropriate in these cases. A simple method of analysis can be used to overcome this problem. Construct a plot of the estimates of the effects on a normal probability scale. The effects that are negligible are normally distributed, with mean zero, and will tend to fall along a straight line on this plot, whereas significant effects will have nonzero means and will not lie along the straight line. We will illustrate this method in the next example.

EXAMPLE 7-6

An article in *Solid State Technology* ("Orthogonal Design for Process Optimization and Its Application in Plasma Etching," May 1987, pp. 127–132) describes the application of factorial designs in developing a nitride etch process on a single-wafer plasma etcher. The process uses C_2F_6 as the reactant gas. It is possible to vary the gas flow, the power applied to the cathode, the pressure in the reactor chamber, and the spacing between the anode and the cathode (gap). Several response variables would usually be of interest in this process, but in this example we will concentrate on etch rate for silicon nitride.

We will use a single replicate of a 2^4 design to investigate this process. Since it is unlikely that the three-factor and four-factor interactions are significant, we will tentatively plan to combine them as an estimate of error. The factor levels used in the design are shown here:

	Design Factor			
Level	Gap (cm)	Pressure (mTorr)	C_2F_6 Flow (SCCM)	Power (w)
Low ($-$)	0.80	450	125	275
High ($+$)	1.20	550	200	325

Table 7-9 presents the data from the 16 runs of the 2^4 design. Table 7-10 is the table of plus and minus signs for the 2^4 design. The signs in the columns of this table can be used to estimate the factor effects. For example, the estimate of factor A is

$$A = \frac{1}{8}[a + ab + ac + abc + ad + abd + acd + abcd - (1) - b$$
$$- c - bc - d - bd - cd - bcd]$$
$$= \frac{1}{8}[669 + 650 + 642 + 635 + 749 + 868 + 860 + 729$$
$$- 550 - 604 - 633 - 601 - 1037 - 1052 - 1075 - 1063]$$
$$= -101.625$$

Table 7-9 The 2^4 Design for the Plasma Etch Experiment

A (gap)	B (pressure)	C (C_2F_6 flow)	D (power)	Etch Rate (Å/min)
−1	−1	−1	−1	550
1	−1	−1	−1	669
−1	1	−1	−1	604
1	1	−1	−1	650
−1	−1	1	−1	633
1	−1	1	−1	642
−1	1	1	−1	601
1	1	1	−1	635
−1	−1	−1	1	1037
1	−1	−1	1	749
−1	1	−1	1	1052
1	1	−1	1	868
−1	−1	1	1	1075
1	−1	1	1	860
−1	1	1	1	1063
1	1	1	1	729

Table 7-10 Contrast Constants for the 2^4 Design

							Factorial Effect								
	A	B	AB	C	AC	BC	ABC	D	AD	BD	ABD	CD	ACD	BCD	ABCD
(1)	−	−	+	−	+	+	−	−	+	+	−	+	−	−	+
a	+	−	−	−	−	+	+	−	−	+	+	+	+	−	−
b	−	+	−	−	+	−	+	−	+	−	+	+	−	+	−
ab	+	+	+	−	−	−	−	−	−	−	−	+	+	+	+
c	−	−	+	+	−	−	+	−	+	+	−	−	+	+	−
ac	+	−	−	+	+	−	−	−	−	+	+	−	−	+	+
bc	−	+	−	+	−	+	−	−	+	−	+	−	+	−	+
abc	+	+	+	+	+	+	+	−	−	−	−	−	−	−	−
d	−	−	+	−	+	+	−	+	−	−	+	−	+	+	−
ad	+	−	−	−	−	+	+	+	+	−	−	−	−	+	+
bd	−	+	−	−	+	−	+	+	−	+	−	−	+	−	+
abd	+	+	+	−	−	−	−	+	+	+	+	−	−	−	−
cd	−	−	+	+	−	−	+	+	−	−	+	+	−	−	+
acd	+	−	−	+	+	−	−	+	+	−	−	+	+	−	−
bcd	−	+	−	+	−	+	−	+	−	+	−	+	−	+	−
abcd	+	+	+	+	+	+	+	+	+	+	+	+	+	+	+

Thus the effect of increasing the gap between the anode and the cathode from 0.80 cm to 1.20 cm is to decrease the mean etch rate by 101.625 Å/min.

It is easy to verify that the complete set of effect estimates is

$$
\begin{array}{llll}
A & = & -101.625 & \qquad B & = & -1.625 \\
AB & = & -7.875 & \qquad C & = & 7.375 \\
AC & = & -24.875 & \qquad BC & = & -43.875 \\
ABC & = & -15.625 & \qquad D & = & 306.125 \\
AD & = & -153.625 & \qquad BD & = & -0.625 \\
ABD & = & 4.125 & \qquad CD & = & -2.125 \\
ACD & = & 5.625 & \qquad BCD & = & -25.375 \\
ABCD & = & -40.125 & & &
\end{array}
$$

The normal probability plot of these effects from the plasma etch experiment is shown in Fig. 7-18. Clearly, the main effects of A and D and the AD interaction are significant, because they fall far from the line passing through the other points. The analysis of variance summarized in Table 7-11 confirms these findings. Note that in the analysis of variance we have pooled the three- and four-factor interactions to form the error mean square. If the normal probability plot had indicated that any of these interactions were important, then we would not have included them in the error term. Consequently,

$$
\hat{\sigma}^2 = 2037.4 \quad \text{and} \quad se(\text{coefficient}) = \frac{1}{2}\sqrt{\frac{2(2037.4)}{16/2}} = 11.28
$$

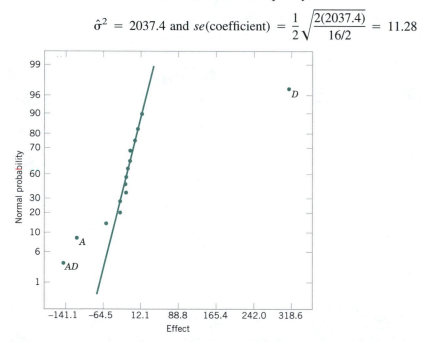

Figure 7-18   Normal probability plot of effects from the plasma etch experiment.

Table 7-11 Analysis for Example 7-6

		Analysis of Variance			
Source	Sum of Squares	Degrees of Freedom	Mean Square	f_0	P-Value
Model	521234	10	52123.40	25.58	0.000
Error	10187	5	2037.40		
Total	531421	15			

Independent Variable	Effect Estimate	Coefficient Estimate	Standard Error of Coefficient	t for H_0 Coefficient $= 0$	P-Value
Intercept		776.06	11.28	68.77	0.000
A	− 101.63	− 50.81	11.28	− 4.50	0.006
B	− 1.63	− 0.81	11.28	− 0.07	0.945
C	7.38	3.69	11.28	0.33	0.757
D	306.12	153.06	11.28	13.56	0.000
AB	− 7.88	− 3.94	11.28	− 0.35	0.741
AC	− 24.87	− 12.44	11.28	− 1.10	0.321
AD	− 153.62	− 76.81	11.28	− 6.81	0.001
BC	− 43.87	− 21.94	11.28	− 1.94	0.109
BD	− 0.62	− 0.31	11.28	− 0.03	0.979
CD	− 2.12	− 1.06	11.28	− 0.09	0.929

Since $A = -101.625$, the effect of increasing the gap between the cathode and anode is to decrease the etch rate. However, $D = 306.125$; thus, applying higher power levels will increase the etch rate. Figure 7-19 is a plot of the AD interaction. This plot indicates that the effect of changing the gap width at low power settings is small, but that increasing the gap at high power settings dramatically reduces the etch rate. High etch rates are obtained at high power settings and narrow gap widths.

The residuals from the experiment can be obtained from the regression model

$$\hat{y} = 776.0625 - \left(\frac{101.625}{2}\right)x_1 + \left(\frac{306.125}{2}\right)x_4 - \left(\frac{153.625}{2}\right)x_1 x_4$$

For example, when both A and D are at the low level, the predicted value is

$$\hat{y} = 776.0625 - \left(\frac{101.625}{2}\right)(-1) + \left(\frac{306.125}{2}\right)(-1) - \left(\frac{153.625}{2}\right)(-1)(-1)$$

$$= 597$$

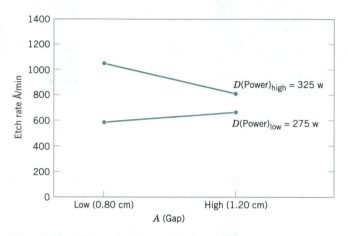

Figure 7-19   *AD* (gap-power) interaction from the plasma etch experiment.

and the four residuals at this treatment combination are

$$e_1 = 550 - 597 = -47$$
$$e_2 = 604 - 597 = 7$$
$$e_3 = 633 - 597 = 36$$
$$e_4 = 601 - 597 = 4$$

The residuals at the other three treatment combinations (*A* high, *D* low), (*A* low, *D* high), and (*A* high, *D* high) are obtained similarly. A normal probability plot of the residuals is shown in Fig. 7-20. The plot is satisfactory. It would also be helpful to plot the residuals versus the predicted values and against each of the factors.

7-7   CENTER POINTS AND BLOCKING IN 2^k DESIGNS

7-7.1   Addition of Center Points

A potential concern in the use of two-level factorial designs is the assumption of linearity in the factor effects. Of course, perfect linearity is unnecessary, and the 2^k system will work quite well even when the linearity assumption holds only approximately. However, there is a method of replicating certain points in the 2^k factorial that will provide protection against curvature as well as allow an independent estimate of error to be obtained. The method consists of adding **center points** to the 2^k design. These consist of n_C replicates run at the point $x_i = 0$, $i = 1, 2, \ldots, k$. One important reason for adding the replicate runs at the design center is that center points do not impact the usual effects estimates in a 2^k design. We assume that the k factors are quantitative. If some of the factors are categorical (such as Tool A and Tool B) the method can be modified.

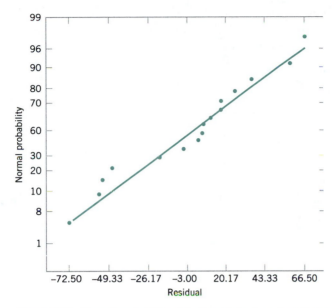

Figure 7-20 Normal probability plot of residuals from the plasma etch experiment.

To illustrate the approach, consider a 2^2 design with one observation at each of the factorial points $(-, -), (+, -), (-, +)$ and $(+, +)$ and n_C observations at the center points $(0, 0)$. Figure 7-21 illustrates the situation. Let $\bar{y}_F$ be the average of the four runs at the four factorial points, and let $\bar{y}_C$ be the average of the n_C run at the center point. If the difference $\bar{y}_F - \bar{y}_C$ is small, then the center points lie on or near the plane passing through the factorial points, and there is no curvature. On the other hand, if $\bar{y}_F - \bar{y}_C$ is large, then curvature is present.

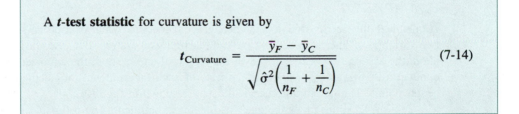

A *t*-test statistic for curvature is given by

$$t_{\text{Curvature}} = \frac{\bar{y}_F - \bar{y}_C}{\sqrt{\hat{\sigma}^2\left(\dfrac{1}{n_F} + \dfrac{1}{n_C}\right)}} \qquad (7\text{-}14)$$

where, in general, n_F is the number of factorial design points. More specifically, when points are added to the center of the 2^k design, then the model we may entertain is

$$Y = \beta_0 + \sum_{j=1}^{k} \beta_j x_j + \sum\sum_{i<j} \beta_{ij} x_i x_j + \sum_{j=1}^{k} \beta_{jj} x_j^2 + \epsilon \qquad (7\text{-}15)$$

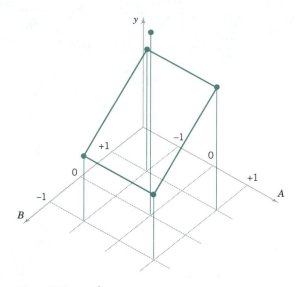

Figure 7-21   A 2^2 design with center points.

where the β_{jj} are pure quadratic effects. The test for curvature actually tests the hypotheses

$$H_0: \sum_{j=1}^{k} \beta_{jj} = 0 \qquad H_1: \sum_{j=1}^{k} \beta_{jj} \neq 0 \qquad (7\text{-}16)$$

Furthermore, if the factorial points in the design are unreplicated, we may use the n_C center points to construct an estimate of error with $n_C - 1$ degrees of freedom. This is referred to as a **pure error** estimate.

EXAMPLE 7-7

A chemical engineer is studying the percent conversion or yield of a process. There are two variables of interest, reaction time and reaction temperature. Because she is uncertain about the assumption of linearity over the region of exploration, the engineer decides to conduct a 2^2 design (with a single replicate of each factorial run) augmented with five center points. The design and the yield data are shown in Fig. 7-22.

Table 7-12 summarizes the analysis for this experiment. The estimate of pure error is calculated from the center points as follows:

$$\hat{\sigma}^2 = \frac{\displaystyle\sum_{\text{center points}} (y_i - \bar{y}_C)^2}{n_C - 1} = \frac{\displaystyle\sum_{i=1}^{5} (y_i - 40.46)^2}{4} = \frac{0.1720}{4} = 0.0430$$

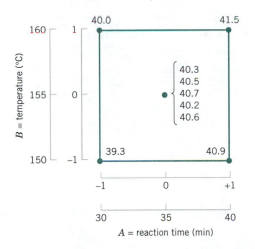

Figure 7-22 The 2^2 design with five center points for Example 7-7.

The average of the points in the factorial portion of the design is $\bar{y}_F = 40.425$, and the average of the points at the center is $\bar{y}_C = 40.46$. The difference $\bar{y}_F - \bar{y}_C = 40.425 - 40.46 = -0.035$ appears to be small. The curvature t ratio is computed from equation 7-14 as follows:

$$t_{\text{Curvature}} = \frac{\bar{y}_F - \bar{y}_C}{\sqrt{\hat{\sigma}^2 \left(\frac{1}{n_F} + \frac{1}{n_C} \right)}} = \frac{-0.035}{\sqrt{0.0430 \left(\frac{1}{4} + \frac{1}{5} \right)}} = 0.252$$

The analysis indicates that there is no evidence of curvature in the response over the region of exploration; that is, the null hypothesis $H_0: \sum_{j=1}^{2} \beta_{jj} = 0$ cannot be rejected.

Table 7-12 displays output from Minitab for this example. The effect of A is $(41.5 + 40.9 - 40.0 - 39.3)/2 = 1.55$ and the other effects are obtained similarly. The pure-error estimate (0.043) agrees with our previous result. Recall from regression modeling that the square of a t-ratio is an F-ratio. Consequently, Minitab uses $0.252^2 = 0.06$ as an F-ratio to obtain an identical test for curvature. The sum of squares for curvature is an intermediate step in the calculation of the F-ratio that equals the square of the t-ratio when the estimate of σ^2 is omitted. That is,

$$SS_{\text{Curvature}} = \frac{(\bar{y}_F - \bar{y}_C)^2}{\frac{1}{n_F} + \frac{1}{n_C}} \tag{7-17}$$

Furthermore, Minitab adds the sum of squares for curvature and for pure error to obtain the residual sum of squares (0.17472) with 5 degrees of freedom. The residual mean

Table 7-12   Analysis for Example 7-7 from Minitab

Factorial Design

Full Factorial Design

Factors:	2	Base Design:	2, 4
Runs:	9	Replicates:	1
Blocks:	none	Center pts (total):	5

All terms are free from aliasing

Fractional Factorial Fit

Estimated Effects and Coefficients for y

Term	Effect	Coef	StDev Coef	T	P
Constant		40.4444	0.06231	649.07	0.000
A	1.5500	0.7750	0.09347	8.29	0.000
B	0.6500	0.3250	0.09347	3.48	0.018
A*B	−0.0500	−0.0250	0.09347	−0.27	0.800

Analysis of Variance for y

Source	DF	Seq SS	Adj SS	Adj MS	F	P
Main Effects	2	2.82500	2.82500	1.41250	40.42	0.001
2-Way Interactions	1	0.00250	0.00250	0.00250	0.07	0.800
Residual Error	5	0.17472	0.17472	0.03494		
Curvature	1	0.00272	0.00272	0.00272	0.06	0.814
Pure Error	4	0.17200	0.17200	0.04300		
Total	8	3.00222				

square (0.03494) is a pooled estimate of σ^2 and it is used in the calculation of the t-ratio for the A, B, and AB effects. The pooled estimate is close to the pure-error estimate in this example because curvature is negligible. If curvature were significant, the pooling would not be appropriate. The estimate of the intercept β_0 (40.444) is the mean of all nine measurements.

7-7.2   Blocking and Confounding

It is often impossible to run all the observations in a 2^k factorial design under homogeneous conditions. Blocking is the design technique that is appropriate for this general situation. However, in many situations the block size is smaller than the number of runs in the complete replicate. In these cases, **confounding** is a useful procedure for running the 2^k

design in 2^p blocks where the number of runs in a block is less than the number of treatment combinations in one complete replicate. The technique causes certain interaction effects to be indistinguishable from blocks or **confounded with blocks**. We will illustrate confounding in the 2^k factorial design in 2^p blocks, where $p < k$.

Consider a 2^2 design. Suppose that each of the $2^2 = 4$ treatment combinations requires 4 hours of laboratory analysis. Thus, 2 days are required to perform the experiment. If days are considered as blocks, then we must assign two of the four treatment combinations to each day.

This design is shown in Fig. 7-23. Note that block 1 contains the treatment combinations (1) and *ab* and that block 2 contains *a* and *b*. The contrasts for estimating the main effects of factors *A* and *B* are

$$\text{Contrast}_A = ab + a - b - (1)$$

$$\text{Contrast}_B = ab + b - a - (1)$$

Note that these contrasts are unaffected by blocking since in each contrast there is one plus and one minus treatment combination from each block. That is, any difference between block 1 and block 2 that increases the readings in one block by an additive constant cancels out. The contrast for the *AB* interaction is

$$\text{Contrast}_{AB} = ab + (1) - a - b$$

Since the two treatment combinations with the plus signs, *ab* and (1), are in block 1 and the two with the minus signs, *a* and *b*, are in block 2, the block effect and the *AB* interaction are identical. That is, the *AB* interaction is confounded with blocks.

The reason for this is apparent from the table of plus and minus signs for the 2^2 design shown in Table 7-3. From the table we see that all treatment combinations that have a plus on *AB* are assigned to block 1, whereas all treatment combinations that have a minus sign on *AB* are assigned to block 2.

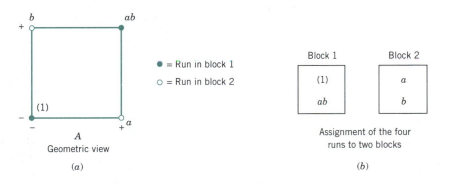

Figure 7-23 A 2^2 design in two blocks. (a) Geometric view. (b) Assignment of the four runs to two blocks.

This scheme can be used to confound any 2^k design in two blocks. As a second example, consider a 2^3 design, run in two blocks. From the table of plus and minus signs, shown in Table 7-7, we assign the treatment combinations that are minus in the ABC column to block 1 and those that are plus in the ABC column to block 2. The resulting design is shown in Fig. 7-24.

EXAMPLE 7-8

An experiment is performed to investigate the effect of four factors on the terminal miss distance of a shoulder-fired ground-to-air-missile. The four factors are target type (A), seeker type (B), target altitude (C), and target range (D). Each factor may be conveniently run at two levels, and the optical tracking system will allow terminal miss distance to be measured to the nearest foot. Two different operators or gunners are used in the flight test and, since there may be differences between operators, the test engineers decided to conduct the 2^4 design in two blocks with $ABCD$ confounded.

The experimental design and the resulting data are shown in Fig. 7-25. The effect estimates obtained from Minitab are shown in Table 7-13. A normal probability plot of the effects in Fig. 7-26 reveals that A (target type), D (target range), AD, and AC have large effects. A confirming analysis of variance, pooling the three-factor interactions as error, is shown in Table 7-14. Since the AC and AD interactions are significant, it is logical to conclude that A (target type), C (target altitude), and D (target range) all have important effects on the miss distance and that there are interactions between target type and altitude and target type and range. Note that the $ABCD$ effect is treated as blocks in this analysis.

It is possible to confound the 2^k design in four blocks of 2^{k-2} observations each. To construct the design, two effects are chosen to confound with blocks. A third effect, the **generalized interaction** of the two effects initially chosen, is also confounded with

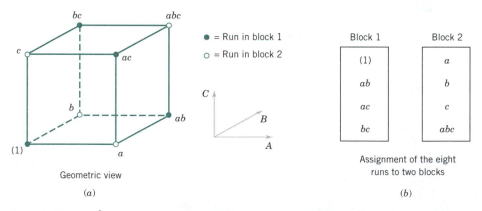

Figure 7-24   The 2^3 design in two blocks with ABC confounded. (a) Geometric view. (b) Assignment of the eight runs to two blocks.

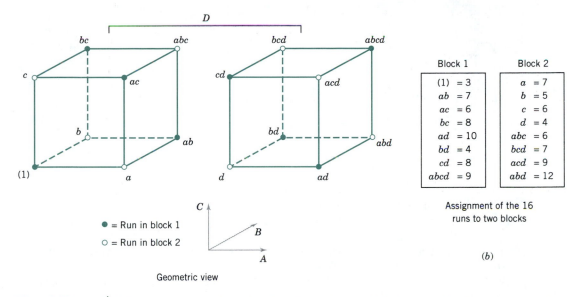

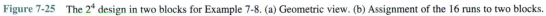

Block 1		Block 2	
(1)	= 3	a	= 7
ab	= 7	b	= 5
ac	= 6	c	= 6
bc	= 8	d	= 4
ad	= 10	abc	= 6
bd	= 4	bcd	= 7
cd	= 8	acd	= 9
abcd	= 9	abd	= 12

Assignment of the 16 runs to two blocks

● = Run in block 1
○ = Run in block 2

Geometric view

(b)

Figure 7-25   The 2^4 design in two blocks for Example 7-8. (a) Geometric view. (b) Assignment of the 16 runs to two blocks.

Table 7-13   Minitab Effect Estimates for Example 7-8

Estimated Effects and Coefficients for Distance		
Term	Effect	Coef
Constant		6.938
Block		0.063
A	2.625	1.312
B	0.625	0.313
C	0.875	0.438
D	1.875	0.938
A*B	−0.125	−0.063
A*C	−2.375	−1.187
A*D	1.625	0.813
B*C	−0.375	−0.188
B*D	−0.375	−0.187
C*D	−0.125	−0.062
A*B*C	−0.125	−0.063
A*B*D	0.875	0.438
A*C*D	−0.375	−0.187
B*C*D	−0.375	−0.187

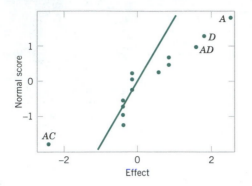

Figure 7-26   Normal probability plot of the effects from Minitab, Example 7-8.

blocks. The generalized interaction of two effects is found by multiplying their respective letters and reducing the exponents modulus 2.

For example, consider the 2^4 design in four blocks. If AC and BD are confounded with blocks, their generalized interaction is $(AC)(BD) = ABCD$. The design is constructed by a partition of the treatments according to the signs of AC and BD. It is easy to verify that the four blocks are

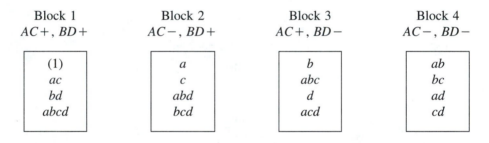

Block 1	Block 2	Block 3	Block 4
$AC+$, $BD+$	$AC-$, $BD+$	$AC+$, $BD-$	$AC-$, $BD-$
(1)	a	b	ab
ac	c	abc	bc
bd	abd	d	ad
abcd	bcd	acd	cd

This general procedure can be extended to confounding the 2^k design in 2^p blocks, where $p < k$. Start by selecting p effects to be confounded, such that no effect chosen is a generalized interaction of the others. Then the blocks can be constructed from the p defining contrasts $L_1, L_2, \ldots, L_p$ that are associated with these effects. In addition to the p effects chosen to be confounded, exactly $2^p - p - 1$ additional effects are confounded with blocks; these are the generalized interactions of the original p effects chosen. Care should be taken so as not to confound effects of potential interest.

For more information on confounding in the 2^k factorial design, refer to Montgomery (1997, Chapter 8). This book contains guidelines for selecting factors to confound with blocks, so that main effects and low-order interactions are not confounded. In particular, the book contains a table of suggested confounding schemes for designs with up to seven factors and a range of block sizes, some of which are as small as two runs.

Table 7-14 Analysis of Variance for Example 7-8

Source of Variation	Sum of Squares	Degrees of Freedom	Mean Square	f_0	P-Value
Blocks ($ABCD$)	0.0625	1	0.0625	0.06	—
A	27.5625	1	27.5625	25.94	0.0070
B	1.5625	1	1.5625	1.47	0.2920
C	3.0625	1	3.0625	2.88	0.1648
D	14.0625	1	14.0625	13.24	0.0220
AB	0.0625	1	0.0625	0.06	—
AC	22.5625	1	22.5625	21.24	0.0100
AD	10.5625	1	10.5625	9.94	0.0344
BC	0.5625	1	0.5625	0.53	—
BD	0.5625	1	0.5625	0.53	—
CD	0.0625	1	0.0625	0.06	—
Error ($ABC + ABD + ACD + BCD$)	4.2500	4	1.0625		
Total	84.9375	15			

EXERCISES FOR SECTIONS 7-5, 7-6, AND 7-7

7-9. An engineer is interested in the effect of cutting speed (A), metal hardness (B), and cutting angle (C) on the life of a cutting tool. Two levels of each factor are chosen, and two replicates of a 2^3 factorial design are run. The tool life data (in hours) are shown in the following table.

Treatment Combination	Replicate	
	I	II
(1)	221	311
a	325	435
b	354	348
ab	552	472
c	440	453
ac	406	377
bc	605	500
abc	392	419

(a) Analyze the data from this experiment using t-ratios with $\alpha = 0.05$.

(b) Find an appropriate regression model that explains tool life in terms of the variables used in the experiment.

(c) Analyze the residuals from this experiment.

7-10. Four factors are thought to influence the taste of a soft-drink beverage: type of sweetener (A), ratio of syrup to water (B), carbonation level (C), and temperature (D). Each factor can be run at two levels, producing a 2^4 design. At each run in the design, samples of the beverage are given to a test panel consisting of 20 people. Each tester assigns the beverage a point score from 1 to 10. Total score is the response variable, and the objective is to find a formulation that maximizes total score. Two replicates of this design are run, and the results are as shown. Analyze the data using t-ratios and draw conclusions. Use $\alpha = 0.05$ in the statistical tests.

Treatment Combination	Replicate	
	I	II
(1)	159	163
a	168	175
b	158	163
ab	166	168
c	175	178
ac	179	183
bc	173	168
abc	179	182
d	164	159
ad	187	189
bd	163	159
abd	185	191
cd	168	174
acd	197	199
bcd	170	174
abcd	194	198

7-11. Consider the experiment in Exercise 7-10. Determine an appropriate model and plot the residuals against the levels of factors A, B, C, and D. Also construct a normal probability plot of the residuals. Comment on these plots and the most important factors influencing taste.

7-12. The data shown here represent a single replicate of a 2^5 design that is used in an experiment to study the compressive strength of concrete. The factors are mix (A), time (B), laboratory (C), temperature (D), and drying time (E).

(1)	= 700		e	= 800
a	= 900		ae	= 1200
b	= 3400		be	= 3500
ab	= 5500		abc	= 6200
c	= 600		ce	= 600
ac	= 1000		ace	= 1200
bc	= 3000		bce	= 3000
abc	= 5300		abce	= 5500
d	= 1000		de	= 1900
ad	= 1100		ade	= 1500
bd	= 3000		bde	= 4000
abd	= 6100		abde	= 6500
cd	= 800		cde	= 1500
acd	= 1100		acde	= 2000
bcd	= 3300		bcde	= 3400
abcd	= 6000		abcde	= 6800

(a) Estimate the factor effects.

(b) Which effects appear important? Use a normal probability plot.

(c) Determine an appropriate model and analyze the residuals from this experiment. Comment on the adequacy of the model.

(d) If it is desirable to maximize the strength, in which direction would you adjust the process variables?

7-13. Consider a famous experiment reported by O. L. Davies (ed.), *The Design and Analysis of Industrial Experiments* (1956). The following data were collected from an unreplicated experiment in which the investigator was interested in determining the effect of four factors on the yield of an isatin derivative used in a fabric-dying process. The four factors are each run at two levels as indicated: (A) acid strength at 87% and 93%, (B) reaction time at 15 min and 30 min, (C) amount of acid 35 ml and 45 ml, and (D) temperature of reaction 60°C and 70°C. The response is the yield of isatin in grams per 100 grams of base material. The data are as follows:

(1)	= 6.08	d	= 6.79
a	= 6.04	ad	= 6.68
b	= 6.53	bd	= 6.73
ab	= 0.43	abd	= 6.08
c	= 6.31	cd	= 6.77
ac	= 6.09	acd	= 6.38
bc	= 6.12	bcd	= 6.49
abc	= 6.36	abcd	= 6.23

(a) Estimate the effects and prepare a normal plot of the effects. Which interaction terms are negligible? Use *t*-ratios to confirm your findings.

(b) Based on your results in part (a), construct a model and analyze the residuals.

7-14. An experiment was run in a semiconductor fabrication plant in an effort to increase yield. Five factors, each at two levels, were studied. The factors (and levels) were A = aperture setting (small, large), B = exposure time (20% below nominal, 20% above nominal), C = development time (30 sec, 45 sec), D = mask dimension (small, large), and E = etch time (14.5 min, 15.5 min). The unreplicated 2^5 design shown here was run.

(1)	= 7	e	= 8
a	= 9	ae	= 12
b	= 34	be	= 35
ab	= 55	abe	= 52
c	= 16	ce	= 15
ac	= 20	ace	= 22
bc	= 40	bce	= 45
abc	= 60	$abce$	= 65
d	= 8	de	= 6
ad	= 10	ade	= 10
bd	= 32	bde	= 30
abd	= 50	$abde$	= 53
cd	= 18	cde	= 15
acd	= 21	$acde$	= 20
bcd	= 44	$bcde$	= 41
$abcd$	= 61	$abcde$	= 63

(a) Construct a normal probability plot of the effect estimates. Which effects appear to be large?

(b) Estimate σ^2 and use t-ratios to confirm your findings for part (a).

(c) Plot the residuals from an appropriate model on normal probability paper. Is the plot satisfactory?

(d) Plot the residuals versus the predicted yields and versus each of the five factors. Comment on the plots.

(e) Interpret any significant interactions.

(f) What are your recommendations regarding process operating conditions?

(g) Project the 2^5 design in this problem into a 2^r for $r < 5$ design in the important factors. Sketch the design and show the average and range of yields at each run. Does this sketch aid in data interpretation?

7-15. Consider the semiconductor experiment in Exercise 7-14. Suppose that a center point (replicated five times) could be added to this design and that the responses at the center are 45, 40, 41, 47, and 43.

(a) Estimate the error using the center points. How does this estimate compare to the estimate obtained in Exercise 7-14?

(b) Calculate the t-ratio for curvature and test at $\alpha = 0.05$.

7-16. Consider the data from Exercise 7-9, replicate I only. Suppose that a center point (with four replicates) is added to these eight runs. The tool life response at the center point is 425, 400, 437, and 418.

(a) Estimate the factor effects.

(b) Estimate pure error using the center points.

(c) Calculate the t-ratio for curvature and test at $\alpha = 0.05$.

(d) Test for main effects and interaction effects, using $\alpha = 0.05$.

(e) Give the regression model and analyze the residuals from this experiment.

7-17. Consider the data from the first replicate of Exercise 7-9. Suppose that these observations could not all be run under the same conditions. Set up a design to run these observations in two blocks of four observations each, with ABC confounded. Analyze the data.

7-18. Consider the data from the first replicate of Exercise 7-10. Construct a design with two blocks of eight observations each, with $ABCD$ confounded. Analyze the data.

7-19. Repeat the previous exercise assuming that four blocks are required. Confound ABD and ABC (and consequently CD) with blocks.

7-20. Construct a 2^5 design in two blocks. Select the *ABCDE* interaction to be confounded with blocks.

7-21. Construct a 2^5 design in four blocks. Select the appropriate effects to confound so that the highest possible interactions are confounded with blocks.

7-22. Consider the data from Exercise 7-13. Construct the design that would have been used to run this experiment in two blocks of eight runs each. Analyze the data and draw conclusions.

7-8 FRACTIONAL REPLICATION OF A 2^k DESIGN

As the number of factors in a 2^k factorial design increases, the number of runs required increases rapidly. For example, a 2^5 requires 32 runs. In this design, only 5 degrees of freedom correspond to main effects, and 10 degrees of freedom correspond to two-factor interactions. Sixteen of the 31 degrees of freedom are used to estimate high-order interactions—that is, three-factor and higher order interactions. Often there is little interest in these high-order interactions, particularly when we first begin to study a process or system. If we can assume that certain high-order interactions are negligible, then a **fractional factorial design** involving fewer than the complete set of 2^k runs can be used to obtain information on the main effects and low-order interactions. In this section, we will introduce fractional replications of the 2^k design.

A major use of fractional factorials is in screening experiments. These are experiments in which many factors are considered with the purpose of identifying those factors (if any) that have large effects. Screening experiments are usually performed in the early stages of a project when it is likely that many of the factors initially considered have little or no effect on the response. The factors that are identified as important are then investigated more thoroughly in subsequent experiments.

7-8.1 One-Half Fraction of a 2^k Design

A one-half fraction of the 2^k design contains 2^{k-1} runs and is often called a 2^{k-1} fractional factorial design. As an example, consider the 2^{3-1} design—that is, a one-half fraction of the 2^3. This design has only four runs, in contrast to the full factorial that would require eight runs. The table of plus and minus signs for the 2^3 design is shown in Table 7-15. Suppose we select the four treatment combinations a, b, c, and abc as our one-half fraction. These treatment combinations are shown in the top half of Table 7-15 and in Fig. 7-27a. We will continue to use both the lowercase letter notation (a, b, c, . . .) and the geometric or plus and minus notation for the treatment combinations.

Note that the 2^{3-1} design is formed by selecting only those treatment combinations that yield a plus on the *ABC* effect. Thus, *ABC* is called the **generator** of this particular fraction. Furthermore, the identity element I is also plus for the four runs, so we call

$$I = ABC$$

the **defining relation** for the design.

Table 7-15 Plus and Minus Signs for the 2^3 Factorial Design

Treatment Combination	Factorial Effect							
	I	*A*	*B*	*C*	*AB*	*AC*	*BC*	*ABC*
a	+	+	−	−	−	−	+	+
b	+	−	+	−	−	+	−	+
c	+	−	−	+	+	−	−	+
abc	+	+	+	+	+	+	+	+
ab	+	+	+	−	+	−	−	−
ac	+	+	−	+	−	+	−	−
bc	+	−	+	+	−	−	+	−
(1)	+	−	−	−	+	+	+	−

The treatment combinations in the 2^{3-1} designs yield three degrees of freedom associated with the main effects. From the upper half of Table 7-15, we obtain the estimates of the main effects as linear combinations of the observations, say,

$$A = \frac{1}{2}[a - b - c + abc]$$
$$B = \frac{1}{2}[-a + b - c + abc]$$
$$C = \frac{1}{2}[-a - b + c + abc]$$

It is also easy to verify that the estimates of the two-factor interactions should be the following linear combinations of the observations:

$$BC = \frac{1}{2}[a - b - c + abc]$$
$$AC = \frac{1}{2}[-a + b - c + abc]$$
$$AB = \frac{1}{2}[-a - b + c + abc]$$

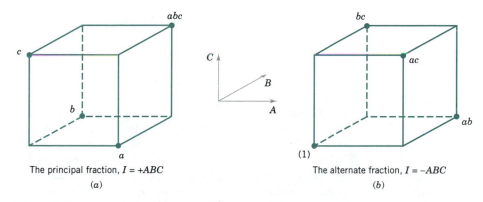

The principal fraction, $I = +ABC$

(a)

The alternate fraction, $I = -ABC$

(b)

Figure 7-27 The one-half fractions of the 2^3 design. (a) The principal fraction, $I = +ABC$. (b) The alternate fraction, $I = -ABC$.

Thus, the linear combination of observations in column A, ℓ_A, estimates both the main effect of A and the BC interaction. That is, the linear combination ℓ_A estimates the sum of these two effects $A + BC$. Similarly, ℓ_B estimates $B + AC$, and ℓ_C estimates $C + AB$. Two or more effects that have this property are called **aliases.** In our 2^{3-1} design, A and BC are aliases, B and AC are aliases, and C and AB are aliases. Aliasing is the direct result of fractional replication. In many practical situations, it will be possible to select the fraction so that the main effects and low-order interactions of interest will be aliased only with high-order interactions (which are probably negligible).

The alias structure for this design is found by using the defining relation $I = ABC$. Multiplying any effect by the defining relation yields the aliases for that effect. In our example, the alias of A is

$$A = A \cdot ABC = A^2BC = BC$$

since $A \cdot I = A$ and $A^2 = I$. The aliases of B and C are

$$B = B \cdot ABC = AB^2C = AC$$

and

$$C = C \cdot ABC = ABC^2 = AB$$

Now suppose that we had chosen the other one-half fraction—that is, the treatment combinations in Table 7-15 associated with minus on ABC. These four runs are shown in the lower half of Table 7-15 and in Fig. 7-27b. The defining relation for this design is $I = -ABC$. The aliases are $A = -BC$, $B = -AC$, and $C = -AB$. Thus, estimates of A, B, and C that result from this fraction really estimate $A - BC$, $B - AC$, and $C - AB$. In practice, it usually does not matter which one-half fraction we select. The fraction with the plus sign in the defining relation is usually called the **principal fraction,** and the other fraction is usually called the **alternate fraction.**

Note that if we had chosen AB as the generator for the fractional factorial, then

$$A = A \cdot AB = B$$

and the two main effects of A and B would be aliased. This typically loses important information.

Sometimes we use **sequences** of fractional factorial designs to estimate effects. For example, suppose we had run the principal fraction of the 2^{3-1} design with generator ABC. However, if after running the principal fraction important effects are aliased, it is possible to estimate them by running the *alternate* fraction. Then, the full factorial design is completed and the effects can be estimated by the usual calculation. Because the experiment has been split over two time periods, it has been confounded with blocks. One might be concerned that changes in the experimental conditions could bias the estimates of the effects. However, it can be shown that if the result of a change in the experimental conditions is to add a constant to all the responses, only the ABC interaction effect is biased as a result of confounding; the remaining effects are not affected. Thus, by combining a sequence of two fractional factorial designs, we can isolate both the main effects and the two-factor interactions. This property makes the fractional factorial design highly useful in experimental problems since we can run sequences of small, efficient experiments, combine information across *several* experiments, and take advantage of learning about the process we are experimenting with as we go along. This is an illustration of the concept of sequential experimentation.

A 2^{k-1} design may be constructed by writing down the treatment combinations for a full factorial with $k - 1$ factors, called the **basic design**, and then adding the kth factor by identifying its plus and minus levels with the plus and minus signs of the highest order interaction. Therefore, a 2^{3-1} fractional factorial is constructed by writing down the basic design as a full 2^2 factorial and then equating factor C with the $\pm AB$ interaction. Thus, to construct the principal fraction, we would use $C = +AB$ as follows:

Basic Design Full 2^2		Fractional Design $2^{3-1}, I = +ABC$		
A	B	A	B	$C = AB$
−	−	−	−	+
+	−	+	−	−
−	+	−	+	−
+	+	+	+	+

To obtain the alternate fraction we would equate the last column to $C = -AB$.

EXAMPLE 7-9

To illustrate the use of a one-half fraction, consider the plasma etch experiment described in Example 7-6. Suppose that we decide to use a 2^{4-1} design with $I = ABCD$ to investigate the four factors gap (A), pressure (B), C_2F_6 flow rate (C), and power setting (D). This design would be constructed by writing down as the basic design a 2^3 in the factors A, B, and C and then setting the levels of the fourth factor $D = ABC$. The design and the etch rate for each trial are shown in Table 7-16. The design is shown graphically in Fig. 7-28.

In this design, the main effects are aliased with the three-factor interactions; note that the alias of A is

$$A \cdot I = A \cdot ABCD$$

$$A = A^2BCD = BCD$$

and similarly

$$B = ACD \qquad C = ABD \qquad D = ABC$$

Table 7-16   The 2^{4-1} Design with Defining Relation $I = ABCD$

A	B	C	D = ABC	Treatment Combination	Etch Rate
−	−	−	−	(1)	550
+	−	−	+	ad	749
−	+	−	+	bd	1052
+	+	−	−	ab	650
−	−	+	+	cd	1075
+	−	+	−	ac	642
−	+	+	−	bc	601
+	+	+	+	abcd	729

The two-factor interactions are aliased with each other. For example, the alias of AB is CD:

$$AB \cdot I = AB \cdot ABCD$$
$$AB = A^2B^2CD = CD$$

The other aliases are

$$AC = BD \quad \text{and} \quad AD = BC$$

The estimates of the main effects and their aliases are found using the four columns of signs in Table 7-16. For example, from column A we obtain the estimated effect as the difference between the averages of the four $+$ runs and the four $-$ runs.

$$\ell_A = A + BCD = \tfrac{1}{4}(-550 + 749 - 1052 + 650 - 1075 + 642 - 601 + 729)$$

$$= -127.00$$

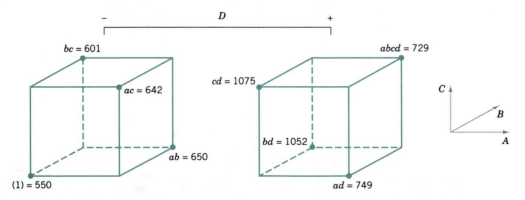

Figure 7-28   The 2^{4-1} design for the experiment of Example 7-9.

The other columns produce

$$\ell_B = B + ACD = 4.00$$

$$\ell_C = C + ABD = 11.50$$

and

$$\ell_D = D + ABC = 290.50$$

Clearly, ℓ_A and ℓ_D are large, and if we believe that the three-factor interactions are negligible, then the main effects A (gap) and D (power setting) significantly affect etch rate.

The interactions are estimated by forming the AB, AC, and AD columns and adding them to the table. For example, the signs in the AB column are $+, -, -, +, +, -, -, +$, and this column produces the estimate

$$\ell_{AB} = AB + CD = \tfrac{1}{4}(550 - 749 - 1052 + 650 + 1075 - 642 - 601 + 729)$$

$$= -10.00$$

From the AC and AD columns we find

$$\ell_{AC} = AC + BD = -25.50$$

$$\ell_{AD} = AD + BC = -197.50$$

The ℓ_{AD} estimate is large; the most straightforward interpretation of the results is that since A and D are large, this is the AD interaction. Thus, the results obtained from the 2^{4-1} design agree with the full factorial results in Example 7-6.

Normality Probability Plots and Residuals

The normal probability plot is useful in assessing the significance of effects from a fractional factorial design, particularly when many effects are to be estimated. We strongly recommend this approach. Residuals can be obtained from a fractional factorial by the regression model method shown previously. These residuals should be graphically analyzed as we have discussed before, both to assess the validity of the underlying model assumptions and to gain additional insight into the experimental situation.

Projection of a 2^{k-1} Design

If one or more factors from a one-half fraction of a 2^k can be dropped, the design will project into a full factorial design. For example, Fig. 7-29 presents a 2^{3-1} design. Note that this design will project into a full factorial in any two of the three original factors. Thus, if we think that at most two of the three factors are important, the 2^{3-1} design is an excellent design for identifying the significant factors. This **projection property** is highly useful in factor screening, for it allows negligible factors to be eliminated, resulting in a stronger experiment in the active factors that remain.

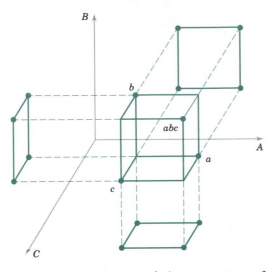

Figure 7-29   Projection of a 2^{3-1} design into three 2^2 designs.

In the 2^{4-1} design used in the plasma etch experiment in Example 7-9, we found that two of the four factors (B and C) could be dropped. If we eliminate these two factors, the remaining columns in Table 7-16 form a 2^2 design in the factors A and D, with two replicates. This design is shown in Fig. 7-30. The main effects of A and D and the strong two-factor AD interaction are clearly evident from this graph.

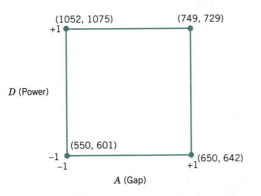

Figure 7-30   The 2^2 design obtained by dropping factors B and C from the plasma etch experiment in Example 7-9.

Design Resolution

The concept of design resolution is a useful way to catalog fractional factorial designs according to the alias patterns they produce. Designs of resolution III, IV, and V are particularly important. The definitions of these terms and an example of each follow.

1. **Resolution III Designs.**   These are designs in which no main effects are aliased with any other main effect, but main effects are aliased with two-factor interactions and some two-factor interactions may be aliased with each other. The 2^{3-1} design with $I = ABC$ is a resolution III design. We usually employ a Roman numeral subscript to indicate design resolution; thus, this one-half fraction is a 2_{III}^{3-1} design.

2. **Resolution IV Designs.**   These are designs in which no main effect is aliased with any other main effect or two-factor interactions, but two-factor interactions are aliased with each other. The 2^{4-1} design with $I = ABCD$ used in Example 7-9 is a resolution IV design (2_{IV}^{4-1}).

3. **Resolution V Designs.**   These are designs in which no main effect or two-factor interaction is aliased with any other main effect or two-factor interaction, but two-factor interactions are aliased with three-factor interactions. A 2^{5-1} design with $I = ABCDE$ is a resolution V design (2_V^{5-1}).

Resolution III and IV designs are particularly useful in factor screening experiments. A resolution IV design provides good information about main effects and will provide some information about all two-factor interactions.

7-8.2 Smaller Fractions: 2^{k-p} Fractional Factorial Design

Although the 2^{k-1} design is valuable in reducing the number of runs required for an experiment, we frequently find that smaller fractions will provide almost as much useful information at even greater economy. In general, a 2^k design may be run in a $1/2^p$ fraction called a 2^{k-p} fractional factorial design. Thus, a 1/4 fraction is called a 2^{k-2} design, a 1/8 fraction is called a 2^{k-3} design, a 1/16 fraction a 2^{k-4} design, and so on.

To illustrate the 1/4 fraction, consider an experiment with six factors and suppose that the engineer is primarily interested in main effects but would also like to get some information about the two-factor interactions. A 2^{6-1} design would require 32 runs and would have 31 degrees of freedom for estimating effects. Since there are only 6 main effects and 15 two-factor interactions, the one-half fraction is inefficient—it requires too many runs. Suppose we consider a 1/4 fraction, or a 2^{6-2} design. This design contains 16 runs and, with 15 degrees of freedom, will allow all 6 main effects to be estimated, with some capability for examining the two-factor interactions.

To generate this design, we would write down a 2^4 design in the factors A, B, C, and D as the basic design, and then add two columns for E and F. To find the new columns we could select the two **design generators** $I = ABCE$ and $I = BCDF$. Thus, column E would be found from $E = ABC$, and column F would be $F = BCD$. That is, columns $ABCE$ and $BCDF$ are equal to the identity column. However, we know that the product of any two columns in the table of plus and minus signs for a 2^k design is simply another column in the table; therefore, the product of $ABCE$ and $BCDF$ that equals $ABCE$ $(BCDF) = AB^2C^2DEF = ADEF$ is also an identity column. Consequently, the **complete defining relation** for the 2^{6-2} design is

$$I = ABCE = BCDF = ADEF$$

We refer to each term in a defining relation (such as $ABCE$ above) as a **word**. To find the alias of any effect, simply multiply the effect by each word in the foregoing defining relation. For example, the alias of A is

$$A = BCE = ABCDF = DEF$$

The complete alias relationships for this design are shown in Table 7-17. In general, the resolution of a 2^{k-p} design is equal to the number of letters in the shortest word in the complete defining relation. Therefore, this is a resolution IV design; main effects are aliased with three-factor and higher interactions, and two-factor interactions are aliased with each other. This design would provide good information on the main effects and would give some idea about the strength of the two-factor interactions. The construction and analysis of the design are illustrated in Example 7-10.

EXAMPLE 7-10

Parts manufactured in an injection-molding process are showing excessive shrinkage, which is causing problems in assembly operations upstream from the injection-molding area. In an effort to reduce the shrinkage, a quality-improvement team has decided to use a designed experiment to study the injection-molding process. The team investigates six factors—mold temperature (A), screw speed (B), holding time (C), cycle time (D), gate size (E), and holding pressure (F)—each at two levels, with the objective of learning how each factor affects shrinkage and obtaining preliminary information about how the factors interact.

Table 7-17   Alias Structure for the 2_{IV}^{6-2} Design with $I = ABCE = BCDF = ADEF$

$A = BCE = DEF = ABCDF$	$AB = CE = ACDF = BDEF$
$B = ACE = CDF = ABDEF$	$AC = BE = ABDF = CDEF$
$C = ABE = BDF = ACDEF$	$AD = EF = BCDE = ABCF$
$D = BCF = AEF = ABCDE$	$AE = BC = DF = ABCDEF$
$E = ABC = ADF = BCDEF$	$AF = DE = BCEF = ABCD$
$F = BCD = ADE = ABCEF$	$BD = CF = ACDE = ABEF$
$ABD = CDE = ACF = BEF$	$BF = CD = ACEF = ABDE$
$ACD = BDE = ABF = CEF$	

Table 7-18 A 2_{IV}^{6-2} Design for the Injection-Molding Experiment in Example 7-10

Run	A	B	C	D	E = ABC	F = BCD	Observed Shrinkage (×10)
1	−	−	−	−	−	−	6
2	+	−	−	−	+	−	10
3	−	+	−	−	+	+	32
4	+	+	−	−	−	+	60
5	−	−	+	−	+	+	4
6	+	−	+	−	−	+	15
7	−	+	+	−	−	−	26
8	+	+	+	−	+	−	60
9	−	−	−	+	−	+	8
10	+	−	−	+	+	+	12
11	−	+	−	+	+	−	34
12	+	+	−	+	−	−	60
13	−	−	+	+	+	−	16
14	+	−	+	+	−	−	5
15	−	+	+	+	−	+	37
16	+	+	+	+	+	+	52

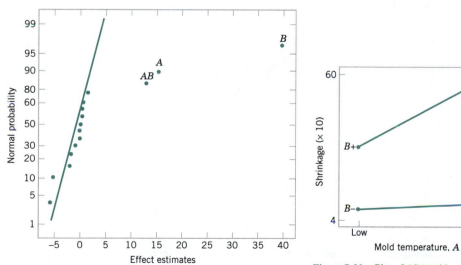

Figure 7-31 Normal probability plot of effects for Example 7-10.

Figure 7-32 Plot of AB (mold temperature-screw speed) interaction for Example 7-10.

The team decides to use a 16-run two-level fractional factorial design for these six factors. The design is constructed by writing down a 2^4 as the basic design in the factors *A, B, C,* and *D* and then setting $E = ABC$ and $F = BCD$ as discussed previously. Table 7-18 shows the design, along with the observed shrinkage ($\times 10$) for the test part produced at each of the 16 runs in the design.

A normal probability plot of the effect estimates from this experiment is shown in Fig. 7-31. Output from Minitab is shown at the end of this chapter in Table 7-33. The only large effects are *A* (mold temperature), *B* (screw speed), and the *AB* interaction. In light of the alias relationship in Table 7-17, it seems reasonable to tentatively adopt these conclusions. The plot of the *AB* interaction in Fig. 7-32 shows that the process is insensitive to temperature if the screw speed is at the low level but sensitive to temperature if the screw speed is at the high level. With the screw speed at a low level, the process should produce an average shrinkage of around 10% regardless of the temperature level chosen.

Based on this initial analysis, the team decides to set both the mold temperature and the screw speed at the low level. This set of conditions should reduce the mean shrinkage of parts to around 10%. However, the variability in shrinkage from part to part is still a potential problem. In effect, the mean shrinkage can be adequately reduced by the above modifications; however, the part-to-part variability in shrinkage over a production run could still cause problems in assembly. One way to address this issue is to see if any of the process factors affect the variability in parts shrinkage.

Figure 7-33 presents the normal probability plot of the residuals. This plot appears satisfactory. The plots of residuals versus each factor were then constructed. One of these plots, that for residuals versus factor *C* (holding time), is shown in Fig. 7-34. The plot

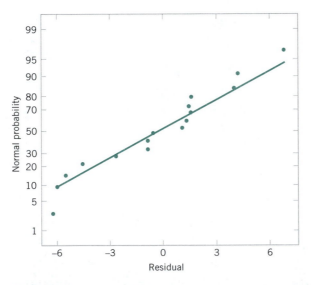

Figure 7-33   Normal probability plot of residuals for Example 7-10.

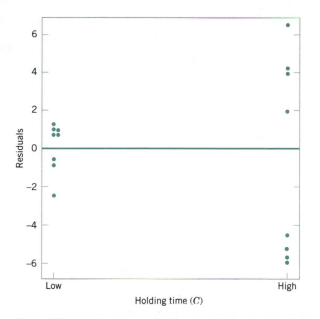

Figure 7-34 Residuals versus holding time (C) for Example 7-10.

reveals much less scatter in the residuals at the low holding time than at the high holding time. These residuals were obtained in the usual way from a model for predicted shrinkage

$$\hat{y} = \hat{\beta}_0 + \hat{\beta}_1 x_1 + \hat{\beta}_2 x_2 + \hat{\beta}_{12} x_1 x_2$$
$$= 27.3125 + 6.9375 x_1 + 17.8125 x_2 + 5.9375 x_1 x_2$$

where x_1, x_2, and $x_1 x_2$ are coded variables that correspond to the factors A and B and the AB interaction. The residuals are then

$$e = y - \hat{y}$$

The regression model used to produce the residuals essentially removes the location effects of A, B, and AB from the data; the residuals therefore contain information about unexplained variability. Figure 7-34 indicates that there is a pattern in the variability and that the variability in the shrinkage of parts may be smaller when the holding time is at the low level.

Figure 7-35 shows the data from this experiment projected onto a cube in the factors A, B, and C. The average observed shrinkage and the range of observed shrinkage are shown at each corner of the cube. From inspection of this figure, we see that running the process with the screw speed (B) at the low level is the key to reducing average parts shrinkage. If B is low, virtually any combination of temperature (A) and holding time (C) will result in low values of average parts shrinkage. However, from examining the

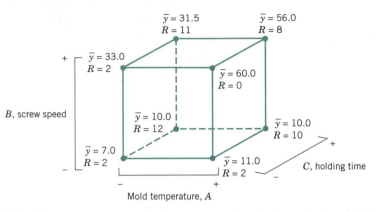

Figure 7-35 Average shrinkage and range of shrinkage in factors *A, B,* and *C* for Example 7-10.

ranges of the shrinkage values at each corner of the cube, it is immediately clear that setting the holding time (*C*) at the low level is the most appropriate choice if we wish to keep the part-to-part variability in shrinkage low during a production run.

The concepts used in constructing the 2^{6-2} fractional factorial design in Example 7-10 can be extended to the construction of any 2^{k-p} fractional factorial design. In general, a 2^k fractional factorial design containing 2^{k-p} runs is called a $1/2^p$ fraction of the 2^k design or, more simply, a 2^{k-p} fractional factorial design. These designs require the selection of *p* independent generators. The defining relation for the design consists of the *p* generators initially chosen and their $2^p - p - 1$ generalized interactions.

The alias structure may be found by multiplying each effect column by the defining relation. Care should be exercised in choosing the generators so that effects of potential interest are not aliased with each other. Each effect has $2^p - 1$ aliases. For moderately large values of *k*, we usually assume higher order interactions (say, third- or fourth-order or higher) to be negligible, and this greatly simplifies the alias structure.

It is important to select the *p* generators for the 2^{k-p} fractional factorial design in such a way that we obtain the best possible alias relationships. A reasonable criterion is to select the generators so that the resulting 2^{k-p} design has the highest possible design resolution. Montgomery (1997) presents a table of recommended generators for 2^{k-p} fractional factorial designs for $k \leq 11$ factors and up to $n \leq 128$ runs. His table is reproduced here as Table 7-19. In this table, the generators are shown with either + or − choices; selection of all generators as + will give a principal fraction, whereas if any generators are − choices, the design will be one of the alternate fractions for the same family. The suggested generators in this table will result in a design of the highest possible resolution. Montgomery (1997) also gives a table of alias relationships for these designs.

Table 7-19   Selected 2^{k-p} Fractional Factorial Designs

Number of Factors k	Fraction	Number of Runs	Design Generators
3	2^{3-1}_{III}	4	$C = \pm AB$
4	2^{4-1}_{IV}	8	$D = \pm ABC$
5	2^{5-1}_{V}	16	$E = \pm ABCD$
	2^{5-2}_{III}	8	$D = \pm AB$ $E = \pm AC$
6	2^{6-1}_{VI}	32	$F = \pm ABCDE$
	2^{6-2}_{IV}	16	$E = \pm ABC$ $F = \pm BCD$
	2^{6-3}_{III}	8	$D = \pm AB$ $E = \pm AC$ $F = \pm BC$
7	2^{7-1}_{VII}	64	$G = \pm ABCDEF$
	2^{7-2}_{IV}	32	$E = \pm ABC$ $G = \pm ABDE$
	2^{7-3}_{IV}	16	$E = \pm ABC$ $F = \pm BCD$ $G = \pm ACD$
	2^{7-4}_{III}	8	$D = \pm AB$ $E = \pm AC$ $F = \pm BC$ $G = \pm ABC$
8	2^{8-2}_{V}	64	$G = \pm ABCD$ $H = \pm ABEF$
	2^{8-3}_{IV}	32	$F = \pm ABC$ $G = \pm ABD$ $H = \pm BCDE$
	2^{8-4}_{IV}	16	$E = \pm BCD$ $F = \pm ACD$ $G = \pm ABC$ $H = \pm ABD$
9	2^{9-2}_{VI}	128	$H = \pm ACDFG$ $J = \pm BCEFG$
	2^{9-3}_{IV}	64	$G = \pm ABCD$ $H = \pm ACEF$ $J = \pm CDEF$
	2^{9-4}_{IV}	32	$F = \pm BCDE$ $G = \pm ACDE$ $H = \pm ABDE$ $J = \pm ABCE$
	2^{9-5}_{III}	16	$E = \pm ABC$ $F = \pm BCD$ $G = \pm ACD$ $H = \pm ABD$ $J = \pm ABCD$

(Continued)

Table 7-19   Selected 2^{k-p} Fractional Factorial Designs *(Continued)*

Number of Factors k	Fraction	Number of Runs	Design Generators
10	2_V^{10-3}	128	$H = \pm ABCG$
			$J = \pm ACDE$
			$K = \pm ACDF$
	2_{IV}^{10-4}	64	$G = \pm BCDF$
			$H = \pm ACDF$
			$J = \pm ABDE$
			$K = \pm ABCE$
	2_{IV}^{10-5}	32	$F = \pm ABCD$
			$G = \pm ABCE$
			$H = \pm ABDE$
			$J = \pm ACDE$
			$K = \pm BCDE$
	2_{III}^{10-6}	16	$E = \pm ABC$
			$F = \pm BCD$
			$G = \pm ACD$
			$H = \pm ABD$
			$J = \pm ABCD$
			$K = \pm AB$
11	2_{IV}^{11-5}	64	$G = \pm CDE$
			$H = \pm ABCD$
			$J = \pm ABF$
			$K = \pm BDEF$
			$L = \pm ADEF$
	2_{IV}^{11-6}	32	$F = \pm ABC$
			$G = \pm BCD$
			$H = \pm CDE$
			$J = \pm ACD$
			$K = \pm ADE$
			$L = \pm BDE$
	2_{III}^{11-7}	16	$E = \pm ABC$
			$F = \pm BCD$
			$G = \pm ACD$
			$H = \pm ABD$
			$J = \pm ABCD$
			$K = \pm AB$
			$L = \pm AC$

Source: Montgomery (1997).

EXAMPLE 7-11

To illustrate the use of Table 7-19, suppose that we have seven factors and that we are interested in estimating the seven main effects and obtaining some insight regarding the two-factor interactions. We are willing to assume that three-factor and higher interactions are negligible. This information suggests that a resolution IV design would be appropriate.

Table 7-19 shows that two resolution IV fractions are available: the 2_{IV}^{7-2} with 32 runs and the 2_{IV}^{7-3} with 16 runs. The aliases involving main effects and two- and three-factor interactions for the 16-run design are presented in Table 7-20. Note that all seven main effects are aliased with three-factor interactions. All the two-factor interactions are aliased in groups of three. Therefore, this design will satisfy our objectives; that is, it will allow the estimation of the main effects, and it will give some insight regarding two-factor interactions. It is not necessary to run the 2_{IV}^{7-2} design, which would require 32 runs. The construction of the 2_{IV}^{7-3} design is shown in Table 7-21. Note that it was constructed by starting with the 16-run 2^4 design in A, B, C, and D as the basic design and then adding the three columns $E = ABC$, $F = BCD$, and $G = ACD$ as suggested in Table 7-19. Thus, the generators for this design are $I = ABCE$, $I = BCDF$, and $I = ACDG$. The complete defining relation is $I = ABCE = BCDF = ADEF = ACDG = BDEG = CEFG = ABFG$. This defining relation was used to produce the aliases in Table 7-20. For example, the alias relationship of A is

$$A = BCE = ABCDF = DEF = CDG = ABDEG = ACEFG = BFG$$

which, if we ignore interactions higher than three factors, agrees with those in Table 7-20.

Table 7-20 Generators, Defining Relation, and Aliases for the 2_{IV}^{7-3} Fractional Factorial Design

Generators and Defining Relation

$$E = ABC \qquad F = BCD \qquad G = ACD$$

$$I = ABCE = BCDF = ADEF = ACDG = BDEG = ABFG = CEFG$$

Aliases

$A = BCE = DEF = CDG = BFG$	$AB = CE = FG$
$B = ACE = CDF = DEG = AFG$	$AC = BE = DG$
$C = ABE = BDF = ADG = EFG$	$AD = EF = CG$
$D = BCF = AEF = ACG = BEG$	$AE = BC = DF$
$E = ABC = ADF = BDG = CFG$	$AF = DE = BG$
$F = BCD = ADE = ABG = CEG$	$AG = CD = BF$
$G = ACD = BDE = ABF = CEF$	$BD = CF = EG$

$$ABD = CDE = ACF = BEF = BCG = AEG = DFG$$

Table 7-21 A 2_{IV}^{7-3} Fractional Factorial Design

	Basic Design						
Run	A	B	C	D	E = ABC	F = BCD	G = ACD
1	−	−	−	−	−	−	−
2	+	−	−	−	+	−	+
3	−	+	−	−	+	+	−
4	+	+	−	−	−	+	+
5	−	−	+	−	+	+	+
6	+	−	+	−	−	+	−
7	−	+	+	−	−	−	+
8	+	+	+	−	+	−	−
9	−	−	−	+	−	+	−
10	+	−	−	+	+	+	−
11	−	+	−	+	+	−	+
12	+	+	−	+	−	−	−
13	−	−	+	+	+	−	−
14	+	−	+	+	−	−	+
15	−	+	+	+	−	+	−
16	+	+	+	+	+	+	+

For seven factors, we can reduce the number of runs even further. The 2^{7-4} design is an eight-run experiment accommodating seven variables. This is a 1/16th fraction and is obtained by first writing down a 2^3 design as the basic design in the factors A, B, and C, and then forming the four new columns from $I = ABD$, $I = ACE$, $I = BCF$, and $I = ABCG$, as suggested in Table 7-19. The design is shown in Table 7-22.

Table 7-22 A 2_{III}^{7-4} Fractional Factorial Design

A	B	C	D = AB	E = AC	F = BC	G = ABC
−	−	−	+	+	+	−
+	−	−	−	−	+	+
−	+	−	−	+	−	+
+	+	−	+	−	−	−
−	−	+	+	−	−	+
+	−	+	−	+	−	−
−	+	+	−	−	+	−
+	+	+	+	+	+	+

The complete defining relation is found by multiplying the generators together two, three, and finally four at a time, producing

$$I = ABD = ACE = BCF = ABCG = BCDE = ACDF = CDG = ABEF$$
$$= BEG = AFG = DEF = ADEG = CEFG = BDFG = ABCDEFG$$

The alias of any main effect is found by multiplying that effect through each term in the defining relation. For example, the alias of A is

$$A = BD = CE = ABCF = BCG = ABCDE = CDF = ACDG$$
$$= BEF = ABEG = FG = ADEF = DEG = ACEFG = ABDFG$$
$$= BCDEFG$$

This design is of resolution III, since the main effect is aliased with two-factor interactions. If we assume that all three-factor and higher interactions are negligible, the aliases of the seven main effects are

$$\ell_A = A + BD + CE + FG$$
$$\ell_B = B + AD + CF + EG$$
$$\ell_C = C + AE + BF + DG$$
$$\ell_D = D + AB + CG + EF$$
$$\ell_E = E + AC + BG + DF$$
$$\ell_F = F + BC + AG + DE$$
$$\ell_G = G + CD + BE + AF$$

This 2_{III}^{7-4} design is called a **saturated fractional factorial,** because all the available degrees of freedom are used to estimate main effects. It is possible to combine sequences of these resolution III fractional factorials to separate the main effects from the two-factor interactions. The procedure is illustrated in Montgomery (1997).

EXERCISES FOR SECTION 7-8

7-23. R. D. Snee ("Experimenting with a Large Number of Variables," in *Experiments in Industry: Design, Analysis and Interpretation of Results,* by R. D. Snee, L. D. Hare, and J. B. Trout, eds., ASQC, 1985) describes an experiment in which a 2^{5-1} design with $I = ABCDE$ was used to investigate the effects of five factors on the color of a chemical product. The factors are A = solvent/reactant, B = catalyst/reactant, C = temperature, D = reactant purity, and E = reactant pH. The results obtained are as follows:

e	$= -0.63$	d	$= 6.79$
a	$= 2.51$	ade	$= 6.47$
b	$= -2.68$	bde	$= 3.45$
abe	$= 1.66$	abd	$= 5.68$
c	$= 2.06$	cde	$= 5.22$
ace	$= 1.22$	acd	$= 4.38$
bce	$= -2.09$	bcd	$= 4.30$
abc	$= 1.93$	$abcde$	$= 4.05$

(a) Write down the complete defining relation and the aliases from the design.

(b) Estimate the effects and prepare a normal probability plot of the effects. Which effects are active?

(c) Interpret the effects and develop an appropriate model for the response.

(d) Plot the residuals from your model against the fitted values. Also construct a normal probability plot of the residuals. Comment on the results.

7-24. Montgomery (1997) describes a 2^{4-1} fractional factorial design used to study four factors in a chemical process. The factors are A = temperature, B = pressure, C = concentration, and D = stirring rate, and the response is filtration rate. The design and the data are shown in Table 7-23.

(a) Write down the complete defining relation and the aliases from the design.

(b) Estimate the effects and prepare a normal probability plot of the effects. Which effects are active?

(c) Interpret the effects and develop an appropriate model for the response.

(d) Plot the residuals from your model against the fitted values. Also construct a normal probability plot of the residuals. Comment on the results.

7-25. An article in *Industrial and Engineering Chemistry* ("More on Planning Experiments to Increase Research Efficiency," 1970, pp. 60–65) uses a 2^{5-2} design to investigate the effect on process yield of A = condensation temperature, B = amount of material 1, C = solvent volume, D = condensation time, and E = amount of material 2. The results obtained are as follows:

$$
\begin{aligned}
e &= 23.2 & cd &= 23.8 \\
ab &= 15.5 & ace &= 23.4 \\
ad &= 16.9 & bde &= 16.8 \\
bc &= 16.2 & abcde &= 18.1
\end{aligned}
$$

(a) Write down the complete defining relation and the aliases from the design. Verify that the design generators used were $I = ACE$ and $I = BDE$.

(b) Estimate the effects and prepare a normal probability plot of the effects. Which effects are active? Verify that the AB and AD interactions are available to use as error.

(c) Interpret the effects and develop an appropriate model for the response.

(d) Plot the residuals from your model against the fitted values. Also construct a normal probability plot of the residuals. Comment on the results.

7-26. Suppose that in Exercise 7-10 it was

Table 7-23 Data for Exercise 7-24

Run	A	B	C	$D = ABC$	Treatment Combination	Filtration Rate
1	−	−	−	−	(1)	45
2	+	−	−	+	ad	100
3	−	+	−	+	bd	45
4	+	+	−	−	ab	65
5	−	−	+	+	cd	75
6	+	−	+	−	ac	60
7	−	+	+	−	bc	80
8	+	+	+	+	$abcd$	96

possible to run only a one-half fraction of replicate I for the 2^4 design. Construct the design and use only the data from the eight runs you have generated to perform the analysis.

7-27. Suppose that in Exercise 7-12 only a one-quarter fraction of the 2^5 design could be run. Construct the design and analyze the data that are obtained by selecting only the response for the eight runs in your design.

7-28. Construct the table of the treatment combinations tested for the 2_{IV}^{6-2} recommended in Table 7-17.

7-29. Construct a 2_{III}^{6-3} fractional factorial design. Write down the aliases, assuming that only main effects and two-factor interactions are of interest.

7-9   RESPONSE SURFACE METHODS AND DESIGNS

Response surface methodology, or RSM, is a collection of mathematical and statistical techniques that are useful for modeling and analysis in applications where a response of interest is influenced by several variables and the objective is to **optimize** this response. For example, suppose that a chemical engineer wishes to find the levels of temperature (x_1) and feed concentration (x_2) that maximize the yield (y) of a process. The process yield is a function of the levels of temperature and feed concentration—say,

$$Y = f(x_1, x_2) + \epsilon \tag{7-18}$$

where ϵ represents the noise or error observed in the response Y. If we denote the expected response by $E(Y) = f(x_1, x_2)$, then the surface represented by $f(x_1, x_2)$ is called a **response surface.**

We may represent the response surface graphically as shown in Fig. 7-36, where $f(x_1, x_2)$ is plotted versus the levels of x_1 and x_2. Note that the response is represented

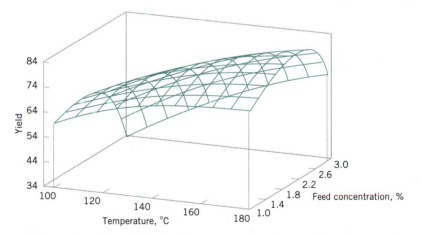

Figure 7-36   A three-dimensional response surface showing the expected yield as a function of temperature and feed concentration.

as a surface plot in a three-dimensional space. To help visualize the shape of a response surface, we often plot the contours of the response surface as shown in Fig. 7-37. In the contour plot, lines of constant response are drawn in the x_1, x_2 plane. Each contour corresponds to a particular height of the response surface. The contour plot is helpful in studying the levels of x_1 and x_2 that result in changes in the shape or height of the response surface.

In most RSM problems, the form of the relationship between the response and the independent variables is unknown. Thus, the first step in RSM is to find a suitable approximation for the true relationship between Y and the independent variables. Usually, a low-order polynomial in some region of the independent variables is employed. If the response is well modeled by a linear function of the independent variables, then the approximating function is the **first-order model**

$$Y = \beta_0 + \beta_1 x_1 + \beta_2 x_2 + \cdots + \beta_k x_k + \epsilon$$

If there is curvature in the system, then a polynomial of higher degree must be used, such as the **second-order model**

$$Y = \beta_0 + \sum_{i=1}^{k} \beta_i x_i + \sum_{i=1}^{k} \beta_{ii} x_i^2 + \sum\sum_{i<j} \beta_{ij} x_i x_j + \epsilon \tag{7-19}$$

Many RSM problems utilize one or both of these approximating polynomials. Of course, it is unlikely that a polynomial model will be a reasonable approximation of the true functional relationship over the entire space of the independent variables, but for a relatively small region they usually work quite well.

The method of least squares, discussed in Chapter 6, is used to estimate the parameters in the approximating polynomials. The response surface analysis is then done in terms of the fitted surface. If the fitted surface is an adequate approximation of the true response

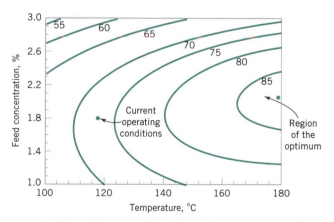

Figure 7-37   A contour plot of the yield response surface in Figure 7-36.

function, then analysis of the fitted surface will be approximately equivalent to analysis of the actual system.

RSM is a **sequential** procedure. Often, when we are at a point on the response surface that is remote from the optimum, such as the current operating conditions in Fig. 7-37, there is little curvature in the system and the first-order model will be appropriate. Our objective here is to lead the experimenter rapidly and efficiently to the general vicinity of the optimum. Once the region of the optimum has been found, a more elaborate model such as the second-order model may be employed, and an analysis may be performed to locate the optimum. From Fig. 7-37, we see that the analysis of a response surface can be thought of as "climbing a hill," where the top of the hill represents the point of maximum response. If the true optimum is a point of minimum response, then we may think of "descending into a valley."

The eventual objective of RSM is to determine the optimum operating conditions for the system or to determine a region of the factor space in which operating specifications are satisfied. Also, note that the word "optimum" in RSM is used in a special sense. The "hill climbing" procedures of RSM guarantee convergence to a local optimum only.

7-9.1 Method of Steepest Ascent

Frequently, the initial estimate of the optimum operating conditions for the system will be far from the actual optimum. In such circumstances, the objective of the experimenter is to move rapidly to the general vicinity of the optimum. We wish to use a simple and economically efficient experimental procedure. When we are remote from the optimum, we usually assume that a first-order model is an adequate approximation to the true surface in a small region of the x's.

The **method of steepest ascent** is a procedure for moving sequentially along the path of steepest ascent—that is, in the direction of the maximum increase in the response. Of course, if **minimization** is desired, then we are talking about the **method of steepest descent**. The fitted first-order model is

$$\hat{y} = \hat{\beta}_0 + \sum_{i=1}^{k} \hat{\beta}_i x_i \tag{7-20}$$

and the first-order response surface—that is, the contours of $\hat{y}$—is a series of parallel lines such as that shown in Fig. 7-38. The direction of steepest ascent is the direction in which $\hat{y}$ increases most rapidly. This direction is normal to the fitted response surface contours. We usually take as the **path of steepest ascent** the line through the center of the region of interest and normal to the fitted surface contours. As shown in Example 7-12, the steps along the path are proportional to the regression coefficients $\{\hat{\beta}_i\}$. The experimenter determines the actual step size based on process knowledge or other practical considerations.

Experiments are conducted along the path of steepest ascent until no further increase in response is observed. Then a new first-order model may be fit, a new direction of steepest ascent determined, and further experiments conducted in that direction until the experimenter feels that the process is near the optimum.

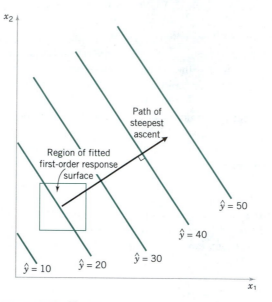

Figure 7-38 First-order response surface and path of steepest ascent.

EXAMPLE 7-12

In Example 7-7 we described an experiment on a chemical process in which two factors, reaction time (x_1) and reaction temperature (x_2), affect the percent conversion or yield (Y). Figure 7-22 shows the 2^2 design plus five center points used in this study. The engineer found that both factors were important, there was no interaction, and there was no curvature in the response surface. Therefore, the first-order model

$$Y = \beta_0 + \beta_1 x_1 + \beta_2 x_2 + \epsilon$$

should be appropriate. Now the effect estimate of time is 1.55 and the effect estimate of temperature is 0.65, and since the regression coefficients $\hat{\beta}_1$ and $\hat{\beta}_2$ are one-half of the corresponding effect estimates, the fitted first-order model is

$$\hat{y} = 40.44 + 0.775x_1 + 0.325x_2$$

Figures 7-39a and b show the contour plot and three-dimensional surface plot of this model. Figure 7-39 also shows the relationship between the **coded variables** x_1 and x_2 (that defined the high and low levels of the factors) and the original variables time (in minutes) and temperature (in °F).

From examining these plots (or the fitted model), we see that to move away from the design center—the point ($x_1 = 0$, $x_2 = 0$)—along the path of steepest ascent, we would move 0.775 unit in the x_1 direction for every 0.325 unit in the x_2 direction. Thus,

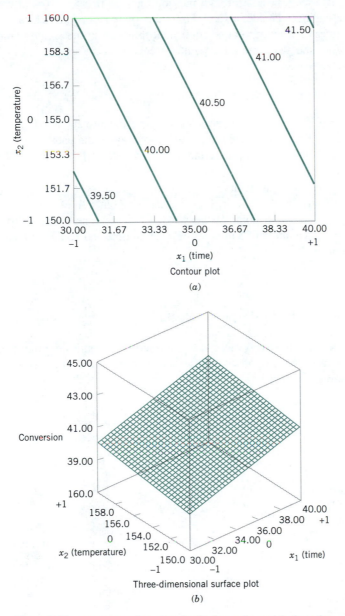

Figure 7-39 Response surface plots for the first-order model of reaction time and temperature.

the path of steepest ascent passes through the point ($x_1 = 0$, $x_2 = 0$) and has a slope 0.325/0.775. The engineer decides to use 5 minutes of reaction time as the basic step size. Now, 5 minutes of reaction time is equivalent to a step in the *coded* variable x_1 of $\Delta x_1 = 1$. Therefore, the steps along the path of steepest ascent are

$$\Delta x_1 = 1.000$$
$$\Delta x_2 = (\hat{\beta}_2/\hat{\beta}_1)\Delta x_1 = (0.325/0.775)\Delta x_1 = 0.42 \qquad (7\text{-}21)$$

A change of $\Delta x_2 = 0.42$ in the coded variable x_2 is equivalent to about 2°F in the original variable temperature. Therefore, the engineer will move along the path of steepest ascent by increasing reaction time by 5 minutes and temperature by 2°F. An actual observation on yield will be determined at each point.

Figure 7-40 shows several points along this path of steepest ascent and the yields actually observed from the process at those points. At points A–D the observed yield increases steadily, but beyond point D the yield decreases. Therefore, steepest ascent would terminate in the vicinity of 55 minutes of reaction time and 163°F with an observed percent conversion of 67%.

7-9.2   Analysis of a Second-Order Response Surface

When the experimenter is relatively close to the optimum, a second-order model is usually required to approximate the response because of curvature in the true response surface. The fitted second-order model is

$$\hat{y} = \hat{\beta}_0 + \sum_{i=1}^{k} \hat{\beta}_i x_i + \sum_{i=1}^{k} \hat{\beta}_{ii} x_i^2 + \sum\sum_{i<j} \hat{\beta}_{ij} x_i x_j \qquad (7\text{-}22)$$

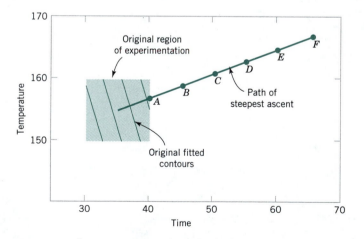

Point A: 40 minutes, 157°F, $y = 40.5$
Point B: 45 minutes, 159°F, $y = 51.3$
Point C: 50 minutes, 161°F, $y = 59.6$
Point D: 55 minutes, 163°F, $y = 67.1$
Point E: 60 minutes, 165°F, $y = 63.6$
Point F: 65 minutes, 167°F, $y = 60.7$

Figure 7-40   Steepest ascent experiment for the first-order model of reaction time and temperature.

where $\hat{\beta}$ denotes the least squares estimate of β. In this section we show how to use this fitted model to find the optimum set of operating conditions for the x's and to characterize the nature of the response surface.

EXAMPLE 7-13

Continuation of Example 7-12

The method of steepest ascent terminated at a reaction time of 55 minutes and a temperature of 163°F. The experimenter decides to fit a second-order model in this region. Table 7-24 and Fig. 7-41 show the experimental design, which consists of a 2^2 design centered at 55 minutes and 165°F, five center points, and four runs along the coordinate axes called axial runs. This type of design is called a **central composite design** (CCD), and it is a very popular design for fitting second-order response surfaces.

Two response variables were measured during this phase of the experiment: percent conversion (yield) and viscosity. The least squares quadratic model for the yield response is

$$\hat{y}_1 = 69.1 + 1.633x_1 + 1.083x_2 - 0.969x_1^2 - 1.219x_2^2 + 0.225x_1x_2$$

The analysis of variance for this model is shown in Table 7-25. Because the coefficient of the x_1x_2 term is not significant, one might choose to remove this term from the model.

Figure 7-42 shows the response surface contour plot and the three-dimensional surface plot for this model. From examination of these plots, the maximum yield is about 70%, obtained at approximately 60 minutes of reaction time and 167°F.

Table 7-24 Central Composite Design for Example 7-13

Observation Number	Time (minutes)	Temperature (°F)	Coded Variables x_1	Coded Variables x_2	Conversion (percent) Response 1	Viscosity (mPa-sec) Response 2
1	50	160	-1	-1	65.3	35
2	60	160	1	-1	68.2	39
3	50	170	-1	1	66	36
4	60	170	1	1	69.8	43
5	48	165	-1.414	0	64.5	30
6	62	165	1.414	0	69	44
7	55	158	0	-1.414	64	31
8	55	172	0	1.414	68.5	45
9	55	165	0	0	68.9	37
10	55	165	0	0	69.7	34
11	55	165	0	0	68.5	35
12	55	165	0	0	69.4	36
13	55	165	0	0	69	37

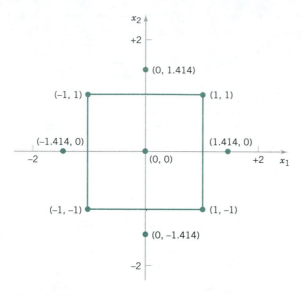

Figure 7-41   Central composite design for Example 7-13.

Table 7-25   Analysis of Variance for the Quadratic Model, Yield Response

Source of Variation	Sum of Squares	Degrees of Freedom	Mean Square	f_0	P-value
Model	45.89	5	9.178	14.93	0.0013
Residual	4.30	7	0.615		
Total	50.19	12			

Independent Variable	Coefficient Estimate	Standard Error	t for H_0 Coefficient $= 0$	P-value
Intercept	69.100	0.351	197.1	0.0000
x_1	1.633	0.277	5.891	0.0006
x_2	1.083	0.277	3.907	0.0058
x_1^2	-0.969	0.297	-3.259	0.0139
x_2^2	-1.219	0.297	-4.100	0.0046
$x_1 x_2$	0.225	0.392	0.5740	0.5839

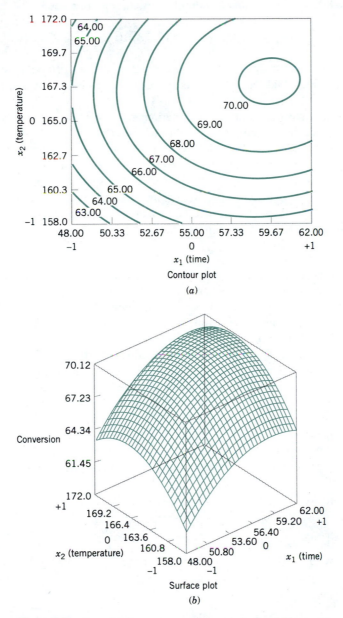

Figure 7-42 Second-order response surface plots in the yield response, Example 7-13.

The viscosity response is adequately described by the first-order model

$$\hat{y}_2 = 37.08 + 3.85x_1 + 3.10x_2$$

Table 7-26 summarizes the analysis of variance for this model. The response surface is shown graphically in Fig. 7-43. Note that viscosity increases as both time and temperature increase.

As in most response surface problems, the experimenter in this example had conflicting objectives regarding the two responses. The objective was to maximize yield, but the acceptable range for viscosity was $38 \le y_2 \le 42$. When there are only a few independent variables, an easy way to solve this problem is to overlay the response surfaces to find the optimum. Figure 7-44 shows the overlay plot of both responses, with the contours $y_1 = 69\%$ conversion, $y_2 = 38$, and $y_2 = 42$ highlighted. The shaded areas on this plot identify infeasible combinations of time and temperature. This graph shows that several combinations of time and temperature will be satisfactory.

Example 7-13 illustrates the use of a central composite design for fitting a second-order response surface model. These designs are widely used in practice because they are relatively efficient with respect to the number of runs required. In general, a CCD in k factors requires 2^k factorial runs, $2k$ axial runs, and at least one center point (3 to 5 center points are typically used). Designs for $k = 2$ and $k = 3$ factors are shown in Fig. 7-45.

The central composite design may be made **rotatable** by proper choice of the axial spacing α in Fig. 7-45. If the design is rotatable, the standard deviation of predicted response $\hat{y}$ is constant at all points that are the same distance from the center of the design. For rotatability, choose $\alpha = (F)^{1/4}$, where F is the number of points in the factorial part of the design (usually $F = 2^k$). For the case of $k = 2$ factors, $\alpha = (2^2)^{1/4} = 1.414$, as was used in the design in Example 7-13. Figure 7-46 presents a contour plot and a surface plot of the standard deviation of prediction for the quadratic model used for the yield response. Note that the contours are concentric circles, implying that yield is predicted

Table 7-26   Analysis of Variance for the First-Order Model, Viscosity Response

Source	Sum of Squares	Degrees of Freedom	Mean Square	f_0	P-value
Model	195.4	2	97.72	15.89	0.0008
Residual	61.5	10	6.15		
Total	256.9	12			

Independent Variable	Coefficient Estimate	Degrees of Freedom	Standard Error of Coefficient	t for H_0 Coefficient $= 0$	P-value
Intercept	37.08	1	0.69	53.91	
x_1	3.85	1	0.88	4.391	0.0014
x_2	3.10	1	0.88	3.536	0.0054

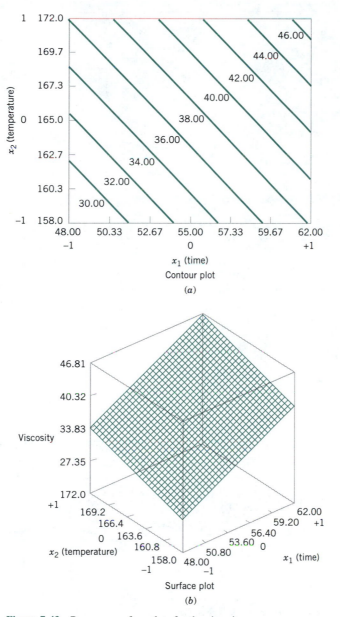

Figure 7-43 Response surface plots for the viscosity response.

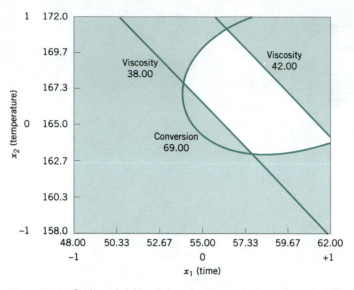

Figure 7-44   Overlay of yield and viscosity response surfaces, Example 7-13.

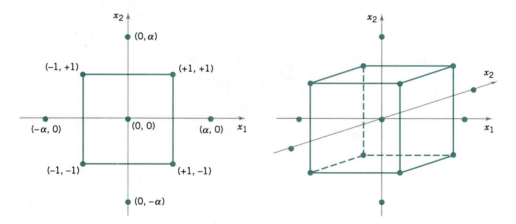

Figure 7-45   Central composite designs for $k = 2$ and $k = 3$.

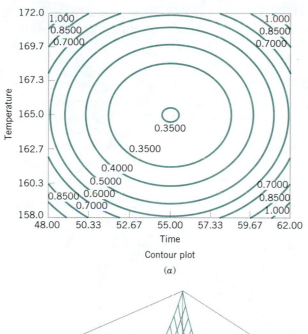

Contour plot

(a)

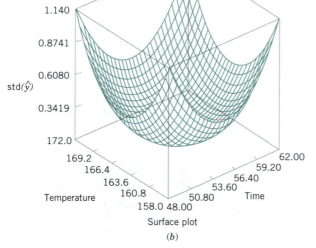

Surface plot

(b)

Figure 7-46 Plots of standard deviation of $\hat{y}$ for a rotatable central composite design.

with equal precision for all points that are the same distance from the center of the design. Also, as one would expect, the precision decreases with increasing distance from the design center.

EXERCISES FOR SECTION 7-9

7-30. An article in *Rubber Age* (Vol. 89, 1961, pp. 453–458) describes an experiment on the manufacture of a product in which two factors were varied. The factors are reaction time (hr) and temperature (°C). These factors are coded as $x_1 = (\text{time} - 12)/8$ and $x_2 = (\text{temperature} - 250)/30$. The following data were observed where y is the yield (in percent).

Run Number	x_1^*	x_2	y
1	−1	0	83.8
2	1	0	81.7
3	0	0	82.4
4	0	0	82.9
5	0	−1	84.7
6	0	1	75.9
7	0	0	81.2
8	−1.414	−1.414	81.3
9	−1.414	1.414	83.1
10	1.414	−1.414	85.3
11	1.414	1.414	72.7
12	0	0	82.0

(a) Plot the points at which the experimental runs were made.

(b) Fit a second-order model to the data. Is the second-order model adequate?

(c) Plot the yield response surface. What recommendations would you make about the operating conditions for this process?

7-31. Consider the experimental design in the accompanying table. This experiment was run in a chemical process.

(a) The response y_1 is the viscosity of the product. Fit an appropriate response surface model.

(b) The response y_2 is the conversion, in grams. Fit an appropriate response surface model.

(c) Where would you recommend that we set x_1, x_2, and x_3 if the objective is to maximize in conversion while keeping viscosity in the range $450 < y_1 < 500$?

x_1	x_2	x_3	y_1	y_2
−1	−1	−1	480	68
0	−1	−1	530	95
1	−1	−1	590	86
−1	0	−1	490	184
0	0	−1	580	220
1	0	−1	660	230
−1	1	−1	490	220
0	1	−1	600	280
1	1	−1	720	310
−1	−1	0	410	134
0	−1	0	450	189
1	−1	0	530	210
−1	0	0	400	230
0	0	0	510	300
1	0	0	590	330
−1	1	0	420	270
0	1	0	540	340
1	1	0	640	380
−1	−1	1	340	164
0	−1	1	390	250
1	−1	1	450	300
−1	0	1	340	250
0	0	1	420	340
1	0	1	520	400
−1	1	1	360	250
0	1	1	470	370
1	1	1	560	440

7-32. A manufacturer of cutting tools has developed two empirical equations for tool life (y_1) and tool cost (y_2). Both models are functions of tool hardness (x_1) and manufacturing time (x_2). The equations are

$$\hat{y}_1 = 10 + 5x_1 + 2x_2$$
$$\hat{y}_2 = 23 + 3x_1 + 4x_2$$

and both equations are valid over the range $-1.5 \leq x_i \leq 1.5$. Suppose that tool life must exceed 12 hours and cost must be below $27.50.

(a) Is there a feasible set of operating conditions?

(b) Where would you run this process?

7-33. An article in *Tappi* (Vol. 43, 1960, pp. 38–44) describes an experiment that investigated the ash value of paper pulp (a measure of inorganic impurities). Two variables, temperature T in degrees Celsius and time t in hours, were studied, and some of the results are shown in the following table. The coded predictor variables shown are

$$x_1 = \frac{(T - 775)}{115} \qquad x_2 = \frac{(t - 3)}{1.5}$$

and the response y is (dry ash value in %) $\times 10^3$.

x_1	x_2	y
−1	−1	211
1	−1	92
−1	1	216
1	1	99
−1.5	0	222
1.5	0	48
0	−1.5	168
0	1.5	179
0	0	122
0	0	175
0	0	157
0	0	146

(a) What type of design has been used in this study? Is the design rotatable?

(b) Fit a quadratic model to the data. Is this model satisfactory?

(c) If it is important to minimize the ash value, where would you run the process?

7-34. In their book *Empirical Model Building and Response Surfaces* (John Wiley, 1987), G. E. P. Box and N. R. Draper describe an experiment with three factors. The data shown in the following

table are a variation of the original experiment on page 247 of their book. Suppose that these data were collected in a semiconductor manufacturing process.

x_1	x_2	x_3	y_1	y_2
−1	−1	−1	24.00	12.49
0	−1	−1	120.33	8.39
1	−1	−1	213.67	42.83
−1	0	−1	86.00	3.46
0	0	−1	136.63	80.41
1	0	−1	340.67	16.17
−1	1	−1	112.33	27.57
0	1	−1	256.33	4.62
1	1	−1	271.67	23.63
−1	−1	0	81.00	0.00
0	−1	0	101.67	17.67
1	−1	0	357.00	32.91
−1	0	0	171.33	15.01
0	0	0	372.00	0.00
1	0	0	501.67	92.50
−1	1	0	264.00	63.50
0	1	0	427.00	88.61
1	1	0	730.67	21.08
−1	−1	1	220.67	133.82
0	−1	1	239.67	23.46
1	−1	1	422.00	18.52
−1	0	1	199.00	29.44
0	0	1	485.33	44.67
1	0	1	673.67	158.21
−1	1	1	176.67	55.51
0	1	1	501.00	138.94
1	1	1	1010.00	142.45

(a) The response y_1 is the average of three readings on resistivity for a single wafer. Fit a quadratic model to this response.

(b) The response y_2 is the standard deviation of the three resistivity measurements. Fit a linear model to this response.

(c) Where would you recommend that we set x_1, x_2, and x_3 if the objective is to hold mean resistivity at 500 and minimize the standard deviation?

7-10   FACTORIAL EXPERIMENTS WITH MORE THAN TWO LEVELS

The 2^k full and fractional factorial designs are usually used in the initial stages of experimentation. After the most important effects have been identified, one might run a factorial experiment with more than two levels in order to obtain details of the relationship between the response and the factors. The basic analysis of variance can be modified to analyze the results from this type of experiment.

The analysis of variance decomposes the total variability in the data into component parts and then compares the various elements in this decomposition. For an experiment with two factors (with a levels for factor A and b levels for factor B) the total variability is measured by the total sum of squares of the observations

$$SS_T = \sum_{i=1}^{a} \sum_{j=1}^{b} \sum_{k=1}^{n} (y_{ijk} - \bar{y}_{...})^2 \tag{7-23}$$

and the sum of squares decomposition follows. The notation is defined in Table 7-27.

The sum of squares identity for a two-factor analysis of variance is

$$\sum_{i=1}^{a} \sum_{j=1}^{b} \sum_{k=1}^{n} (y_{ijk} - \bar{y}_{...})^2 = bn \sum_{i=1}^{a} (\bar{y}_{i..} - \bar{y}_{...})^2$$

$$+ an \sum_{j=1}^{b} (\bar{y}_{.j.} - \bar{y}_{...})^2$$

$$+ n \sum_{i=1}^{a} \sum_{j=1}^{b} (\bar{y}_{ij.} - \bar{y}_{i..} - \bar{y}_{.j.} + \bar{y}_{...})^2$$

$$+ \sum_{i=1}^{a} \sum_{j=1}^{b} \sum_{k=1}^{n} (y_{ijk} - \bar{y}_{ij.})^2 \tag{7-24}$$

The sum of squares identity may be written symbolically as

$$SS_T = SS_A + SS_B + SS_{AB} + SS_E$$

corresponding to each of the terms in equation 7-24. There are $abn - 1$ total degrees of freedom. The main effects A and B have $a - 1$ and $b - 1$ degrees of freedom, whereas the interaction effect AB has $(a - 1)(b - 1)$ degrees of freedom. Within each of the ab cells in Table 7-26, there are $n - 1$ degrees of freedom for the n replicates, and observations in the same cell can differ only because of random error. Therefore, there are

Table 7-27 Data Arrangement for a Two-Factor Factorial Design

		Factor B				
		1	2	...	b	Totals	Averages
Factor A	1	$y_{111}, y_{112},$ $\ldots, y_{11n}$	$y_{121}, y_{122},$ $\ldots, y_{12n}$		$y_{1b1}, y_{1b2},$ $\ldots, y_{1bn}$	$y_{1..}$	$\bar{y}_{1..}$
	2	$y_{211}, y_{212},$ $\ldots, y_{21n}$	$y_{221}, y_{222},$ $\ldots, y_{22n}$		$y_{2b1}, y_{2b2},$ $\ldots, y_{2bn}$	$y_{2..}$	$\bar{y}_{2..}$
	$\vdots$						
	a	$y_{a11}, y_{a12},$ $\ldots, y_{a1n}$	$y_{a21}, y_{a22},$ $\ldots, y_{a2n}$		$y_{ab1}, y_{ab2},$ $\ldots, y_{abn}$	$y_{a..}$	$\bar{y}_{a..}$
Totals		$y_{.1.}$	$y_{.2.}$		$y_{.b.}$	$y_{...}$	
Averages		$\bar{y}_{.1.}$	$\bar{y}_{.2.}$		$\bar{y}_{.b.}$		$\bar{y}_{...}$

$ab(n-1)$ degrees of freedom for error. Therefore, the degrees of freedom are partitioned according to

$$abn - 1 = (a - 1) + (b - 1) + (a - 1)(b - 1) + ab(n - 1)$$

If we divide each of the sum of squares by the corresponding number of degrees of freedom, we obtain the mean squares for A, B, the interaction, and error:

$$MS_A = \frac{SS_A}{a - 1} \qquad MS_B = \frac{SS_B}{b - 1}$$

$$MS_{AB} = \frac{SS_{AB}}{(a - 1)(b - 1)} \qquad MS_E = \frac{SS_E}{ab(n - 1)} \tag{7-25}$$

To test that the row, column, and interaction effects are zero, we would use the ratios

$$F_0 = \frac{MS_A}{MS_E} \qquad F_0 = \frac{MS_B}{MS_E} \qquad F_0 = \frac{MS_{AB}}{MS_E} \tag{7-26}$$

respectively. Each is compared to an F distribution with $a - 1$, $b - 1$, and $(a - 1) \times (b - 1)$ degrees of freedom in the numerator and $ab(n - 1)$ degrees of freedom in the denominator. This analysis is summarized in Table 7-28.

It is usually best to conduct the test for interaction first and then to evaluate the main effects. If interaction is not significant, interpretation of the tests on the main effects is straightforward. However, when interaction is significant the main effects of the factors involved in the interaction may not have much practical interpretative value. Knowledge of the interaction is usually more important than knowledge about the main effects.

Table 7-28 Analysis of Variance Table for a Two-Factor Factorial

Source of Variation	Sum of Squares	Degrees of Freedom	Mean Square	F_0
A treatments	SS_A	$a - 1$	$MS_A = \dfrac{SS_A}{a - 1}$	$\dfrac{MS_A}{MS_E}$
B treatments	SS_B	$b - 1$	$MS_B = \dfrac{SS_B}{b - 1}$	$\dfrac{MS_B}{MS_E}$
Interaction	SS_{AB}	$(a - 1)(b - 1)$	$MS_{AB} = \dfrac{SS_{AB}}{(a - 1)(b - 1)}$	$\dfrac{MS_{AB}}{MS_E}$
Error	SS_E	$ab(n - 1)$	$MS_E = \dfrac{SS_E}{ab(n - 1)}$	
Total	SS_T	$abn - 1$		

EXAMPLE 7-14

Aircraft primer paints are applied to aluminum surfaces by two methods, dipping and spraying. The purpose of the primer is to improve paint adhesion, and some parts can be primed using either application method. The process engineering group responsible for this operation is interested in learning whether three different primers differ in their adhesion properties. A factorial experiment was performed to investigate the effect of paint primer type and application method on paint adhesion. Three specimens were painted with each primer using each application method, a finish paint was applied, and the adhesion force was measured. The data from the experiment are shown in Table 7-29. The circled numbers in the cells are the cell totals $y_{ij.}$. The sums of squares required to perform the analysis of variance are computed by computer software and summarized in Table 7-30. The experimenter has decided to use $\alpha = 0.05$. Since $f_{0.05,2,12} = 3.89$ and $f_{0.05,1,12} = 4.75$, we conclude that the main effects of primer type and application method affect adhesion force. Furthermore, since $1.5 < f_{0.05,2,12}$, there is no indication of interaction between these factors. The last column of Table 7-30 shows the P-value

Table 7-29 Adhesion Force Data for Example 7-14 for Primer Type ($i = 1, 2, 3$) and Application Method ($j = 1, 2$) with $n = 3$ Replicates

	Primer Type	Dipping	Spraying	Totals $y_{i..}$
	1	4.0, 4.5, 4.3	5.4, 4.9, 5.6	28.7
	2	5.6, 4.9, 5.4	5.8, 6.1, 6.3	34.1
	3	3.8, 3.7, 4.0	5.5, 5.0, 5.0	27.0
Totals	$y_{.j.}$	40.2	49.6	$y_{...} = 89.8$

Table 7-30 Analysis of Variance for Example 7-14

Source of Variation	Sum of Squares	Degrees of Freedom	Mean Square	f_0	P-value
Primer types	4.58	2	2.29	28.63	2.7 × E-5
Application methods	4.91	1	4.91	61.38	5.0 × E-7
Interaction	0.24	2	0.12	1.50	0.2621
Error	0.99	12	0.08		
Total	10.72	17			

for each F-ratio. Note that the P-values for the two test statistics for the main effects are considerably less than 0.05, whereas the P-value for the test statistic for the interaction is greater than 0.05.

A graph of the cell adhesion force averages $\{\bar{y}_{ij.}\}$ versus levels of primer type for each application method is shown in Fig. 7-47. The averages are available in the computer output in Table 7-32. The no-interaction conclusion is obvious in this graph, because the two curves are nearly parallel. Furthermore, since a large response indicates greater adhesion force, we conclude that spraying is the best application method and that primer type 2 is most effective.

Model Adequacy Checking

Just as in the other experiments discussed in this chapter, the residuals from a factorial experiment play an important role in assessing model adequacy. In general, the residuals from a two-factor factorial are

$$e_{ijk} = y_{ijk} - \bar{y}_{ij.}$$

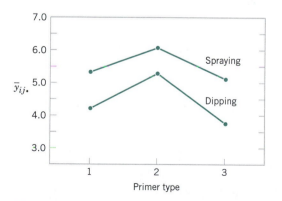

Figure 7-47 Graph of average adhesion force versus primer types for both application methods.

Table 7-31   Residuals for the Aircraft Primer Experiment in Example 7-14

Primer Type	Application Method	
	Dipping	Spraying
1	−0.27, 0.23, 0.03	0.10, −0.40, 0.30
2	0.30, −0.40, 0.10	−0.27, 0.03, 0.23
3	−0.03, −0.13, 0.17	0.33, −0.17, −0.17

That is, the residuals are just the difference between the observations and the corresponding cell averages. If interaction is negligible, then the cell averages could be replaced by a better predictor, but we only consider the simpler case.

Table 7-31 presents the residuals for the aircraft primer paint data in Example 7-14. The normal probability plot of these residuals is shown in Fig. 7-48. This plot has tails that do not fall exactly along a straight line passing through the center of the plot, indicating some potential problems with the normality assumption, but the deviation from normality does not appear severe. Figures 7-49 and 7-50 plot the residuals versus the levels of primer types and application methods, respectively. There is some indication that primer type 3 results in slightly lower variability in adhesion force than the other two primers. The graph of residuals versus fitted values in Fig. 7-51 does not reveal any unusual or diagnostic pattern.

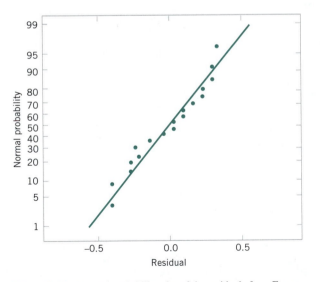

Figure 7-48   Normal probability plot of the residuals from Example 7-14.

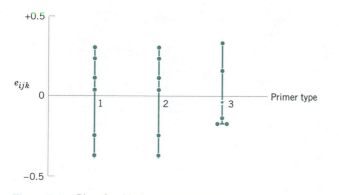

Figure 7-49 Plot of residuals versus primer type.

Computer Output

Table 7-32 shows some of the output from the analysis of variance procedure in Minitab for the aircraft primer paint experiment in Example 7-14.

This means table presents the sample means by primer type, by application method, and by cell (AB). The standard error for each mean is computed as $\sqrt{MS_E/m}$, where m is the number of observations in each sample mean. For example, each cell has $m = 3$ observations, so the standard error of a cell mean is $\sqrt{MS_E/3} = \sqrt{0.0822/3} = 0.1655$. A 95% confidence interval can be determined from the mean plus or minus the standard error times the multiplier $t_{0.025,12} = 2.179$. Minitab (and many other programs) will also produce the residual plots and interaction plot shown previously.

Table 7-33 shows output from Minitab for the fractional factorial experiment in Example 7-10. The design generators and aliases are displayed when the design is created in Minitab. The effects and ANOVA table are displayed when the design is analyzed. The F-tests are not shown because effects are not yet pooled into an estimate of error.

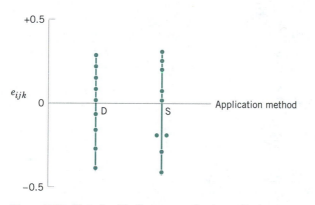

Figure 7-50 Plot of residuals versus application method.

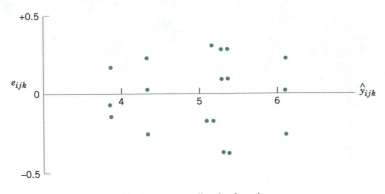

Figure 7-51   Plot of residuals versus predicted values $\hat{y}_{ijk}$.

Table 7-32   Analysis of Variance from Minitab for Example 7-14

Analysis of Variance (Balanced Designs)

Factor	Type	Levels	Values		
primer	fixed	3	1	2	3
method	fixed	2	1	2	

Analysis of Variance for y

Source	DF	SS	MS	F	P
primer	2	4.5811	2.2906	27.86	0.000
method	1	4.9089	4.9089	59.70	0.000
primer*method	2	0.2411	0.1206	1.47	0.269
Error	12	0.9867	0.0822		
Total	17	10.7178			

Means

primer	N	y
1	6	4.7833
2	6	5.6833
3	6	4.5000

method	N	y
1	9	4.4667
2	9	5.5111

primer	method	N	y
1	1	3	4.2667
1	2	3	5.3000
2	1	3	5.3000
2	2	3	6.0667
3	1	3	3.8333
3	2	3	5.1667

Table 7-33   Analysis from Minitab for Fractional Factorial Experiment in Example 7-10

Factorial Design

Fractional Factorial Design

Factors:	6	Base Design:	6, 16	Resolution: IV
Runs:	16	Replicates:	1	Fraction: 1/4
Blocks:	none	Center pts (total):	0	

Design Generators: E = ABC   F = BCD

Alias Structure

I + ABCE + ADEF + BCDF

A + BCE + DEF + ABCDF
B + ACE + CDF + ABDEF
C + ABE + BDF + ACDEF
D + AEF + BCF + ABCDE
E + ABC + ADF + BCDEF
F + ADE + BCD + ABCEF
AB + CE + ACDF + BDEF
AC + BE + ABDF + CDEF
AD + EF + ABCF + BCDE
AE + BC + DF + ABCDEF
AF + DE + ABCD + BCEF
BD + CF + ABEF + ACDE
BF + CD + ABDE + ACEF
ABD + ACF + BEF + CDE
ABF + ACD + BDE + CEF

Fractional Factorial Fit

Estimated Effects and Coefficients for y

Term	Effect	Coef
Constant		27.313
A	13.875	6.938
B	35.625	17.812
C	−0.875	−0.438
D	1.375	0.687
E	0.375	0.187
F	0.375	0.188
A∗B	11.875	5.938
A∗C	−1.625	−0.813
A∗D	−5.375	−2.688
A∗E	−1.875	−0.937
A∗F	0.625	0.313
B∗D	−0.125	−0.062
B∗F	−0.125	−0.062
A∗B∗D	0.125	0.062
A∗B∗F	−4.875	−2.437

(Continued)

Table 7-33 Analysis from Minitab for Fractional Factorial Experiment in Example 7-10 (*Continued*)

Analysis of Variance for y

Source	DF	Seq SS	Adj SS	Adj MS	F	P
Main Effects	6	5858.37	5858.37	976.40	**	
2-Way Interactions	7	705.94	705.94	100.85	**	
3-Way Interactions	2	95.12	95.12	47.56	**	
Residual Error	0	0.00	0.00	0.00		
Total	15	6659.44				

EXERCISES FOR SECTION 7-10

7-35. Consider the experiment in Exercise 7-2. Suppose that the experiment was actually carried out on three types of drying times and two types of paint. The data are:

Paint	Drying Time (min)		
	20	25	30
1	74	73	78
	64	61	85
	50	44	92
2	99	98	66
	86	73	45
	68	88	85

(a) Perform the analysis of variance with α = 0.05. What is your conclusion about the significance of the interaction effect?

(b) Assess the adequacy of the model by analyzing the residuals. What is your conclusion?

(c) If the smaller values are desirable, what levels of the factors do you recommend to obtain the necessary surface finish?

7-36. Consider the experiment in Exercise 7-4. Suppose that the experiment was actually carried out on three levels of temperature and two positions. The data are:

Position	Temperature (°C)		
	800	825	850
1	570	1063	565
	565	1080	510
	583	1043	590
2	528	988	526
	547	1026	538
	521	1004	532

(a) Perform the analysis of variance with α = 0.05. What is your conclusion about the significance of the interaction effect?

(b) Assess the adequacy of the model by analyzing the residuals. What is your conclusion?

(c) If higher density values are desirable, what levels of the factors do you recommend?

7-37. The percentage of hardwood concentration in raw pulp, the freeness, and the cooking time of the pulp are being investigated for their effects on the strength of paper. The data from a three-factor factorial experiment are shown in Table 7-34.

Table 7-34   Data for Exercise 7-37

Percentage of Hardwood Concentration	Cooking Time 1.5 Hours			Cooking Time 2.0 Hours		
	Freeness			Freeness		
	350	500	650	350	500	650
10	96.6	97.7	99.4	98.4	99.6	100.6
	96.0	96.0	99.8	98.6	100.4	100.9
15	98.5	96.0	98.4	97.5	98.7	99.6
	97.2	96.9	97.6	98.1	96.0	99.0
20	97.5	95.6	97.4	97.6	97.0	98.5
	96.6	96.2	98.1	98.4	97.8	99.8

(a) Use a statistical software package to perform the analysis of variance. Use $\alpha = 0.05$.

(b) Find P-values for the F-ratios in part (a) and interpret your results.

(c) The residuals are found by $e_{ijkl} = y_{ijkl} - \bar{y}_{ijk\cdot}$. Graphically analyze the residuals from this experiment.

7-38. The quality control department of a fabric finishing plant is studying the effect of several factors on dyeing for a blended cotton/synthetic cloth used to manufacture shirts. Three operators, three cycle times, and two temperatures were selected, and three small specimens of cloth were dyed under each set of conditions. The finished cloth was compared to a standard, and a numerical score was assigned. The results are shown in Table 7-35.

(a) Perform the analysis of variance with $\alpha = 0.05$. Interpret your results.

(b) The residuals may be obtained from $e_{ijkl} = y_{ijkl} - \bar{y}_{ijk\cdot}$. Graphically analyze the residuals from this experiment.

Table 7-35   Data for Exercise 7-38

	Temperature					
	300°			350°		
	Operator			Operator		
Cycle Time	1	2	3	1	2	3
40	23	27	31	24	38	34
	24	28	32	23	36	36
	25	26	28	28	35	39
50	36	34	33	37	34	34
	35	38	34	39	38	36
	36	39	35	35	36	31
60	28	35	26	26	36	28
	24	35	27	29	37	26
	27	34	25	25	34	34

SUPPLEMENTAL EXERCISES

7-39. An article in *Process Engineering* (No. 71, 1992, pp. 46–47) presents a two-factor factorial experiment used to investigate the effect of pH and catalyst concentration on product viscosity (cSt). The data are as follows.

		Catalyst Concentration	
		2.5	2.7
pH	5.6	192, 199, 189, 198	178, 186, 179, 188
	5.9	185, 193, 185, 192	197, 196, 204, 204

(a) Test for main effects and interactions using $\alpha = 0.05$. What are your conclusions?

(b) Graph the interaction and discuss the information provided by this plot.

(c) Analyze the residuals from this experiment.

7-40. Heat treating of metal parts is a widely used manufacturing process. An article in the *Journal of Metals* (Vol. 41, 1989) describes an experiment to investigate flatness distortion from heat treating for three types of gears and two heat-treating times. Some of the data are as follows.

	Time (minutes)	
Gear Type	90	120
20-tooth	0.0265	0.0560
	0.0340	0.0650
24-tooth	0.0430	0.0720
	0.0510	0.0880

(a) Is there any evidence that flatness distortion is different for the different gear types? Is there any indication that heat treating time affects the flatness distortion? Do these factors interact? Use $\alpha = 0.05$.

(b) Construct graphs of the factor effects that aid in drawing conclusions from this experiment.

(c) Analyze the residuals from this experiment. Comment on the validity of the underlying assumptions.

7-41. An article in the *Textile Research Institute Journal* (Vol. 54, 1984, pp. 171–179) reported the results of an experiment that studied the effects of treating fabric with selected inorganic salts on the flammability of the material. Two application levels of each salt were used, and a vertical burn test was used on each sample. (This finds the temperature at which each sample ignites.) The burn test data are shown in Table 7-36.

(a) Test for differences between salts, application levels, and interactions. Use $\alpha = 0.01$.

(b) Draw a graph of the interaction between salt and application level. What conclusions can you draw from this graph?

(c) Analyze the residuals from this experiment.

7-42. An article in the *IEEE Transactions on Components, Hybrids, and Manufacturing Technology* (Vol. 15, 1992) describes an experiment for investigating a method for aligning optical chips onto circuit boards. The method involves placing solder bumps onto the bottom of the chip. The experiment used two solder bump sizes and two alignment methods. The response variable is alignment accuracy (μm). The data are as follows.

Solder Bump Size (diameter in μm)	Alignment Method	
	1	2
75	4.60	1.05
	4.53	1.00
130	2.33	0.82
	2.44	0.95

(a) Is there any indication that either solder bump size or alignment method affects the alignment accuracy? Is there any evidence of interaction between these factors? Use $\alpha = 0.05$.

Table 7-36 Data for Exercise 7-41

Level	Salt					
	Untreated	$MgCl_2$	NaCl	$CaCO_3$	$CaCl_2$	Na_2CO_3
1	812	752	739	733	725	751
	827	728	731	728	727	761
	876	764	726	720	719	755
2	945	794	741	786	756	910
	881	760	744	771	781	854
	919	757	727	779	814	848

(b) What recommendations would you make about this process?

(c) Analyze the residuals from this experiment. Comment on model adequacy.

7-43. An article in *Solid State Technology* (Vol. 29, 1984, pp. 281–284) describes the use of factorial experiments in photolithography, an important step in the process of manufacturing integrated circuits. The variables in this experiment (all at two levels) are prebake temperature (A), prebake time (B), and exposure energy (C), and the response variable is delta-line width, the difference between the line on the mask and the printed line on the device. The data are as follows: $(1) = -2.30$, $a = -9.87$, $b = -18.20$, $ab = -30.20$, $c = -23.80$, $ac = -4.30$, $bc = -3.80$, and $abc = -14.70$.

(a) Estimate the factor effects.

(b) Suppose that a center point is added to this design and four replicates are obtained: -10.50, -5.30, -11.60, and -7.30. Calculate an estimate of experimental error.

(c) Test the significance of main effects, interactions, and curvature. At $\alpha = 0.05$, what conclusions can you draw?

(d) What model would you recommend for predicting the delta-line width response, based on the results of this experiment?

(e) Analyze the residuals from this experiment, and comment on model adequacy.

7-44. An article in the *Journal of Coatings Technology* (Vol. 60, 1988, pp. 27–32) describes a 2^4 factorial design used for studying a silver automobile basecoat. The response variable is distinctness of image (DOI). The variables used in the experiment are

$A =$ Percent of polyester by weight of polyester/melamine (low value = 50%, high value = 70%)

$B =$ Percent cellulose acetate butyrate carboxylate (low value = 15%, high value = 30%)

$C =$ Percent aluminum stearate (low value = 1%, high value = 3%)

$D =$ Percent acid catalyst (low value = 0.25%, high value = 0.50%)

The responses are $(1) = 63.8$, $a = 77.6$, $b = 68.8$, $ab = 76.5$, $c = 72.5$, $ac = 77.2$, $bc = 77.7$, $abc = 84.5$, $d = 60.6$, $ad = 64.9$, $bd = 72.7$, $abd = 73.3$, $cd = 68.0$, $acd = 76.3$, $bcd = 76.0$, and $abcd = 75.9$.

(a) Estimate the factor effects.

(b) From a normal probability plot of the effects, identify a tentative model for the data from this experiment.

(c) Using the apparently negligible factors as an estimate of error, test for significance of the factors identified in part (b). Use $\alpha = 0.05$.

(d) What model would you use to describe the process, based on this experiment? Interpret the model.

(e) Analyze the residuals from the model in part (d) and comment on your findings.

7-45. An article in the *Journal of Manufacturing Systems* (Vol. 10, 1991, pp. 32–40) describes an experiment to investigate the effect of four factors P = waterjet pressure, F = abrasive flow rate, G = abrasive grain size, and V = jet traverse speed on the surface roughness of a waterjet cutter. A 2^4 design with seven center points is shown in Table 7-37.

(a) Estimate the factor effects.

(b) Form a tentative model by examining a normal probability plot of the effects.

(c) Is the model in part (b) a reasonable description of the process? Use $\alpha = 0.05$.

(d) Interpret the results of this experiment.

(e) Analyze the residuals from this experiment.

7-46. Construct a 2_{IV}^{4-1} design for the problem in Exercise 7-44. Select the data for the

Table 7-37 Data for Exercise 7-45

Run	V (in/min)	F (lb/min)	P (kpsi)	G (mesh no.)	Surface Roughness (μm)
1	6	2.0	38	80	104
2	2	2.0	38	80	98
3	6	2.0	30	80	103
4	2	2.0	30	80	96
5	6	1.0	38	80	137
6	2	1.0	38	80	112
7	6	1.0	30	80	143
8	2	1.0	30	80	129
9	6	2.0	38	170	88
10	2	2.0	38	170	70
11	6	2.0	30	170	110
12	2	2.0	30	170	110
13	6	1.0	38	170	102
14	2	1.0	38	170	76
15	6	1.0	30	170	98
16	2	1.0	30	170	68
17	4	1.5	34	115	95
18	4	1.5	34	115	98
19	4	1.5	34	115	100
20	4	1.5	34	115	97
21	4	1.5	34	115	94
22	4	1.5	34	115	93
23	4	1.5	34	115	91

eight runs that would have been required for this design. Analyze these runs and compare your conclusions to those obtained in Exercise 7-44 for the full factorial.

7-47. Construct a 2_{IV}^{4-1} design for the problem in Exercise 7-45. Select the data for the eight runs that would have been required for this design, plus the center points. Analyze these data and compare your conclusions to those obtained in Exercise 7-45 for the full factorial.

7-48. Construct a 2_{IV}^{8-4} design in 16 runs. What are the alias relationships in this design?

7-49. Construct a 2_{III}^{5-2} design in eight runs. What are the alias relationships in this design?

7-50. In a process development study on yield, four factors were studied, each at two levels: time (A), concentration (B), pressure (C), and temperature (D). A single replicate of a 2^4 design was run, and the resulting data are shown in Table 7-38.

(a) Plot the effect estimates on a normal probability scale. Which factors appear to have large effects?
(b) Conduct an analysis of variance using the normal probability plot in part (a) for guidance in forming an error term. What are your conclusions?
(c) Analyze the residuals from this experiment. Does your analysis indicate any potential problems?
(d) Can this design be collapsed into a 2^3 design with two replicates? If so, sketch the design with the average and range of yield shown at each point in the cube. Interpret the results.

7-51. Consider the experiment described in Exercise 7-50. Find 95% confidence intervals on the factor effects that appear important. Use the normal probability plot to provide guidance concerning the effects that can be combined to provide an estimate of error.

Table 7-38 Data for Exercise 7-50

Run Number	Actual Run Order	A	B	C	D	Yield (lbs)	Factor Levels		
								Low (−)	High (+)
1	5	−	−	−	−	12	A (h)	2.5	3
2	9	+	−	−	−	18	B (%)	14	18
3	8	−	+	−	−	13	C (psi)	60	80
4	13	+	+	−	−	16	D (°C)	225	250
5	3	−	−	+	−	17			
6	7	+	−	+	−	15			
7	14	−	+	+	−	20			
8	1	+	+	+	−	15			
9	6	−	−	−	+	10			
10	11	+	−	−	+	25			
11	2	−	+	−	+	13			
12	15	+	+	−	+	24			
13	4	−	−	+	+	19			
14	16	+	−	+	+	21			
15	10	−	+	+	+	17			
16	12	+	+	+	+	23			

7-52. An article in the *Journal of Quality Technology* (Vol. 17, 1985, pp. 198–206) describes the use of a replicated fractional factorial to investigate the effect of five factors on the free height of leaf springs used in an automotive application. The factors are A = furnace temperature, B = heating time, C = transfer time, D = hold down time, and E = quench oil temperature. The data are shown in the following table.

A	B	C	D	E	Free Height		
−	−	−	−	−	7.78	7.78	7.81
+	−	−	+	−	8.15	8.18	7.88
−	+	−	+	−	7.50	7.56	7.50
+	+	−	−	−	7.59	7.56	7.75
−	−	+	+	−	7.54	8.00	7.88
+	−	+	−	−	7.69	8.09	8.06
−	+	+	−	−	7.56	7.52	7.44
+	+	+	+	−	7.56	7.81	7.69
−	−	−	−	+	7.50	7.56	7.50
+	−	−	+	+	7.88	7.88	7.44
−	+	−	+	+	7.50	7.56	7.50
+	+	−	−	+	7.63	7.75	7.56
−	−	+	+	+	7.32	7.44	7.44
+	−	+	−	+	7.56	7.69	7.62
−	+	+	−	+	7.18	7.18	7.25
+	+	+	+	+	7.81	7.50	7.59

(a) What is the generator for this fraction? Write out the alias structure.
(b) Analyze the data. What factors influence mean free height?
(c) Calculate the range of free height for each run. Is there any indication that any of these factors affects variability in free height?
(d) Analyze the residuals from this experiment and comment on your findings.

7-53. An article in *Rubber Chemistry and Technology* (Vol. 47, 1974, pp. 825–836) describes an experiment that studies the Mooney viscosity of rubber to several variables, including silica filler (parts per hundred) and oil filler (parts per hundred). Some of the data from this experiment are shown here, where

$$x_1 = \frac{\text{silica} - 60}{15} \qquad x_2 = \frac{\text{oil} - 21}{15}$$

Coded Levels		
x_1	x_2	y
−1	−1	13.71
1	−1	14.15
−1	1	12.87
1	1	13.53
−1.4	·0	12.99
1.4	0	13.89
0	−1.4	14.16
0	1.4	12.90
0	0	13.75
0	0	13.66
0	0	13.86
0	0	13.63
0	0	13.74

Fit a quadratic model to these data. What values of x_1 and x_2 will maximize the Mooney viscosity?

Team Exercise

7-54. The project consists of planning, designing, conducting and analyzing an experiment, using appropriate experimental design principles. The context of the project experiment is limited only by your imagination. Students have conducted experiments directly connected to their own research interests, a project that they are involved with at work (something for the industrial participants, or the part-timers in industry to think about), or if all else fails, you could conduct a "household" experiment (such as how do varying factors such as type of cooking oil, amount of oil, cooking temperature, pan type, brand of popcorn, etc. affect the yield and taste of popcorn).

The major requirement is that the experiment must involve at least three factors. Each of the interim steps requires information about the problem, the factors, the responses that will be observed, and the specific details of the design that will be used. Your final report should include a clear statement of objectives, the procedures and techniques used, appropriate analyses, and specific conclusions that state what you learned from the experiment.

Statistical

Quality

Control

CHAPTER OUTLINE

8-1 QUALITY IMPROVEMENT AND STATISTICS

The quality of products and services has become a major decision factor in most businesses today. Regardless of whether the consumer is an individual, a corporation, a military defense program, or a retail store, when the consumer is making purchase decisions, he or she is likely to consider quality to be equal in importance to cost and schedule. Consequently, **quality improvement** has become a major concern to many U.S. corpora-

448

tions. This chapter is about **statistical quality control,** a collection of tools that are essential in quality-improvement activities.

Quality means **fitness for use.** For example, you or I may purchase automobiles that we expect to be free of manufacturing defects and that should provide reliable and economical transportation, a retailer buys finished goods with the expectation that they are properly packaged and arranged for easy storage and display, and a manufacturer buys raw material and expects to process it with no rework or scrap. In other words, all consumers expect that the products and services they buy will meet their requirements. Those requirements define fitness for use.

Quality or fitness for use is determined through the interaction of **quality of design** and **quality of conformance.** By quality of design we mean the different grades or levels of performance, reliability, serviceability, and function that are the result of deliberate engineering and management decisions. By quality of conformance, we mean the systematic **reduction of variability** and **elimination of defects** until every unit produced is identical and defect-free.

Some confusion exists in our society about quality improvement; some people still think that it means gold-plating a product or spending more money to develop a product or process. This thinking is wrong. Quality improvement means the systematic **elimination of waste.** Examples of waste include scrap and rework in manufacturing, inspection and test, errors on documents (such as engineering drawings, checks, purchase orders, and plans), customer complaint hotlines, warranty costs, and the time required to do things over again that could have been done right the first time. A successful quality-improvement effort can eliminate much of this waste and lead to lower costs, higher productivity, increased customer satisfaction, increased business reputation, higher market share, and ultimately higher profits for the company.

Statistical methods play a vital role in quality improvement. Some applications are outlined next:

1. In product design and development, statistical methods, including designed experiments, can be used to compare different materials, components, or ingredients, and to help determine both system and component tolerances. This application can significantly lower development costs and reduce development time.

2. Statistical methods can be used to determine the capability of a manufacturing process. Statistical process control can be used to systematically improve a process by reducing variability.

3. Experimental design methods can be used to investigate improvements in the process. These improvements can lead to higher yields and lower manufacturing costs.

4. Life testing provides reliability and other performance data about the product. This can lead to new and improved designs and products that have longer useful lives and lower operating and maintenance costs.

Some of these applications have been illustrated in earlier chapters of this book. It is essential that engineers, scientists, and managers have an in-depth understanding of these statistical tools in any industry or business that wants to be a high-quality, low-cost producer. In this chapter we provide an introduction to the basic methods of statistical quality control that, along with experimental design, form the basis of a successful quality-improvement effort.

8-2   STATISTICAL QUALITY CONTROL

The field of statistical quality control can be broadly defined as those statistical and engineering methods that are used in measuring, monitoring, controlling, and improving quality. Statistical quality control is a relatively new field, dating back to the 1920s. Dr. Walter A. Shewhart of the Bell Telephone Laboratories was one of the early pioneers of the field. In 1924 he wrote a memorandum showing a modern control chart, one of the basic tools of statistical process control. Harold F. Dodge and Harry G. Romig, two other Bell System employees, provided much of the leadership in the development of statistically based sampling and inspection methods. The work of these three men forms much of the basis of the modern field of statistical quality control. World War II saw the widespread introduction of these methods to U.S. industry. Dr. W. Edwards Deming and Dr. Joseph M. Juran have been instrumental in spreading statistical quality-control methods since World War II.

The Japanese have been particularly successful in deploying statistical quality-control methods and have used statistical methods to gain significant advantage over their competitors. In the 1970s American industry suffered extensively from Japanese (and other foreign) competition; that has led, in turn, to renewed interest in statistical quality-control methods in the United States. Much of this interest focuses on *statistical process control* and *experimental design.* Many U.S. companies have begun extensive programs to implement these methods in their manufacturing, engineering, and other business organizations.

8-3   STATISTICAL PROCESS CONTROL

It is impractical to inspect quality into a product; the product must be built right the first time. The manufacturing process must therefore be stable or repeatable and capable of operating with little variability around the target or nominal dimension. Online statistical process control is a powerful tool for achieving process stability and improving capability through the reduction of variability.

It is customary to think of **statistical process control (SPC)** as a set of problem-solving tools that may be applied to any process. The major tools of SPC[1] are

1.   Histogram
2.   Pareto chart

---

[1] Some prefer to include the experimental design methods discussed in Chapter 7 as part of the SPC toolkit. We did not do so, because we think of SPC as an online approach to quality improvement using techniques founded on passive observation of the process, whereas design of experiments is an active approach in which deliberate changes are made to the process variables. As such, designed experiments are often referred to as offline quality control.

   3. Cause-and-effect diagram
   4. Defect-concentration diagram
   5. Control chart
   6. Scatter diagram
   7. Check sheet

Although these tools are an important part of SPC, they comprise only the technical aspect of the subject. An equally important element of SPC is attitude—a desire of all individuals in the organization for continuous improvement in quality and productivity through the systematic reduction of variability. The control chart is the most powerful of the SPC tools. For complete discussion of these methods, see Montgomery (1996).

8-4 INTRODUCTION TO CONTROL CHARTS

8-4.1 Basic Principles

In any production process, regardless of how well-designed or carefully maintained it is, a certain amount of inherent or natural variability will always exist. This natural variability or "background noise" is the cumulative effect of many small, essentially unavoidable causes. When the background noise in a process is relatively small, we usually consider it an acceptable level of process performance. In the framework of statistical quality control, this natural variability is often called a "stable system of chance causes." A process that is operating with only **chance causes** of variation present is said to be in statistical control. In other words, the chance causes are an inherent part of the process.

Other kinds of variability may occasionally be present in the output of a process. This variability in key quality characteristics usually arises from three sources: improperly adjusted machines, operator errors, or defective raw materials. Such variability is generally large when compared to the background noise, and it usually represents an unacceptable level of process performance. We refer to these sources of variability that are not part of the chance cause pattern as **assignable causes.** A process that is operating in the presence of assignable causes is said to be out of control.[2]

Production processes will often operate in the in-control state, producing acceptable product for relatively long periods of time. Occasionally, however, assignable causes will occur, seemingly at random, resulting in a "shift" to an out-of-control state where a large proportion of the process output does not conform to requirements. A major objective of statistical process control is to quickly detect the occurrence of assignable causes or process shifts so that investigation of the process and corrective action may be undertaken before many nonconforming units are manufactured. The control chart is an online process-monitoring technique widely used for this purpose.

Recall the following from Chapter 1: Figs. 1-14(a) and (b) illustrate that adjustments to common causes of variation increase the variation of a process, whereas Figs. 1-15(a)

---

[2] The terminology *chance* and *assignable* causes was developed by Dr. Walter A. Shewhart. Today, some writers use *common* cause instead of *chance* cause and *special* cause instead of *assignable* cause.

and (b) illustrate that actions should be taken in response to assignable causes of variation. Control charts may also be used to estimate the parameters of a production process and, through this information, to determine the capability of a process to meet specifications. The control chart can also provide information that is useful in improving the process. Finally, remember that the eventual goal of statistical process control is the *elimination of variability in the process.* Although it may not be possible to eliminate variability completely, the control chart helps reduce it as much as possible.

A typical control chart is shown in Fig. 8-1, which is a graphical display of a quality characteristic that has been measured or computed from a sample versus the sample number or time. Often, the samples are selected at periodic intervals such as every hour. The chart contains a center line (CL) that represents the average value of the quality characteristic corresponding to the in-control state. (That is, only chance causes are present.) Two other horizontal lines, called the upper control limit (UCL) and the lower control limit (LCL), are also shown on the chart. These control limits are chosen so that if the process is in control, nearly all of the sample points will fall between them. In general, as long as the points plot within the control limits, the process is assumed to be in control, and no action is necessary. However, a point that plots outside of the control limits is interpreted as evidence that the process is out of control, and investigation and corrective action are required to find and eliminate the assignable cause or causes responsible for this behavior. The sample points on the control chart are usually connected with straight-line segments, so that it is easier to visualize how the sequence of points has evolved over time.

Even if all the points plot inside the control limits, if they behave in a systematic or nonrandom manner, then this is an indication that the process is out of control. For example, if 18 of the last 20 points plotted above the center line but below the upper control limit and only two of these points plotted below the center line but above the lower control limit, we would be very suspicious that something was wrong. If the process is in control, all the plotted points should have an essentially random pattern. Methods designed to find sequences or nonrandom patterns can be applied to control charts as an

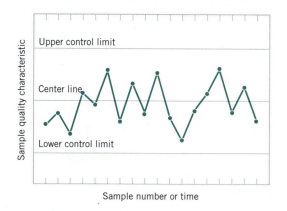

Figure 8-1   A typical control chart.

aid in detecting out-of-control conditions. A particular nonrandom pattern usually appears on a control chart for a reason, and if that reason can be found and eliminated, process performance can be improved.

There is a close connection between control charts and hypothesis testing. Essentially, the control chart is a test of the hypothesis that the process is in a state of statistical control. A point plotting within the control limits is equivalent to failing to reject the hypothesis of statistical control, and a point plotting outside the control limits is equivalent to rejecting the hypothesis of statistical control.

We may give a general *model* for a control chart.

General Model for a Control Chart

Let W be a sample statistic that measures some quality characteristic of interest, and suppose that the mean of W is μ_W and the standard deviation of W is σ_W.[3] Then the center line, the upper control limit, and the lower control limit become

$$UCL = \mu_W + k\sigma_W$$
$$CL = \mu_W$$
$$LCL = \mu_W - k\sigma_W \qquad (8\text{-}1)$$

where k is the "distance" of the control limits from the center line, expressed in standard deviation units.

A common choice is $k = 3$. This general theory of control charts was first proposed by Dr. Walter A. Shewhart, and control charts developed according to these principles are often called **Shewhart control charts.**

The control chart is a device for describing exactly what is meant by statistical control; as such, it may be used in a variety of ways. In many applications, it is used for online process monitoring. That is, sample data are collected and used to construct the control chart, and if the sample values of $\bar{x}$ (say) fall within the control limits and do not exhibit any systematic pattern, we say the process is in control at the level indicated by the chart. Note that we may be interested here in determining *both* whether the past data came from a process that was in control and whether future samples from this process indicate statistical control.

The most important use of a control chart is to *improve* the process. We have found that, generally,

1.   Most processes do not operate in a state of statistical control.

---

[3]Note that "sigma" refers to the standard deviation of the statistic plotted on the chart (i.e., σ_w), not the standard deviation of the quality characteristic.

2. Consequently, the routine and attentive use of control charts will identify assignable causes. If these causes can be eliminated from the process, variability will be reduced and the process will be improved.

This process-improvement activity using the control chart is illustrated in Fig. 8-2. Note that:

3. The control chart will only *detect* assignable causes. Management, operator, and engineering *action* will usually be necessary to eliminate the assignable cause. An action plan for responding to control chart signals is vital.

In identifying and eliminating assignable causes, it is important to find the underlying **root cause** of the problem and to attack it. A cosmetic solution will not result in any real, long-term process improvement. Developing an effective system for corrective action is an essential component of an effective SPC implementation.

We may also use the control chart as an *estimating device*. That is, from a control chart that exhibits statistical control, we may estimate certain process parameters, such as the mean, standard deviation, and fraction nonconforming or fallout. These estimates may then be used to determine the *capability* of the process to produce acceptable products. Such **process capability studies** have considerable impact on many management decision problems that occur over the product cycle, including make-or-buy decisions, plant and process improvements that reduce process variability, and contractual agreements with customers or suppliers regarding product quality.

Control charts may be classified into two general types. Many quality characteristics can be measured and expressed as numbers on some continuous scale of measurement. In such cases, it is convenient to describe the quality characteristic with a measure of central tendency and a measure of variability. Control charts for central tendency and

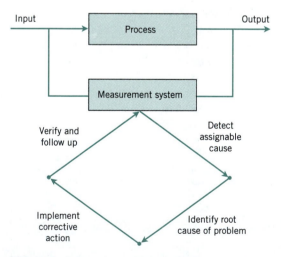

Figure 8-2 Process improvement using the control chart.

variability are collectively called **variables control charts.** The $\overline{X}$ chart is the most widely used chart for monitoring central tendency, whereas charts based on either the sample range or the sample standard deviation are used to control process variability. Many quality characteristics are not measured on a continuous scale or even a quantitative scale. In these cases, we may judge each unit of product as either conforming or nonconforming on the basis of whether or not it possesses certain attributes, or we may count the number of nonconformities (defects) appearing on a unit of product. Control charts for such quality characteristics are called **attributes control charts.**

Control charts have had a long history of use in industry. There are at least five reasons for their popularity.

1. **Control charts are a proven technique for improving productivity.** A successful control chart program will reduce scrap and rework, which are the primary productivity killers in *any* operation. If you reduce scrap and rework, then productivity increases, cost decreases, and production capacity (measured in the number of *good* parts per hour) increases.

2. **Control charts are effective in defect prevention.** The control chart helps keep the process in control, which is consistent with the "do it right the first time" philosophy. It is never cheaper to sort out the "good" units from the "bad" later on than it is to build it right initially. If you do not have effective process control, you are paying someone to make a nonconforming product.

3. **Control charts prevent unnecessary process adjustments.** A control chart can distinguish between background noise and abnormal variation; no other device, including a human operator, is as effective in making this distinction. If process operators adjust the process based on periodic tests unrelated to a control chart program, they will often overreact to the background noise and make unneeded adjustments. These unnecessary adjustments can usually result in a deterioration of process performance. In other words, the control chart is consistent with the "if it isn't broken, don't fix it" philosophy.

4. **Control charts provide diagnostic information.** Frequently, the pattern of points on the control chart will contain information that is of diagnostic value to an experienced operator or engineer. This information allows the operator to implement a change in the process that will improve its performance.

5. **Control charts provide information about process capability.** The control chart provides information about the value of important process parameters and their stability over time, which allows an estimate of process capability to be made. This information is of tremendous use to product and process designers.

Control charts are among the most effective management control tools, and they are as important as cost controls and material controls. Modern computer technology has made it easy to implement control charts in any type of process, for data collection and analysis can be performed on a microcomputer or a local area network terminal in real-time, online at the work center.

8-4.2   Design of a Control Chart

To illustrate these ideas, we give a simplified example of a control chart. In manufacturing automobile engine piston rings, the inside diameter of the rings is a critical quality characteristic. The process mean inside ring diameter is 74 mm, and it is known that the standard deviation of ring diameter is 0.01 mm. A control chart for average ring diameter is shown in Fig. 8-3. Every hour a random sample of five rings is taken, the average ring diameter of the sample (say, $\bar{x}$) is computed, and $\bar{x}$ is plotted on the chart. Because this control chart utilizes the sample mean $\bar{X}$ to monitor the process mean, it is usually called an $\bar{X}$ control chart. Note that all the points fall within the control limits, so the chart indicates that the process is in statistical control.

Consider how the control limits were determined. The process average is 74 mm, and the process standard deviation is $\sigma = 0.01$ mm. Now if samples of size $n = 5$ are taken, the standard deviation of the sample average $\bar{X}$ is

$$\sigma_{\bar{X}} = \frac{\sigma}{\sqrt{n}} = \frac{0.01}{\sqrt{5}} = 0.0045$$

Therefore, if the process is in control with a mean diameter of 74 mm, by using the central limit theorem to assume that $\bar{X}$ is approximately normally distributed, we would expect approximately $100(1 - \alpha)\%$ of the sample mean diameters $\bar{X}$ to fall between $74 + z_{\alpha/2}(0.0045)$ and $74 - z_{\alpha/2}(0.0045)$. As discussed above, we customarily choose the constant $z_{\alpha/2}$ to be 3, so that the upper and lower control limits become

$$UCL = 74 + 3(0.0045) = 74.0135$$

and

$$LCL = 74 - 3(0.0045) = 73.9865$$

as shown on the control chart. These are the three-sigma control limits referred to above.

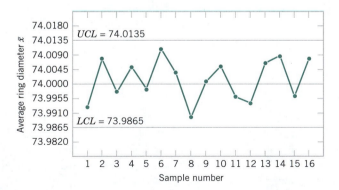

Figure 8-3 $\bar{X}$ control chart for piston ring diameter.

Note that the use of three-sigma limits implies that $\alpha = 0.0027$; that is, the probability that the point plots outside the control limits when the process is in control is 0.0027. The width of the control limits is inversely related to the sample size n for a given multiple of sigma. Choosing the control limits is equivalent to setting up the critical region for testing the hypothesis

$$H_0: \mu = 74$$
$$H_1: \mu \neq 74$$

where $\sigma = 0.01$ is known. Essentially, the control chart tests this hypothesis repeatedly at different points in time.

In designing a control chart, we must specify both the sample size to use and the frequency of sampling. In general, larger samples will make it easier to detect small shifts in the process. When choosing the sample size, we must keep in mind the size of the shift that we are trying to detect. If we are interested in detecting a relatively large process shift, then we use smaller sample sizes than those that would be employed if the shift of interest were relatively small.

We must also determine the frequency of sampling. The most desirable situation from the point of view of detecting shifts would be to take large samples very frequently; however, this is usually not economically feasible. The general problem is one of *allocating sampling effort*. That is, either we take small samples at short intervals or larger samples at longer intervals. Current industry practice tends to favor smaller, more frequent samples, particularly in high-volume manufacturing processes, or where a great many types of assignable causes can occur. Furthermore, as automatic sensing and measurement technology develops, it is becoming possible to greatly increase frequencies. Ultimately, every unit can be tested as it is manufactured. This capability will not eliminate the need for control charts because the test system does not prevent defects. The increased data will increase the effectiveness of process control and improve quality.

8-4.3 Rational Subgroups

A fundamental idea in the use of control charts is to collect sample data according to what Shewhart called the **rational subgroup** concept. Generally, this means that subgroups or samples should be selected so that to the extent possible, the variability of the observations within a subgroup include all the chance or natural variability and exclude the assignable variability. Then, the control limits will represent bounds for all the chance variability and not the assignable variability. Consequently, assignable causes will tend to generate points that are outside of the control limits, whereas chance variability will tend to generate points that are within the control limits.

When control charts are applied to production processes, the time order of production is a logical basis for rational subgrouping. Even though time order is preserved, it is still possible to form subgroups erroneously. If some of the observations in the subgroup are taken at the end of one 8-hour shift and the remaining observations are taken at the start of the next 8-hour shift, then any differences between shifts might not be detected. Time order is frequently a good basis for forming subgroups because it allows us to detect assignable causes that occur over time.

Two general approaches to constructing rational subgroups are used. In the first approach, each subgroup consists of units that were produced at the same time (or as close together as possible). This approach is used when the primary purpose of the control chart is to detect process shifts. It minimizes variability due to assignable causes *within* a sample, and it maximizes variability *between* samples if assignable causes are present. It also provides better estimates of the standard deviation of the process in the case of variables control charts. This approach to rational subgrouping essentially gives a "snapshot" of the process at each point in time where a sample is collected.

In the second approach, each sample consists of units of product that are representative of *all* units that have been produced since the last sample was taken. Essentially, each subgroup is a *random sample* of *all* process output over the sampling interval. This method of rational subgrouping is often used when the control chart is employed to make decisions about the acceptance of all units of product that have been produced since the last sample. In fact, if the process shifts to an out-of-control state and then back in control again *between* samples, it is sometimes argued that the first method of rational subgrouping defined above will be ineffective against these types of shifts, and so the second method must be used.

When the rational subgroup is a random sample of all units produced over the sampling interval, considerable care must be taken in interpreting the control charts. If the process mean drifts between several levels during the interval between samples, the range of observations within the sample may consequently be relatively large. It is the within-sample variability that determines the width of the control limits on an $\overline{X}$ chart, so this practice will result in wider limits on the $\overline{X}$ chart. This makes it harder to detect shifts in the mean. In fact, we can often make *any* process appear to be in statistical control just by stretching out the interval between observations in the sample. It is also possible for shifts in the process average to cause points on a control chart for the range or standard deviation to plot out of control, even though no shift in process variability has taken place.

There are other bases for forming rational subgroups. For example, suppose a process consists of several machines that pool their output into a common stream. If we sample from this common stream of output, it will be very difficult to detect whether or not some of the machines are out of control. A logical approach to rational subgrouping here is to apply control chart techniques to the output for each individual machine. Sometimes this concept needs to be applied to different heads on the same machine, different workstations, different operators, and so forth.

The rational subgroup concept is very important. The proper selection of samples requires careful consideration of the process, with the objective of obtaining as much useful information as possible from the control chart analysis.

8-4.4 Analysis of Patterns on Control Charts

A control chart may indicate an out-of-control condition either when one or more points fall beyond the control limits, or when the plotted points exhibit some nonrandom pattern

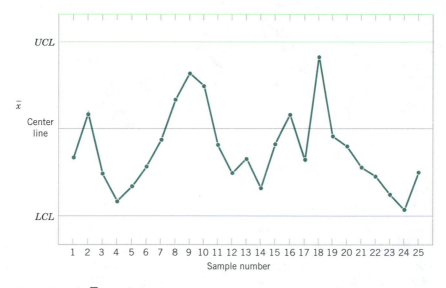

Figure 8-4   An $\overline{X}$ control chart.

of behavior. For example, consider the $\overline{X}$ chart shown in Fig. 8-4. Although all 25 points fall within the control limits, the points do not indicate statistical control because their pattern is very nonrandom in appearance. Specifically, we note that 19 of the 25 points plot below the center line, while only 6 of them plot above. If the points are truly random, we should expect a more even distribution of them above and below the center line. We also observe that following the fourth point, five points in a row increase in magnitude. This arrangement of points is called a **run.** Since the observations are increasing, we could call it a run up; similarly, a sequence of decreasing points is called a run down. This control chart has an unusually long run up (beginning with the fourth point) and an unusually long run down (beginning with the eighteenth point).

In general, we define a run as a sequence of observations of the same type. In addition to runs up and runs down, we could define the types of observations as those above and below the center line, respectively, so that two points in a row above the center line would be a run of length 2.

A run of length 8 or more points has a very low probability of occurrence in a random sample of points. Consequently, any type of run of length 8 or more is often taken as a signal of an out-of-control condition. For example, 8 consecutive points on one side of the center line will indicate that the process is out of control.

Although runs are an important measure of nonrandom behavior on a control chart, other types of patterns may also indicate an out-of-control condition. For example, consider the $\overline{X}$ chart in Fig. 8-5. Note that the plotted sample averages exhibit a cyclic behavior, yet they all fall within the control limits. Such a pattern may indicate a problem with the process, such as operator fatigue, raw material deliveries, and heat or stress buildup. The yield may be improved by eliminating or reducing the sources of variability causing this

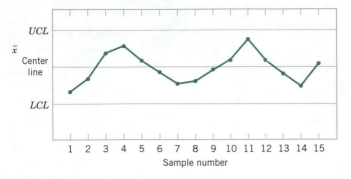

Figure 8-5   An $\overline{X}$ chart with a cyclic pattern.

cyclic behavior (see Fig. 8-6). In Fig. 8-6, *LSL* and *USL* denote the lower and upper specification limits of the process. These limits represent bounds within which acceptable product must fall and they are often based on customer requirements.

The problem is one of **pattern recognition**—that is, recognizing systematic or nonrandom patterns on the control chart and identifying the reason for this behavior. The ability to interpret a particular pattern in terms of assignable causes requires experience and knowledge of the process. That is, we must not only know the statistical principles of control charts, but we must also have a good understanding of the process.

The Western Electric Handbook (1956) suggests a set of decision rules for detecting nonrandom patterns on control charts. These rules are as follows:

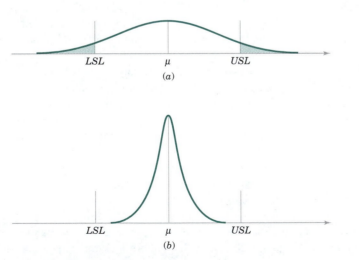

Figure 8-6   (a) Variability with the cyclic pattern. (b) Variability with the cyclic pattern eliminated.

The **Western Electric rules** would conclude that the process is out of control if either

1. One point plots outside three-sigma control limits.
2. Two out of three consecutive points plot beyond a two-sigma limit.
3. Four out of five consecutive points plot at a distance of one sigma or beyond from the center line.
4. Eight consecutive points plot on one side of the center line.

We have found these rules very effective in practice for enhancing the sensitivity of control charts. Rules 2 and 3 apply to one side of the center line at a time. That is, a point above the *upper* two-sigma limit followed immediately by a point below the *lower* two-sigma limit would not signal an out-of-control alarm.

Figure 8-7 shows an $\overline{X}$ control chart for the piston ring process with the one-sigma, two-sigma, and three-sigma limits used in the Western Electric procedure. Note that these inner limits, sometimes called **warning limits**, partition the control chart into three zones A, B, and C on each side of the center line. Consequently, the Western Electric rules are sometimes called the **zone rules** for control charts. Note that the last four points fall in zone B or beyond. Thus, since four of five consecutive points exceed the one-sigma limit, the Western Electric procedure will conclude that the pattern is nonrandom and the process is out of control.

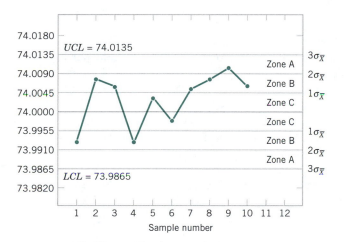

Figure 8-7 The Western Electric zone rules.

8-5 $\overline{X}$ AND R CONTROL CHARTS

When dealing with a quality characteristic that can be expressed as a measurement, it is customary to monitor both the mean value of the quality characteristic and its variability. Control over the average quality is exercised by the control chart for averages, usually called the $\overline{X}$ chart. Process variability can be controlled by either a range chart (R chart) or a standard deviation chart (S chart), depending on how the population standard deviation is estimated. We will discuss only the R chart.

Suppose that the process mean and standard deviation μ and σ are known and that we can assume that the quality characteristic has a normal distribution. Consider the $\overline{X}$ chart. As discussed previously, we can use μ as the center line for the control chart, and we can place the upper and lower three-sigma limits at $UCL = \mu + 3\sigma/\sqrt{n}$ and $LCL = \mu - 3\sigma/\sqrt{n}$, respectively.

When the parameters μ and σ are unknown, we usually estimate them on the basis of preliminary samples, taken when the process is thought to be in control. We recommend the use of at least 20 to 25 preliminary samples. Suppose m preliminary samples are available, each of size n. Typically, n will be 4, 5, or 6; these relatively small sample sizes are widely used and often arise from the construction of rational subgroups.

Grand Mean

Let the sample mean for the ith sample be $\overline{X}_i$. Then we estimate the mean of the population, μ, by the *grand mean*

$$\overline{\overline{X}} = \frac{1}{m} \sum_{i=1}^{m} \overline{X}_i \tag{8-2}$$

Thus, we may take $\overline{\overline{X}}$ as the center line on the $\overline{X}$ control chart.

We may estimate σ from either the standard deviation or the range of the observations within each sample. Since it is more frequently used in practice, we confine our discussion to the range method. The sample size is relatively small, so there is little loss in efficiency in estimating σ from the sample ranges.

The relationship between the range R of a sample from a normal population with known parameters and the standard deviation of that population is needed. Since R is a random variable, the quantity $W = R/\sigma$, called the relative range, is also a random variable. The parameters of the distribution of W have been determined for any sample size n.

The mean of the distribution of W is called d_2, and a table of d_2 for various n is given in Table VI of Appendix A. The standard deviation of W is called d_3. Because $R = \sigma W$,

$$\mu_R = d_2\sigma$$
$$\sigma_R = d_3\sigma$$

Average Range and Estimator of σ

Let R_i be the range of the ith sample, and let

$$\overline{R} = \frac{1}{m} \sum_{i=1}^{m} R_i \tag{8-3}$$

be the average range. Then $\overline{R}$ is an estimator of μ_R and an estimator of σ is

$$\hat{\sigma} = \frac{\overline{R}}{d_2} \tag{8-4}$$

Therefore, we may use as our upper and lower control limits for the $\overline{X}$ chart

$$UCL = \overline{\overline{X}} + \frac{3}{d_2\sqrt{n}}\overline{R}$$

$$LCL = \overline{\overline{X}} - \frac{3}{d_2\sqrt{n}}\overline{R} \tag{8-5}$$

Define the constant

$$A_2 = \frac{3}{d_2\sqrt{n}} \tag{8-6}$$

Now once we have computed the sample values $\overline{\overline{x}}$ and $\overline{r}$, the parameters of the $\overline{X}$ control chart may be defined as follows.

The $\overline{X}$ Control Chart

The center line and upper and lower control limits for an $\overline{X}$ control chart are

$$UCL = \overline{\overline{x}} + A_2\overline{r}$$
$$CL = \overline{\overline{x}}$$
$$LCL = \overline{\overline{x}} - A_2\overline{r} \tag{8-7}$$

where the constant A_2 is tabulated for various sample sizes in Appendix A Table VI.

The parameters of the R chart may also be easily determined. The center line will obviously be $\overline{R}$. To determine the control limits, we need an estimate of σ_R, the standard deviation of R. Once again, assuming that the process is in control, the distribution of the relative range, W, will be useful. Because σ is unknown, we may estimate σ_R as

$$\hat{\sigma}_R = d_3 \frac{\overline{R}}{d_2}$$

and we would use as the upper and lower control limits on the R chart

$$UCL = \overline{R} + \frac{3d_3}{d_2}\overline{R} = \left(1 + \frac{3d_3}{d_2}\right)\overline{R}$$

$$LCL = \overline{R} - \frac{3d_3}{d_2}\overline{R} = \left(1 - \frac{3d_3}{d_2}\right)\overline{R}$$

Setting $D_3 = 1 - 3d_3/d_2$ and $D_4 = 1 + 3d_3/d_2$ leads to the following definition.

The R Control Chart

The center line and upper and lower control limits for an R chart are

$$UCL = D_4\overline{r}$$
$$CL = \overline{r} \tag{8-8}$$
$$LCL = D_3\overline{r}$$

where $\overline{r}$ is the sample average range, and the constants D_3 and D_4 are tabulated for various sample sizes in Appendix A Table VI.

When preliminary samples are used to construct limits for control charts, these limits are customarily treated as trial values. Therefore, the m sample means and ranges should be plotted on the appropriate charts, and any points that exceed the control limits should be investigated. If assignable causes for these points are discovered, they should be eliminated and new limits for the control charts determined. In this way, the process may be eventually brought into statistical control and its inherent capabilities assessed. Other changes in process centering and dispersion may then be contemplated. Also, we often study the R chart first because if the process variability is not constant over time the control limits calculated for the $\overline{X}$ chart can be misleading.

EXAMPLE 8-1

A component part for a jet aircraft engine is manufactured by an investment casting process. The vane opening on this casting is an important functional parameter of the part. We will illustrate the use of $\overline{X}$ and R control charts to assess the statistical stability of this process. Table 8-1 presents 20 samples of five parts each. The values given in the table have been coded by using the last three digits of the dimension; that is, 31.6 should be 0.50316 inch.

The quantities $\overline{\overline{x}} = 33.3$ and $\overline{r} = 5.8$ are shown at the foot of Table 8-1. The value of A_2 for samples of size 5 is $A_2 = 0.577$. Then the trial control limits for the $\overline{X}$ chart are

$$\overline{\overline{x}} \pm A_2\overline{r} = 33.32 \pm (0.577)(5.8) = 33.32 \pm 3.35$$

or

$$UCL = 36.67$$
$$LCL = 29.97$$

For the R chart, the trial control limits are

$$UCL = D_4\overline{r} = (2.115)(5.8) = 12.27$$
$$LCL = D_3\overline{r} = (0)(5.8) = 0$$

The $\overline{X}$ and R control charts with these trial control limits are shown in Fig. 8-8. Note that samples 6, 8, 11, and 19 are out of control on the $\overline{X}$ chart and that sample 9 is out of control on the R chart. (These points are labeled with a "1" because they violate the

Table 8-1   Vane-Opening Measurements

Sample Number	x_1	x_2	x_3	x_4	x_5	$\overline{x}$	r
1	33	29	31	32	33	31.6	4
2	33	31	35	37	31	33.4	6
3	35	37	33	34	36	35.0	4
4	30	31	33	34	33	32.2	4
5	33	34	35	33	34	33.8	2
6	38	37	39	40	38	38.4	3
7	30	31	32	34	31	31.6	4
8	29	39	38	39	39	36.8	10
9	28	33	35	36	43	35.0	15
10	38	33	32	35	32	34.0	6
11	28	30	28	32	31	29.8	4
12	31	35	35	35	34	34.0	4
13	27	32	34	35	37	33.0	10
14	33	33	35	37	36	34.8	4
15	35	37	32	35	39	35.6	7
16	33	33	27	31	30	30.8	6
17	35	34	34	30	32	33.0	5
18	32	33	30	30	33	31.6	3
19	25	27	34	27	28	28.2	9
20	35	35	36	33	30	33.8	6
						$\overline{\overline{x}} = 33.32$	$\overline{r} = 5.8$

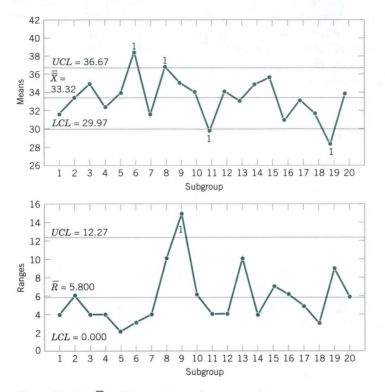

Figure 8-8 The $\overline{X}$ and R control charts for vane opening.

first Western Electric rule.) Suppose that all of these assignable causes can be traced to a defective tool in the wax-molding area. We should discard these five samples and recompute the limits for the $\overline{X}$ and R charts. These new revised limits are, for the $\overline{X}$ chart,

$$UCL = \overline{\overline{x}} + A_2 \overline{r} = 33.21 + (0.577)(5.0) = 36.10$$
$$LCL = \overline{\overline{x}} - A_2 \overline{r} = 33.21 - (0.577)(5.0) = 30.33$$

and for the R chart,

$$UCL = D_4 \overline{r} = (2.115)(5.0) = 10.57$$
$$LCL = D_3 \overline{r} = (0)(5.0) = 0$$

The revised control charts are shown in Fig. 8-9. Note that we have treated the first 20 preliminary samples as **estimation data** with which to establish control limits. These limits can now be used to judge the statistical control of future production. As each new sample becomes available, the values of $\overline{x}$ and r should be computed and plotted on the control charts. It may be desirable to revise the limits periodically, even if the process remains stable. The limits should always be revised when process improvements are made.

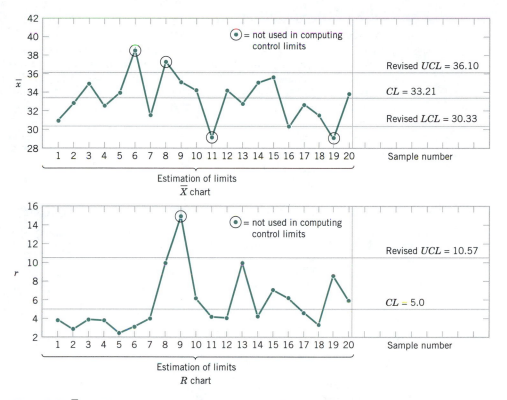

Figure 8-9 $\overline{X}$ and R control charts for vane opening, revised limits.

Computer Construction of $\overline{X}$ and R Control Charts

Many computer programs construct $\overline{X}$ and R control charts. Figures 8-8 and 8-9 show charts similar to those produced by Minitab for the vane-opening data in Example 8-1. This program will allow the user to select any multiple of sigma as the width of the control limits and utilize the Western Electric rules to detect out-of-control points. The program will also prepare a summary report as in Table 8-2 and exclude subgroups from the calculation of the control limits.

Table 8-2 Summary Report from Minitab for the Vane-Opening Data in Example 8-1

Test Results for Xbar Chart
TEST 1. One point more than 3.00 sigmas from center line.
Test Failed at points: 6 8 11 19

Test Results for R Chart
TEST 1. One point more than 3.00 sigmas from center line.
Test Failed at points: 9

EXERCISES FOR SECTION 8-5

8-1. An extrusion die is used to produce aluminum rods. The diameter of the rods is a critical quality characteristic. The following table shows values for 20 samples of three rods each. Specifications on the rods are 0.4030 ± 0.0010 inch. The values given are the last three digits of the measurement; that is, 36 is read as 0.4036.

Sample	Observation		
	1	2	3
1	36	33	34
2	30	34	31
3	33	32	29
4	35	30	34
5	33	31	33
6	32	34	33
7	27	36	35
8	32	36	41
9	32	33	39
10	36	40	37
11	20	30	33
12	30	32	38
13	34	35	30
14	36	39	37
15	38	33	34
16	33	43	35
17	36	39	37
18	35	34	31
19	36	33	37
20	34	33	31

(a) Using all the data, find trial control limits for $\overline{X}$ and R charts, construct the chart, and plot the data.

(b) Use the trial control limits from part (a) to identify out-of-control points. If necessary, revise your control limits, assuming that any samples that plot outside the control limits can be eliminated.

8-2. Twenty samples of size 4 are drawn from a process at 1-hour intervals, and the following data are obtained:

$$\sum_{i=1}^{20} \overline{x}_i = 378.50 \qquad \sum_{i=1}^{20} r_i = 7.80$$

Find trial control limits for $\overline{X}$ and R charts.

8-3. The overall length of a skew used in a knee replacement device is monitored using $\overline{X}$ and R charts. The following table gives the length for 20 samples of size 4. (Measurements are coded from 2.00 mm; that is, 15 is 2.15 mm.)

Sample	Observation			
	1	2	3	4
1	16	18	15	13
2	16	15	17	16
3	15	16	20	16
4	14	16	14	12
5	14	15	13	16
6	16	14	16	15
7	16	16	14	15
8	17	13	17	16
9	15	11	13	16
10	15	18	14	13
11	14	14	15	13
12	15	13	15	16
13	13	17	16	15
14	11	14	14	21
15	14	15	14	13
16	18	15	16	14
17	14	16	19	16
18	16	14	13	19
19	17	19	17	13
20	12	15	12	17

(a) Using all the data, find trial control limits for $\overline{X}$ and R charts, construct the chart, and plot the data.

(b) Use the trial control limits from part (a) to identify out-of-control points. If necessary, revise your control limits, assuming that any samples that plot outside the control limits can be eliminated.

8-4. Samples of size $n = 6$ are collected from a process every hour. After 20 samples have been collected, we calculate $\overline{\overline{x}} = 20.0$ and $\overline{r}/d_2 = 1.4$. Find trial control limits for $\overline{X}$ and R charts.

8-5. Control charts for $\overline{X}$ and R are to be set up for an important quality characteristic. The sample size is $n = 4$, and $\overline{x}$ and r

are computed for each of 25 preliminary samples. The summary data are

$$\sum_{i=1}^{25} \overline{x}_i = 7657 \qquad \sum_{i=1}^{25} r_i = 1180$$

(a) Find trial control limits for $\overline{X}$ and R charts.

(b) Assuming that the process is in control, estimate the process mean and standard deviation.

8-6. The thickness of a metal part is an important quality parameter. Data on thickness (in inches) are given here, for 25 samples of five parts each.

Sample Number	x_1	x_2	x_3	x_4	x_5
1	0.0629	0.0636	0.0640	0.0635	0.0640
2	0.0630	0.0631	0.0622	0.0625	0.0627
3	0.0628	0.0631	0.0633	0.0633	0.0630
4	0.0634	0.0630	0.0631	0.0632	0.0633
5	0.0619	0.0628	0.0630	0.0619	0.0625
6	0.0613	0.0629	0.0634	0.0625	0.0628
7	0.0630	0.0639	0.0625	0.0629	0.0627
8	0.0628	0.0627	0.0622	0.0625	0.0627
9	0.0623	0.0626	0.0633	0.0630	0.0624
10	0.0631	0.0631	0.0633	0.0631	0.0630
11	0.0635	0.0630	0.0638	0.0635	0.0633
12	0.0623	0.0630	0.0630	0.0627	0.0629
13	0.0635	0.0631	0.0630	0.0630	0.0630
14	0.0645	0.0640	0.0631	0.0640	0.0642
15	0.0619	0.0644	0.0632	0.0622	0.0635
16	0.0631	0.0627	0.0630	0.0628	0.0629
17	0.0616	0.0623	0.0631	0.0620	0.0625
18	0.0630	0.0630	0.0626	0.0629	0.0628
19	0.0636	0.0631	0.0629	0.0635	0.0634
20	0.0640	0.0635	0.0629	0.0635	0.0634
21	0.0628	0.0625	0.0616	0.0620	0.0623
22	0.0615	0.0625	0.0619	0.0619	0.0622
23	0.0630	0.0632	0.0630	0.0631	0.0630
24	0.0635	0.0629	0.0635	0.0631	0.0633
25	0.0623	0.0629	0.0630	0.0626	0.0628

(a) Using all the data, find trial control limits for $\overline{X}$ and R charts, construct the chart, and plot the data. Is the process in statistical control?

(b) Use the trial control limits from part (a) to identify out-of-control points.

List the sample numbers of the out-of-control points.

8-7. The copper content of a plating bath is measured three times per day, and the results are reported in ppm. The values for 25 days are shown in the following table.

Sample	Observation		
	1	2	3
1	5.10	6.10	5.50
2	5.70	5.59	5.29
3	6.31	5.00	6.07
4	6.83	8.10	7.96
5	5.42	5.29	6.71
6	7.03	7.29	7.54
7	6.57	5.89	7.08
8	5.96	7.52	7.29
9	8.15	6.69	6.06
10	6.11	5.14	6.68
11	6.49	5.68	5.51
12	5.12	4.26	4.49
13	5.59	5.21	4.94
14	7.59	7.93	6.90
15	6.72	6.79	5.23
16	6.30	5.37	7.08
17	6.33	6.33	5.80
18	6.91	6.05	6.03
19	8.05	6.52	8.51
20	6.39	5.07	6.86
21	5.63	6.42	5.39
22	6.51	6.90	7.40
23	6.91	6.87	6.83
24	6.28	6.09	6.71
25	5.07	7.17	6.11

(a) Using all the data, find trial control limits for $\overline{X}$ and R charts, construct the chart, and plot the data. Is the process in statistical control?

(b) If necessary, revise the control limits computed in part (a), assuming any samples that plot outside the control limits can be eliminated. Continue to eliminate points outside the control limits and revise, until all points plot between control limits.

8-6   CONTROL CHARTS FOR INDIVIDUAL MEASUREMENTS

In many situations, the sample size used for process control is $n = 1$; that is, the sample consists of an individual unit. Some examples of these situations are as follows.

1. Automated inspection and measurement technology is used, and every unit manufactured is analyzed.

2. The production rate is very slow, and it is inconvenient to allow sample sizes of $n > 1$ to accumulate before being analyzed.

3. Repeat measurements on the process differ only because of laboratory or analysis error, as in many chemical processes.

4. In process plants, such as papermaking, measurements on some parameters such as coating thickness *across* the roll will differ very little and produce a standard deviation that is much too small if the objective is to control coating thickness *along* the roll.

In such situations, the **control chart for individuals** is useful. The control chart for individuals uses the **moving range** of two successive observations to estimate the process variability. The moving range is defined as $MR_i = |X_i - X_{i-1}|$. It is also possible to establish a control chart on the moving range. The parameters for these charts are defined as follows.

Control Chart for Individuals

The center line and upper and lower control limits for a control chart for individuals are

$$UCL = \bar{x} + 3\frac{\overline{mr}}{d_2}$$

$$CL = \bar{x} \qquad\qquad (8\text{-}9)$$

$$LCL = \bar{x} - 3\frac{\overline{mr}}{d_2}$$

and for a control chart for moving ranges

$$UCL = D_4\overline{mr}$$

$$CL = \overline{mr}$$

$$LCL = D_3\overline{mr}$$

The procedure is illustrated in the following example.

Table 8-3   Chemical Process Concentration Measurements

Observation	Concentration x	Moving Range mr
1	102.0	
2	94.8	7.2
3	98.3	3.5
4	98.4	0.1
5	102.0	3.6
6	98.5	3.5
7	99.0	0.5
8	97.7	1.3
9	100.0	2.3
10	98.1	1.9
11	101.3	3.2
12	98.7	2.6
13	101.1	2.4
14	98.4	2.7
15	97.0	1.4
16	96.7	0.3
17	100.3	3.6
18	101.4	1.1
19	97.2	4.2
20	101.0	3.8
	$\bar{x} = 99.1$	$\overline{mr} = 2.59$

EXAMPLE 8-2

Table 8-3 shows 20 observations on concentration for the output of a chemical process. The observations are taken at 1-hour intervals. If several observations are taken at the same time, the observed concentration reading will differ only because of measurement error. Since the measurement error is small, only one observation is taken each hour.

To set up the control chart for individuals, note that the sample average of the 20 concentration readings is $\bar{x} = 99.1$ and that the moving ranges of two observations are shown in the last column of Table 8-3. The average of the 19 moving ranges is $\overline{mr} = 2.59$. To set up the moving-range chart, we note that $D_3 = 0$ and $D_4 = 3.267$ for $n = 2$. Therefore, the moving-range chart has center line $\overline{mr} = 2.59$, $LCL = 0$, and $UCL = D_4\overline{mr} = (3.267)(2.59) = 8.46$. The control chart is shown as the lower control chart in Fig. 8-10. This control chart was constructed by Minitab. Because no points exceed the upper control limit, we may now set up the control chart for individual concentration measurements. If a moving range of $n = 2$ observations is used, then $d_2 = 1.128$. For the data in Table 8-3 we have

$$UCL = \bar{x} + 3\frac{\overline{mr}}{d_2} = 99.1 + 3\frac{2.59}{1.128} = 105.99$$

$$CL = \bar{x} = 99.1$$

$$LCL = \bar{x} - 3\frac{\overline{mr}}{d_2} = 99.1 - 3\frac{2.59}{1.128} = 92.21$$

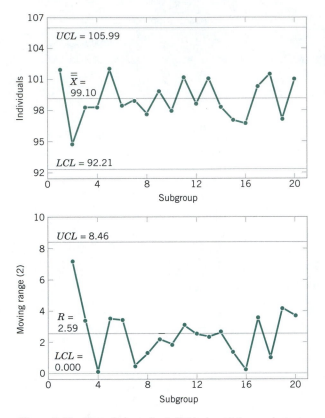

Figure 8-10 Control charts for individuals and the moving range (from Minitab) for the chemical process concentration data in Example 8-2.

The control chart for individual concentration measurements is shown as the upper control chart in Fig. 8-10. There is no indication of an out-of-control condition.

The chart for individuals can be interpreted much like an ordinary $\overline{X}$ control chart. A shift in the process average will result in either a point (or points) outside the control limits, or a pattern consisting of a run on one side of the center line.

Some care should be exercised in interpreting patterns on the moving-range chart. The moving ranges are correlated, and this correlation may often induce a pattern of runs or cycles on the chart. The individual measurements are assumed to be uncorrelated, however, and any apparent pattern on the individuals' control chart should be carefully investigated.

The control chart for individuals is very insensitive to small shifts in the process mean. For example, if the size of the shift in the mean is one standard deviation, the average number of points to detect this shift is 43.9. This result is shown later in the chapter. Although the performance of the control chart for individuals is much better for large shifts, in many situations the shift of interest is not large and more rapid shift detection is desirable. In these cases, we recommend the *cumulative sum control chart* or an *exponentially weighted moving-average chart* (Montgomery, 1996).

Some individuals have suggested that limits narrower than three-sigma be used on the chart for individuals in order to enhance its ability to detect small process shifts. This is a dangerous suggestion, for narrower limits will dramatically increase false alarms such that the charts may be ignored and become useless. If you are interested in detecting small shifts, use the cumulative sum or exponentially weighted moving-average control chart referred to above.

EXERCISES FOR SECTION 8-6

8-8. Twenty successive hardness measurements are made on a metal alloy, and the data are shown in the following table.

Observation	Hardness
1	54
2	52
3	54
4	52
5	51
6	55
7	54
8	62
9	49
10	54
11	49
12	53
13	55
14	54
15	56
16	52
17	55
18	51
19	55
20	52

(a) Using all the data, compute trial control limits for individual observations and moving-range $n = 2$ charts. Construct the chart and plot the data. Determine whether the process is in statistical control. If not, assume that assignable causes can be found to eliminate these samples and revise the control limits.

(b) Estimate the process mean and standard deviation for the in-control process.

8-9. The purity of a chemical product is measured hourly. Purity determinations for the last 24 hours are shown in the following table.

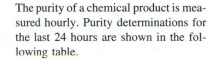

Observation	Purity
1	81
2	83
3	82
4	80
5	84
6	76
7	83
8	85
9	79
10	82
11	75
12	80
13	83
14	86
15	84
16	85
17	81
18	83
19	77
20	82
21	75
22	83
23	85
24	86

(a) Using all the data, compute trial control limits for individual observations and moving-range $n = 2$ charts. Construct the chart and plot the data. Determine whether the process is in statistical control. If not, assume that assignable causes can be found to eliminate these samples and revise the control limits.

(b) Estimate the process mean and standard deviation for the in-control process.

8-10. The diameter of individual holes is measured in consecutive order by an automatic sensor. The results of measuring 25 holes are as follows.

Sample	Diameter
1	14.06
2	23.70
3	15.10
4	22.46
5	35.26
6	22.74
7	20.14
8	11.62
9	10.21
10	8.29
11	16.49
12	15.34
13	14.08
14	20.68
15	16.33
16	16.29
17	9.59
18	15.83
19	15.65
20	19.80
21	21.64
22	28.38
23	21.58
24	8.38
25	17.00

(a) Using all the data, compute trial control limits for individual observations and moving-range $n = 2$ charts. Construct the control chart and plot the data. Determine whether the process is in statistical control. If not, assume that assignable causes can be found to eliminate these samples and revise the control limits.

(b) Estimate the process mean and standard deviation for the in-control process.

8-11. The viscosity of a chemical intermediate is measured every hour. Twenty samples consisting of a single observation are as follows.

Sample	Viscosity
1	378
2	438
3	487
4	515
5	485
6	474
7	486
8	548
9	502
10	440
11	462
12	502
13	449
14	470
15	501
16	470
17	512
18	530
19	462
20	491

(a) Using all the data, compute trial control limits for individual observations and moving-range $n = 2$ charts. Determine whether the process is in statistical control. If not, assume that assignable causes can be found to eliminate these samples and revise the control limits.

(b) Estimate the process mean and standard deviation for the in-control process.

8-7 PROCESS CAPABILITY

It is usually necessary to obtain some information about the **capability** of the process—that is, the performance of the process when it is operating in control. Two graphical tools, the **tolerance chart** (or tier chart) and the **histogram,** are helpful in assessing process capability. The tolerance chart for all 20 samples from the vane-manufacturing process is shown in Fig. 8-11. The specifications on vane opening are 0.5030 ± 0.0010 in. In terms of the coded data, the upper specification limit is $USL = 40$ and the lower specification limit is $LSL = 20$, and these limits are shown on the chart in Fig. 8-11. Each measurement is plotted on the tolerance chart. Measurements from the same subgroup are connected with lines. The tolerance chart is useful in revealing patterns over time in the individual measurements, or it may show that a particular value of $\bar{x}$ or r was produced by one or two unusual observations in the sample. For example, note the two unusual observations in sample 9 and the single unusual observation in sample 8. Note also that it is appropriate to plot the

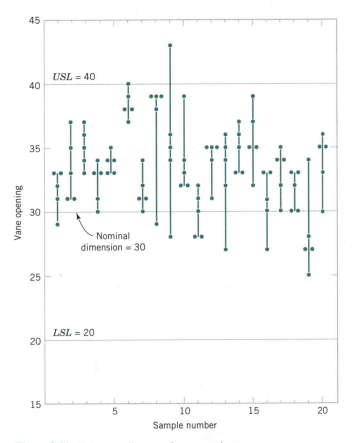

Figure 8-11   Tolerance diagram of vane openings.

specification limits on the tolerance chart, since it is a chart of individual measurements. **It is never appropriate to plot specification limits on a control chart or to use the specifications in determining the control limits.** Specification limits and control limits are unrelated. Finally, note from Fig. 8-11 that the process is running off-center from the nominal dimension of 30 (or 0.5030 inch).

The histogram for the vane-opening measurements is shown in Fig. 8-12. The observations from samples 6, 8, 9, 11, and 19 (corresponding to out of-control points on either the $\overline{X}$ or R chart) have been deleted from this histogram. The general impression from examining this histogram is that the process is capable of meeting the specification but that it is running off-center.

Another way to express process capability is in terms of an index that is defined as follows.

The **process capability ratio** (*PCR*) is

$$PCR = \frac{USL - LSL}{6\sigma} \qquad (8\text{-}10)$$

The numerator of *PCR* is the width of the specifications. The limits 3σ on either side of the process mean are sometimes called **natural tolerance limits,** for these represent limits

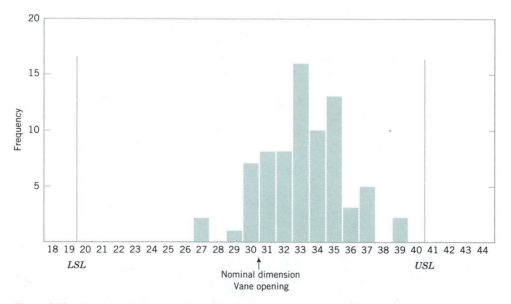

Figure 8-12   Histogram for vane opening.

that an in-control process should meet with most of the units produced. Consequently, 6σ is often referred to as the width of the process. For the vane opening, where our sample size is 5, we could estimate σ as

$$\hat{\sigma} = \frac{\bar{r}}{d_2} = \frac{5.0}{2.326} = 2.15$$

Therefore, the *PCR* is estimated to be

$$\widehat{PCR} = \frac{USL - LSL}{6\hat{\sigma}}$$

$$= \frac{40 - 20}{6(2.15)}$$

$$= 1.55$$

The *PCR* has a natural interpretation: $(1/PCR)100$ is simply the percentage of the specifications' width used by the process. Thus, the vane-opening process uses approximately $(1/1.55)100 = 64.5\%$ of the specifications' width.

Figure 8-13a shows a process for which the *PCR* exceeds unity. Since the process natural tolerance limits lie inside the specifications, very few defective or nonconforming units will be produced. If $PCR = 1$, as shown in Fig. 8-13b, more nonconforming units result. In fact, for a normally distributed process, if $PCR = 1$, the fraction nonconforming is 0.27%, or 2700 parts per million. Finally, when the *PCR* is less than unity, as in Fig. 8-13c, the process is very yield-sensitive and a large number of nonconforming units will be produced.

The definition of the *PCR* given in equation 8-10 implicitly assumes that the process is centered at the nominal dimension. If the process is running off-center, its **actual capability** will be less than indicated by the *PCR*. It is convenient to think of *PCR* as a measure of **potential capability**—that is, capability with a centered process. If the process is not centered, then a measure of actual capability is often used. This ratio, called PCR_k, is defined next.

$$PCR_k = \min\left[\frac{USL - \mu}{3\sigma}, \frac{\mu - LSL}{3\sigma}\right] \qquad (8\text{-}11)$$

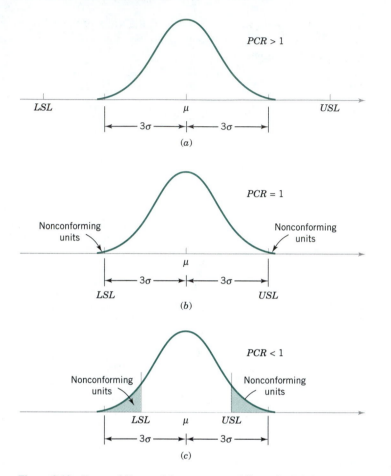

Figure 8-13   Process fallout and the process capability ratio (*PCR*).

In effect, PCR_k is a one-sided process capability ratio that is calculated relative to the specification limit nearest to the process mean. For the vane-opening process, we find that the estimate of the process capability ratio PCR_k is

$$\widehat{PCR_k} = \min\left[\frac{USL - \bar{\bar{x}}}{3\hat{\sigma}}, \ \frac{\bar{\bar{x}} - LSL}{3\hat{\sigma}}\right]$$

$$= \min\left[\frac{40 - 33.19}{3(2.15)} = 1.06, \ \frac{33.19 - 20}{3(2.15)} = 2.04\right]$$

$$= 1.06$$

Note that if $PCR = PCR_k$, the process is centered at the nominal dimension. Since $\widehat{PCR_k} = 1.06$ for the vane-opening process and $\widehat{PCR} = 1.55$, the process is obviously running off-center, as was first noted in Figs. 8-11 and 8-12. This off-center operation was ultimately traced to an oversized wax tool. Changing the tooling resulted in a substantial improvement in the process (Montgomery, 1996).

The fractions of nonconforming output (or fallout) below the lower specification limit and above the upper specification limit are often of interest. Suppose that the output from a normally distributed process in statistical control is denoted as X. The fractions are determined from

$$P(X < LSL) = P(Z < (LSL - \mu)/\sigma)$$
$$P(X > USL) = P(Z > (USL - \mu)/\sigma)$$

EXAMPLE 8-3

For an electronic manufacturing process, a current has specifications of 100 ± 10 milliamperes. The process mean μ and standard deviation σ are 107.0 and 1.5, respectively. Consequently,

$$PCR = (110 - 90)/(6 \cdot 1.5) = 2.22 \quad \text{and} \quad PCR_k = (110 - 107)/(3 \cdot 1.5) = 0.67$$

The small PCR_k indicates that the process is likely to produce currents outside of the specification limits. From the normal distribution in Appendix A Table I,

$$P(X < LSL) = P(Z < (90 - 107)/1.5) = P(Z < -11.33) = 0$$
$$P(X > USL) = P(Z > (110 - 107)/1.5) = P(Z > 2) = 0.023$$

For this example, the relatively large probability of exceeding the USL is a warning of potential problems with this criterion even if none of the measured observations in a preliminary sample exceed this limit. We emphasize that the fraction-nonconforming calculation assumes that the observations are normally distributed and the process is in control. Departures from normality can seriously affect the results. The calculation should be interpreted as an approximate guideline for process performance. To make matters worse, μ and σ need to be estimated from the data available and a small sample size can result in poor estimates that further degrade the calculation.

Montgomery (1996) provides guidelines on appropriate values of the PCR and a table relating fallout for a normally distributed process in statistical control to the value of PCR. Many U.S. companies use $PCR = 1.33$ as a minimum acceptable target and $PCR = 1.66$ as a minimum target for strength, safety, or critical characteristics. Some companies require that internal processes and those at suppliers achieve a $PCR_k = 2.0$. Figure 8-14 illustrates a process with $PCR = PCR_k = 2.0$. Assuming a normal distribution, the calculated fallout for this process is 0.0018 parts per million. A process with $PCR_k = 2.0$ is referred to as a **six-sigma process** because the distance from the process mean to the nearest specification is six standard deviations. The reason that such a large process capability is often required is that it is difficult to maintain a process mean at the

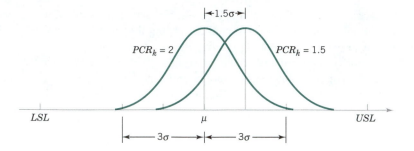

Figure 8-14 Mean of a six-sigma process shifts by 1.5 standard deviations.

center of the specifications for long periods of time. A common model that is used to justify the importance of a six-sigma process is illustrated by referring to Fig. 8-14. If the process mean shifts off-center by 1.5 standard deviations, the PCR_k decreases to $4.5\sigma/3\sigma = 1.5$. Assuming a normally distributed process, the fallout of the shifted process is **3.4 parts per million.** Consequently, the mean of a six-sigma process can shift 1.5 standard deviations from the center of the specifications and still maintain a fallout of 3.4 parts per million.

In addition, some U.S. companies, particularly the automobile industry, have adopted the Japanese terminology $C_p = PCR$ and $C_{pk} = PCR_k$. Because C_p has another meaning in statistics (in multiple regression) we prefer the traditional notation PCR and PCR_k.

We repeat that process capability calculations are meaningful only for stable processes; that is, processes that are in control. A process capability ratio indicates whether or not the natural or chance variability in a process is acceptable relative to the specifications.

EXERCISES FOR SECTION 8-7

8-12. A normally distributed process uses 66.7% of the specification band. It is centered at the nominal dimension, located halfway between the upper and lower specification limits.
  (a) Estimate PCR and PCR_k. Interpret these ratios.
  (b) What fallout level (fraction defective) is produced?

8-13. Reconsider Exercise 8-1. Use the revised control limits and process estimates.
  (a) Estimate PCR and PCR_k. Interpret these ratios.
  (b) What percentage of defectives is being produced by this process?

8-14. Reconsider Exercise 8-2, where the specification limits are 18.50 ± 0.50.

  (a) What conclusions can you draw about the ability of the process to operate within these limits? Estimate the percentage of defective items that will be produced.
  (b) Estimage PCR and PCR_k. Interpret these ratios.

8-15. Reconsider Exercise 8-3. Using the process estimates, what is the fallout level if the coded specifications are 15 ± 4 mm? Estimate PCR and interpret this ratio.

8-16. A normally distributed process uses 85% of the specification band. It is centered at the nominal dimension, located halfway between the upper and lower specification limits.
  (a) Estimate PCR and PCR_k. Interpret these ratios.

(b) What fallout level (fraction defective) is produced?

8-17. Reconsider Exercise 8-5. Suppose that the quality characteristic is normally distributed with specification at 300 ± 40. What is the fallout level? Estimate PCR and PCR_k and interpret these ratios.

8-18. Reconsider Exercise 8-4. Assuming that both charts exhibit statistical control and that the process specifications are at 20 ± 5, estimate PCR and PCR_k and interpret these ratios.

8-19. Reconsider Exercise 8-7. Given that the specifications are at 6.0 ± 1.0, estimate PCR and PCR_k for the in-control process and interpret these ratios.

8-20. Reconsider 8-6. What are the natural tolerance limits of this process?

8-21. Reconsider 8-11. The viscosity specifications are at 500 ± 25. Calculate estimates of the process capability ratios PCR and PCR_k for the in-control process using all the data and provide an interpretation.

8-8 ATTRIBUTE CONTROL CHARTS

8-8.1 P Chart (Control Chart for Proportions) and nP Chart

Often it is desirable to classify a product as either defective or nondefective on the basis of comparison with a standard. This classification is usually done to achieve economy and simplicity in the inspection operation. For example, the diameter of a ball bearing may be checked by determining whether it will pass through a gauge consisting of circular holes cut in a template. This kind of measurement would be much simpler than directly measuring the diameter with a device such as a micrometer. Control charts for attributes are used in these situations. Attribute control charts often require a considerably larger sample size than do their measurements counterparts. In this section, we will discuss the **fraction-defective control chart,** or *P* chart. Sometimes the *P* chart is called the **control chart for fraction nonconforming.**

Suppose *D* is the number of defective units in a random sample of size *n*. We assume that *D* is a binomial random variable with unknown parameter *p*. The fraction defective

$$\hat{P} = \frac{D}{n}$$

of each sample is plotted on the chart. Furthermore, the variance of the statistic $\hat{P}$ is

$$\sigma_{\hat{P}}^2 = \frac{p(1-p)}{n}$$

Therefore, a *P* chart for fraction defective could be constructed using *p* as the center line and control limits at

$$UCL = p + 3\sqrt{\frac{p(1-p)}{n}}$$

$$LCL = p - 3\sqrt{\frac{p(1-p)}{n}} \tag{8-12}$$

However, the true process fraction defective is almost always unknown and must be estimated using the data from preliminary samples.

Suppose that m preliminary samples each of size n are available, and let D_i be the number of defectives in the ith sample. The $\hat{P}_i = D_i/n$ is the sample fraction defective in the ith sample. The average fraction defective is

$$\overline{P} = \frac{1}{m} \sum_{i=1}^{m} \hat{P}_i = \frac{1}{mn} \sum_{i=1}^{m} D_i \tag{8-13}$$

Now $\overline{P}$ may be used as an estimator of p in the center line and control limit calculations.

The P Chart

The center line and upper and lower control limits for the P chart are

$$UCL = \overline{p} + 3 \sqrt{\frac{\overline{p}(1 - \overline{p})}{n}}$$

$$CL = \overline{p}$$

$$LCL = \overline{p} - 3 \sqrt{\frac{\overline{p}(1 - \overline{p})}{n}} \tag{8-14}$$

where $\overline{p}$ is the observed value of the average fraction defective.

These control limits are based on the normal approximation to the binomial distribution. When p is small, the normal approximation may not always be adequate. In such cases, we may use control limits obtained directly from a table of binomial probabilities. If $\overline{p}$ is small, the lower control limit may be a negative number. If this should occur, it is customary to consider zero as the lower control limit.

EXAMPLE 8-4

Suppose we wish to construct a fraction-defective control chart for a ceramic substrate production line. We have 20 preliminary samples, each of size 100; the number of defectives in each sample is shown in Table 8-4. Assume that the samples are numbered

Table 8-4   Number of Defectives in Samples of 100 Ceramic
Substrates

Sample	No. of Defectives	Sample	No. of Defectives
1	44	11	36
2	48	12	52
3	32	13	35
4	50	14	41
5	29	15	42
6	31	16	30
7	46	17	46
8	52	18	38
9	44	19	26
10	48	20	30

in the sequence of production. Note that $\bar{p} = (800/2000) = 0.40$; therefore, the trial parameters for the control chart are

$$UCL = 0.40 + 3\sqrt{\frac{(0.40)(0.60)}{100}} = 0.55$$

$$CL = 0.40$$

$$LCL = 0.40 - 3\sqrt{\frac{(0.40)(0.60)}{100}} = 0.25$$

The control chart is shown in Fig. 8-15. All samples are in control. If they were not, we would search for assignable causes of variation and revise the limits accordingly. This chart can be used for controlling future production.

Although this process exhibits statistical control, its defective rate ($\bar{p} = 0.40$) is very poor. We should take appropriate steps to investigate the process to determine why such a large number of defective units are being produced. Defective units should be analyzed to determine the specific types of defects present. Once the defect types are known, process

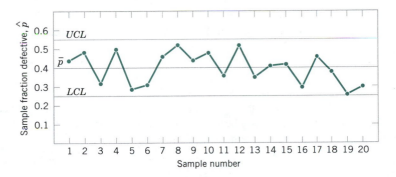

Figure 8-15 P chart for a ceramic substrate.

changes should be investigated to determine their impact on defect levels. Designed experiments may be useful in this regard.

Computer software also produces an **nP chart.** This is simply a control chart of $n\hat{P} = D$, the number of defectives in a sample. The points, center line, and control limits for this chart are multiples (times n) of the corresponding elements of a P chart. The use of an nP chart avoids the fractions in a P chart.

The *nP* Chart

The center line and upper and lower control limits for the nP chart are

$$UCL = n\bar{p} + 3\sqrt{n\bar{p}(1-\bar{p})}$$
$$CL = n\bar{p}$$
$$LCL = n\bar{p} - 3\sqrt{n\bar{p}(1-\bar{p})}$$

where $\bar{p}$ is the observed value of the average fraction defective.

For the data in Example 8-4 the center line is $n\bar{p} = 100(0.4) = 40$ and the upper and lower control limits for the nP chart are $UCL = 100(0.4) + \sqrt{100(0.4)(0.6)} = 44.9$ and $LCL = 100(0.4) - \sqrt{100(0.4)(0.6)} = 35.1$. The number of defectives in Table 8-4 would be plotted on such a chart.

8-8.2 U Chart (Control Chart for Average Number of Defects per Unit) and C Chart

It is sometimes necessary to monitor the number of defects in a unit of product rather than the fraction defective. Suppose that in the production of cloth it is necessary to control the number of defects per yard, or that in assembling an aircraft wing the number of missing rivets must be controlled. In these situations we may use the control chart for defects per unit or the *U* **chart.** Many defects-per-unit situations can be modeled by the Poisson distribution. If each sample consists of n units and there are C total defects in the sample, then

$$U = \frac{C}{n}$$

is the average number of defects per unit. A U chart may be constructed for such data. If there are m samples, and the number of defects in these samples is $C_1, C_2, \ldots, C_m$, then the estimator of the average number of defects per unit is

$$\bar{U} = \frac{1}{m}\sum_{i=1}^{m} U_i = \frac{1}{mn}\sum_{i=1}^{m} C_i \tag{8-15}$$

The parameters of the U chart are defined as follows.

The U Chart

The center line and upper and lower control limits on the U chart are

$$UCL = \bar{u} + 3\sqrt{\frac{\bar{u}}{n}}$$

$$CL = \bar{u}$$

$$LCL = \bar{u} - 3\sqrt{\frac{\bar{u}}{n}} \qquad (8\text{-}16)$$

where $\bar{u}$ is the average number of defects per unit.

If the number of defects in a unit is a Poisson random variable with parameter λ, then the mean and variance of this distribution are both λ. Each point on the chart is U, the average number of defects per unit. Therefore, the mean of U is λ and the variance of U is λ/n. Usually λ is unknown and $\overline{U}$ is the estimator of λ that is used to set the control limits.

These control limits are based on the normal approximation to the Poisson distribution. When λ is small, the normal approximation may not always be adequate. In such cases, we may use control limits obtained directly from a table of Poisson probabilities. If $\bar{u}$ is small, the lower control limit may be a negative number. If this should occur, it is customary to consider zero as the lower control limit.

EXAMPLE 8-5

Printed circuit boards are assembled by a combination of manual assembly and automation. A flow solder machine is used to make the mechanical and electrical connections of the leaded components to the board. The boards are run through the flow solder process almost continuously, and every hour five boards are selected and inspected for process-control purposes. The number of defects in each sample of five boards is noted. Results for 20 samples are shown in Table 8-5.

The center line for the U chart is

$$\bar{u} = \frac{1}{20}\sum_{i=1}^{20} u_i = \frac{32}{20} = 1.6$$

and the upper and lower control limits are

$$UCL = \bar{u} + 3\sqrt{\frac{\bar{u}}{n}} = 1.6 + 3\sqrt{\frac{1.6}{5}} = 3.3$$

$$LCL = \bar{u} - 3\sqrt{\frac{\bar{u}}{n}} = 1.6 - 3\sqrt{\frac{1.6}{5}} = 0$$

The control chart is plotted in Fig. 8-16. Because LCL is negative, it is set to zero. From the control chart in Fig. 8-16, we see that the process is in control. However, eight defects per group of five circuit boards is too many (about $8/5 = 1.6$ defects/board), and the process needs improvement. An investigation needs to be made of the specific types of defects found on the printed circuit boards. This will usually suggest potential avenues for process improvement.

Computer software also produces a **C chart.** This is simply a control chart of C, the total of defects in a sample. The use of a C chart avoids the fractions that can occur in a U chart.

The C Chart

The center line and upper and lower control limits for the C chart are

$$UCL = \bar{c} + 3\sqrt{\bar{c}}$$
$$CL = \bar{c} \qquad\qquad (8\text{-}17)$$
$$LCL = \bar{c} - 3\sqrt{\bar{c}}$$

where $\bar{c}$ is the average number of defects in a sample.

Table 8-5 Number of Defects in Samples of Five Printed Circuit Boards

Sample	Number of Defects c_i	Defects per Unit u_i	Sample	Number of Defects c_i	Defects per Unit u_i
1	6	1.2	11	9	1.8
2	4	0.8	12	15	3.0
3	8	1.6	13	8	1.6
4	10	2.0	14	10	2.0
5	9	1.8	15	8	1.6
6	12	2.4	16	2	0.4
7	16	3.2	17	7	1.4
8	2	0.4	18	1	0.2
9	3	0.6	19	7	1.4
10	10	2.0	20	13	2.6

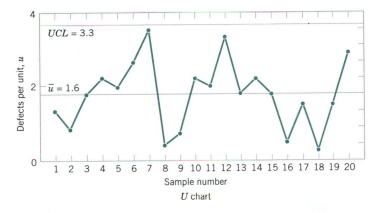

Figure 8-16 *U* chart of defects per unit on printed circuit boards.

For the data in Example 8-5

$$\bar{c} = \frac{1}{20} \sum_{i=1}^{20} c_i = \left(\frac{1}{20}\right) 160 = 8$$

and the upper and lower control limits for the *C* chart are $UCL = 8 + 3\sqrt{8} = 16.5$ and $LCL = 8 - 3\sqrt{8} = -0.5$, which is set to zero. The number of defects in Table 8-5 would be plotted on such a chart.

EXERCISES FOR SECTION 8-8

8-22. Suppose the following number of defects has been found in successive samples of size 100 (read down):

6	6	10
7	9	8
3	6	7
9	10	7
6	9	7
9	2	6
4	8	14
14	4	18
3	8	13
5	10	6

(a) Using all the data, compute trial control limits for a fraction-defective control chart, construct the chart, and plot the data.

(b) Determine whether the process is in statistical control. If not, assume that assignable causes can be found and out-of-control points eliminated. Revise the control limits.

8-23. The following represent the number of solder defects observed on 24 samples of four printed circuit boards: 7, 6, 8, 10, 24, 6, 5, 4, 8, 11, 15, 8, 4, 16, 11, 12, 8, 6, 5, 9, 7, 14, 8, 21.

(a) Using all the data, compute trial control limits for a U control chart, construct the chart, and plot the data.

(b) Can we conclude that the process is in control using a U chart? If not, assume that assignable causes can be found, list points, and revise the control limits.

8-24.   The following represent the number of defects per 1000 feet in rubber-covered wire: 1, 1, 3, 7, 8, 10, 5, 13, 0, 19, 24, 6, 9, 11, 15, 8, 3, 6, 7, 4, 9, 20, 11, 7, 18, 10, 6, 4, 0, 9, 7, 3, 1, 8, 12. Do the data come from a controlled process?

8-25.   Consider the data in Exercise 8-23. Set up a C chart for this process. Compare it to the U chart in Exercise 8-23. Comment on your findings.

8-26.   The following are the numbers of defective solder joints found during successive samples of 500 solder joints.

Day	No. of Defectives	Day	No. of Defectives
1	106	12	37
2	116	13	25
3	164	14	88
4	89	15	101
5	99	16	64
6	40	17	51
7	112	18	74
8	36	19	71
9	69	20	43
10	74	21	80
11	42		

(a) Using all the data, compute trial control limits for both a P chart and an nP chart, construct the charts, and plot the data.

(b) Determine whether the process is in statistical control. If not, assume that assignable causes can be found and out-of-control points eliminated. Revise the control limits.

8-9   CONTROL CHART PERFORMANCE

Specifying the control limits is one of the critical decisions that must be made in designing a control chart. By moving the control limits further from the center line, we decrease the risk of a type I error—that is, the risk of a point falling beyond the control limits, indicating an out-of-control condition when no assignable cause is present. However, widening the control limits will also increase the risk of a type II error—that is, the risk of a point falling between the control limits when the process is really out of control. If we move the control limits closer to the center line, the opposite effect is obtained: The risk of type I error is increased, while the risk of type II error is decreased.

The control limits on a Shewhart control chart are customarily located a distance of plus or minus three standard deviations of the variable plotted on the chart from the center line; that is, the constant k in equation 8-1 should be set equal to 3. These limits are called **three-sigma control limits.**

A way to evaluate decisions regarding sample size and sampling frequency is through the **average run length (ARL)** of the control chart. Essentially, the ARL is the average number of points that must be plotted before a point indicates an out-of-control condition. For any Shewhart control chart, the ARL can be calculated from the mean of a geometric random variable (Montgomery, 1996) as

$$ARL = \frac{1}{p} \tag{8-18}$$

where p is the probability that any point exceeds the control limits. Thus, for an $\overline{X}$ chart with three-sigma limits, $p = 0.0027$ is the probability that a single point falls outside the limits when the process is in control, so

$$\text{ARL} = \frac{1}{p} = \frac{1}{0.0027} \cong 370$$

is the average run length of the $\overline{X}$ chart when the process is in control. That is, even if the process remains in control, an out-of-control signal will be generated every 370 points, on the average.

Consider the piston ring process discussed earlier, and suppose we are sampling every hour. Thus, we will have a **false alarm** about every 370 hours on the average. Suppose that we are using a sample size of $n = 5$ and that when the process goes out of control the mean shifts to 74.0135 mm. Then, the probability that $\overline{X}$ falls between the control limits of Fig. 8-3 is equal to

$$P[73.9865 \leq \overline{X} \leq 74.0135 \text{ when } \mu = 74.0135]$$

$$= P\left[\frac{73.9865 - 74.0135}{0.0045} \leq Z \leq \frac{74.0135 - 74.0135}{0.0045}\right]$$

$$= P[-6 \leq Z \leq 0] = 0.5$$

Therefore, p in equation 8-17 is 0.50, and the out-of-control ARL is

$$\text{ARL} = \frac{1}{p} = \frac{1}{0.5} = 2$$

That is, the control chart will require two samples to detect the process shift, on the average, so 2 hours will elapse between the shift and its detection (*again on the average*). Suppose this approach is unacceptable, because production of piston rings with a mean diameter of 74.0135 mm results in excessive scrap costs and delays final engine assembly. How can we reduce the time needed to detect the out-of-control condition? One method is to sample more frequently. For example, if we sample every half hour, then only 1 hour will elapse (on the average) between the shift and its detection. The second possibility is to increase the sample size. For example, if we use $n = 10$, then the control limits in Fig. 8-3 narrow to 73.9905 and 74.0095. The probability of $\overline{X}$ falling between the control limits when the process mean is 74.0135 mm is approximately 0.1, so that $p = 0.9$, and the out-of-control ARL is

$$\text{ARL} = \frac{1}{p} = \frac{1}{0.9} = 1.11$$

Table 8-6   Average Run Length (ARL) for an $\overline{X}$ Chart
With Three-Sigma Control Limits

Magnitude of Process Shift	ARL $n = 1$	ARL $n = 4$
0	370.4	370.4
0.5σ	155.2	43.9
1.0σ	43.9	6.3
1.5σ	15.0	2.0
2.0σ	6.3	1.2
3.0σ	2.0	1.0

Thus, the larger sample size would allow the shift to be detected about twice as quickly as the old one. If it became important to detect the shift in the first hour after it occurred, two control chart designs would work:

Design 1	Design 2
Sample size: $n = 5$	Sample size: $n = 10$
Sampling frequency: every half hour	Sampling frequency: every hour

Table 8-6 provides average run lengths for an $\overline{X}$ chart with three-sigma control limits. The average run lengths are calculated for shifts in the process mean from 0 to 3.0σ and for sample sizes of $n = 1$ and $n = 4$ by using $1/p$, where p is the probability that a point plots outside of the control limits. Figure 8-17 illustrates a shift in the process mean of 2σ.

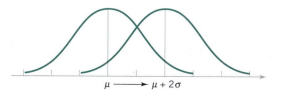

Figure 8-17   Process mean shift of 2σ.

EXERCISES FOR SECTION 8-9

8-27. Consider the $\overline{X}$ control chart in Fig. 8-3. Suppose that the mean shifts to 74.010 mm.
   (a) What is the probability that this shift will be detected on the next sample?
   (b) What is the ARL after the shift?

8-28. An $\overline{X}$ chart uses samples of size 6. The center line is at 100, and the upper and lower three-sigma control limits are at 106 and 94, respectively.
   (a) What is the process σ?
   (b) Suppose the process mean shifts to 105. Find the probability that this shift will be detected on the next sample.
   (c) Find the ARL to detect the shift in part (b).

8-29. Consider the revised $\overline{X}$ control chart in Exercise 8-1 with $\hat{\sigma} = 2.922$, $UCL = 39.34$, $LCL = 29.22$ and $n = 3$. Suppose that the mean shifts to 38.
   (a) What is the probability that this shift will be detected on the next sample?
   (b) What is the ARL after the shift?

8-30. Consider the $\overline{X}$ control chart in Exercise 8-2 with $\bar{r} = 0.39$, $UCL = 19.209$, $LCL = 18.641$ and $n = 4$. Suppose that the mean shifts to 18.7.
   (a) What is the probability that this shift will be detected on the next sample?
   (b) What is the ARL after the shift?

8-31. Consider the revised $\overline{X}$ control chart in Exercise 8-3 with $\bar{r} = 3.895$, $UCL = 17.98$, $LCL = 12.31$ and $n = 4$. Suppose that the mean shifts to 14.5.
   (a) What is the probability that this shift will be detected on the next sample?
   (b) What is the ARL after the shift?

8-32. Consider the $\overline{X}$ control chart in Exercise 8-4 with $\hat{\sigma} = 1.40$, $UCL = 21.71$, $LCL = 18.29$ and $n = 6$. Suppose that the mean shifts to 19.
   (a) What is the probability that this shift will be detected on the next sample?
   (b) What is the ARL after the shift?

8-33. Consider the $\overline{X}$ control chart in Exercise 8-5 with $\bar{r} = 47.2$, $UCL = 340.69$, $LCL = 271.87$ and $n = 4$. Suppose that the mean shifts to 300.
   (a) What is the probability that this shift will be detected on the next sample?
   (b) What is the ARL after the shift?

8-34. Consider the revised $\overline{X}$ control chart in Exercise 8-6 with $\hat{\sigma} = 0.00024$, $UCL = 0.06331$, $LCL = 0.06266$ and $n = 5$. Suppose that the mean shifts to 0.06280.
   (a) What is the probability that this shift will be detected on the next sample?
   (b) What is the ARL after the shift?

8-35. Consider the revised $\overline{X}$ control chart in Exercise 8-7 with $\hat{\sigma} = 0.671$, $UCL = 7.385$, $LCL = 5.061$ and $n = 3$. Suppose that the mean shifts to 5.5.
   (a) What is the probability that this shift will be detected on the next sample?
   (b) What is the ARL after the shift?

SUPPLEMENTAL EXERCISES

8-36. The diameter of fuse pins used in an aircraft engine application is an important quality characteristic. Twenty-five samples of three pins each are as follows (in mm).

Sample Number	Diameter		
1	64.030	64.002	64.019
2	63.995	63.992	64.001
3	63.988	64.024	64.021
4	64.002	63.996	63.993
5	63.992	64.007	64.015
6	64.009	63.994	63.997
7	63.995	64.006	63.994
8	63.985	64.003	63.993
9	64.008	63.995	64.009
10	63.998	74.000	63.990
11	63.994	63.998	63.994
12	64.004	64.000	64.007
13	63.983	64.002	63.998
14	64.006	63.967	63.994
15	64.012	64.014	63.998
16	64.000	63.984	64.005
17	63.994	64.012	63.986
18	64.006	64.010	64.018
19	63.984	64.002	64.003
20	64.000	64.010	64.013
21	63.988	64.001	64.009
22	64.004	63.999	63.990
23	64.010	63.989	63.990
24	64.015	64.008	63.993
25	63.982	63.984	63.995

   (a) Set up $\overline{X}$ and R charts for this process. If necessary, revise limits so that no observations are out-of-control.
   (b) Estimate the process mean and standard deviation.
   (c) Suppose the process specifications are at 64 ± 0.02. Calculate an estimate of PCR. Does the process meet a minimum capability level of $PCR \geq 1.33$?
   (d) Calculate an estimate of PCR_k. Use

this ratio to draw conclusions about process capability.

(e) To make this process a six-sigma process, the variance σ^2 would have to be decreased such that $PCR_k = 2.0$. What should this new variance value be?

(f) Suppose the mean shifts to 64.005. What is the probability that this shift will be detected on the next sample? What is the ARL after the shift?

8-37. Plastic bottles for liquid laundry detergent are formed by blow-molding. Twenty samples of $n = 100$ bottles are inspected in time order of production, and the number defective in each sample is reported. The data are shown here.

Sample	Number Defective
1	9
2	11
3	10
4	8
5	3
6	8
7	8
8	10
9	3
10	5
11	9
12	8
13	8
14	8
15	6
16	10
17	17
18	11
19	9
20	10

(a) Set up a P chart for this process. Is the process in statistical control?

(b) Suppose that instead of $n = 100$, $n = 200$. Use the data given to set up a P chart for this process. Is the process in statistical control?

(c) Compare your control limits for the

P charts in parts (a) and (b). Explain why they differ. Also, explain why your assessment about statistical control differs for the two sizes of n.

8-38. Cover cases for a personal computer are manufactured by injection molding. Samples of five cases are taken from the process periodically, and the number of defects is noted. Twenty-five samples are shown here.

Sample	No. of Defects
1	3
2	2
3	0
4	1
5	4
6	3
7	2
8	4
9	1
10	0
11	2
12	3
13	2
14	8
15	0
16	2
17	4
18	3
19	5
20	0
21	2
22	1
23	9
24	3
25	2

(a) Using all the data, find trial control limits for this U chart for the process.

(b) Use the trial control limits from part (a) to identify out-of-control points. If necessary, revise your control limits.

(c) Suppose that instead of samples of 5 cases, the sample size was 10. Repeat parts (a) and (b). Explain how this change alters your answers to parts (a) and (b).

8-39. Consider the data in Exercise 8-38.

(a) Using all the data, find trial control limits for a C chart for this process.

(b) Use the trial control limits of part (a) to identify out-of-control points. If necessary, revise your control limits.

(c) Suppose that instead of samples of 5 cases, the sample was 10 cases. Repeat parts (a) and (b). Explain how this alters your answers to parts (a) and (b).

8-40. Suppose that a process is in control and an $\overline{X}$ chart is used with a sample size of 4 to monitor the process. Suddenly there is a mean shift of 1.75.

(a) If three-sigma control limits are in use on the $\overline{X}$ chart, what is the probability that this shift will remain undetected for three consecutive samples?

(b) If two-sigma control limits are in use on the $\overline{X}$ chart, what is the probability that this shift will remain undetected for three consecutive samples?

(c) Compare your answers to parts (a) and (b) and explain why they differ. Also, which limits you would recommend using and why?

8-41. Consider the control chart for individuals with three-sigma limits.

(a) Suppose that a shift in the process mean of magnitude σ occurs. Verify that the ARL for detecting the shift is ARL = 43.9.

(b) Find the ARL for detecting a shift of magnitude 2σ in the process mean.

(c) Find the ARL for detecting a shift of magnitude 3σ in the process mean.

(d) Compare your answers to parts (a), (b), and (c) and explain why the ARL for detection is decreasing as the magnitude of the shift increases.

8-42. Consider a control chart for individuals, applied to a continuous 24-hour chemi-

cal process with observations taken every hour.

(a) If the chart has three-sigma limits, verify that the in-control ARL is ARL = 370. How many false alarms would occur each 30-day month, on the average, with this chart?

(b) Suppose that the chart has two-sigma limits. Does this reduce the ARL for detecting a shift in the mean of magnitude σ? (Recall that the ARL for detecting this shift with three-sigma limits is 43.9.)

(c) Find the in-control ARL if two-sigma limits are used on the chart. How many false alarms would occur each month with this chart? Is this in-control ARL performance satisfactory? Explain your answer.

8-43. The depth of a keyway is an important part quality characteristic. Samples of size $n = 5$ are taken every 4 hours from the process and 20 samples are given as follows.

Sample	\multicolumn{5}{c}{Observation}				
	1	2	3	4	5
1	139.9	138.8	139.85	141.1	139.8
2	140.7	139.3	140.55	141.6	140.1
3	140.8	139.8	140.15	141.9	139.9
4	140.6	141.1	141.05	141.2	139.6
5	139.8	138.9	140.55	141.7	139.6
6	139.8	139.2	140.55	141.2	139.4
7	140.1	138.8	139.75	141.2	138.8
8	140.3	140.6	140.65	142.5	139.9
9	140.1	139.1	139.05	140.5	139.1
10	140.3	141.1	141.25	142.6	140.9
11	138.4	138.1	139.25	140.2	138.6
12	139.4	139.1	139.15	140.3	137.8
13	138.0	137.5	138.25	141.0	140.0
14	138.0	138.1	138.65	139.5	137.8
15	141.2	140.5	141.45	142.5	141.0
16	141.2	141.0	141.95	141.9	140.1
17	140.2	140.3	141.45	142.3	139.6
18	139.6	140.3	139.55	141.7	139.4
19	136.2	137.2	137.75	138.3	137.7
20	138.8	137.7	140.05	140.8	138.9

(a) Using all the data, find trial control limits for $\overline{X}$ and R charts. Is the process in control?

(b) Use the trial control limits from part (a) to identify out-of-control points. If necessary, revise your control limits. Then, estimate the process standard deviation.

(c) Suppose that the specifications are at 140 ± 2. Using the results from part (b), what statements can you make about process capability? Compute estimates of the appropriate process capability ratios.

(d) To make this process a "six-sigma process," the variance σ^2 would have to be decreased such that $PCR_k = 2.0$. What should this new variance value be?

(e) Suppose the mean shifts to 139.7. What is the probability that this shift will be detected on the next sample? What is the ARL after the shift?

8-44. A process is controlled by a P chart using samples of size 100. The center line on the chart is 0.05.

(a) What is the probability that the control chart detects a shift to 0.07 on the first sample following the shift?

(b) What is the probability that the control chart does not detect a shift to 0.07 on the first sample following the shift but does detect it on the second sample?

(c) Suppose that instead of a shift in the mean to 0.07, the mean shifts to 0.10. Repeat parts (a) and (b).

(d) Compare your answers for a shift to 0.07 and for a shift to 0.10. Explain why they differ. Also, explain why a shift to 0.10 is easier to detect.

8-45. Suppose the average number of defects in a unit is known to be 8. If the mean number of defects in a unit shifts to 16, what is the probability that it will be detected by the U chart on the first sample following the shift

(a) if the sample size is $n = 4$?

(b) if the sample size is $n = 10$?
Use a normal approximation for U.

8-46. Suppose the average number of defects in a unit is known to be 10. If the mean number of defects in a unit shifts to 14, what is the probability that it will be detected by the U chart on the first sample following the shift

(a) if the sample size is $n = 1$?

(b) if the sample size is $n = 5$?
Use a normal approximation for U.

8-47. Suppose that an $\overline{X}$ control chart with two-sigma limits is used to control a process. Find the probability that a false out-of-control signal will be produced on the next sample. Compare this with the corresponding probability for the chart with three-sigma limits and discuss. Comment on when you would prefer to use two-sigma limits instead of three-sigma limits.

8-48. Consider an $\overline{X}$ control chart with k-sigma control limits. Develop a general expression for the probability that a point will plot outside the control limits when the process mean has shifted by δ units from the center line.

8-49. Consider the $\overline{X}$ control chart with two-sigma limits in Exercise 8-47.

(a) Find the probability of no signal on the first sample but a signal on the second.

(b) What is the probability that there will not be a signal in three samples?

8-50. Suppose a process has a $PCR = 2$, but the mean is exactly three standard deviations above the upper specification limit. What is the probability of making a product outside the specification limits?

Team Exercise

8-51. Obtain time-ordered data from a process of interest. Use the data to construct appropriate control charts and comment on the control of the process. Can you make any recommendations to improve the process? If appropriate, calculate appropriate measures of process capability.

APPENDICES

Appendix A

Statistical

Tables and

Charts

Table I   Cumulative Standard Normal Distribution

$$\Phi(z) = P(Z \le z) = \int_{-\infty}^{z} \frac{1}{\sqrt{2\pi}} e^{\frac{-u^2}{2}} \, du$$

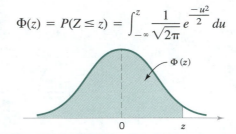

z	−0.09	−0.08	−0.07	−0.06	−0.05	−0.04	−0.03	−0.02	−0.01	−0.00	z
−3.9	0.000033	0.000034	0.000036	0.000037	0.000039	0.000041	0.000042	0.000044	0.000046	0.000048	−3.9
−3.8	0.000050	0.000052	0.000054	0.000057	0.000059	0.000062	0.000064	0.000067	0.000069	0.000072	−3.8
−3.7	0.000075	0.000078	0.000082	0.000085	0.000088	0.000092	0.000096	0.000100	0.000104	0.000108	−3.7
−3.6	0.000112	0.000117	0.000121	0.000126	0.000131	0.000136	0.000142	0.000147	0.000153	0.000159	−3.6
−3.5	0.000165	0.000172	0.000179	0.000185	0.000193	0.000200	0.000208	0.000216	0.000224	0.000233	−3.5
−3.4	0.000242	0.000251	0.000260	0.000270	0.000280	0.000291	0.000302	0.000313	0.000325	0.000337	−3.4
−3.3	0.000350	0.000362	0.000376	0.000390	0.000404	0.000419	0.000434	0.000450	0.000467	0.000483	−3.3
−3.2	0.000501	0.000519	0.000538	0.000557	0.000577	0.000598	0.000619	0.000641	0.000664	0.000687	−3.2
−3.1	0.000711	0.000736	0.000762	0.000789	0.000816	0.000845	0.000874	0.000904	0.000935	0.000968	−3.1
−3.0	0.001001	0.001035	0.001070	0.001107	0.001144	0.001183	0.001223	0.001264	0.001306	0.001350	−3.0
−2.9	0.001395	0.001441	0.001489	0.001538	0.001589	0.001641	0.001695	0.001750	0.001807	0.001866	−2.9
−2.8	0.001926	0.001988	0.002052	0.002118	0.002186	0.002256	0.002327	0.002401	0.002477	0.002555	−2.8
−2.7	0.002635	0.002718	0.002803	0.002890	0.002980	0.003072	0.003167	0.003264	0.003364	0.003467	−2.7
−2.6	0.003573	0.003681	0.003793	0.003907	0.004025	0.004145	0.004269	0.004396	0.004527	0.004661	−2.6
−2.5	0.004799	0.004940	0.005085	0.005234	0.005386	0.005543	0.005703	0.005868	0.006037	0.006210	−2.5
−2.4	0.006387	0.006569	0.006756	0.006947	0.007143	0.007344	0.007549	0.007760	0.007976	0.008198	−2.4
−2.3	0.008424	0.008656	0.008894	0.009137	0.009387	0.009642	0.009903	0.010170	0.010444	0.010724	−2.3
−2.2	0.011011	0.011304	0.011604	0.011911	0.012224	0.012545	0.012874	0.013209	0.013553	0.013903	−2.2
−2.1	0.014262	0.014629	0.015003	0.015386	0.015778	0.016177	0.016586	0.017003	0.017429	0.017864	−2.1
−2.0	0.018309	0.018763	0.019226	0.019699	0.020182	0.020675	0.021178	0.021692	0.022216	0.022750	−2.0
−1.9	0.023295	0.023852	0.024419	0.024998	0.025588	0.026190	0.026803	0.027429	0.028067	0.028717	−1.9
−1.8	0.029379	0.030054	0.030742	0.031443	0.032157	0.032884	0.033625	0.034379	0.035148	0.035930	−1.8
−1.7	0.036727	0.037538	0.038364	0.039204	0.040059	0.040929	0.041815	0.042716	0.043633	0.044565	−1.7
−1.6	0.045514	0.046479	0.047460	0.048457	0.049471	0.050503	0.051551	0.052616	0.053699	0.054799	−1.6
−1.5	0.055917	0.057053	0.058208	0.059380	0.060571	0.061780	0.063008	0.064256	0.065522	0.066807	−1.5
−1.4	0.068112	0.069437	0.070781	0.072145	0.073529	0.074934	0.076359	0.077804	0.079270	0.080757	−1.4
−1.3	0.082264	0.083793	0.085343	0.086915	0.088508	0.090123	0.091759	0.093418	0.095098	0.096801	−1.3
−1.2	0.098525	0.100273	0.102042	0.103835	0.105650	0.107488	0.109349	0.111233	0.113140	0.115070	−1.2
−1.1	0.117023	0.119000	0.121001	0.123024	0.125072	0.127143	0.129238	0.131357	0.133500	0.135666	−1.1
−1.0	0.137857	0.140071	0.142310	0.144572	0.146859	0.149170	0.151505	0.153864	0.156248	0.158655	−1.0
−0.9	0.161087	0.163543	0.166023	0.168528	0.171056	0.173609	0.176185	0.178786	0.181411	0.184060	−0.9
−0.8	0.186733	0.189430	0.192150	0.194894	0.197662	0.200454	0.203269	0.206108	0.208970	0.211855	−0.8
−0.7	0.214764	0.217695	0.220650	0.223627	0.226627	0.229650	0.232695	0.235762	0.238852	0.241964	−0.7
−0.6	0.245097	0.248252	0.251429	0.254627	0.257846	0.261086	0.264347	0.267629	0.270931	0.274253	−0.6
−0.5	0.277595	0.280957	0.284339	0.287740	0.291160	0.294599	0.298056	0.301532	0.305026	0.308538	−0.5
−0.4	0.312067	0.315614	0.319178	0.322758	0.326355	0.329969	0.333598	0.337243	0.340903	0.344578	−0.4
−0.3	0.348268	0.351973	0.355691	0.359424	0.363169	0.366928	0.370700	0.374484	0.378281	0.382089	−0.3
−0.2	0.385908	0.389739	0.393580	0.397432	0.401294	0.405165	0.409046	0.412936	0.416834	0.420740	−0.2
−0.1	0.424655	0.428576	0.432505	0.436441	0.440382	0.444330	0.448283	0.452242	0.456205	0.460172	−0.1
0.0	0.464144	0.468119	0.472097	0.476078	0.480061	0.484047	0.488033	0.492022	0.496011	0.500000	0.0

Table 1 Cumulative Standard Normal Distribution (*continued*)

$$\Phi(z) = P(Z \le z) = \int_{-\infty}^{z} \frac{1}{\sqrt{2\pi}} e^{\frac{-u^2}{2}} du$$

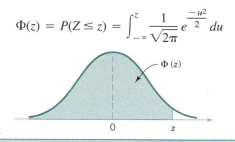

z	0.00	0.01	0.02	0.03	0.04	0.05	0.06	0.07	0.08	0.09	z
0.0	0.500000	0.503989	0.507978	0.511967	0.515953	0.519939	0.523922	0.527903	0.531881	0.535856	0.0
0.1	0.539828	0.543795	0.547758	0.551717	0.555760	0.559618	0.563559	0.567495	0.571424	0.575345	0.1
0.2	0.579260	0.583166	0.587064	0.590954	0.594835	0.598706	0.602568	0.606420	0.610261	0.614092	0.2
0.3	0.617911	0.621719	0.625516	0.629300	0.633072	0.636831	0.640576	0.644309	0.648027	0.651732	0.3
0.4	0.655422	0.659097	0.662757	0.666402	0.670031	0.673645	0.677242	0.680822	0.684386	0.687933	0.4
0.5	0.691462	0.694974	0.698468	0.701944	0.705401	0.708840	0.712260	0.715661	0.719043	0.722405	0.5
0.6	0.725747	0.729069	0.732371	0.735653	0.738914	0.742154	0.745373	0.748571	0.751748	0.754903	0.6
0.7	0.758036	0.761148	0.764238	0.767305	0.770350	0.773373	0.776373	0.779350	0.782305	0.785236	0.7
0.8	0.788145	0.791030	0.793892	0.796731	0.799546	0.802338	0.805106	0.807850	0.810570	0.813267	0.8
0.9	0.815940	0.818589	0.821214	0.823815	0.826391	0.828944	0.831472	0.833977	0.836457	0.838913	0.9
1.0	0.841345	0.843752	0.846136	0.848495	0.850830	0.853141	0.855428	0.857690	0.859929	0.862143	1.0
1.1	0.864334	0.866500	0.868643	0.870762	0.872857	0.874928	0.876976	0.878999	0.881000	0.882977	1.1
1.2	0.884930	0.886860	0.888767	0.890651	0.892512	0.894350	0.896165	0.897958	0.899727	0.901475	1.2
1.3	0.903199	0.904902	0.906582	0.908241	0.909877	0.911492	0.913085	0.914657	0.916207	0.917736	1.3
1.4	0.919243	0.920730	0.922196	0.923641	0.925066	0.926471	0.927855	0.929219	0.930563	0.931888	1.4
1.5	0.933193	0.934478	0.935744	0.936992	0.938220	0.939429	0.940620	0.941792	0.942947	0.944083	1.5
1.6	0.945201	0.946301	0.947384	0.948449	0.949497	0.950529	0.951543	0.952540	0.953521	0.954486	1.6
1.7	0.955435	0.956367	0.957284	0.958185	0.959071	0.959941	0.960796	0.961636	0.962462	0.963273	1.7
1.8	0.964070	0.964852	0.965621	0.966375	0.967116	0.967843	0.968557	0.969258	0.969946	0.970621	1.8
1.9	0.971283	0.971933	0.972571	0.973197	0.973810	0.974412	0.975002	0.975581	0.976148	0.976705	1.9
2.0	0.977250	0.977784	0.978308	0.978822	0.979325	0.979818	0.980301	0.980774	0.981237	0.981691	2.0
2.1	0.982136	0.982571	0.982997	0.983414	0.983823	0.984222	0.984614	0.984997	0.985371	0.985738	2.1
2.2	0.986097	0.986447	0.986791	0.987126	0.987455	0.987776	0.988089	0.988396	0.988696	0.988989	2.2
2.3	0.989276	0.989556	0.989830	0.990097	0.990358	0.990613	0.990863	0.991106	0.991344	0.991576	2.3
2.4	0.991802	0.992024	0.992240	0.992451	0.992656	0.992857	0.993053	0.993244	0.993431	0.993613	2.4
2.5	0.993790	0.993963	0.994132	0.994297	0.994457	0.994614	0.994766	0.994915	0.995060	0.995201	2.5
2.6	0.995339	0.995473	0.995604	0.995731	0.995855	0.995975	0.996093	0.996207	0.996319	0.996427	2.6
2.7	0.996533	0.996636	0.996736	0.996833	0.996928	0.997020	0.997110	0.997197	0.997282	0.997365	2.7
2.8	0.997445	0.997523	0.997599	0.997673	0.997744	0.997814	0.997882	0.997948	0.998012	0.998074	2.8
2.9	0.998134	0.998193	0.998250	0.998305	0.998359	0.998411	0.998462	0.998511	0.998559	0.998605	2.9
3.0	0.998650	0.998694	0.998736	0.998777	0.998817	0.998856	0.998893	0.998930	0.998965	0.998999	3.0
3.1	0.999032	0.999065	0.999096	0.999126	0.999155	0.999184	0.999211	0.999238	0.999264	0.999289	3.1
3.2	0.999313	0.999336	0.999359	0.999381	0.999402	0.999423	0.999443	0.999462	0.999481	0.999499	3.2
3.3	0.999517	0.999533	0.999550	0.999566	0.999581	0.999596	0.999610	0.999624	0.999638	0.999650	3.3
3.4	0.999663	0.999675	0.999687	0.999698	0.999709	0.999720	0.999730	0.999740	0.999749	0.999758	3.4
3.5	0.999767	0.999776	0.999784	0.999792	0.999800	0.999807	0.999815	0.999821	0.999828	0.999835	3.5
3.6	0.999841	0.999847	0.999853	0.999858	0.999864	0.999869	0.999874	0.999879	0.999883	0.999888	3.6
3.7	0.999892	0.999896	0.999900	0.999904	0.999908	0.999912	0.999915	0.999918	0.999922	0.999925	3.7
3.8	0.999928	0.999931	0.999933	0.999936	0.999938	0.999941	0.999943	0.999946	0.999948	0.999950	3.8
3.9	0.999952	0.999954	0.999956	0.999958	0.999959	0.999961	0.999963	0.999964	0.999966	0.999967	3.9

$t_{1-\alpha} = -t_\alpha$

Table II Percentage Points $t_{\alpha,v}$ of the t Distribution

Is σ unknown?
Is $n < 30$?
$t = \dfrac{\bar{X} - \mu}{s/\sqrt{n}}$

area $-t_{1-\alpha}$ 0 $t_{\alpha,v}$ α right hand tails

df = n - 1   degrees of freedom

sample size of 6

v \ α	.40	.25	.10	.05	.025	.01	.005	.0025	.001	.0005
1	.325	1.000	3.078	6.314	12.706	31.821	63.657	127.32	318.31	636.62
2	.289	.816	1.886	2.920	4.303	6.965	9.925	14.089	23.326	31.598
3	.277	.765	1.638	2.353	3.182	4.541	5.841	7.453	10.213	12.924
4	.271	.741	1.533	2.132	2.776	3.747	4.604	5.598	7.173	8.610
5	.267	.727	1.476	2.015	2.571	3.365	4.032	4.773	5.893	6.869
6	.265	.718	1.440	1.943	2.447	3.143	3.707	4.317	5.208	5.959
7	.263	.711	1.415	1.895	2.365	2.998	3.499	4.029	4.785	5.408
8	.262	.706	1.397	1.860	2.306	2.896	3.355	3.833	4.501	5.041
9	.261	.703	1.383	1.833	2.262	2.821	3.250	3.690	4.297	4.781
10	.260	.700	1.372	1.812	2.228	2.764	3.169	3.581	4.144	4.587
11	.260	.697	1.363	1.796	2.201	2.718	3.106	3.497	4.025	4.437
12	.259	.695	1.356	1.782	2.179	2.681	3.055	3.428	3.930	4.318
13	.259	.694	1.350	1.771	2.160	2.650	3.012	3.372	3.852	4.221
14	.258	.692	1.345	1.761	2.145	2.624	2.977	3.326	3.787	4.140
15	.258	.691	1.341	1.753	2.131	2.602	2.947	3.286	3.733	4.073
16	.258	.690	1.337	1.746	2.120	2.583	2.921	3.252	3.686	4.015
17	.257	.689	1.333	1.740	2.110	2.567	2.898	3.222	3.646	3.965
18	.257	.688	1.330	1.734	2.101	2.552	2.878	3.197	3.610	3.922
19	.257	.688	1.328	1.729	2.093	2.539	2.861	3.174	3.579	3.883
20	.257	.687	1.325	1.725	2.086	2.528	2.845	3.153	3.552	3.850
21	.257	.686	1.323	1.721	2.080	2.518	2.831	3.135	3.527	3.819
22	.256	.686	1.321	1.717	2.074	2.508	2.819	3.119	3.505	3.792
23	.256	.685	1.319	1.714	2.069	2.500	2.807	3.104	3.485	3.767
24	.256	.685	1.318	1.711	2.064	2.492	2.797	3.091	3.467	3.745
25	.256	.684	1.316	1.708	2.060	2.485	2.787	3.078	3.450	3.725
26	.256	.684	1.315	1.706	2.056	2.479	2.779	3.067	3.435	3.707
27	.256	.684	1.314	1.703	2.052	2.473	2.771	3.057	3.421	3.690
28	.256	.683	1.313	1.701	2.048	2.467	2.763	3.047	3.408	3.674
29	.256	.683	1.311	1.699	2.045	2.462	2.756	3.038	3.396	3.659
30	.256	.683	1.310	1.697	2.042	2.457	2.750	3.030	3.385	3.646
40	.255	.681	1.303	1.684	2.021	2.423	2.704	2.971	3.307	3.551
60	.254	.679	1.296	1.671	2.000	2.390	2.660	2.915	3.232	3.460
120	.254	.677	1.289	1.658	1.980	2.358	2.617	2.860	3.160	3.373
∞	.253	.674	1.282	1.645	1.960	2.326	2.576	2.807	3.090	3.291

v = degrees of freedom.

Table III   Percentage Points $\chi^2_{\alpha,\nu}$ of the Chi-Square Distribution

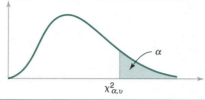

α ν	.995	.990	.975	.950	.900	.500	.100	.050	.025	.010	.005
1	.00+	.00+	.00+	.00+	.02	.45	2.71	3.84	5.02	6.63	7.88
2	.01	.02	.05	.10	.21	1.39	4.61	5.99	7.38	9.21	10.60
3	.07	.11	.22	.35	.58	2.37	6.25	7.81	9.35	11.34	12.84
4	.21	.30	.48	.71	1.06	3.36	7.78	9.49	11.14	13.28	14.86
5	.41	.55	.83	1.15	1.61	4.35	9.24	11.07	12.83	15.09	16.75
6	.68	.87	1.24	1.64	2.20	5.35	10.65	12.59	14.45	16.81	18.55
7	.99	1.24	1.69	2.17	2.83	6.35	12.02	14.07	16.01	18.48	20.28
8	1.34	1.65	2.18	2.73	3.49	7.34	13.36	15.51	17.53	20.09	21.96
9	1.73	2.09	2.70	3.33	4.17	8.34	14.68	16.92	19.02	21.67	23.59
10	2.16	2.56	3.25	3.94	4.87	9.34	15.99	18.31	20.48	23.21	25.19
11	2.60	3.05	3.82	4.57	5.58	10.34	17.28	19.68	21.92	24.72	26.76
12	3.07	3.57	4.40	5.23	6.30	11.34	18.55	21.03	23.34	26.22	28.30
13	3.57	4.11	5.01	5.89	7.04	12.34	19.81	22.36	24.74	27.69	29.82
14	4.07	4.66	5.63	6.57	7.79	13.34	21.06	23.68	26.12	29.14	31.32
15	4.60	5.23	6.27	7.26	8.55	14.34	22.31	25.00	27.49	30.58	32.80
16	5.14	5.81	6.91	7.96	9.31	15.34	23.54	26.30	28.85	32.00	34.27
17	5.70	6.41	7.56	8.67	10.09	16.34	24.77	27.59	30.19	33.41	35.72
18	6.26	7.01	8.23	9.39	10.87	17.34	25.99	28.87	31.53	34.81	37.16
19	6.84	7.63	8.91	10.12	11.65	18.34	27.20	30.14	32.85	36.19	38.58
20	7.43	8.26	9.59	10.85	12.44	19.34	28.41	31.41	34.17	37.57	40.00
21	8.03	8.90	10.28	11.59	13.24	20.34	29.62	32.67	35.48	38.93	41.40
22	8.64	9.54	10.98	12.34	14.04	21.34	30.81	33.92	36.78	40.29	42.80
23	9.26	10.20	11.69	13.09	14.85	22.34	32.01	35.17	38.08	41.64	44.18
24	9.89	10.86	12.40	13.85	15.66	23.34	33.20	36.42	39.36	42.98	45.56
25	10.52	11.52	13.12	14.61	16.47	24.34	34.28	37.65	40.65	44.31	46.93
26	11.16	12.20	13.84	15.38	17.29	25.34	35.56	38.89	41.92	45.64	48.29
27	11.81	12.88	14.57	16.15	18.11	26.34	36.74	40.11	43.19	46.96	49.65
28	12.46	13.57	15.31	16.93	18.94	27.34	37.92	41.34	44.46	48.28	50.99
29	13.12	14.26	16.05	17.71	19.77	28.34	39.09	42.56	45.72	49.59	52.34
30	13.79	14.95	16.79	18.49	20.60	29.34	40.26	43.77	46.98	50.89	53.67
40	20.71	22.16	24.43	26.51	29.05	39.34	51.81	55.76	59.34	63.69	66.77
50	27.99	29.71	32.36	34.76	37.69	49.33	63.17	67.50	71.42	76.15	79.49
60	35.53	37.48	40.48	43.19	46.46	59.33	74.40	79.08	83.30	88.38	91.95
70	43.28	45.44	48.76	51.74	55.33	69.33	85.53	90.53	95.02	100.42	104.22
80	51.17	53.54	57.15	60.39	64.28	79.33	96.58	101.88	106.63	112.33	116.32
90	59.20	61.75	65.65	69.13	73.29	89.33	107.57	113.14	118.14	124.12	128.30
100	67.33	70.06	74.22	77.93	82.36	99.33	118.50	124.34	129.56	135.81	140.17

ν = degrees of freedom.

Table IV Percentage Points $f_{\alpha,u,v}$ of the F Distribution

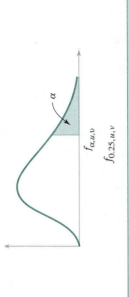

$f_{\alpha,u,v}$

$f_{0.25,u,v}$

		Degrees of freedom for the numerator (u)																	
v	1	2	3	4	5	6	7	8	9	10	12	15	20	24	30	40	60	120	∞
1	5.83	7.50	8.20	8.58	8.82	8.98	9.10	9.19	9.26	9.32	9.41	9.49	9.58	9.63	9.67	9.71	9.76	9.80	9.85
2	2.57	3.00	3.15	3.23	3.28	3.31	3.34	3.35	3.37	3.38	3.39	3.41	3.43	3.43	3.44	3.45	3.46	3.47	3.48
3	2.02	2.28	2.36	2.39	2.41	2.42	2.43	2.44	2.44	2.44	2.45	2.46	2.46	2.46	2.47	2.47	2.47	2.47	2.47
4	1.81	2.00	2.05	2.06	2.07	2.08	2.08	2.08	2.08	2.08	2.08	2.08	2.08	2.08	2.08	2.08	2.08	2.08	2.08
5	1.69	1.85	1.88	1.89	1.89	1.89	1.89	1.89	1.89	1.89	1.89	1.89	1.88	1.88	1.88	1.88	1.87	1.87	1.87
6	1.62	1.76	1.78	1.79	1.79	1.78	1.78	1.78	1.77	1.77	1.77	1.76	1.76	1.75	1.75	1.75	1.74	1.74	1.74
7	1.57	1.70	1.72	1.72	1.71	1.71	1.70	1.70	1.70	1.69	1.68	1.68	1.67	1.67	1.66	1.66	1.65	1.65	1.65
8	1.54	1.66	1.67	1.66	1.66	1.65	1.64	1.64	1.63	1.63	1.62	1.62	1.61	1.60	1.60	1.59	1.59	1.58	1.58
9	1.51	1.62	1.63	1.63	1.62	1.61	1.60	1.60	1.59	1.59	1.58	1.57	1.56	1.56	1.55	1.54	1.54	1.53	1.53
10	1.49	1.60	1.60	1.59	1.59	1.58	1.57	1.56	1.56	1.55	1.54	1.53	1.52	1.52	1.51	1.51	1.50	1.49	1.48
11	1.47	1.58	1.58	1.57	1.56	1.55	1.54	1.53	1.53	1.52	1.51	1.50	1.49	1.49	1.48	1.47	1.47	1.46	1.45
12	1.46	1.56	1.56	1.55	1.54	1.53	1.52	1.51	1.51	1.50	1.49	1.48	1.47	1.46	1.45	1.45	1.44	1.43	1.42
13	1.45	1.55	1.55	1.53	1.52	1.51	1.50	1.49	1.49	1.48	1.47	1.46	1.45	1.44	1.43	1.42	1.42	1.41	1.40
14	1.44	1.53	1.53	1.52	1.51	1.50	1.49	1.48	1.47	1.46	1.45	1.44	1.43	1.42	1.41	1.41	1.40	1.39	1.38
15	1.43	1.52	1.52	1.51	1.49	1.48	1.47	1.46	1.46	1.45	1.44	1.43	1.41	1.41	1.40	1.39	1.38	1.37	1.36
16	1.42	1.51	1.51	1.50	1.48	1.47	1.46	1.45	1.44	1.44	1.43	1.41	1.40	1.39	1.38	1.37	1.36	1.35	1.34
17	1.42	1.51	1.50	1.49	1.47	1.46	1.45	1.44	1.43	1.43	1.41	1.40	1.39	1.38	1.37	1.36	1.35	1.34	1.33
18	1.41	1.50	1.49	1.48	1.46	1.45	1.44	1.43	1.42	1.42	1.40	1.39	1.38	1.37	1.36	1.35	1.34	1.33	1.32
19	1.41	1.49	1.49	1.47	1.46	1.44	1.43	1.42	1.41	1.41	1.40	1.38	1.37	1.36	1.35	1.34	1.33	1.32	1.30
20	1.40	1.49	1.48	1.47	1.45	1.44	1.43	1.42	1.41	1.40	1.39	1.37	1.36	1.35	1.34	1.33	1.32	1.31	1.29
21	1.40	1.48	1.48	1.46	1.44	1.43	1.42	1.41	1.40	1.39	1.38	1.37	1.35	1.34	1.33	1.32	1.31	1.30	1.28
22	1.40	1.48	1.47	1.45	1.44	1.42	1.41	1.40	1.39	1.39	1.37	1.36	1.34	1.33	1.32	1.31	1.30	1.29	1.28
23	1.39	1.47	1.47	1.45	1.43	1.42	1.41	1.40	1.39	1.38	1.37	1.35	1.34	1.33	1.32	1.31	1.30	1.28	1.27
24	1.39	1.47	1.46	1.44	1.43	1.41	1.40	1.39	1.38	1.38	1.36	1.35	1.33	1.32	1.31	1.30	1.29	1.28	1.26
25	1.39	1.47	1.46	1.44	1.42	1.41	1.40	1.39	1.38	1.37	1.36	1.34	1.33	1.32	1.31	1.29	1.28	1.27	1.25
26	1.38	1.46	1.45	1.44	1.42	1.41	1.39	1.38	1.37	1.37	1.35	1.34	1.32	1.31	1.30	1.29	1.28	1.26	1.25
27	1.38	1.46	1.45	1.43	1.42	1.40	1.39	1.38	1.37	1.36	1.35	1.33	1.32	1.31	1.30	1.28	1.27	1.26	1.24
28	1.38	1.46	1.45	1.43	1.41	1.40	1.39	1.38	1.36	1.36	1.34	1.33	1.31	1.30	1.29	1.28	1.27	1.25	1.24
29	1.38	1.45	1.45	1.43	1.41	1.40	1.38	1.37	1.36	1.35	1.34	1.32	1.31	1.30	1.29	1.27	1.26	1.25	1.23
30	1.38	1.45	1.44	1.42	1.41	1.39	1.38	1.37	1.36	1.35	1.34	1.32	1.30	1.29	1.28	1.27	1.26	1.24	1.23
40	1.36	1.44	1.42	1.40	1.39	1.37	1.36	1.35	1.33	1.33	1.31	1.30	1.28	1.26	1.25	1.24	1.22	1.21	1.19
60	1.35	1.42	1.41	1.38	1.37	1.35	1.33	1.32	1.31	1.30	1.29	1.27	1.25	1.24	1.22	1.21	1.19	1.17	1.15
120	1.34	1.40	1.39	1.37	1.35	1.33	1.31	1.30	1.28	1.28	1.26	1.24	1.22	1.21	1.19	1.18	1.16	1.13	1.10
∞	1.32	1.39	1.37	1.35	1.33	1.31	1.29	1.28	1.27	1.25	1.24	1.22	1.19	1.18	1.16	1.14	1.12	1.08	1.00

Degrees of freedom for the denominator (v)

Table IV Percentage Points of the F Distribution (*continued*)

$$f_{0.10,u,v}$$

v	\multicolumn{19}{c}{Degrees of freedom for the numerator (u)}
	1	2	3	4	5	6	7	8	9	10	12	15	20	24	30	40	60	120	∞
1	39.86	49.50	53.59	55.83	57.24	58.20	58.91	59.44	59.86	60.19	60.71	61.22	61.74	62.00	62.26	62.53	62.79	63.06	63.33
2	8.53	9.00	9.16	9.24	9.29	9.33	9.35	9.37	9.38	9.39	9.41	9.42	9.44	9.45	9.46	9.47	9.47	9.48	9.49
3	5.54	5.46	5.39	5.34	5.31	5.28	5.27	5.25	5.24	5.23	5.22	5.20	5.18	5.18	5.17	5.16	5.15	5.14	5.13
4	4.54	4.32	4.19	4.11	4.05	4.01	3.98	3.95	3.94	3.92	3.90	3.87	3.84	3.83	3.82	3.80	3.79	3.78	3.76
5	4.06	3.78	3.62	3.52	3.45	3.40	3.37	3.34	3.32	3.30	3.27	3.24	3.21	3.19	3.17	3.16	3.14	3.12	3.10
6	3.78	3.46	3.29	3.18	3.11	3.05	3.01	2.98	2.96	2.94	2.90	2.87	2.84	2.82	2.80	2.78	2.76	2.74	2.72
7	3.59	3.26	3.07	2.96	2.88	2.83	2.78	2.75	2.72	2.70	2.67	2.63	2.59	2.58	2.56	2.54	2.51	2.49	2.47
8	3.46	3.11	2.92	2.81	2.73	2.67	2.62	2.59	2.56	2.54	2.50	2.46	2.42	2.40	2.38	2.36	2.34	2.32	2.29
9	3.36	3.01	2.81	2.69	2.61	2.55	2.51	2.47	2.44	2.42	2.38	2.34	2.30	2.28	2.25	2.23	2.21	2.18	2.16
10	3.29	2.92	2.73	2.61	2.52	2.46	2.41	2.38	2.35	2.32	2.28	2.24	2.20	2.18	2.16	2.13	2.11	2.08	2.06
11	3.23	2.86	2.66	2.54	2.45	2.39	2.34	2.30	2.27	2.25	2.21	2.17	2.12	2.10	2.08	2.05	2.03	2.00	1.97
12	3.18	2.81	2.61	2.48	2.39	2.33	2.28	2.24	2.21	2.19	2.15	2.10	2.06	2.04	2.01	1.99	1.96	1.93	1.90
13	3.14	2.76	2.56	2.43	2.35	2.28	2.23	2.20	2.16	2.14	2.10	2.05	2.01	1.98	1.96	1.93	1.90	1.88	1.85
14	3.10	2.73	2.52	2.39	2.31	2.24	2.19	2.15	2.12	2.10	2.05	2.01	1.96	1.94	1.91	1.89	1.86	1.83	1.80
15	3.07	2.70	2.49	2.36	2.27	2.21	2.16	2.12	2.09	2.06	2.02	1.97	1.92	1.90	1.87	1.85	1.82	1.79	1.76
16	3.05	2.67	2.46	2.33	2.24	2.18	2.13	2.09	2.06	2.03	1.99	1.94	1.89	1.87	1.84	1.81	1.78	1.75	1.72
17	3.03	2.64	2.44	2.31	2.22	2.15	2.10	2.06	2.03	2.00	1.96	1.91	1.86	1.84	1.81	1.78	1.75	1.72	1.69
18	3.01	2.62	2.42	2.29	2.20	2.13	2.08	2.04	2.00	1.98	1.93	1.89	1.84	1.81	1.78	1.75	1.72	1.69	1.66
19	2.99	2.61	2.40	2.27	2.18	2.11	2.06	2.02	1.98	1.96	1.91	1.86	1.81	1.79	1.76	1.73	1.70	1.67	1.63
20	2.97	2.59	2.38	2.25	2.16	2.09	2.04	2.00	1.96	1.94	1.89	1.84	1.79	1.77	1.74	1.71	1.68	1.64	1.61
21	2.96	2.57	2.36	2.23	2.14	2.08	2.02	1.98	1.95	1.92	1.87	1.83	1.78	1.75	1.72	1.69	1.66	1.62	1.59
22	2.95	2.56	2.35	2.22	2.13	2.06	2.01	1.97	1.93	1.90	1.86	1.81	1.76	1.73	1.70	1.67	1.64	1.60	1.57
23	2.94	2.55	2.34	2.21	2.11	2.05	1.99	1.95	1.92	1.89	1.84	1.80	1.74	1.72	1.69	1.66	1.62	1.59	1.55
24	2.93	2.54	2.33	2.19	2.10	2.04	1.98	1.94	1.91	1.88	1.83	1.78	1.73	1.70	1.67	1.64	1.61	1.57	1.53
25	2.92	2.53	2.32	2.18	2.09	2.02	1.97	1.93	1.89	1.87	1.82	1.77	1.72	1.69	1.66	1.63	1.59	1.56	1.52
26	2.91	2.52	2.31	2.17	2.08	2.01	1.96	1.92	1.88	1.86	1.81	1.76	1.71	1.68	1.65	1.61	1.58	1.54	1.50
27	2.90	2.51	2.30	2.17	2.07	2.00	1.95	1.91	1.87	1.85	1.80	1.75	1.70	1.67	1.64	1.60	1.57	1.53	1.49
28	2.89	2.50	2.29	2.16	2.06	2.00	1.94	1.90	1.87	1.84	1.79	1.74	1.69	1.66	1.63	1.59	1.56	1.52	1.48
29	2.89	2.50	2.28	2.15	2.06	1.99	1.93	1.89	1.86	1.83	1.78	1.73	1.68	1.65	1.62	1.58	1.55	1.51	1.47
30	2.88	2.49	2.28	2.14	2.03	1.98	1.93	1.88	1.85	1.82	1.77	1.72	1.67	1.64	1.61	1.57	1.54	1.50	1.46
40	2.84	2.44	2.23	2.09	2.00	1.93	1.87	1.83	1.79	1.76	1.71	1.66	1.61	1.57	1.54	1.51	1.47	1.42	1.38
60	2.79	2.39	2.18	2.04	1.95	1.87	1.82	1.77	1.74	1.71	1.66	1.60	1.54	1.51	1.48	1.44	1.40	1.35	1.29
120	2.75	2.35	2.13	1.99	1.90	1.82	1.77	1.72	1.68	1.65	1.60	1.55	1.48	1.45	1.41	1.37	1.32	1.26	1.19
∞	2.71	2.30	2.08	1.94	1.85	1.77	1.72	1.67	1.63	1.60	1.55	1.49	1.42	1.38	1.34	1.30	1.24	1.17	1.00

Degrees of freedom for the denominator (v)

Table IV Percentage Points of the F Distribution (*continued*)

$$f_{0.05,u,v}$$

v \ u	1	2	3	4	5	6	7	8	9	10	12	15	20	24	30	40	60	120	∞
1	161.4	199.5	215.7	224.6	230.2	234.0	236.8	238.9	240.5	241.9	243.9	245.9	248.0	249.1	250.1	251.1	252.2	253.3	254.3
2	18.51	19.00	19.16	19.25	19.30	19.33	19.35	19.37	19.38	19.40	19.41	19.43	19.45	19.45	19.46	19.47	19.48	19.49	19.50
3	10.13	9.55	9.28	9.12	9.01	8.94	8.89	8.85	8.81	8.79	8.74	8.70	8.66	8.64	8.62	8.59	8.57	8.55	8.53
4	7.71	6.94	6.59	6.39	6.26	6.16	6.09	6.04	6.00	5.96	5.91	5.86	5.80	5.77	5.75	5.72	5.69	5.66	5.63
5	6.61	5.79	5.41	5.19	5.05	4.95	4.88	4.82	4.77	4.74	4.68	4.62	4.56	4.53	4.50	4.46	4.43	4.40	4.36
6	5.99	5.14	4.76	4.53	4.39	4.28	4.21	4.15	4.10	4.06	4.00	3.94	3.87	3.84	3.81	3.77	3.74	3.70	3.67
7	5.59	4.74	4.35	4.12	3.97	3.87	3.79	3.73	3.68	3.64	3.57	3.51	3.44	3.41	3.38	3.34	3.30	3.27	3.23
8	5.32	4.46	4.07	3.84	3.69	3.58	3.50	3.44	3.39	3.35	3.28	3.22	3.15	3.12	3.08	3.04	3.01	2.97	2.93
9	5.12	4.26	3.86	3.63	3.48	3.37	3.29	3.23	3.18	3.14	3.07	3.01	2.94	2.90	2.86	2.83	2.79	2.75	2.71
10	4.96	4.10	3.71	3.48	3.33	3.22	3.14	3.07	3.02	2.98	2.91	2.85	2.77	2.74	2.70	2.66	2.62	2.58	2.54
11	4.84	3.98	3.59	3.36	3.20	3.09	3.01	2.95	2.90	2.85	2.79	2.72	2.65	2.61	2.57	2.53	2.49	2.45	2.40
12	4.75	3.89	3.49	3.26	3.11	3.00	2.91	2.85	2.80	2.75	2.69	2.62	2.54	2.51	2.47	2.43	2.38	2.34	2.30
13	4.67	3.81	3.41	3.18	3.03	2.92	2.83	2.77	2.71	2.67	2.60	2.53	2.46	2.42	2.38	2.34	2.30	2.25	2.21
14	4.60	3.74	3.34	3.11	2.96	2.85	2.76	2.70	2.65	2.60	2.53	2.46	2.39	2.35	2.31	2.27	2.22	2.18	2.13
15	4.54	3.68	3.29	3.06	2.90	2.79	2.71	2.64	2.59	2.54	2.48	2.40	2.33	2.29	2.25	2.20	2.16	2.11	2.07
16	4.49	3.63	3.24	3.01	2.85	2.74	2.66	2.59	2.54	2.49	2.42	2.35	2.28	2.24	2.19	2.15	2.11	2.06	2.01
17	4.45	3.59	3.20	2.96	2.81	2.70	2.61	2.55	2.49	2.45	2.38	2.31	2.23	2.19	2.15	2.10	2.06	2.01	1.96
18	4.41	3.55	3.16	2.93	2.77	2.66	2.58	2.51	2.46	2.41	2.34	2.27	2.19	2.15	2.11	2.06	2.02	1.97	1.92
19	4.38	3.52	3.13	2.90	2.74	2.63	2.54	2.48	2.42	2.38	2.31	2.23	2.16	2.11	2.07	2.03	1.98	1.93	1.88
20	4.35	3.49	3.10	2.87	2.71	2.60	2.51	2.45	2.39	2.35	2.28	2.20	2.12	2.08	2.04	1.99	1.95	1.90	1.84
21	4.32	3.47	3.07	2.84	2.68	2.57	2.49	2.42	2.37	2.32	2.25	2.18	2.10	2.05	2.01	1.96	1.92	1.87	1.81
22	4.30	3.44	3.05	2.82	2.66	2.55	2.46	2.40	2.34	2.30	2.23	2.15	2.07	2.03	1.98	1.94	1.89	1.84	1.78
23	4.28	3.42	3.03	2.80	2.64	2.53	2.44	2.37	2.32	2.27	2.20	2.13	2.05	2.01	1.96	1.91	1.86	1.81	1.76
24	4.26	3.40	3.01	2.78	2.62	2.51	2.42	2.36	2.30	2.25	2.18	2.11	2.03	1.98	1.94	1.89	1.84	1.79	1.73
25	4.24	3.39	2.99	2.76	2.60	2.49	2.40	2.34	2.28	2.24	2.16	2.09	2.01	1.96	1.92	1.87	1.82	1.77	1.71
26	4.23	3.37	2.98	2.74	2.59	2.47	2.39	2.32	2.27	2.22	2.15	2.07	1.99	1.95	1.90	1.85	1.80	1.75	1.69
27	4.21	3.35	2.96	2.73	2.57	2.46	2.37	2.31	2.25	2.20	2.13	2.06	1.97	1.93	1.88	1.84	1.79	1.73	1.67
28	4.20	3.34	2.95	2.71	2.56	2.45	2.36	2.29	2.24	2.19	2.12	2.04	1.96	1.91	1.87	1.82	1.77	1.71	1.65
29	4.18	3.33	2.93	2.70	2.55	2.43	2.35	2.28	2.22	2.18	2.10	2.03	1.94	1.90	1.85	1.81	1.75	1.70	1.64
30	4.17	3.32	2.92	2.69	2.53	2.42	2.33	2.27	2.21	2.16	2.09	2.01	1.93	1.89	1.84	1.79	1.74	1.68	1.62
40	4.08	3.23	2.84	2.61	2.45	2.34	2.25	2.18	2.12	2.08	2.00	1.92	1.84	1.79	1.74	1.69	1.64	1.58	1.51
60	4.00	3.15	2.76	2.53	2.37	2.25	2.17	2.10	2.04	1.99	1.92	1.84	1.75	1.70	1.65	1.59	1.53	1.47	1.39
120	3.92	3.07	2.68	2.45	2.29	2.17	2.09	2.02	1.96	1.91	1.83	1.75	1.66	1.61	1.55	1.50	1.43	1.35	1.25
∞	3.84	3.00	2.60	2.37	2.21	2.10	2.01	1.94	1.88	1.83	1.75	1.67	1.57	1.52	1.46	1.39	1.32	1.22	1.00

Degrees of freedom for the numerator (*u*)

Degrees of freedom for the denominator (*v*)

A-8

Table IV Percentage Points of the F Distribution (continued)

$$f_{0.025,u,v}$$

Degrees of freedom for the numerator (u)

v \ u	1	2	3	4	5	6	7	8	9	10	12	15	20	24	30	40	60	120	∞
1	647.8	799.5	864.2	899.6	921.8	937.1	948.2	956.7	963.3	968.6	976.7	984.9	993.1	997.2	1001	1006	1010	1014	1018
2	38.51	39.00	39.17	39.25	39.30	39.33	39.36	39.37	39.39	39.40	39.41	39.43	39.45	39.46	39.46	39.47	39.48	39.49	39.50
3	17.44	16.04	15.44	15.10	14.88	14.73	14.62	14.54	14.47	14.42	14.34	14.25	14.17	14.12	14.08	14.04	13.99	13.95	13.90
4	12.22	10.65	9.98	9.60	9.36	9.20	9.07	8.98	8.90	8.84	8.75	8.66	8.56	8.51	8.46	8.41	8.36	8.31	8.26
5	10.01	8.43	7.76	7.39	7.15	6.98	6.85	6.76	6.68	6.62	6.52	6.43	6.33	6.28	6.23	6.18	6.12	6.07	6.02
6	8.81	7.26	6.60	6.23	5.99	5.82	5.70	5.60	5.52	5.46	5.37	5.27	5.17	5.12	5.07	5.01	4.96	4.90	4.85
7	8.07	6.54	5.89	5.52	5.29	5.12	4.99	4.90	4.82	4.76	4.67	4.57	4.47	4.42	4.36	4.31	4.25	4.20	4.14
8	7.57	6.06	5.42	5.05	4.82	4.65	4.53	4.43	4.36	4.30	4.20	4.10	4.00	3.95	3.89	3.84	3.78	3.73	3.67
9	7.21	5.71	5.08	4.72	4.48	4.32	4.20	4.10	4.03	3.96	3.87	3.77	3.67	3.61	3.56	3.51	3.45	3.39	3.33
10	6.94	5.46	4.83	4.47	4.24	4.07	3.95	3.85	3.78	3.72	3.62	3.52	3.42	3.37	3.31	3.26	3.20	3.14	3.08
11	6.72	5.26	4.63	4.28	4.04	3.88	3.76	3.66	3.59	3.53	3.43	3.33	3.23	3.17	3.12	3.06	3.00	2.94	2.88
12	6.55	5.10	4.47	4.12	3.89	3.73	3.61	3.51	3.44	3.37	3.28	3.18	3.07	3.02	2.96	2.91	2.85	2.79	2.72
13	6.41	4.97	4.35	4.00	3.77	3.60	3.48	3.39	3.31	3.25	3.15	3.05	2.95	2.89	2.84	2.78	2.72	2.66	2.60
14	6.30	4.86	4.24	3.89	3.66	3.50	3.38	3.29	3.21	3.15	3.05	2.95	2.84	2.79	2.73	2.67	2.61	2.55	2.49
15	6.20	4.77	4.15	3.80	3.58	3.41	3.29	3.20	3.12	3.06	2.96	2.86	2.76	2.70	2.64	2.59	2.52	2.46	2.40
16	6.12	4.69	4.08	3.73	3.50	3.34	3.22	3.12	3.05	2.99	2.89	2.79	2.68	2.63	2.57	2.51	2.45	2.38	2.32
17	6.04	4.62	4.01	3.66	3.44	3.28	3.16	3.06	2.98	2.92	2.82	2.72	2.62	2.56	2.50	2.44	2.38	2.32	2.25
18	5.98	4.56	3.95	3.61	3.38	3.22	3.10	3.01	2.93	2.87	2.77	2.67	2.56	2.50	2.44	2.38	2.32	2.26	2.19
19	5.92	4.51	3.90	3.56	3.33	3.17	3.05	2.96	2.88	2.82	2.72	2.62	2.51	2.45	2.39	2.33	2.27	2.20	2.13
20	5.87	4.46	3.86	3.51	3.29	3.13	3.01	2.91	2.84	2.77	2.68	2.57	2.46	2.41	2.35	2.29	2.22	2.16	2.09
21	5.83	4.42	3.82	3.48	3.25	3.09	2.97	2.87	2.80	2.73	2.64	2.53	2.42	2.37	2.31	2.25	2.18	2.11	2.04
22	5.79	4.38	3.78	3.44	3.22	3.05	2.93	2.84	2.76	2.70	2.60	2.50	2.39	2.33	2.27	2.21	2.14	2.08	2.00
23	5.75	4.35	3.75	3.41	3.18	3.02	2.90	2.81	2.73	2.67	2.57	2.47	2.36	2.30	2.24	2.18	2.11	2.04	1.97
24	5.72	4.32	3.72	3.38	3.15	2.99	2.87	2.78	2.70	2.64	2.54	2.44	2.33	2.27	2.21	2.15	2.08	2.01	1.94
25	5.69	4.29	3.69	3.35	3.13	2.97	2.85	2.75	2.68	2.61	2.51	2.41	2.30	2.24	2.18	2.12	2.05	1.98	1.91
26	5.66	4.27	3.67	3.33	3.10	2.94	2.82	2.73	2.65	2.59	2.49	2.39	2.28	2.22	2.16	2.09	2.03	1.95	1.88
27	5.63	4.24	3.65	3.31	3.08	2.92	2.80	2.71	2.63	2.57	2.47	2.36	2.25	2.19	2.13	2.07	2.00	1.93	1.85
28	5.61	4.22	3.63	3.29	3.06	2.90	2.78	2.69	2.61	2.55	2.45	2.34	2.23	2.17	2.11	2.05	1.98	1.91	1.83
29	5.59	4.20	3.61	3.27	3.04	2.88	2.76	2.67	2.59	2.53	2.43	2.32	2.21	2.15	2.09	2.03	1.96	1.89	1.81
30	5.57	4.18	3.59	3.25	3.03	2.87	2.75	2.65	2.57	2.51	2.41	2.31	2.20	2.14	2.07	2.01	1.94	1.87	1.79
40	5.42	4.05	3.46	3.13	2.90	2.74	2.62	2.53	2.45	2.39	2.29	2.18	2.07	2.01	1.94	1.88	1.80	1.72	1.64
60	5.29	3.93	3.34	3.01	2.79	2.63	2.51	2.41	2.33	2.27	2.17	2.06	1.94	1.88	1.82	1.74	1.67	1.58	1.48
120	5.15	3.80	3.23	2.89	2.67	2.52	2.39	2.30	2.22	2.16	2.05	1.94	1.82	1.76	1.69	1.61	1.53	1.43	1.31
∞	5.02	3.69	3.12	2.79	2.57	2.41	2.29	2.19	2.11	2.05	1.94	1.83	1.71	1.64	1.57	1.48	1.39	1.27	1.00

Degrees of freedom for the denominator (v)

Table IV Percentage Points of the F Distribution (*continued*)

$$f_{0.01,u,v}$$

Degrees of freedom for the numerator (u)

v	1	2	3	4	5	6	7	8	9	10	12	15	20	24	30	40	60	120	∞
1	4052	4999.5	5403	5625	5764	5859	5928	5982	6022	6056	6106	6157	6209	6235	6261	6287	6313	6339	6366
2	98.50	99.00	99.17	99.25	99.30	99.33	99.36	99.37	99.39	99.40	99.42	99.43	99.45	99.46	99.47	99.47	99.48	99.49	99.50
3	34.12	30.82	29.46	28.71	28.24	27.91	27.67	27.49	27.35	27.23	27.05	26.87	26.69	26.60	26.50	26.41	26.32	26.22	26.13
4	21.20	18.00	16.69	15.98	15.52	15.21	14.98	14.80	14.66	14.55	14.37	14.20	14.02	13.93	13.84	13.75	13.65	13.56	13.46
5	16.26	13.27	12.06	11.39	10.97	10.67	10.46	10.29	10.16	10.05	9.89	9.72	9.55	9.47	9.38	9.29	9.20	9.11	9.02
6	13.75	10.92	9.78	9.15	8.75	8.47	8.26	8.10	7.98	7.87	7.72	7.56	7.40	7.31	7.23	7.14	7.06	6.97	6.88
7	12.25	9.55	8.45	7.85	7.46	7.19	6.99	6.84	6.72	6.62	6.47	6.31	6.16	6.07	5.99	5.91	5.82	5.74	5.65
8	11.26	8.65	7.59	7.01	6.63	6.37	6.18	6.03	5.91	5.81	5.67	5.52	5.36	5.28	5.20	5.12	5.03	4.95	4.86
9	10.56	8.02	6.99	6.42	6.06	5.80	5.61	5.47	5.35	5.26	5.11	4.96	4.81	4.73	4.65	4.57	4.48	4.40	4.31
10	10.04	7.56	6.55	5.99	5.64	5.39	5.20	5.06	4.94	4.85	4.71	4.56	4.41	4.33	4.25	4.17	4.08	4.00	3.91
11	9.65	7.21	6.22	5.67	5.32	5.07	4.89	4.74	4.63	4.54	4.40	4.25	4.10	4.02	3.94	3.86	3.78	3.69	3.60
12	9.33	6.93	5.95	5.41	5.06	4.82	4.64	4.50	4.39	4.30	4.16	4.01	3.86	3.78	3.70	3.62	3.54	3.45	3.36
13	9.07	6.70	5.74	5.21	4.86	4.62	4.44	4.30	4.19	4.10	3.96	3.82	3.66	3.59	3.51	3.43	3.34	3.25	3.17
14	8.86	6.51	5.56	5.04	4.69	4.46	4.28	4.14	4.03	3.94	3.80	3.66	3.51	3.43	3.35	3.27	3.18	3.09	3.00
15	8.68	6.36	5.42	4.89	4.56	4.32	4.14	4.00	3.89	3.80	3.67	3.52	3.37	3.29	3.21	3.13	3.05	2.96	2.87
16	8.53	6.23	5.29	4.77	4.44	4.20	4.03	3.89	3.78	3.69	3.55	3.41	3.26	3.18	3.10	3.02	2.93	2.84	2.75
17	8.40	6.11	5.18	4.67	4.34	4.10	3.93	3.79	3.68	3.59	3.46	3.31	3.16	3.08	3.00	2.92	2.83	2.75	2.65
18	8.29	6.01	5.09	4.58	4.25	4.01	3.84	3.71	3.60	3.51	3.37	3.23	3.08	3.00	2.92	2.84	2.75	2.66	2.57
19	8.18	5.93	5.01	4.50	4.17	3.94	3.77	3.63	3.52	3.43	3.30	3.15	3.00	2.92	2.84	2.76	2.67	2.58	2.49
20	8.10	5.85	4.94	4.43	4.10	3.87	3.70	3.56	3.46	3.37	3.23	3.09	2.94	2.86	2.78	2.69	2.61	2.52	2.42
21	8.02	5.78	4.87	4.37	4.04	3.81	3.64	3.51	3.40	3.31	3.17	3.03	2.88	2.80	2.72	2.64	2.55	2.46	2.36
22	7.95	5.72	4.82	4.31	3.99	3.76	3.59	3.45	3.35	3.26	3.12	2.98	2.83	2.75	2.67	2.58	2.50	2.40	2.31
23	7.88	5.66	4.76	4.26	3.94	3.71	3.54	3.41	3.30	3.21	3.07	2.93	2.78	2.70	2.62	2.54	2.45	2.35	2.26
24	7.82	5.61	4.72	4.22	3.90	3.67	3.50	3.36	3.26	3.17	3.03	2.89	2.74	2.66	2.58	2.49	2.40	2.31	2.21
25	7.77	5.57	4.68	4.18	3.85	3.63	3.46	3.32	3.22	3.13	2.99	2.85	2.70	2.62	2.54	2.45	2.36	2.27	2.17
26	7.72	5.53	4.64	4.14	3.82	3.59	3.42	3.29	3.18	3.09	2.96	2.81	2.66	2.58	2.50	2.42	2.33	2.23	2.13
27	7.68	5.49	4.60	4.11	3.78	3.56	3.39	3.26	3.15	3.06	2.93	2.78	2.63	2.55	2.47	2.38	2.29	2.20	2.10
28	7.64	5.45	4.57	4.07	3.75	3.53	3.36	3.23	3.12	3.03	2.90	2.75	2.60	2.52	2.44	2.35	2.26	2.17	2.06
29	7.60	5.42	4.54	4.04	3.73	3.50	3.33	3.20	3.09	3.00	2.87	2.73	2.57	2.49	2.41	2.33	2.23	2.14	2.03
30	7.56	5.39	4.51	4.02	3.70	3.47	3.30	3.17	3.07	2.98	2.84	2.70	2.55	2.47	2.39	2.30	2.21	2.11	2.01
40	7.31	5.18	4.31	3.83	3.51	3.29	3.12	2.99	2.89	2.80	2.66	2.52	2.37	2.29	2.20	2.11	2.02	1.92	1.80
60	7.08	4.98	4.13	3.65	3.34	3.12	2.95	2.82	2.72	2.63	2.50	2.35	2.20	2.12	2.03	1.94	1.84	1.73	1.60
120	6.85	4.79	3.95	3.48	3.17	2.96	2.79	2.66	2.56	2.47	2.34	2.19	2.03	1.95	1.86	1.76	1.66	1.53	1.38
∞	6.63	4.61	3.78	3.32	3.02	2.80	2.64	2.51	2.41	2.32	2.18	2.04	1.88	1.79	1.70	1.59	1.47	1.32	1.00

Degrees of freedom for the denominator (v)

Chart V   Operating Characteristic Curves for the *t*-Test

β

$$d = \frac{\left| \mu_0 - \mu_1 \right|}{\sigma}$$

can substitute S as

$\hat{\sigma}$

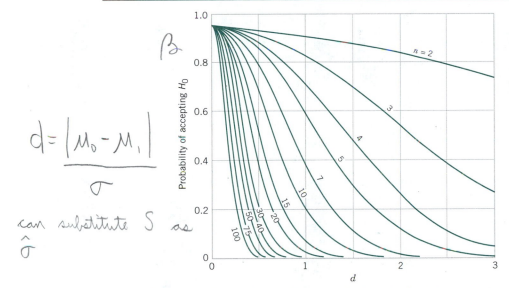

(*a*) *OC* curves for different values of n for the two-sided t-test for a level of significance $\alpha = 0.05$.

β

β

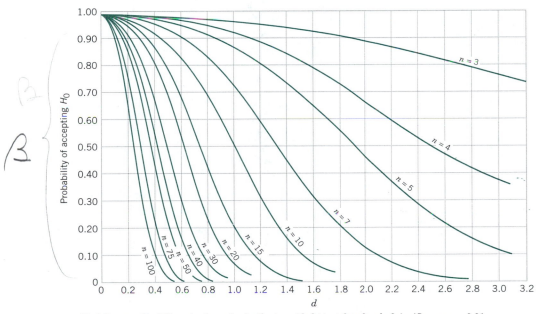

(*b*) *OC* curves for different values of n for the two-sided t-test for a level of significance $\alpha = 0.01$.

Source: These charts are reproduced with permission from "Operating Characteristics for the Common Statistical Tests of Significance," by C. L. Ferris, F. E. Grubbs, and C. L. Weaver, *Annals of Mathematical Statistics,* June 1946, and from *Engineering Statistics,* 2nd Edition, by A. H. Bowker and G. J. Lieberman, Prentice-Hall, 1972.

Chart V   Operating Characteristic Curves for the *t*-Test (*continued*)

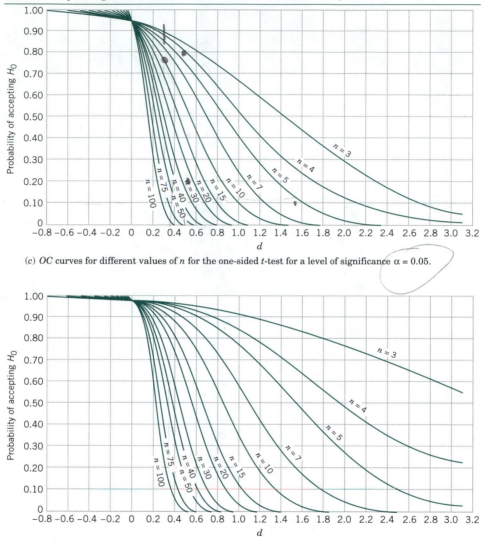

(*c*) *OC* curves for different values of *n* for the one-sided *t*-test for a level of significance $\alpha = 0.05$.

(*d*) *OC* curves for different values of *n* for the one-sided *t*-test for a level of significance $\alpha = 0.01$.

Table VI Factors for Constructing Variables Control Charts

| | $\overline{X}$ **Chart** | | | *R* **Chart** | | |
| | Factors for Control Limits | | | Factors for Control Limits | | |
n^a	A_1	A_2	d_2	D_3	D_4	n
2	3.760	1.880	1.128	0	3.267	2
3	2.394	1.023	1.693	0	2.575	3
4	1.880	.729	2.059	0	2.282	4
5	1.596	.577	2.326	0	2.115	5
6	1.410	.483	2.534	0	2.004	6
7	1.277	.419	2.704	.076	1.924	7
8	1.175	.373	2.847	.136	1.864	8
9	1.094	.337	2.970	.184	1.816	9
10	1.028	.308	3.078	.223	1.777	10
11	.973	.285	3.173	.256	1.744	11
12	.925	.266	3.258	.284	1.716	12
13	.884	.249	3.336	.308	1.692	13
14	.848	.235	3.407	.329	1.671	14
15	.816	.223	3.472	.348	1.652	15
16	.788	.212	3.532	.364	1.636	16
17	.762	.203	3.588	.379	1.621	17
18	.738	.194	3.640	.392	1.608	18
19	.717	.187	3.689	.404	1.596	19
20	.697	.180	3.735	.414	1.586	20
21	.679	.173	3.778	.425	1.575	21
22	.662	.167	3.819	.434	1.566	22
23	.647	.162	3.858	.443	1.557	23
24	.632	.157	3.895	.452	1.548	24
25	.619	.153	3.931	.459	1.541	25

[a] $n > 25$: $A_1 = 3/\sqrt{n}$ where n = number of observations in sample.

Appendix B
Bibliography

INTRODUCTORY WORKS AND GRAPHICAL METHODS

Chambers, J., Cleveland, W., Kleiner, B., and P. Tukey (1983), *Graphical Methods for Data Analysis,* Wadsworth & Brooks/Cole, Pacific Grove, CA. A very well-written presentation of graphical methods in statistics.

Freedman, D., Pisani, R., Purves R., and A. Adbikari (1991), *Statistics,* 2nd ed., Norton, New York. An excellent introduction to statistical thinking, requiring minimal mathematical background.

Hoaglin, D., Mosteller, F., and J. Tukey (1983), *Understanding Robust and Exploratory Data Analysis,* John Wiley & Sons, New York. Good discussion and illustration of techniques such as stem-and-leaf displays and box plots.

Tanur, J., et al. (eds.) (1989), *Statistics: A Guide to the Unknown,* 3rd ed., Wadsworth & Brooks/ Cole, Pacific Grove, CA. Contains a collection of short nonmathematical articles describing different applications of statistics.

Tukey, J. (1977), *Exploratory Data Analysis,* Addison-Wesley, Reading, MA. Introduces many new descriptive and analytical methods. Not extremely easy to read.

PROBABILITY

Derman, C., Gleser, L., and I. Olkin (1980), *Probability Models and Applications,* Macmillan, New York. A comprehensive treatment of probability at a higher mathematical level than this book.

Hoel, P. G., Port, S. C., and C. J. Stone (1971), *Introduction to Probability Theory,* Houghton Mifflin, Boston. A well-written and comprehensive treatment of probability theory and the standard discrete and continuous distributions.

Mosteller, F., Rourke, R., and G. Thomas (1970), *Probability with Statistical Applications,* 2nd ed., Addison-Wesley, Reading, MA. A precalculus introduction to probability with many excellent examples.

Ross, S. (1985), *A First Course in Probability,* 3rd ed., Macmillan, New York. More mathematically sophisticated than this book, but has many excellent examples and exercises.

ENGINEERING STATISTICS

Devore, J. L., (1995), *Probability and Statistics for Engineering and the Sciences,* 4th ed., Wadsworth & Brooks/Cole, Pacific Grove, CA. A more comprehensive book that covers many of the same topics as this text, but at a slightly higher mathematical level. Many of the examples and exercises involve applications to biological and life sciences.

Hines, W. W., and D. C. Montgomery (1990), *Probability and Statistics in Engineering and Management Science,* 3rd ed., John Wiley & Sons, New York. Covers many of the same topics as this book. More emphasis on probability and a higher mathematical level.

Ross, S. (1987), *Introduction to Probability and Statistics for Engineers and Scientists,* John Wiley & Sons, New York. More tightly written and mathematically oriented than this book, but contains some good examples.

Walpole, R. E., Myers, R. H., and S. L. Myers (1998), *Probability and Statistics for Engineers and Scientists,* 6th ed., Macmillan, New York. A very well-written book at about the same level as this one.

Montgomery, D. C., and G. C. Runger (1999), *Applied Statistics and Probability for Engineers,* 2nd ed., John Wiley & Sons, New York. A more comprehensive book on engineering statistics at about the same level as this one.

EMPIRICAL MODEL BUILDING

Daniel, C., and F. Wood (1980), *Fitting Equations to Data,* 2nd ed., John Wiley & Sons, New York. An excellent reference containing many insights on data analysis.

Draper, N., and H. Smith (1998), *Applied Regression Analysis,* 3rd ed., John Wiley & Sons, New York. A comprehensive book on regression written for statistically oriented readers.

Montgomery, D. C., and E. A. Peck (1992), *Introduction to Linear Regression Analysis,* 2nd ed., John Wiley & Sons, New York. A comprehensive book on regression written for engineers and physical scientists.

Myers, R. H. (1990), *Classical and Modern Regression with Applications,* 2nd ed., PWS-Kent, Boston. Contains many examples with annotated SAS output. Very well written.

DESIGN OF EXPERIMENTS

Box, G. E. P., Hunter, W. G., and J. S. Hunter (1978), *Statistics for Experimenters,* John Wiley & Sons, New York. An excellent introduction to the subject for those readers desiring a statistically oriented treatment. Contains many useful suggestions for data analysis.

Montgomery, D. C., (1997), *Design and Analysis of Experiments,* 4th ed., John Wiley & Sons, New York. Written at the same level as the Box, Hunter, and Hunter book, but focused on engineering applications.

STATISTICAL QUALITY CONTROL AND RELATED METHODS

Duncan, A. J. (1974), *Quality Control and Industrial Statistics,* 4th ed., Richard D. Irwin, Homewood, Illinois. A classic book on the subject.

Grant, E. L., and R. S. Leavenworth (1988), *Statistical Quality Control,* 6th ed., McGraw-Hill, New York. One of the first books on the subject; contains many good examples.

John, P. W. M. (1990), *Statistical Methods in Engineering and Quality Improvement,* John Wiley & Sons, New York. Not a methods book, but a well-written presentation of statistical methodology for quality improvement.

Montgomery, D. C. (1996), *Introduction to Statistical Quality Control,* 3rd ed., John Wiley & Sons, New York. A modern comprehensive treatment of the subject written at the same level as this book.

Ryan, T. P. (2000), *Statistical Methods for Quality Improvement,* 2nd ed., John Wiley & Sons, New York. Gives broad coverage of the field, with some emphasis on newer techniques.

Western Electric Company (1956), *Statistical Quality Control Handbook,* Western Electric Company, Inc., Indianapolis, Indiana. An oldie but a goodie.

Appendix C
Answers to Selected
Problems

CHAPTER 2

Section 2-1

2-1.   4.375, 4.658

2-3.   1288.4, 15.8

2-5.   43.975, 12.294

Section 2-2

2-13. $M = 1436.5$, $Q_1 = 1097.8$, $Q_3 = 1735.0$

2-15. $M = 89.250$, $Q_1 = 86.100$, $Q_3 = 93.125$

Section 2-4

2-23.   **a.** $\bar{x} = 4$ **b.** $s^2 = 0.867$, $s = 0.931$

2-25.   **a.** $\bar{x} = 952.44$, $s^2 = 9.55$, $s = 3.09$ **b.** $M = 953$, Any increase

2-27.   **a.** $\bar{x} = 48.125$, $M = 49$ **b.** $s^2 = 7.247$, $s = 2.692$
         d. 5th Percentile: 44, 95th Percentile: 52

Supplemental Exercises

2-35.   **a.** $s^2 = 19.89\Omega^2$, $s = 4.46\Omega$ **b.** $\bar{x} = 41.5$, $s^2 = 19.89\Omega^2$, $s = 4.46\Omega$
         c. No effect   **d.** $s^2 = 1989\Omega^2$, $s = 44.6\Omega$

2-39.   **a.** Sample 1 Range $= 4$, Sample 2 Range $= 4$, Yes
         b. Sample 1 $s = 1.604$, Sample 2 $s = 1.852$, No

2-41.   **b.** $\bar{x} = 9.3$, $s = 4.56$

CHAPTER 3

Section 3-2

3-1.   Continuous

3-3.   Continuous

3-5.   Discrete

3-7.   Discrete

Section 3-3

3-9.   **a.** Yes   **b.** 0.6   **c.** 0.4   **d.** 1

3-11.   **a.** 0.7   **b.** 0.9   **c.** 0.2   **d.** 0.5

3-13.   **a.** 0.55   **b.** 0.95   **c.** 0.50

Section 3-4

3-15.   **a.** $k = 3/64$, $E(X) = 3$, $V(X) = 0.6$
        b. $k = 1/6$, $E(X) = 1.22$, $V(X) = 0.284$
        c. $k = 1$, $E(X) = 1$, $V(X) = 1$

3-17.   **a.** 1   **b.** 0.8647   **c.** 0.8647   **d.** 0.1353   **e.** $x = 9$

3-19.   **a.** 0.7165   **b.** 0.2031   **c.** 0.6321   **d.** $x = 316.2$
        e. $E(X) = 77.865$, $V(X) = 31627$

3-21.   **a.** 0.5   **b.** 0.5   **c.** 0.2   **d.** 0.4

3-23.   **b.** 2.0   **c.** 0.96   **d.** 0.0204

3-25.   **a.** 0.8   **b.** 0.5

3-27.   **a.** 0   **b.** 0.75   **c.** 0.5   **d.** $E(X) = 3.33$, $V(X) = 0.22$

Section 3-5

3-29.   **a.** $z = 0$ **b.** $z = -3.09$ **c.** $z = -1.18$ **d.** $z = -1.11$ **e.** $z = 1.75$

3-31.   **a.** 0.9773   **b.** 0.8413   **c.** 0.6827   **d.** 0.9973   **e.** 0.4773   **f.** 0.4987

3-33.   **a.** 0.9773   **b.** 0.9998   **c.** 0.4773   **d.** 0.8413   **e.** 0.6853

3-35.   **a.** 0.9938   **b.** 0.1359   **c.** $x = 5835.51$

3-37.   **a.** 0.0082   **b.** 0.7211   **c.** $x = 0.5641$

3-39.   **a.** 12.309   **b.** 12.155

3-41.   **a.** 0.1587   **b.** 90   **c.** 0.9973

3-43.   **a.** 0.8186   **b.** 72.85

3-45.   **a.** 0.0668   **b.** 0.8664   **c.** 0.000214

Section 3-6

3-47.   Normally distributed

3-49.   Variability is greatly reduced.

Section 3-7

3-51.   **a.** 0.433   **b.** 0.409   **c.** 0.316   **d.** $E(X) = 3.319, V(X) = 3.7212$

3-53.   **a.** 4/7   **b.** 3/7   **c.** $E(X) = 11/7, V(X) = 26/49$

3-55.   **a.** 0.17   **b.** 0.10   **c.** 0.91   **d.** $E(X) = 9.92, V(X) = 1.954$

Section 3-8

3-75.   **a.** Reasonable   **b.** Not Reasonable   **c.** Not Reasonable   **d.** Not Reasonable
        e. Not Reasonable   **f.** Reasonable   **g.** Reasonable
        h. Not Reasonable   **i.** Not Reasonable

3-59.   **a.** 0.2461   **b.** 0.0547   **c.** 0.0108   **d.** 0.3223

3-61.   **a.** 0.0015   **b.** 0.9298   **c.** 0   **d.** 0.5852

3-63.   0.0043

3-65.   **a.** $n = 50, p = 0.1$ **b.** 0.1117   **c.** 0

3-67.   **a.** 0.9961   **b.** 0.0039   **c.** $E(X) = 112.5 \; \sigma_x = 3.354$

3-69.   0.0595

Section 3-9

3-71.   **a.** 0.7408   **b.** 0.9997   **c.** 0   **d.** 0.0333

3-73. $E(X) = V(X) = 4.017$

3-75.   **a.** 0.0844   **b.** 0.0103   **c.** 0.0185   **d.** 0.1251

3-77.   **a.** 0   **b.** 0.6321

3-79.   **a.** 0.7261   **b.** 0.0731

3-81.   **a.** 0.2941

3-83.   **a.** 0.0076   **b.** 0.1462

3-85.   **a.** 0.3679   **b.** 0.0489   **c.** 0.0183   **d.** 14.979

3-87.   **a.** 30 seconds   **b.** 30 seconds   **c.** 1.5 minutes

3-89.   **a.** 0.3012   **b.** 0.7981

3-91.   **a.** 0.1353   **b.** 0.2707   **c.** 5 miles

3-93.   **a.** 0.3679   **b.** 0.3679   **c.** 2 hours

3-95.   **a.** 0.1218 years or 1.4 months   **b.** $\lambda = 33.3$ years

Section 3-10

3-97.   **a.** 0.075   **b.** 0.851
3-99.   **a.** 0.129   **b.** 0.488
3-101.   0.839
3-103.   **a.** 0.119   **b.** 0.079   **c.** 0.118   **d.** 0.995   **e.** 0.983

Section 3-11

3-105.   **a.** 0.247   **b.** 0.764   **c.** 0.574   **d.** 0.185
3-107.   **a.** 0.372   **b.** 0.140   **c.** 0.344   **d.** 0.578
3-109.   **a.** 0.840   **b.** 0.403   **c.** 0
3-111.   **a.** 0.032   **b.** 0.973
3-113.   **a.** 0.874   **b.** 0.126
3-115.   **a.** 0.15   **b.** 0.08   **c.** 0.988

Section 3-12

3-117.   **a.** 46   **b.** 40   **c.** 0.041   **d.** 0.171
3-119.   **a.** $E(T) = 3$mm, $\sigma_T = 0.141$ mm   **b.** 0.0169
3-121.   **a.** $E(D) = 6$mm, $\sigma_D = 0.123$ mm   **b.** 0.207
3-123.   **a.** 0.002   **b.** 6
3-125.   0.431
3-127.   **a.** 0.176   **b.** 0.824   **c.** 0.0005
3-129.   **a.** 0.079   **b.** 0.104   **c.** 0.187

Supplemental Exercises

3-131.   **a.** 0.777   **b.** 0.777   **c.** 0.173   **d.** 0   **e.** 0.050
3-133.   **a.** 0.6   **b.** 0.8   **c.** 0.7   **d.** 3.9   **e.** 3.09
3-135.   The competitor's tube
3-137.   **a.** 0.130   **b.** 0.897   **c.** 42.5   **d.** 0.151, 0.736
3-139.   **a.** Exponential with a mean of 12 minutes   **b.** 0.287   **c.** 0.341   **d.** 0.436
3-141.   **a.** Exponential with a mean of 100   **b.** 0.632   **c.** 0.135   **d.** 0.607
3-143.   **a.** 0.098   **b.** 0.0006   **c.** 0.00005
3-145.   **a.** 33.3 psi   **b.** 22.36 psi
3-147.   **a.** $E(W + X + Y) = 240$, $V(W + X + Y) = 0.42$ **b.** 0.001

3-149. **a.** 0.008   **b.** 0

3-151. **a.** 18 ppm   **b.** 3.4 ppm

3-153. **a.** 0.919   **b.** 0.402   **c.** Machine 1 is preferable   **d.** 0.252

3-155. **a.** No   **b.** No

3-157. 5.74×10^{-14}

3-159. 1×10^{-8}

3-161. **a.** 0.898, 0.098, 0.005, 0.102
        b. 0.853, 0.145, 0.0025, 0.147
        c. Improve series component

3-163. **a.** No   **b.** Data appear normal   **c.** 312.825

3-165. **a.** Fit is adequate   **b.** 0.38   **c.** 8.65

CHAPTER 4

Section 4-2

4-1. $\overline{X}_1$ is the better estimator

4-5.   0.5

4-9.   **b.** $-\dfrac{\sigma^2}{n}$ **c.** decreases

Section 4-3

4-11. **a.** 0.013   **b.** 0.068

4-13. 13.69

4-15. **a.** 0.024   **b.** 0.012   **c.** 0

4-17. **a.** 0.057   **b.** 0.057

4-19. **a.** 0.023   **b.** 0.023

4-21. **a.** 0.058   **b.** 0.265

4-23. 8.85, 9.16

Section 4-4

4-25. **a.** Do not reject H_0 **b.** 0.274   **c.** 5   **d.** 0.681   **e.** (87.85, 93.11)

4-27. **a.** Reject H_0 **b.** 0.001   **c.** (74.035, 74.037)   **d.** 74.036

4-29. **a.** Reject H_0 **b.** 0   **c.** (3237.53, 3273.31)   **d.** (3231.96, 3278.88)

4-31. 16

Section 4-5

4-33.   **a.** Do not reject H_0 **b.** No   **c.** (59321, 63663)

4-35.   **a.** Reject H_0 **b.** $0.025 < P$-value < 0.50 **c.** $5522.3 \leq \mu$

4-37.   **a.** Do not reject H_0 **b.** (2231.0, 2271.8)   **c.** $2234.8 \leq \mu$

4-39.   **a.** Reject H_0, $0.0005 < P$-value < 0.001 **b.** (4.025, 4.092)

4-41.   **a.** Reject H_0 **b.** Reject H_0 **c.** Yes, Power $= 1$ **d.** (1.091, 1.105)

Section 4-6

4-43.   **a.** Do not reject H_0 **b.** $0.5 < P$-value < 0.9 **c.** 0.012

4-45.   **a.** Do not reject H_0 **b.** $0.50 \leq P$-value ≤ 0.90 **c.** 2214

4-47.   **a.** Reject H_0 **b.** (0.31, 0.46)

Section 4-7

4-49.   622

4-51.   666

4-53.   2401

4-55.   **a.** 0.8288   **b.** 4543

4-57.   **a.** Do not reject H_0 **b.** 0.732   **c.** 0.1314   **d.** 473

4-59.   **a.** 0.085   **b.** 0

Section 4-9

4-61.   **a.** Do not reject H_0 **b.** 0.7185

4-63.   **a.** Do not reject H_0 **b.** 0.5475

4-65.   **a.** Reject H_0 **b.** 0.0155

Supplemental Exercises

4-69.   Symmetric interval is shorter

4-71.   **a.** $0.05 < P(\chi^2_{15} \leq 7.68) < 0.10$
        b. $0.01 < P(\chi^2_{29} \leq 14.85) < 0.025$
        c. $P(\chi^2_{70} \leq 35.84) < 0.005$

4-73.   **b.** 17   **c.** (17, 33.25)   **d.** 343.7   **e.** (28.23, 343.74)
        f. (16.82, 28.98); (15.80, 192.44)   **g.** (16.88, 33.12); (28.16, 342.92)

4-77.   **a.** 0.452   **b.** 0.102   **c.** 0.014

4-79.   **a.** Reject H_0 **b.** 0   **c.** $590.44 \leq \mu$ **d.** (153.63, 712.24)

4-81.   **b.** Reject H_0, Normality

4-83.   **a.** Do not reject   **b.** Reject

4-85.   **a.** Do not reject H_0, Normality

4-87.   **a.** Do not reject H_0

4-89.   **a.** (0.554, 0.720)   **b.** (0.538, 0.734)   **d.** Yes

4-91.   **a.** 0.026   **b.** 0.131   **c.** 0.869

4-93.   **a.** Do not reject H_0 **b.** 0.641

CHAPTER 5

Section 5-2

5-1.   **a.** Do not reject H_0 **b.** 0.3222   **c.** 0.977   **d.** (−0.0098, 0.0298)   **e.** 12

5-3.   **a.** Reject H_0 **b.** 0   **c.** 0.2483   **d.** (−8.21, −4.49)

5-5.   **a.** Reject H_0 **b.** 0   **c.** 9

5-7.   8

5-9.   (−21.08, 7.72)

5-11.   **a.** Reject H_0 **b.** 0.02872

Section 5-3

5-13.   **a.** Do not reject H_0 **b.** P-value > 0.80 **c.** (−0.394, 0.494)

5-15.   **a.** Reject H_0 **b.** (−5.151, −1.209)

5-17.   **a.** Reject H_0 **b.** $0.01 < P$-value < 0.02 **c.** (−0.749, −0.111)

5-19.   **a.** Reject H_0 **b.** $0.001 < P$-value < 0.005 **c.** (0.045, 0.240)

5-21.   38

5-23.   Do not reject H_0; $0.40 < P$-value

5-25.   (5.14, 11.82); Lower limit: 5.71

Section 5-4

5-31.   **a.** (−1.216, 2.550)   **b.** Normality assumption holds

5-33.   **a.** Do not reject H_0 **b.** (−13.790, 2.810)

5-35.   Reject H_0

5-37. $n = 5$, Yes

Section 5-5

5-39.   **a.** 1.47   **b.** 2.19   **c.** 4.41   **d.** 0.595   **e.** 0.439   **f.** 0.297

5-41.   Do not reject H_0

5-43.   **a.** (0.0844, 8.9200)   **b.** (0.145, 13.400)   **c.** Lower limit: 0.582

5-45.   Do not reject H_0

5-47.   Reject H_0

5-49.   (0.255, 2.000)

Section 5-6

5-51.   **a.** 0.819   **b.** 383

5-53.   Reject H_0

5-55. $(-0.045, 0.021)$

Section 5-8

5-57.   **a.** Reject H_0

5-59.   **a.** Reject H_0 **b.** P-value $= 0$

5-61.   **a.** Reject H_0

5-63.   **a.** Do not reject H_0

5-65.   **a.** Do not reject H_0

Supplemental Exercises

5-67.   **b.** Do not reject H_0

5-69.   **a.** (0.1583, 5.1570)   **b.** Variances do not differ significantly

5-71.   **a.** $(-5.362, -0.638)$ **b.** $(-6.340, 0.340)$

5-73.   **a.** Reject H_0 **b.** Reject H_0

5-75.   **a.** Do not reject H_0 **b.** Do not reject H_0 **c.** Conclusions are the same
        d. Do not reject H_0

5-77.   **a.** Reject H_0 **b.** Conclusions are the same

5-79.   21

5-81.   **a.** 16   **b.** $n = 5$ is sufficient

5-83.   **a.** Answers given in bold.

Source of Variation	Sum of Squares	Degrees of Freedom	Mean Square	F_0
Treatments	95.129	**4**	**23.782**	**4.112**
Error	**86.752**	**15**	**5.783**	
Total	**181.881**	19		

b. Do not reject H_0

5-85. **a.** Normality is valid   **b.** P-value > 0.80, Do not reject H_0 **c.** 40

5-87. **a.** Reject H_0 **b.** P-value $= 0.002$

CHAPTER 6

Section 6-2

6-1. **a.** $\hat{y} = 48.013 - 2.330x$ **b.** 37.994   **c.** 39.392   **d.** 6.708

6-3. **a.** $\hat{y} = 0.4631476 + 0.0074902x$ **b.** 0.00749

6-5. **a.** $\hat{y} = 0.0249 + 0.1290x$ **b.** 0.055   **c.** 0.054

6-7. **a.** $\hat{\beta}_0 = -1.9122$, $\hat{\beta}_1 = 0.0931$, $\hat{\beta}_2 = 0.2532$
b. 29.37

6-9. **a.** $\hat{y} = 108 + 2x_1$ $\hat{y} = 101 + 5.5x_1$ for $x_2 = 2$
$\hat{y} = 132 + 2x_1$ $\hat{y} = 119 + 17.5x_1$ for $x_2 = 8$
b. 2, no   **c.** $x_2 = 5$, 11.5; $x_2 = 2$, 5.5; $x_2 = 8$, 17.5; Yes

6-11. **a.** $\hat{y} = -102.713 + 0.605x_1 + 8.924x_2 + 1.438x_3 + 0.0136x_4$ **b.** 287.6

Sections 6-3 and 6-4

6-13. **a.** Reject H_0 **b.** 0.008

6-15. **a.** Reject H_0 **b.** Reject H_0 **c.** 0.000000342

6-17. **a.** Reject H_0, P-value $\cong 0$ **b.** 8.06
c. Reject H_0, P-value $\cong 0$

6-19. **a.** Do not reject H_0, P-value $= 0.0303$ **b.** 242.716
c. Do not reject H_0 for any regressor

6-21. **a.** Do not reject H_0 **b.** Do not reject H_0
c. Interaction model: 561.28, No interaction: 650.14

Sections 6-5 and 6-6

6-23. **a.** $(-2.92, -1.74)$ **b.** $(46.71, 49.31)$ **c.** $(41.285, 43.095)$ **d.** $(39.096, 45.284)$

6-25. **a.** $(-0.00961, -0.00444)$ **b.** $(16.2448, 27.3318)$
c. $(7.914, 10.372)$ **d.** $(4.072, 14.214)$

6-27. **a.** $(9.101, 9.315)$ **b.** $(-11.622, -1.049)$
c. $(498.720, 501.528)$ **d.** $(495.573, 504.674)$

6-29. **a.** β_1: $(-0.2672, 1.478)$ β_2: $(-3.614, 21.4610)$
β_3: $(-4.219, 7.094)$ β_4: $(-1.722, 1.749)$

b. $(263.78, 311.34)$ **c.** $(243.711, 331.413)$

6-31. **a.** $(14.4015, 22.07)$ **b.** $(77.74, 105.10)$ **c.** $(82.27, 100.58)$

Section 6-7

6-33.   **d.** 98.6%

6-35.   **a.** $R^2 = 0.86179$, adj $R^2 = 0.76965$
      b. $R^2 = 0.9205$, adj $R^2 = 0.8011$

6-37.   **b.** $R^2 = 0.9937$, adj $R^2 = 0.9925$

      c. $\hat{y}^* = 6.225 - 0.1665x_1 - 0.000228x_2 + 0.1573x_3$

Supplemental Exercises

6-43.   **a.** Average size $= 0.12$ **b.** Points 17, 18

6-45.   **a.** Reject H_0, P-value $\cong 0$
      b. Do not reject H_0: $\beta_3 = 0$
        Reject H_0: $\beta_4 = 0$
        Do not reject H_0: $\beta_5 = 0$

6-47.   **b.** $\hat{y} = 0.4705 + 20.5673x$ **c.** 21.038   **d.** 1.6629   **f.** $\hat{y} = 21.03146x$

6-49.   **b.** $\hat{y} = 2625.39 - 36.962x$ **c.** 1886.154

6-53.   Reject H_0, P-value $\cong 0$

CHAPTER 7

Section 7-4

7-1.   **a.**

Term	Coeff	SE(coeff)
Material	9.313	7.730
Temperature	-33.938	7.730
Mat*Temp	4.678	7.730

      c. Only Temperature is significant
      d. Material: (12.294, 49.546)
        Temperature: $(-98.796, 39.956)$
        Material*Temperature: $(-21.546, 40.294)$

7-3.   **a.**

Term	Coeff	SE(coeff)
PVP	7.8250	0.5744
Time	3.3875	0.5744
PVP*Time	-2.7125	0.5744

      c. All are significant
      d. PVP: (13.352, 17.948)   Time: (4.477, 9.073)
        PVP*Time: $(-7.723, -3.127)$

7-5. **a. Term**

Term	Coeff	SE(coeff)
Temp	−0.875	0.7181
%Copper	3.625	0.7181
Temp*Copper	1.375	0.7181

c. %Copper is significant

d. Temperature: (−4.622, 1.122)   %Copper: (4.378, 10.122)
Temperature*%Copper: (−0.122, 5.622)

7-7. **a. Term**

Term	Coeff	SE(coeff)
Doping	−0.2425	−0.07622
Anneal	1.015	0.07622
Doping*Anneal	0.1675	0.07622

c. Polysilicon doping and Anneal are both significant.

d. Polysilicon doping: (−0.7898, −0.1802)   Anneal: (1.725, 2.34)
Polysilicon doping*Anneal: (0.0302, 0.6398)

Sections 7-5, 7-6, and 7-7

7-9. **a. Term**

Term	Effect	Coeff
A	18.25	9.13
B	84.25	42.13
C	71.75	35.88
AB	−11.25	−5.62
AC	−119.25	−59.62
BC	−24.25	−12.12
ABC	−34.75	−17.38

b. Tool Life = 413 + 9.1 A + 42.1 B + 35.9 C − 59.6 AC

7-11. Score = 175 + 8.50 A + 5.44 C + 4.19 D + 4.56 AD

7-13. **a.** No terms are significant.
b. There is no appropriate model.

7-15. **a.** $\hat{\sigma}^2 = 8.2$
b. $t = -9.20$, curvature is not significant.

7-17. **a.** Block 1: (1) ab ac bc; Block 2: a b c abc;
None of the factors or interactions appear to be significant using only Replicate 1.

7-19. Block 1: (1) acd bcd ab; Block 2: a, b, cd, abcd; Block 3: d ac bc abd;
Block 4: c ad bd abc
The factors A, C, and D and the interactions AD and ACD are significant.

7-21. ABC, CDE

Section 7-8

7-23. **a.** $I = ABCDE$
b.

Term	A	B	C	D	E	AB	AC	AD	AE	BC	BD	BE
Effect	1.435	−1.47	−0.273	4.545	−0.703	1.15	−0.91	−1.23	0.427	0.293	0.12	0.16

c. color $= 2.77 + 0.718A - 0.732B + 2.27D + 0.575AB - 0.615AD$

7-25. **a.** $I = ACE = BDE = ABCD$
b. Estimated Effects

Term	A	B	C	D	E	AB	AD
Effect	−1.525	−5.175	2.275	−0.675	2.275	1.825	−1.275

c. Factor B is significant (using AB and AD for error)

7-27. Strength $= 3025 + 725\,A + 1825\,B + 875\,D + 325\,E$

Section 7-9

7-31. **a.** $y_1 = 499.26 - 85x_1 + 35x_2 - 71.67x_3 + 25.83x_1x_2$
b. $y_2 = 299 + 50.89x_1 + 75.78x_2 + 59.5x_3 - 17.33x_1^2 - 34x_2^2 - 17.17x_3^2 + 13.33x_1x_2 + 26.83x_1x_3 - 17.92x_2x_3$

7-33. **a.** Central composite design, not rotatable
b. Quadratic model not reasonable
c. Increase x_1

Section 7-11

7-35. **a.** Interaction is significant
c. Paint type $= 1$, Drying time $= 25$ minutes.

7-37. **a.**

Source	F
Hardwood	7.55
cooktime	31.31
freeness	19.71
Hardwood*cooktime	2.89
Hardwood*freeness	2.94
cooktime*freeness	0.95
Hardwood*cooktime*freeness	0.94

b. Hardwood, cooktime, freeness, and hardwood*freeness are significant.

Supplemental Exercises

7-39. **a.**

Source	pH	Catalyst	pH*Catalyst
t_0	2.54	-0.05	5.02

pH and pH*Catalyst are significant

c. Viscosity $= 191.563 + 2.937$ pH $- 0.062$ Catalyst $+ 5.812$ pH*Catalyst

7-41. **a.**

Source	Level	Salt	Level*Salt
f_0	63.24	39.75	5.29

Level, Salt, and Level*Salt are all significant.

b. Application Level 1 increases flammability average.

7-43. **a.**

Term	A	B	C	AB	AC	BC	ABC
Effect	-2.74	-6.66	3.49	-8.71	7.04	11.46	-6.49

b. $\hat{\sigma}_c^2 = 8.40$

c. Two-way interactions are significant (specifically BC interaction)

d. deltaline $= -11.8 - 1.37$ A $- 3.33$ B $+ 1.75$ C $- 4.35$ AB $+ 3.52$ AC $+ 5.73$ BC

7-45. **a.**

Term	Effect	Term	Effect
V	15.75	FP	-6.00
F	-10.75	FG	19.25
P	-8.75	PG	-3.75
G	-25.00	VFP	1.25
VF	-8.00	VFG	-1.50
VP	3.00	VPG	0.50
VG	2.75		

b. V, F, P, G, VF, and FGP are possibly significant

Term	V	F	P	G	FPG	VF
f_0	2.71	-1.85	-1.51	-4.31	-2.15	-1.38

V, G, and FPG are significant.

SurfaceRough $= 101 + 7.88$ V $- 5.37$ F $- 4.38$ P $- 12.5$ G $- 6.25$ fpg

7-47.

Term	V	F	P	G	VF	VP	VG
t_0	54.38	-0.94	-3.62	-3.62	-2.53	4.21	-0.84

Factors F, P, G, VF, and VP are significant.

7-53. Model: $y = 13.728 + 0.2966x_1 - 0.4052x_2 - 0.1240x_1^2 - 0.079x_1^2$

Maximum viscosity is 14.425 using $x_1 = 1.19$ and $x_2 = -2.56$

CHAPTER 8

8-1. **a.** $\bar{x}$ Chart: $UCL = 39.42$, $LCL = 28.48$ R Chart: $UCL = 13.77$, $LCL = 0$

b. $\bar{x}$ Chart: $UCL = 39.34$, $LCL = 29.22$ R Chart: $UCL = 12.74$, $LCL = 0$

8-3.   **a.** $\bar{x}$ Chart: $UCL = 18.20, LCL = 12.08$ R Chart: $UCL = 9.581, LCL = 0$
        b. $\bar{x}$ Chart: $UCL = 17.98, LCL = 12.31$ R Chart: $UCL = 8.885, LCL = 0$

8-5.   **a.** $\bar{x}$ Chart: $UCL = 340.69, LCL = 271.87$
        R Chart: $UCL = 107.71, LCL = 0$
        b. $\hat{\mu} = \bar{\bar{x}} = 306.28$ $\hat{\sigma} = \dfrac{\bar{r}}{d_2} = \dfrac{47.2}{2.059} = 22.92$

8-7.   **a.** $\bar{x}$ Chart: $UCL = 7.511, LCL = 5.137$ R Chart: $UCL = 2.986, LCL = 0$
        b. $\bar{x}$ Chart: $UCL = 7.385, LCL = 5.061$ R Chart: $UCL = 2.924, LCL = 0$

8-9.   **a.** x Chart: $UCL = 92.19, LCL = 71.144$ MR Chart: $UCL = 12.93, LCL = 0$
        b. $\hat{\mu} = \bar{x} = 81.67$ $\hat{\sigma} = 3.51$

8-11.   **a.** x Chart: $UCL = 580.2, LCL = 380$ MR Chart: $UCL = 123, LCL = 0$
        Revised Limits:
        x Chart: $UCL = 582.3, LCL = 388$ MR Chart: $UCL = 118.9, LCL = 0$
        b. $\hat{\mu} = \bar{x} = 485.5$ $\hat{\sigma} = 32.26$

Section 8-7

8-13.   **a.** $PCR = 1.141, PCR_k = 0.5932$ **b.** 0.0376

8-15.   Fallout is 0.034937, $PCR = 0.705$. The process capability appears to be poor.

8-17.   Fallout is 0.0925, $PCR = 0.582, PCR_k = 0.490$

8-19. $PCR = 0.497, PCR_k = 0.386$

8-21. $PCR = 0.258, PCR_k = 0.108$

8-23.   **a.** $UCL = 5.380, LCL = 0.3167$
        b. Revised limits: $UCL = 5.237, LCL = 0.2625$

8-25. $UCL = 21.52, LCL = 1.267$
        Revised limits: $UCL = 20.95, LCL = 1.050$

Section 8-9

8-27.   **a.** 0.2177   **b.** 4.6

8-29.   **a.** 0.2148   **b.** 4.6

8-31.   **a.** 0.0103   **b.** 97.1

8-33.   **a.** 0.00734   **b.** 136.24

8-35.   **a.** 0.1292   **b.** 7.74

Supplemental Exercises

8-37. $UCL = 0.1694, LCL = 0.0016$ Revised limits: $UCL = 0.0856, LCL = 0$

8-39. $UCL = 7.514, LCL = 0$ Revised limits: $UCL = 6.509, LCL = 0$

8-41. **a.** 43.9   **b.** 6.30   **c.** 2.00

8-43. **a.** $\bar{x}$ Chart: $UCL = 141.2, LCL = 138.6$ R Chart: $UCL = 4.832, LCL = 0$

     b. Revised limits: $\bar{x}$ Chart: $UCL = 141.5, LCL = 138.8$
       R Chart: $UCL = 4.922, LCL = 0$
     $\hat{\sigma} = 1$
     c. $PCR = 0.67, PCR_k = 0.60$ **d.** Decrease to $\hat{\sigma}^2 = 0.135$ **e.** $ARL = 43.96$

8-45. **a.** 0.96995   **b.** 1

8-47. 0.0455

8-49. **a.** 0.0434   **b.** 0.8696

Index

Summary of Two-Sample Hypothesis Testing Procedures

Handwritten annotation (top): $\Delta_0 = \mu_1 - \mu_2$

Case	Null Hypothesis	Test Statistic	Alternative Hypothesis	Criteria for Rejection	OC Curve Parameter	OC Curve Appendix A Chart IV
1.	$H_0: \mu_1 = \mu_2$; σ_1^2 and σ_2^2 known	$z_0 = \dfrac{\bar{x}_1 - \bar{x}_2 - \Delta_0}{\sqrt{\dfrac{\sigma_1^2}{n_1} + \dfrac{\sigma_2^2}{n_2}}}$	$H_1: \mu_1 \neq \mu_2$ $H_1: \mu_1 > \mu_2$ $H_1: \mu_1 < \mu_2$	$|z_0| > z_{\alpha/2}$ $z_0 > z_\alpha$ $z_0 < -z_\alpha$	— — —	— — —
2.	$H_0: \mu_1 = \mu_2$; $\sigma_1^2 = \sigma_2^2$ unknown	$t_0 = \dfrac{\bar{x}_1 - \bar{x}_2 - \Delta_0}{s_p\sqrt{\dfrac{1}{n_1} + \dfrac{1}{n_2}}}$	$H_1: \mu_1 \neq \mu_2$ $H_1: \mu_1 > \mu_2$ $H_1: \mu_1 < \mu_2$	$|t_0| > t_{\alpha/2,\,n_1+n_2-2}$ $t_0 > t_{\alpha,\,n_1+n_2-2}$ $t_0 < -t_{\alpha,\,n_1+n_2-2}$	$d = |\mu - \mu_0|/2\sigma$ $d = (\mu - \mu_0)/2\sigma$ $d = (\mu_0 - \mu)/2\sigma$	a, b c, d c, d
3.	$H_0: \mu_1 = \mu_2$; $\sigma_1^2 \neq \sigma_2^2$ unknown	$t_0 = \dfrac{\bar{x}_1 - \bar{x}_2}{\sqrt{\dfrac{s_1^2}{n_1} + \dfrac{s_2^2}{n_2}}}$	$H_1: \mu_1 \neq \mu_2$ $H_1: \mu_1 > \mu_2$ $H_1: \mu_1 < \mu_2$	$|t_0| > t_{\alpha/2,\,v}$ $t_0 > t_{\alpha,\,v}$ $t_0 < -t_{\alpha,\,v}$	— — —	— — —
4.	Paired data $H_0: \mu_D = 0$	$t_0 = \dfrac{\bar{d} - \Delta_0}{s_d/\sqrt{n}}$	$H_1: \mu_d \neq 0$ $H_1: \mu_d > 0$ $H_1: \mu_d < 0$	$|t_0| > t_{\alpha/2,\,n-1}$ $t_0 > t_{\alpha,\,n-1}$ $t_0 < -t_{\alpha,\,n-1}$	— — —	— — —
5.	$H_0: \sigma_1^2 = \sigma_2^2$	$f_0 = s_1^2/s_2^2$	$H_1: \sigma_1^2 \neq \sigma_2^2$ $H_1: \sigma_1^2 > \sigma_2^2$	$f_0 > f_{\alpha/2,\,n_1-1,\,n_2-1}$ or $f_0 < f_{1-\alpha/2,\,n_1-1,\,n_2-1}$ $f_0 > f_{\alpha,\,n_1-1,\,n_2-1}$	—	—
6.	$H_0: p_1 = p_2$	$z_0 = \dfrac{\hat{p}_1 - \hat{p}_2}{\sqrt{\hat{p}(1 - \hat{p})\left[\dfrac{1}{n_1} + \dfrac{1}{n_2}\right]}}$	$H_1: p_1 \neq p_2$ $H_1: p_1 > p_2$ $H_1: p_1 < p_2$	$|z_0| > z_{\alpha/2}$ $z_0 > z_\alpha$ $z_0 < -z_\alpha$	— — —	— — —

For Case 3:

$$v = \frac{\left(\dfrac{s_1^2}{n_1} + \dfrac{s_2^2}{n_2}\right)^2}{\dfrac{(s_1^2/n_1)^2}{n_1+1} + \dfrac{(s_2^2/n_2)^2}{n_2+1}} - 2$$

Handwritten annotations:

Pooled variance:
$$S_p^2 = \frac{(n_1-1)S_1^2 + (n_2-1)S_2^2}{n_1+n_2-2}$$

$\mu_1 - \mu_2 = 0$

Adjusts crit. region ←